D1526479

Lecture Notes in Physics

The Editorial Policy for Proceedings

The series Lecture Notes in Physics reports new developments in physical research and teaching – quickly, informally, and at a high level. The proceedings to be considered for publication in this series should be limited to only a few areas of research, and these should be closely related to each other. The contributions should be of a high standard and should avoid lengthy redraftings of papers already published or about to be published elsewhere. As a whole, the proceedings should aim for a balanced presentation of the theme of the conference including a description of the techniques used and enough motivation for a broad readership. It should not be assumed that the published proceedings must reflect the conference in its entirety. (A listing or abstracts of papers presented at the meeting but not included in the proceedings could be added as an appendix.)
When applying for publication in the series Lecture Notes in Physics the volume's editor(s) should submit sufficient material to enable the series editors and their referees to make a fairly accurate evaluation (e.g. a complete list of speakers and titles of papers to be presented and abstracts). If, based on this information, the proceedings are (tentatively) accepted, the volume's editor(s), whose name(s) will appear on the title pages, should select the papers suitable for publication and have them refereed (as for a journal) when appropriate. As a rule discussions will not be accepted. The series editors and Springer-Verlag will normally not interfere with the detailed editing except in fairly obvious cases or on technical matters.
Final acceptance is expressed by the series editor in charge, in consultation with Springer-Verlag only after receiving the complete manuscript. It might help to send a copy of the authors' manuscripts in advance to the editor in charge to discuss possible revisions with him. As a general rule, the series editor will confirm his tentative acceptance if the final manuscript corresponds to the original concept discussed, if the quality of the contribution meets the requirements of the series, and if the final size of the manuscript does not greatly exceed the number of pages originally agreed upon. The manuscript should be forwarded to Springer-Verlag shortly after the meeting. In cases of extreme delay (more than six months after the conference) the series editors will check once more the timeliness of the papers. Therefore, the volume's editor(s) should establish strict deadlines, or collect the articles during the conference and have them revised on the spot. If a delay is unavoidable, one should encourage the authors to update their contributions if appropriate. The editors of proceedings are strongly advised to inform contributors about these points at an early stage.
The final manuscript should contain a table of contents and an informative introduction accessible also to readers not particularly familiar with the topic of the conference. The contributions should be in English. The volume's editor(s) should check the contributions for the correct use of language. At Springer-Verlag only the prefaces will be checked by a copy-editor for language and style. Grave linguistic or technical shortcomings may lead to the rejection of contributions by the series editors. A conference report should not exceed a total of 500 pages. Keeping the size within this bound should be achieved by a stricter selection of articles and not by imposing an upper limit to the length of the individual papers. Editors receive jointly 30 complimentary copies of their book. They are entitled to purchase further copies of their book at a reduced rate. As a rule no reprints of individual contributions can be supplied. No royalty is paid on Lecture Notes in Physics volumes. Commitment to publish is made by letter of interest rather than by signing a formal contract. Springer-Verlag secures the copyright for each volume.

The Production Process

The books are hardbound, and the publisher will select quality paper appropriate to the needs of the author(s). Publication time is about ten weeks. More than twenty years of experience guarantee authors the best possible service. To reach the goal of rapid publication at a low price the technique of photographic reproduction from a camera-ready manuscript was chosen. This process shifts the main responsibility for the technical quality considerably from the publisher to the authors. We therefore urge all authors and editors of proceedings to observe very carefully the essentials for the preparation of camera-ready manuscripts, which we will supply on request. This applies especially to the quality of figures and halftones submitted for publication. In addition, it might be useful to look at some of the volumes already published. As a special service, we offer free of charge LATEX and TEX macro packages to format the text according to Springer-Verlag's quality requirements. We strongly recommend that you make use of this offer, since the result will be a book of considerably improved technical quality. To avoid mistakes and time-consuming correspondence during the production period the conference editors should request special instructions from the publisher well before the beginning of the conference. Manuscripts not meeting the technical standard of the series will have to be returned for improvement.

For further information please contact Springer-Verlag, Physics Editorial Department V, Tiergartenstrasse 17, D-69121 Heidelberg, FRG

M. Dienes M. Month B. Strasser S. Turner (Eds.)

Frontiers of Particle Beams: Factories with $e^+ e^-$ Rings

Proceedings of a Topical Course
Held by the Joint US-CERN School
on Particle Accelerators at Benalmádena, Spain
29 October - 4 November 1992

Springer-Verlag

Berlin Heidelberg New York
London Paris Tokyo
Hong Kong Barcelona
Budapest

Editors

M. Dienes
205 E. Harmon. Apt. 908
Las Vegas, NV 89109, USA

M. Month
US Particle Accelerator School
Fermilab, P. O. Box 500
Batavia, IL 60510, USA

B. Strasser
Chemin de la Léchère
F-74160 St. Julien-en-Genevois, France

S. Turner
CERN Accelerator School
CH-1211 Geneva 23, Switzerland

ISBN 3-540-56588-4 Springer-Verlag Berlin Heidelberg New York
ISBN 0-387-56588-4 Springer-Verlag New York Berlin Heidelberg

Printed in Germany

58/3140-543210 - Printed on acid-free paper

Pstack
SLIMON
2-3-94

JOINT US-CERN SCHOOL ON PARTICLE ACCELERATORS
FRONTIERS OF PARTICLE BEAMS: FACTORIES WITH $e^+ e^-$ RINGS

Members of the Organizing Institutions

CERN Accelerator School
Head: E.J.N. Wilson
S. Turner
S. von Wartburg

US Particle Accelerator School
Director: M. Month
M. Paul
S. Winchester
M. Dienes
A. Gonzales

Universidad de Sevilla
Vice Rector: M. Lozano

Members of the Programme Committees

Y. Baconnier
P.J. Bryant
A. Chao
S. Chattopadhyay
R. Cooper
J. Gareyte
G. Jackson
R. Jameson
J. Jowett
M. Month
J.-M. Quesada
R. Ruth
J. Seeman
R. Siemann
S. Tazzari
S. Turner
W. Weng
H. Wiedemann
F. Willeke
C. Willmott
P. Wilson
M. Zisman

Local Organising Committee

M. Paul
J.-M. Quesada
R. Rubio-Marin
S. Turner
S. von Wartburg
S. Winchester

Editors of the Proceedings

M. Dienes
M. Month
B. Strasser
S. Turner

Sponsors

CERN
US Department of Energy
National Science Foundation
Comisión Interministerial de Ciencia y Tecnología
Centro para el Desarollo Tecnológico e Industrial
Universidad de Sevilla

PREFACE

Since 1985, the CERN Accelerator School (CAS) and the US Particle Accelerator School (USPAS) have jointly organized a series of specialized courses. The first of these, on the topic of nonlinear dynamics and its relationship to accelerator physics, was held in Santa Margherita di Pula, Sardinia. The second was held on South Padre Island, Texas, in 1986 with the title "Frontiers of Particle Beams" and covered the latest concepts in the physics and technology of particle accelerators. Next came the course held in Anacapri, Isola di Capri in 1988. At this point "Frontiers of Particle Beams" was adopted as the series title and this particular course concentrated on the techniques of observing, diagnosing, and correcting the properties of particle beams in accelerators and storage rings. The fourth course in the series took place on Hilton Head Island, South Carolina, in 1990 and dealt with the limitations on beam performance resulting from the higher and higher beam intensities becoming available. The proceedings of each of these joint schools were published in the 'Lecture Notes in Physics' series of Springer-Verlag as volumes 247, 296, 343, and 400 respectively.

The present volume is the proceedings of the latest of these joint schools, held in Benalmádena, Spain. This course dealt with the design and development of high performance "factories" using $e^+ e^-$ colliders. Topics covered were: physics motivation, overall design of factories and their detectors, high luminosity, injection, short bunches, instabilities, feedback, beam-beam interaction, lattice and interaction-region design, special schemes, RF, vacuum, ion clearing and background.

Our host on this occasion was the University of Seville; on behalf of the CERN and US Accelerator Schools we express our sincere appreciation to its Vice Rector, M. Lozano, to J.-M. Quesada, J.A. Rubio and C. Willmott for their support. Special thanks are also due to the CAS Advisory Committee and the USPAS Executive Board for their continuing support, as well as to the Programme and Local Organizing Committees for their dedicated efforts. As usual the lecturers deserve all our praise for accepting a very exacting task and for the serious effort they put into preparing and presenting the various topics as well as writing the chapters presented in this volume. Finally we thank all the participants for their attendance at the course and their lively interest during the lectures and in the ensuing discussions.

Geneva, Switzerland — S. Turner, CAS

Chicago, Illinois — M. Month, USPAS

September 1993

CONTENTS

Basic e^+e^- Collider Physics

David L. Rubin

Cornell University, Ithaca, New York 14853, USA

In the most basic colliding-beam storage ring bunches of conterrotating electrons and positrons share a common equilibrium orbit. The evenly spaced bunches collide at $2n_b$ locations around the ring, where n_b is the number of bunches in each beam. At each collision point the machine optics are manipulated to minimize both the spot size and the effect of the beam-beam interaction. The separated function guide field is characterized by alternate gradient focusing.

1 Equations of Motion

To the magnetic guide field there corresponds a Lorentz force

$$\mathbf{F} = q\mathbf{v} \times \mathbf{B} \ , \tag{1}$$

yielding identical trajectories for conterrotating and oppositely charged beams, at least in so far as the particle energy is independent of azimuthal location around the ring[1]. Then the beam trajectories are described by the equation of motion

$$m\frac{d^2\mathbf{x}}{dt^2} = q\mathbf{v}\times\mathbf{B} \ . \tag{2}$$

The field $\mathbf{B}$ depends on the azimuthal position s along the magnetic axis of the machine. We define a cylindrical coordinate system in which x and y correspond to horizontal and vertical displacement with respect to the reference orbit. The reference orbit is circular. The radius of curvature is determined by the part of the field that is nonzero on the reference orbit, such as the uniform field of a bending magnet. The magnetic field can then be expanded about the reference

[1] Synchrotron radiation and RF acceleration both break the symmetry, leading to finite transverse displacement of the electron and positron orbits. The beams tend to have minimum energy at opposite ends of the accelerating cavities. Nevertheless, if the interaction points are located symmetrically with respect to the cavities, and if they are points of vanishing dispersion, head-on collision of the electrons and positrons is certain.

orbit. If we define the coefficients of the terms linear in x and y as $k_x(s)$ and $k_y(s)$ respectively, the equations of motion become

$$\frac{d^2x}{ds^2} = k_x(s)x + F_x(s) \tag{3}$$

$$\frac{d^2y}{ds^2} = k_y(s)y + F_y(s) \ , \tag{4}$$

where the time t has been replaced by position along the trajectory s, as the dependent variable, and $k_x = -\frac{1}{R^2} + \frac{q}{\gamma mc}\frac{\partial B_y(0)}{\partial x}$ with $\frac{1}{R} = -\frac{q}{\gamma mc}B_y(0)$ and $k_y = -\frac{q}{\gamma mc}\frac{\partial B_x(0)}{\partial y}$. The functions $F_x(s)$ and $F_y(s)$ are meant to include all but the linear terms in the effective force. In the ideal linear lattice $F_x = F_y = 0$ [9].

Then the change of variables:

$$x = \sqrt{\beta_x(s)}u_x \tag{5}$$

$$\phi(s) = \frac{1}{Q_x}\int^s \frac{ds'}{\beta_x(s')} \tag{6}$$

yields

$$\frac{d^2u_x}{d\phi^2} + Q_x^2 u_x = 0 \ , \tag{7}$$

and similarly for motion in the vertical direction. $\beta(s)$ is periodic in s, that is $\beta(s + C) = \beta(s)$, where C is the circumference of the machine and

$$2\pi\int_s^{s+C} \frac{ds'}{\beta_x(s')} = Q_x \tag{8}$$

so that ϕ advances from 0 to 2π in one full turn. The distribution of quadrupole magnets determines the frequency Q_x and Q_y of transverse oscillations.

There is of course a third such equation describing longitudinal or energy oscillations. The so-called synchrotron frequency depends on the time dependence of the accelerating voltage and the energy dependence of the revolution frequency.

1.1 Nonlinearities and Resonances

The field errors, magnetic nonlinearities, beam-beam interaction etc. are included in the term $F(s)$ which leads to the modification of the simple harmonic oscillations as follows:

$$\frac{d^2u_x}{d\phi^2} + Q_x^2 u_x = Q_x^2\beta_x^{3/2}(s)F_x(s) \ . \tag{9}$$

The driving term $F(s)$ is necessarily periodic in s and can be written as a Fourier expansion $F(s) \to \sum a_m e^{im\phi(s)}$. If for example the perturbing field is simply a dipole kick sufficiently well localized that it can be described as a delta function, we can write, using (6),

$$F_x(s) = \delta(s)\Delta\theta = \frac{d\phi}{ds}\delta(\phi) = \frac{1}{Q\beta_x}\frac{\Delta\theta}{2\pi}\sum_{m=-\infty}^{\infty} e^{im\phi(s)} \ . \tag{10}$$

$\Delta\theta$ is the angular kick imparted to the beam by the dipole in question. Then (9) becomes

$$\frac{d^2u_x}{d\phi^2} + Q_x^2 u_x = Q_x\beta_x^{1/2}\frac{\Delta\theta}{2\pi}\sum_{m=-\infty}^{\infty} e^{im\phi} \ . \tag{11}$$

The driven oscillator (11) has resonances whenever $Q^2 - m^2 = 0$.

In general, in addition to its periodicity with respect to s, the function F may depend on the phase-space coordinates of the particle $(x, x', y, y', l, \delta p/p)$. Such dependence leads to higher-order resonances. For example, the force associated with a one-dimensional, localized sextupole field is $F(x, s) = S(s)x^2$, where $S(s) = \frac{1}{2}\frac{\partial^2 B_y(s)}{\partial x^2}\frac{q}{\gamma mc}$. We suppose that the field is uniform over the length l of the magnet so that the net kick due to the sextupole is $\Delta\theta = Sx^2 l$. If as before the s dependence is represented as a delta function, then the equation of motion is

$$\frac{d^2u_x}{d\phi^2} + Q_x^2 u_x = Q_x\beta_x^{1/2}\frac{Sx^2 l}{2\pi}\sum_{m=-\infty}^{\infty} e^{im\phi} \ . \tag{12}$$

We solve for the motion perturbatively, replacing x by its unperturbed value namely $x = a\sqrt{\beta}\cos(Q_x\phi)$. Then (12) becomes

$$\frac{d^2u_x}{d\phi^2} + Q_x^2 u_x = Q_x\beta^{3/2}\frac{a^2 S}{2\pi}\frac{1}{2}\sum_{-\infty}^{\infty}\left[e^{i(m+2Q_x)\phi} + e^{i(m-2Q_x)\phi} + 2e^{im\phi}\right] \ . \tag{13}$$

The beam responds resonantly when $(m \pm 2Q_x)^2 - Q_x^2 = 0$, or when $Q_x = m/3$.

In general, if $F(s)$ is independent of x, resonances occur for $Q_x = m$. If $F(s, x) \sim x$ there are resonances whenever $2Q_x = m$, and for $F(s, x) \sim x^2$ at $3Q_x = m$ etc. For the most general case, the perturbation F is a nonlinear function of x, y and z and the resonance condition is $pQ_x + rQ_y + nQ_z = m$ where p, r, n and m are integers. In an electron machine the radiation damping ameliorates the effects of all but relatively low order resonances. Nevertheless, great care is required in the choice of the operating point as we shall see.

1.2 Tune Plane

The resonant structure of the "tune plane" is apparent in a detailed measure of the beam lifetime, beam tails, or vertical beam size as a function of the machine tune. Such a scan in the vicinity of the CESR operating point is shown in Figure 1. In this scan the vertical beam size is measured at each of 3600 tunes. Each measurement appears at the corresponding point in the grid. In CESR, the revolution frequency is 390.1kHz. The plotted range in horizontal

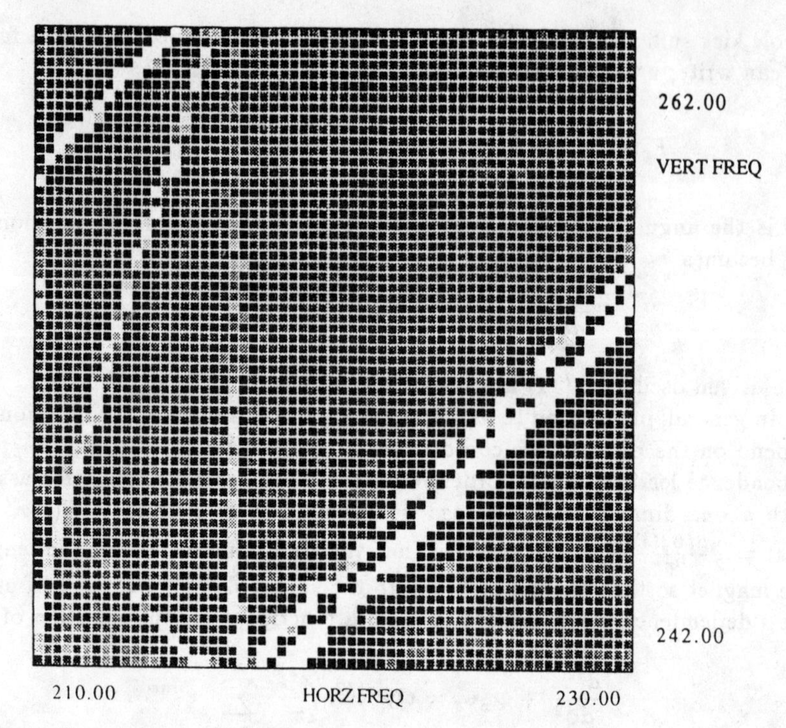

Fig. 1. Scan of vertical beam size as a function of horizontal and vertical tune appears in the upper plot. The synchrotron tune $Q_s = 0.06$. The CESR operating point is near the center of the scan.

fractional tune is from $Q_x = 210./390.1$ to $230./390.1$ $(0.538 < Q_x < 0.590)$, and in vertical tune $0.620 < Q_y < 0.672$. The step size in both horizontal and vertical is $\Delta Q = 8.5 \times 10^{-4}$. The synchrotron tune $Q_s = 0.06$ is fixed throughout the scan.

The most prominent line, across the lower right-hand corner of the plot, is the first synchrobetatron sideband off of the transverse coupling resonance, $-Q_x + Q_y - Q_s = 0$. The second sideband, $-Q_x + Q_y - 2Q_s = 0$, is in the upper left corner. The resonance that is independent of the vertical tune is a sideband off of the horizontal half integer, $2Q_y - 2Q_s = 0$. The remaining two lines satisy $3Q_x - Q_y = 1$, and $Q_x + Q_y - 3Q_s = 1$ respectively. It is left to the reader to determine which is which. It happens that colliding-beam performance is optimum when the single-beam tunes, that is the tunes of the σ mode, are $f_x \sim 218kHz$ and $f_y \sim 254kHz$.

The synchrobetatron coupling is due to the energy dependence of the optical functions[2] and/or the dispersion in the RF accelerating cavity. The dispersion is

[2] The focusing function $k(s)$ that appears in equation (1) will in general depend on the beam energy and the beam energy oscillates at the synchrotron frequency.

the energy dependence of the transverse displacement of the beam. Because the dispersion is nonzero in the CESR cavities, the RF kick, which also depends on energy, is coupled linearly to the horizontal displacement. Higher-order transverse resonances are due to the cumulative effect of the machine sextupoles. Indeed, single-beam resonance structure is well understood in terms of known coupling mechanisms and nonlinearities, and simulations yield scans that bear a remarkable resemblance to the observed tune dependence of beam size shown in Figure 1.

1.3 Matrix Formalism

In view of (9) it is obviously of great interest to compute $\beta(s)$. In principle we might compute $\beta(s)$ by noting first of all that the solution to the homogenous equation of motion is $u_x = a\cos\phi$, where a is an arbitrary constant, and then that $x = a\sqrt{\beta}\cos\phi$. Then substitution into (3) and (4) with $F_x = F_y = 0$ and using (6), yields a differential equation for $\beta(s)$. Given the periodicity of $\beta(s)$, it is uniquely defined by the differential equation.

In practice it is more convenient to compute $\beta(s)$ by way of a matrix formalism. In so far as the guide field consists of distinct multipoles, the focusing function $k(s)$ in equations (3) and (4) is piecewise constant. That is, k is uniform through each quadrupole, bend and drift. Therefore (3) and (4) can be solved directly for the motion within each element. In particular

$$x(s) = A\cos(\sqrt{k}s) + B\sin(\sqrt{k}s) \tag{14}$$

$$x'(s) = -A\sqrt{k}\sin(\sqrt{k}s) + B\sqrt{k}\sin(\sqrt{k}s)\ . \tag{15}$$

The constants A and B are determined by specifying the values of x and x' at some s. If we write that $(x(0), x'(0)) = (x_0, x_0')$ at the entrance to an element of length l then

$$x(l) = x_0\cos(\sqrt{k}l) + \frac{x_0'}{\sqrt{k}}\sin(\sqrt{k}l) \tag{16}$$

$$x'(l) = -x_0\sqrt{k}\sin(\sqrt{k}l) + x_0'\sqrt{k}\cos(\sqrt{k}l)\ , \tag{17}$$

and in matrix notation

$$\begin{pmatrix} x(l) \\ x'(l) \end{pmatrix} = M\begin{pmatrix} x_0 \\ x_0' \end{pmatrix} = \begin{pmatrix} \cos(\sqrt{k}l) & \frac{1}{\sqrt{k}}\sin(\sqrt{k}l) \\ -\sqrt{k}\sin(\sqrt{k}l) & \cos(\sqrt{k}l) \end{pmatrix}\begin{pmatrix} x_0 \\ x_0' \end{pmatrix}\ . \tag{18}$$

Each linear machine element can be described by a similar unit determinant matrix, and the mapping of the phase-space vector (x, x') through a sequence of elements, by the product of the intermediate matrices.

Now consider propogation of the vector [2]

$$\mathbf{x}_1 \rightarrow \mathbf{x}_2 = M\mathbf{x}_1\ , \tag{19}$$

where $x_1 = a\sqrt{\beta}_1 \cos\phi_1$. Differentiation with respect to s yields $x_1' = a\frac{1}{2}\frac{\beta_1'}{\sqrt{\beta_1}}\cos\phi_1 - a\frac{1}{\sqrt{\beta_1}}\sin\phi_1$. Then equation (18) becomes

$$\begin{pmatrix} a\sqrt{\beta}_2\cos\phi_2 \\ -a\frac{\alpha_2}{\sqrt{\beta_2}}\cos\phi_2 - a\frac{1}{\sqrt{\beta_2}}\sin\phi_2 \end{pmatrix} = M \begin{pmatrix} a\sqrt{\beta}_1\cos\phi_1 \\ -a\frac{\alpha_1}{\sqrt{\beta_1}}\cos\phi_1 - a\frac{1}{\sqrt{\beta_1}}\sin\phi_1 \end{pmatrix}. \tag{20}$$

Note that we have defined $\alpha \equiv -\frac{1}{2}\beta'$. Now if we define the matrix

$$G_i = \begin{pmatrix} \sqrt{\beta}_i & 0 \\ -\frac{\alpha_i}{\sqrt{\beta_i}} & -\frac{1}{\sqrt{\beta_i}} \end{pmatrix} \tag{21}$$

so that

$$\begin{pmatrix} x \\ x' \end{pmatrix} = G\begin{pmatrix} \cos\phi \\ \sin\phi \end{pmatrix}, \tag{22}$$

then (20) can be rewritten in a most suggestive form, namely

$$G_2\begin{pmatrix} a\cos\phi_2 \\ a\sin\phi_2 \end{pmatrix} = MG_1\begin{pmatrix} a\cos\phi_1 \\ a\sin\phi_1 \end{pmatrix}. \tag{23}$$

Apparently

$$G_2^{-1}MG_1 = \begin{pmatrix} \cos\Delta\phi & -\sin\Delta\phi \\ \sin\Delta\phi & \cos\Delta\phi \end{pmatrix}. \tag{24}$$

Here $\Delta\phi = \phi_2 - \phi_1$. Finally, (24) can be solved for M, and M can be written in terms of the twiss parameters α, β, and ϕ:[3]

$$M = G_2\begin{pmatrix} \cos\Delta\phi & \sin\Delta\phi \\ -\sin\Delta\phi & \cos\Delta\phi \end{pmatrix}G_1^{-1} \tag{25}$$

$$= \begin{pmatrix} \sqrt{\frac{\beta_2}{\beta_1}}[\cos\Delta\phi + \alpha_1\sin\Delta\phi] & \sqrt{\beta_2\beta_1}\cos\Delta\phi \\ \frac{1}{\sqrt{\beta_1\beta_2}}[\Delta\alpha\cos\Delta\phi - (1+\alpha_2\alpha_1)\sin\Delta\phi] & \sqrt{\frac{\beta_1}{\beta_2}}[\cos\Delta\phi - \alpha_2\sin\Delta\phi] \end{pmatrix}, \tag{26}$$

where $\Delta\alpha = \alpha_1 - \alpha_2$. And if M is the full-turn matrix, then $\beta_1 = \beta_2 = \beta$, $\alpha_1 = \alpha_2 = \alpha$ and $\Delta\phi = \mu = 2\pi Q$, and

$$M(s+C,s) = \begin{pmatrix} \cos\mu + \alpha\sin\mu & \beta\sin\mu \\ -\gamma\sin\mu & \cos\mu - \alpha\sin\mu \end{pmatrix}. \tag{27}$$

($\gamma \equiv (1+\alpha^2)/\beta$.) The twiss parameters at any point in the machine are thus determined by computation of the full-turn transfer matrix at that point.

---

[3] Note that for a 2×2 matrix T, $T^{-1} = T^{\dagger}/\det T$ and if $T = \begin{pmatrix} a & b \\ c & d \end{pmatrix}$, then $T^{\dagger} = \begin{pmatrix} d & -b \\ -c & a \end{pmatrix}$.

1.4 Propogating Twiss Parameters

The reader may be amused by a further extension of the formalism that permits propogation of the twiss parameters through a sequence of elements. Returning to (22), it is clear that,for any phase space vector $\mathbf{x} = (x, x')$, $G^{-1}\begin{pmatrix} x \\ x' \end{pmatrix} = a\begin{pmatrix} \cos\phi \\ \sin\phi \end{pmatrix}$ so that the quantity $(G_{11}^{-1}x+G_{12}^{-1}x')^2+(G_{21}^{-1}x+G_{22}^{-1}x')^2 = a^2(\cos^2\phi+\sin^2\phi)$ is an invariant. Substitution of the elements of G^{-1} identified in (21) yields $a^2 = \gamma x^2 + 2\alpha x x' + \beta x'^2 \equiv \boldsymbol{\gamma}\cdot\mathbf{X}$ where the three vectors $\boldsymbol{\gamma} = (\gamma, 2\alpha, \beta)$ and $\mathbf{X} = (x^2, xx', x'^2)$. The scalar product is an invariant. Since we know how x and x' propogate we can construct a 3×3 matrix $\mathcal{M}$ corresponding to propogation of $\mathbf{X}$. The reader can verify that

$$\mathcal{M} = \begin{pmatrix} M_{11}^2 & 2M_{11}M_{12} & M_{12}^2 \\ M_{11}M_{21} & (M_{11}M_{22}+M_{12}M_{21}) & M_{12}M_{22} \\ M_{21}^2 & 2M_{21}M_{22} & M_{22}^2 \end{pmatrix} . \tag{28}$$

Then $\mathbf{X}_2 = \mathcal{M}\mathbf{X}_1$. We would like to determine the corresponding matrix $\mathcal{N}$ for $\boldsymbol{\gamma}$. Since $\boldsymbol{\gamma}_2\cdot\mathbf{X}_2 = \boldsymbol{\gamma}_1\cdot\mathbf{X}_1 = \boldsymbol{\gamma}_1\cdot\mathcal{M}^{-1}\mathcal{M}\mathbf{X}_1 = \boldsymbol{\gamma}_1\mathcal{M}^{-1}\mathbf{X}_2$, it is obvious that $\boldsymbol{\gamma}_2 = (\mathcal{M}^{-1})^T\boldsymbol{\gamma}_1$ and $\mathcal{N} = (\mathcal{M}^{-1})^T$. $\mathcal{N}$ is constructed by replacing the elements of M with those of its inverse in (28) and transposing, yielding[4]

$$\mathcal{N} = \begin{pmatrix} M_{22}^2 & -M_{22}M_{21} & M_{21}^2 \\ -2M_{22}M_{12} & (M_{11}M_{22}+M_{12}M_{21}) & -2M_{21}M_{11}^2 \\ M_{12}^2 & -M_{12}M_{11} & M_{11}^2 \end{pmatrix} . \tag{29}$$

As an example consider the propogation of the twiss parameters through a field-free region. Using (18) to obtain the elements of the matrix M through the field-free region we find that

$$\mathcal{N}(s) = \begin{pmatrix} 1 & 0 & 0 \\ -2s & 1 & 0 \\ s^2 & -s & 1 \end{pmatrix} . \tag{30}$$

At a point of minimum β, in a field-free region like that near the interaction point, $\boldsymbol{\gamma}_0 = (1/\beta_0, 0, \beta_0)$ and at a distance s from the minimum

$$\boldsymbol{\gamma}(s) = \mathcal{N}\boldsymbol{\gamma}(0) = (1/\beta_0, -2s/\beta_0, \beta_0 + s^2/\beta_0) . \tag{31}$$

[4] In the case of coupled motion with transport described by 4×4 matrices, the procedure is readily generalized; γ becomes a 10-component vector and $\mathcal{N}$ a 10×10 matrix.

2 Luminosity

The objective of the basic collider is of course luminosity, which depends on the beam parameters as

$$L = \frac{n_b f_{rev} N_b^2}{4\pi\sigma_x\sigma_y} . \tag{32}$$

n_b is the number of bunches, N_b the number of particles per bunch, f_{rev} the revolution frequency and σ_x and σ_y the transverse beam size at the collision point. That is as long as the bunch length is small compared to height and/or width. It is clear that for a given beam current, L is a maximum when the cross-sectional area of the bunch is a minimum. The minimum bunch size is limited by the beam-beam interaction.

2.1 Beam-Beam Interaction

The dependence of the vertical kick experienced by a particle in one beam, as a function of its displacement from the centroid of the opposing beam, is indicated in Figure 2. The force is linear in displacement as long as that displacement is

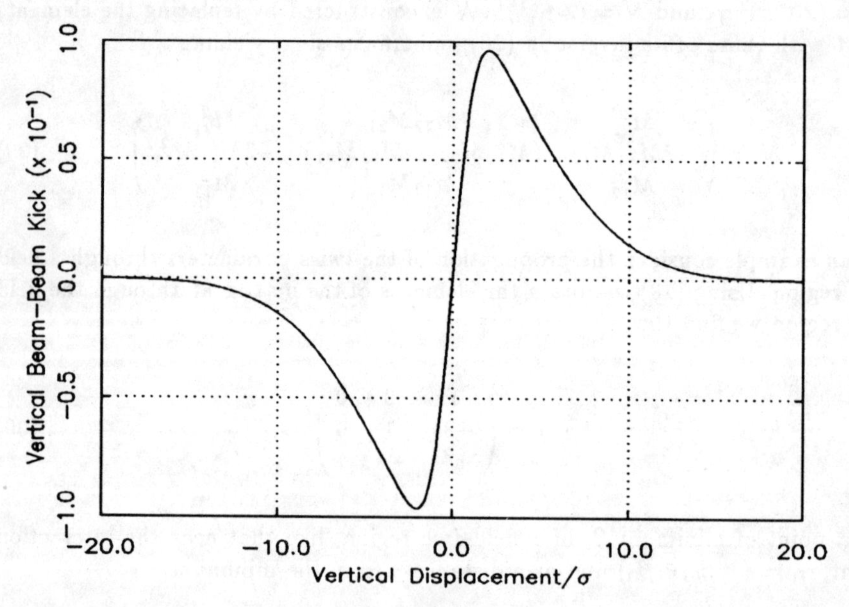

Fig. 2. The vertical beam-beam kick as a function of vertical displacement from the center of the opposing bunch for an aspect ratio $\sigma_x/\sigma_y = 10$. The kick is in arbitrary units and the displacement in units of the rms vertical beam size.

small, giving rise to the linear beam-beam tune shift. The focal length associated with the presumed Gaussian charge distribution in the linear regime is

$$\frac{1}{f_i} = \frac{2N_b r_e}{\gamma(\sigma_x + \sigma_y)\sigma_i}, \tag{33}$$

where $i = x, y$ and r_e is the classical radius of the electron. And the corresponding beam-beam tune shift is

$$\Delta Q_i = \frac{\beta_i^*}{4\pi f_i} \ . \tag{34}$$

The field becomes increasingly nonlinear for $y \geq \sigma_y$. The beam-beam interaction is the most important source of nonlinearity in the machine, or at least it should be. Presumably it is the nonlinearities that will limit the luminosity per unit current. Machine nonlinearities, that is, sextupole-induced nonlinearities can be controlled by careful design. The beam-beam-induced nonlinearities are beyond our control. Let us consider some of the dynamical implications of the beam-beam interaction.

2.2 Beam-beam coupling

In the event of conterrotating beams in collision we need to write an equation of motion for each. The resulting coupled equations are

$$\frac{d^2u_1}{d\phi^2} + Q_1^2 u_1 = Q_1^2 \beta^{*\frac{3}{2}} F_2(x_1 - x_2) \tag{35}$$

$$\frac{d^2u_2}{d\phi^2} + Q_2^2 u_2 = Q_2^2 \beta^{*\frac{3}{2}} F_1(x_2 - x_1) \ . \tag{36}$$

u_1 and u_2 are horizontal coordinates for the motion of beams 1 and 2 respectively. F_1 is the force on beam 2 due to beam 1 and F_2 the force on beam 1 due to beam 2. The beam-beam coupling yields normal modes A and B with tunes Q_A and Q_B such that $Q_A - Q_B \propto \Delta Q$, with ΔQ given in (34). In the limit of a purely linear beam-beam kick, the coupled equations can be solved by making the substitution $u_1 = u_A + u_B$ and $u_2 = u_A - u_B$. Then the tune of mode A, the σ mode, is the same as the unperturbed tune Q, and mode B, the π mode is shifted such that $Q_B - Q_A = 2\Delta Q$, for ΔQ defined above. That is, the normal-mode splitting is just twice the linear beam-beam tune shift.

The beam-beam coupling can also be evaluated in a matrix formalism. In the absence of the beam-beam coupling, the phase-space coordinates x and x' are mapped through a single turn by a 2×2 matrix M, so that $\mathbf{x}_{n+1} = M\mathbf{x}_n$ and $\mathbf{x} = (x, x')$. If the eigenvalues λ of M have unit magnitude, then the motion is stable. The eigenvalues have the form $\lambda = e^{\pm i\mu}$ and the tune $Q = \mu/2\pi$.

In order to add the linear beam-beam coupling we need to generalize the phase-space vector to accomodate the coordinates of both beams. So we define $\mathbf{X} = (x_1, x_1', x_2, x_2')$, where subscripts 1 and 2 indicate beams 1 and 2 respectively. Then propagation through a full turn is described by the 4×4 matrix

$T = \begin{pmatrix} M & 0 \\ 0 & M \end{pmatrix}$. The beam-beam kick is given by the

$$B = \begin{pmatrix} C & D \\ D & C \end{pmatrix} = \begin{pmatrix} 1 & 0 & 0 & 0 \\ 1/f & 1 & -1/f & 0 \\ 0 & 0 & 1 & 0 \\ -1/f & 0 & 1/f & 1 \end{pmatrix} . \tag{37}$$

Then according to B, the kick received by beam 1 is $\Delta x_1' = (x_1 - x_2)/f$, etc. and f is the focal length given above. The full-turn matrix

$$F = BT = \begin{pmatrix} CM & DM \\ DM & CM \end{pmatrix} \tag{38}$$

and the eigenvalues of F yield the normal-mode tunes. In particular

$$\cos\mu_A - \cos\mu_B = [(\cos\mu_1 - \cos\mu_2)^2 + |DM + (DM)^\dagger|]^{\frac{1}{2}} . \tag{39}$$

The technique is readily generalized to include the effects of multiple beam-beam interactions in a single turn.

Of course the beam-beam kick is linear only over a rather limited range. For particles with amplitudes that sample the nonlinear fields of the opposing bunch, the kick falls rapidly with displacement and the measured normal-mode splitting is somewhat less than that predicted by the purely linear model [1] [3]. Apparently the region of the tune plane clear of resonances must expand with the beam-beam interaction to accomodate the splitting of the normal modes.

2.3 Saturation of the Tune-Shift Parameter

In the limit where $\sigma_y \ll \sigma_x$ the beam-beam tune shift parameter

$$\xi_v \equiv \Delta Q = \frac{\beta^*}{2\pi} \frac{2 I_b r_e}{e f_{rev} \sigma_x \sigma_y} \quad \text{and} \quad L = \frac{n_b \gamma I_b \xi_v}{2\beta^* e r_e} . \tag{40}$$

I_b is the bunch current, e the charge of the electron, and we have made use of (32-34). The strength of the beam-beam interaction increases with bunch current, as the charge per unit area of the bunch. Resonant response of one beam to the other eventually leads to an increase in the area of the bunch that precludes further growth of the bunch charge density. The tune-shift parameter is saturated. Beyond saturation luminosity increases linearly with current. The maximum luminosity per unit current obtains above saturation and so that is clearly the regime in which to operate [4]. It is evidently desirable to remain in saturation throughout the course of a fill. Data collected at the Cornell Electron Storage Ring operation for high energy physics, and shown in Figures 3 and 4, indicate both saturation of the tune-shift parameter and the linear dependence of luminosity on current.

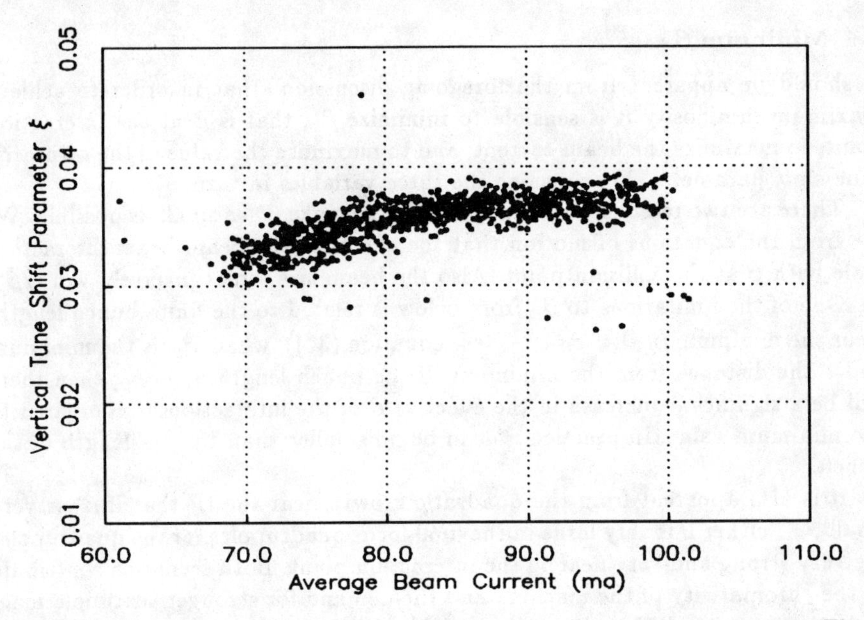

Fig. 3. Beam-beam tune-shift parameter as a function of average beam current. There are seven bunches in each beam. Data collected during a 24-hour period of operation for high energy physics are shown.

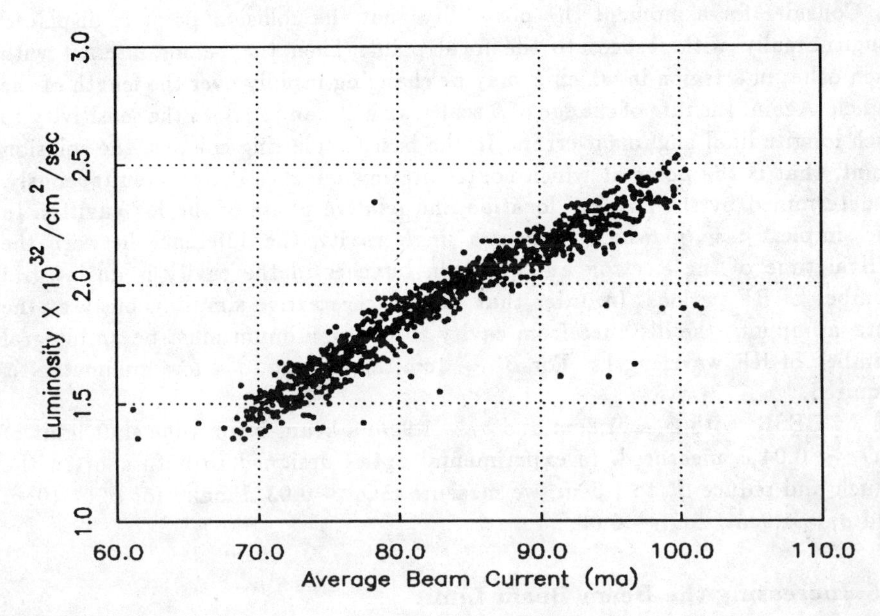

Fig. 4. Luminosity as a function of average beam current.

2.4 Minimum β

It should be apparent from the foregoing discussion that in order to achieve maximum luminosity it is sensible to minimize β^*, that is β at the interaction point, to maximize the beam current, and to maximize the value of the saturated tune-shift parameter. We consider the three variables in turn.

There are two reasons why it is attractive to make β^* as small as possible. We see from the equations of motion that the effects of the beam-beam interaction scale with β at the collision point. Also the beam size scales inversely wih $\sqrt{\beta^*}$.

One of the limitations to β^* from below is related to the finite bunch length. Near the minimum $\beta(s) = \beta_0 + \frac{s^2}{\beta_0}$ (see equation (31)) where β_0, is the minimum and s the distance from the minimum. If the bunch length $\sigma_z \sim \beta_0$, then there will be a significant increase in the effective β of the interaction as compared to the minimum value. In practice β^* can be no smaller than the length of the bunch.

It is also apparent from the quadratic growth near the IP that, if β^* is very small, either β is very large in the final focus quadrupoles, or the quadrupoles are very strong and very near to the interaction point. Both scenarios contribute to the chromaticity of the machine and the demand for stronger sextupole magnets. And as noted above, sextupole fields drive all manner of nonlinear resonances. Also, the interaction region tends to be cluttered with particle detectors and there is not always adequate real estate for optimal placement of such quadrupoles.

There are other practical, rather than fundamental limits on the minimum β. Consider for a moment the possibility that the collision point is displaced longitudinally with respect to the focal point. Then the beams interact with each other in a region in which β may be changing rapidly over the length of the bunch. Again, the rate of change of β scales with $\frac{1}{\beta_0}$, and so does the sensitivity to such longitudinal alignment errors. In the basic single-ring collider, the collision point, that is the point at which conterrotating bunches arrive simultaneously, is determined by the physical location and relative phase of the RF cavities. In the simplest case, in which there is a single cavity, the difference between the arrival time of the electron and positron bunches in the cavity is an integral number of RF periods. In order that the bunches arrive simultaneously at the beta minimum, the distance from cavity to that minimum must be an integral number of RF wavelengths. For $\beta^* \sim 1cm$, alignment to a few millimeters is required.

At CESR, with $\beta^* = 1.8cm$ and $\sigma_l \sim 1.9cm$ a beam-beam tune shift limit of $\Delta Q_v \sim 0.04$ is measured. In experimental optics designed to both shorten the bunch and reduce β^* to $1.3cm$, we measure $\Delta Q_v \sim 0.03$. Finally for $\beta^* \sim 10cm$ and $\sigma_l = 1.9cm$, $\Delta Q_v \sim 0.06$.

2.5 Increasing the Beam-Beam Limit

Referring again to our expression for luminosity (40), it is clear that maximizing the limiting beam-beam tune shift is of great interest. Our discussion of beam-beam coupling suggested that nonlinear resonances are largely responsible for limiting the bunch charge density, and that the beam-beam interaction itself is the source of many such resonances. In the basic collider with a single bunch in each beam, the conterrotating bunches collide at two diametrically opposed points. The equation of motion including the effects of the beam-beam interaction at both points is

$$\frac{d^2u}{d\phi^2} + Q^2 = Q^2\beta(s)^{\frac{3}{2}}\left(f_1(\mathbf{x}_1, s_1) + f_2(\mathbf{x}_2, s_2)\right) \ . \tag{41}$$

Taking advantage of the periodicity in s we have

$$\frac{d^2u}{d\phi^2} + Q^2 = Q^2\beta(\phi)^{\frac{3}{2}}\left(\sum_m a_m^1(\mathbf{x})e^{im(\phi-\phi_1)} + \sum_m a_m^2(\mathbf{x})e^{im(\phi-\phi_2)}\right) \ . \tag{42}$$

If the twiss parameters at the two collision points are identical, $f_1 = f_2$, and $a_m^1 = a_m^2 = a_m$,

$$\frac{d^2u}{d\phi^2} + Q^2 = Q^2\beta(\phi)^{\frac{3}{2}}\sum_m a_m(\mathbf{x})e^{im\phi}\left(e^{-im(\phi_1+\phi_2)}\right) \tag{43}$$

$$= Q^2\beta(\phi)^{\frac{3}{2}}\sum_m a_m(\mathbf{x})e^{im\phi}e^{-i\frac{m}{2}\phi^+}2\cos\frac{m}{2}\phi^- \ , \tag{44}$$

where $\phi^\pm = \phi_1 \pm \phi_2$. If the betatron phase advance separating the points is identically $2\pi(Q/2)$ then $\phi^- = \phi_1 - \phi_2 = \pi$. All m odd terms vanish and m even terms increase by a factor of two. The resonance condition is $pQ = 2m$. We might equivalently imagine that a single turn of the machine is two turns of a simpler machine with tune $Q \to Q_{\frac{1}{2}} = Q/2$. In the smaller machine there is a single beam-beam interaction with resonances at all m, $pQ_{\frac{1}{2}} = m$. It is easy to show with the help of (43,44) that for each resonance $pQ = 2m$ there is a corresponding resonance $pQ_{\frac{1}{2}} = m$ of equal strength. In such an idealized case of symmetrically placed collision points, the resonance behavior is determined not by the full-turn phase advance but rather by the phase advance from one point to the next, that is by $Q_{\frac{1}{2}}$. Multiple collision points are in such circumstance no cause for alarm as long as the point $Q_x/N, Q_y/N$ in the tune plane is clear of dangerous resonances. N is the number of interaction points per turn.

In practice it is very difficult to achieve perfect symmetry, both in terms of phase advance and interaction point parameters. Furthermore, resonances are excited by machine hardware, such as RF cavities and sextupole magnets. All such hardware must be symmetrically placed in order to ensure that $Q_{\frac{1}{2}}$ and not Q determine the resonance behavior. As an example, suppose that there are two interaction points and only a single RF cavity. Synchrobetatron resonances associated with cavity field nonlinearities will be circumscribed by the full-turn tune. Beam-beam resonances will depend on the half-turn tune. While it may be desirable to operate with phase advance just above the half integer for best beam-beam performance, in a machine with two interaction points that implies a full-turn tune just above the integer, very near to the synchrobetatron coupling resonance.

We conclude that while in principle performance is very nearly independent of numbers of collision points, the requisite optical symmetry is difficult to achieve and at very least is an expensive constraint on the machine design. The result of asymmetries is excitation of additional resonances and the narrowing of available good operating regions of the tune plane, and ultimately reduced beam-beam tune shift. Of course it is also much easier to keep one group of experimentalists happy than two.

In so far as radiation damping determines the equilibrium beam size it is desirable to have fewer collision points. The radiation damping decrement δ is the ratio of the time between collisions to the radiation damping time. A machine with one rather than two beam-beam collisions per turn has twice the damping decrement between collisions. On the other hand, that decrement is typically very small, of order two parts in 10^5 at CESR, and it is not clear how a factor of two is relevant. There is however some evidence that the limiting tune shift paramter $\xi_v \sim \sqrt{\delta}$ [4] [8].

2.6 Multiple Bunches

Luminosity is expected to scale with the number of bunches in each beam at fixed bunch current. As noted above, the number of collision points in the ring is twice the number of bunches in the beam. And there can be no longer any doubt that the consequence of those parasitic collisions will be a degradation of the beam-beam limiting tune shift. At best we suffer for having complicated the tune-plane. And unless we are prepared to install a low-beta insert at each of the new crossing points, the beam-beam current limit will fall precipitously. It is therefore attractive to consider separation of the beams at the parasitic crossing points, and this is typically accomplished with electrostatic deflectors. The most straightforward implementation is to locate separators $\pm\frac{1}{4}$ wavelengths from the crossing point, generating a half wavelength closed bump. Since the beam width tends to be many times the beam height, separation based on beam size is simpler in the vertical than in the horizontal plane. For the same reason, it is somewhat easier to build vertical than horizontal separators. The disadvantage of vertical separation is that vertical displacement in sextupole magnets introduces transverse coupling. If the vertical displacement is opposite for the two beams, then the transverse coupling is differential and therefore difficult to correct. Vertical separation is most usefully employed in regions devoid of sextupoles.

As the number of bunches in each beam and the number of parasitic crossing points is increased, it is advantageous to generate with a minimal set of electrostatic deflectors, a differential closed-orbit distortion that yields separation at multiple crossing points. In the "pretzel" scheme the closed orbits are off axis through a large fraction of the machine arcs and the sextupole-induced coupling noted above renders vertical separation a dubious proposition. At CESR, two symmetrically placed and powered pairs of horizontal deflectors separate the seven-bunch beams at the thirteen parasitic crossing points as shown in Figure 5.

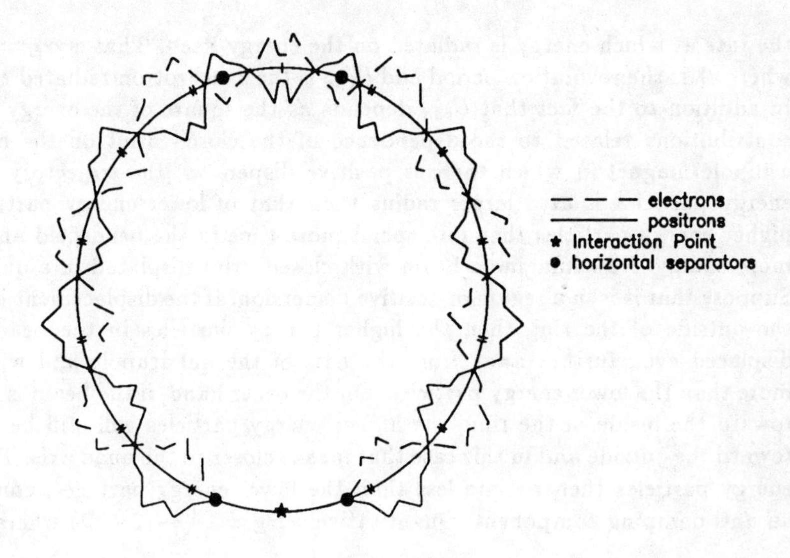

Fig. 5. Electron and positron closed orbits separate the beams at the parasitic crossing points with seven bunches in each beam. The crossing points are indicated by the tick marks in the plot. The multiplicity of crossing points within each lobe of the orbit is a result of the unequal bunch spacing.

Optical Effects of Differential Displacement in Quadrupoles. Because the orbits of the electrons and positrons are distinct through most of the quadrupole magnets, and nearly all of the nonlinear elements in the storage ring, there will in general be differences in the electron and positron beam parameters that increase with separator voltage. The exclusive use of magnetic elements for the guide field in the simplest machine assured commonality of closed orbits and beam optics. The introduction of the differential distortion by way of electrostatic elements spoils that symmetry, and it is compensated only by employment of specific design criteria.

In general both quadrupole and sextupole magnets can contribute to differences in the characteristics of the off-axis electron and positron beams. The effective field of a quadrupole magnet on the closed orbit is that of a bend with radius of curvature inversely proportional to the displacement. Electrons and positrons thus experience equal but opposite fields, and there will be asymmetric contributions to the dispersion function.

The radiation damping times will also be different for electrons and positrons. The damping rate α_E for energy oscillations is determined by the dependence of the rate at which energy is radiated on the energy itself. That is $\alpha_E = \frac{1}{T_0}\frac{dU_{rad}}{dE}$, where T_0 is the revolution period and U_{rad} is the synchrotron radiated energy [6]. In addition to the fact that U_{rad} depends as the square of the energy there are contributions related to the dependence of the closed orbit on the energy. In a dipole magnet in which there is positive dispersion, the trajectory of higher energy particles is at a larger radius than that of lower energy particles. The higher energy particles therefore spend more time in the bend field and radiate more energy. Now imagine a beam with closed orbit displaced in a quadrupole. Suppose that it is in a region of positive dispersion. If the displacement is towards the outside of the ring then the higher energy particles in the beam will be displaced even further away from the axis of the quadrupole and will radiate more than the lower energy particles. On the other hand, if the beam is displaced toward the inside of the ring, the higher energy particles will still be displaced toward the outside and in this case that means closer to the quad axis. The higher energy particles then radiate less than the lower energy particles, contributing an anti-damping component. Quantitatively $\alpha_E = \frac{U_0}{2T_0E_0}(2+\mathcal{D})$ where

$$\mathcal{D} = \frac{\oint \eta G(G^2+2k)ds}{\oint G^2 ds} , \tag{45}$$

$G(s) = 1/\rho(s)$ and $\rho(s)$ is the radius of curvature of the on-energy particle in the dipole field and k is defined in (3) and (4). In the event of displaced orbits in quadrupoles $\rho(s) = \frac{1}{x(s)k(s)}$ and (45) becomes

$$\mathcal{D} = \frac{\sum_i \eta_i(G_i^3 + 2x_ik_i^2)l_i}{\sum_i(G_i^2+(x_ik_i)^2)l_i} \tag{46}$$

where the integral is relaced by a sum over discrete elements and l_i is the length of each. To lowest order, neglecting the dependence of the dispersion on the particle species and noting that $x_i^+ = -x_i^-$,

$$\mathcal{D}_+ - \mathcal{D}_- = \frac{2\sum_i \eta_i(2x_ik_i^2)l_i}{\sum_i G_i^2 l_i} . \tag{47}$$

x_i^+ and x_i^- are the displacements of the positrons and electrons respectively.

Since the damping of the longitudinal oscillations is different for electrons and positrons, so is the damping of horizontal oscillations. As a consequence of the differences in damping times and dispersion, the beam emittances will also be unequal.

Effects of Differential Displacement in Sextupole Fields. The differential displacement of the beams in the sextupoles is perhaps even more serious than in the quadrupoles. A sextupole behaves like a quadrupole with focal length inversely proportional to the displacement from the axis. But the quadrupole distribution determines the β function and the tune. As a result, electrons and positrons may well have different tunes and β^* in a machine with pretzeled orbits. Sometimes it is possible to eliminate differential effects by exploiting various symmetries. In CESR for example, which is symmetric about the diameter that includes the interaction point, an antisymmetric closed-orbit distortion yields equal tunes and emittances, β^*, and chromaticity, but in general unequal dispersion and $\beta(s)$ throughout the arcs and nonzero β' at the interaction point.

Evaluation of Optical Parameters of Displaced Beams. In any event, the optics can be designed to minimize the electron and positron differences as long as we know how to calculate those differences, and it is to that end that we now turn our attention. We showed earlier the relationship between the elements of the linear map and the twiss parameters β, β' and the phase advance ϕ. If the closed orbit is coincident with the longitudinal axis of the machine elements, then the full-turn linear map is simply the product of the individual linear transfer matrices. (Remember that for nonlinear elements such as sextupoles the linear part of the transfer matrix is simply that of a field-free region.) The matrix describes the linear dependence of the phase-space coordinates of the trajectory as it leaves the element, on the phase-space coordinates at its entrance.

In a nonlinear element the linear part of the mapping of the phase space will in general depend on the displacement of the trajectory from the magnetic axis. Therefore the linear part of the mapping depends on the closed orbit. We can compute the linear part of the mapping, that is the Jacobian of the map, but first we must identify the coordinates of the closed orbit.

If we write the dependence of phase-space coordinates of the trajectory as it exits the magnet or sequence of magnets on the initial coordinates as $x_i^{out}(x_j^{in})$ then the Jacobian

$$J_{ij} = \frac{\partial x_i^{out}}{\partial x_j^{in}} . \tag{48}$$

Here x_i^{out} and x_j^{in} refer to the displacements $x, x', y, y' \ldots$ from the closed orbit. If all of the machine elements are strictly linear then $J_{ij} = M_{ij}$ where M is the product of the individual transfer matrices. In the presence of nonlinear elements, the derivatives J_{ij} can be computed numerically with a tracking code. If the mapping is through a full turn then there exists a closed orbit and its coordinates can be computed as well. If the motion is confined to a plane, so that $\mathbf{x} = (x, x')$, then it is necessary to compute $\Delta \mathbf{x}^{out}$ as a function of three distinct initial values $\Delta \mathbf{x}^{in}$ to establish both the location of the closed orbit and the matrix J. This is accomplished by tracking from the phase-space coordinate $\mathbf{x}_0 + \Delta \mathbf{x}^{in}$ through the full turn to find $\mathbf{x}_0 + \Delta \mathbf{x}^{out}$. $\mathbf{x}_0$ is the coordinate of the closed orbit. The numerically computed Jacobian is the Jacobian of partial derivatives in the limit where $\Delta \mathbf{x}^{in} \to 0$. Since $\Delta \mathbf{x}^{in}$ is the displacement with respect to the closed orbit and the closed orbit is in general not well established apriori, iteration may be required to obtain sufficient accuracy.

For motion that may include transverse coupling $\mathbf{x} = (x, x', y, y')$, five distinct and nondegenerate trajectories establish the 4×4 Jacobian matrix and the closed orbit. In order to determine the linear full-turn map we suppose that initial and final coordinates are related as follows :

$$\Delta \mathbf{x}_\alpha^{out} = J \Delta \mathbf{x}_\alpha^{in} \tag{49}$$

$$= \mathbf{x}_\alpha^{out} - \mathbf{x}^0 = J(\mathbf{x}_\alpha^{in} - \mathbf{x}^0) \ . \tag{50}$$

Then

$$\mathbf{x}_\alpha^{out} = J\mathbf{x}_\alpha^{in} - (J - I)\mathbf{x}_\alpha^0 = J\mathbf{x}_\alpha^{in} + \mathbf{z}^0 \ , \tag{51}$$

where $\mathbf{z}^0 = (I - J)\mathbf{x}^0$. The suffix α runs from one to five and indicates the five initial sets of coordinates x_α^{in} mapped to the five final vectors x_α^{out}, and $x_{i\alpha}^{out}$ indicates the ith component of the αth vector. That is $x_{i\alpha}^{out} = J_{ij}x_{j\alpha} + z_i^0$. Now define the matrices

$$Y = \begin{pmatrix} x_{11} & x_{12} & x_{13} & x_{14} & x_{15} \\ x_{21} & x_{22} & x_{23} & x_{24} & x_{25} \\ x_{31} & x_{32} & x_{33} & x_{34} & x_{35} \\ x_{41} & x_{42} & x_{43} & x_{44} & x_{45} \\ 1 & 1 & 1 & 1 & 1 \end{pmatrix} , \tag{52}$$

and

$$H = \begin{pmatrix} & & & & z_1^0 \\ & & J_{ij} & & z_2^0 \\ & & & & z_3^0 \\ & & & & z_4^0 \\ 0 & 0 & 0 & 0 & 1 \end{pmatrix} , \tag{53}$$

so that (51) becomes

$$Y^{out} = HY^{in} \ . \tag{54}$$

Then $H = Y^{out}[Y^{in}]^{-1}$ from which the Jacobian and the coordinate of the closed orbit can be extracted [7] .The twiss parameters in turn are defined by the Jacobian. Once the closed orbit has been established the linear mapping between any two points in the lattice can be calculated in this fashion, yielding $\beta(s)$ and $\phi(s)$ along the trajectory of the closed orbit.

We now are in a position to compute the distinct properties of the electron and positron beams, and so we can at least in principle arrange the optical elements to minimize the differences. The most significant include differences in beam energies, damping times, and β^*.

Separated Orbits and Guide Field Errors. Of course our ability to compute and compensate effects of displaced orbits is ultimately limited by our knowledge of the details of the guide field. In particular, multipole errors can introduce peculiar effects in the company of separated orbits.

Quadrupole errors introduce a differential kick that can result in differential horizontal displacement of the beams at the interaction point with calamitous impact on luminosity. Elimination of such differences depends on a capability to reliably measure relative displacement of the beams. Generally the collinearity of the beams at the IP can be restored by a straightforward adjustment of the electrostatic separator voltages.

Skew quadrupole errors introduce a vertical kick that is proportional to the horizontal displacement of the closed orbit. The resulting vertical separation at the IP is similarly removed if vertical electrostatic deflectors are available.

Lacking vertical deflectors there is no alternative but to use skew quadrupole trims to make corrections. This can prove a particularly unwieldy scenario. The skew fields couple horizontal and vertical oscillations. Skew quadrupoles located in regions where there is a large differential horizontal displacement will simultanously affect both the vertical closed orbit and the vertical beam size. Optimization of luminosity depends critically on careful tuning of skew quadrupoles. In a machine with displaced orbits, tuning is greatly facilitated if there are skew quadrupoles located in regions where the orbits coincide.

3 Crossing Angle

In a multiple-bunch scheme in which beams collide head-on, the collinearity through the interaction region is of great practical value. It permits a direct determination of the relative dispacement of the beams at the IP, and provides a stretch of guide field in which skew quads can be implemented with abandon. But the collinearity ultimately limits spacing of the bunches in each beam. To be precise, if the collinear region extends a distance s from the IP, then the distance between the bunches in each beam must be at least $2s$. Otherwise, the electrons and positrons will collide at the parasitic crossing point nearest the IP. We can accomodate beams with very closely spaced bunches in a single ring if we introduce a small horizontal crossing angle at the interaction point as shown in Figure 6. The crossing angle required to provide adequate separation at the first parasitic crossing point depends in detail on the bunch spacing, and $\beta(s)$ in the interaction region. In typical low-β optics with a bunch separation of 8.4 meters, a crossing half angle of $\theta_c \sim \pm 2mrad$ is sufficient.

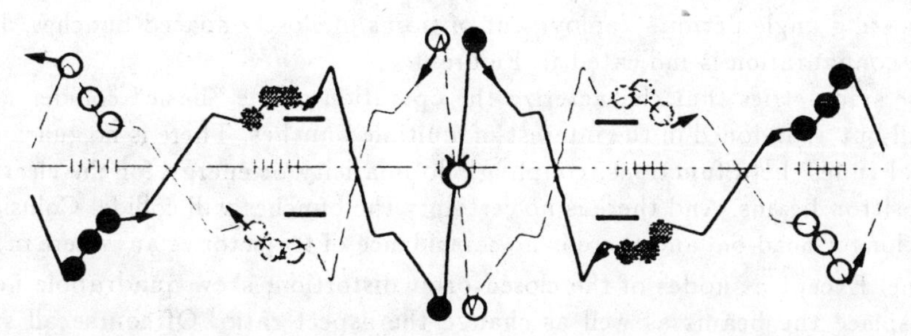

Fig. 6. Horizontal crossing angle at the interaction point. Note the separation of the incoming positron bunch (open circle) and the outgoing electron bunch (solid circle). The collision point is at the center of the plot. The three-bunch trains are 85 meters apart. Spacing of the bunches within each train is 8.4 meters.

3.1 Dependence of Beam-Beam Tune Shift on Crossing Angle

With the introduction of the crossing angle the nonlinear beam-beam kick gains an additional dependence on the longitudinal displacement of the particle from the center of the bunch and a synchrobetatron coupling results. The effect of the coupling is characterized by the "badness" parameter $\kappa = \frac{\theta_c}{(\sigma_x/\sigma_z)}$ [5].Measurements at CESR indicate a soft dependence of the beam-beam tune-shift parameter on the crossing angle as shown in Figure 7 [10]. The implementation

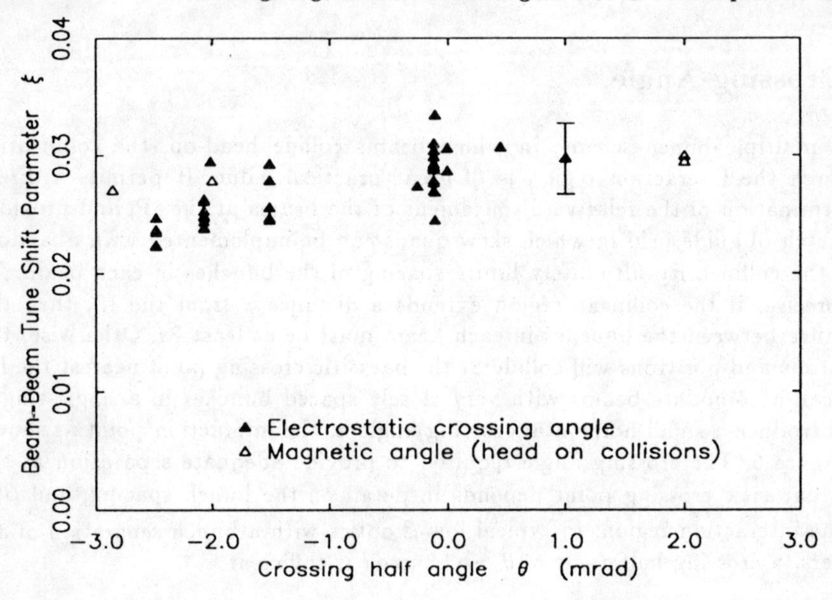

Fig. 7. The vertical tune-shift parameter as a function of crossing angle for average beam currents greater than $9ma/beam$. Open triangles are for head-on collisions, but with beams traversing the interaction point at the angle indicated. The open triangles thus give some notion of the effect of the large displacement in the IR optics independent of the crossing angle.

of a crossing angle permits deployment of trains of closely spaced bunches. Just such a configuration is indicated in Figure 8.

The symmetries that characterize the operation of the "basic" collider have been all but abandoned in the interest of multiple bunches. There is no guarantee of equal tunes, beta functions, coupling, chromaticity or energy for the electron and positron beams. And there is no certainty the bunches will collide. Collisions are no longer head-on, and there is no coincidence of trajectories anywhere in the machine. Except at nodes of the closed-orbit distortion, skew quadrupole fields will displace the beams as well as change the aspect ratio. Of course all such effects can be corrected in principle with the help of suitable combinations of trim magnets. But the tuning by the accelerator operator that is such a critical aspect of performance, is enormously complicated, and good diagnostics become increasingly important.

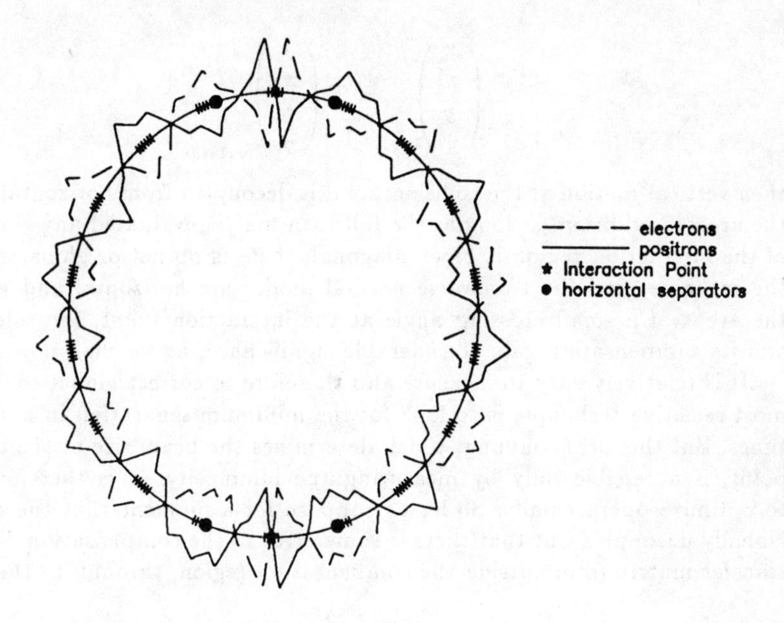

Fig. 8. The crossing angle at the IP is $\pm 2.1mrad$ and the peak horizontal displacement in the horizontally focusing interaction region quadrupole is about 2cm. The tick marks indicate the parasitic crossing points of the 9 trains of three bunches each. The bunches are spaced 8.4 meters apart within each train. The interaction point is at the bottom of the plot. Nine trains does not lend itself to equal spacings in CESR injector and storage ring, thus leading to the inconsistency in the number of crossing points from one lobe to the next.

3.2 Solenoid Compensation

Most collider detectors depend on the magnetic field of a solenoid in order to measure the momentum of the products of e^+e^- annihilation. The field of the solenoid generates transverse coupling in the stored beams. The radial fringe field at each end of the solenoid imparts a kick with amplitude proportional to the displacement of the beam from the magnet axis, and in a direction perpendicular to that displacement. The beam follows a circular path about the longitudinal field of the solenoid. Compensation of the transverse coupling is usually accomplished by a small rotation of the interaction region quadrupoles about their longitudinal axis. The rotation angles are chosen so that the mapping of the transverse phase space, through the compensation region to the interaction point, is block diagonal. Then horizontal emittance does not contribute to vertical beam size at the interaction point. That is, if the transport through the region is described by

$$T = \begin{pmatrix} M & 0 \\ 0 & N \end{pmatrix} \tag{55}$$

where M and N are 2×2 matrices, and

$$\begin{pmatrix} x \\ x' \\ y \\ y' \end{pmatrix}_{IP} = T \begin{pmatrix} x \\ x' \\ y \\ y' \end{pmatrix}_{outside} \tag{56}$$

then vertical motion at the collision point is decoupled from horizontal motion in the arcs. In addition, as long as the full-turn matrix evaluated anywhere outside of the interaction region is block diagonal, there is no net or global coupling in the machine, and the transverse normal modes are horizontal and vertical. In the event of a small crossing angle at the interaction point, the solenoid field and its compensation gain considerable significance, as we shall now see.

It is relatively easy to measure and therefore to correct global coupling. The most sensitive technique is to look for the minimum separation in normal-mode tunes. But the local coupling, which determines the beam size at the interaction point, is accessible only by measuring the luminosity. It is therefore difficult to optimize operationally. So let us suppose for a moment that the machine is globally decoupled but that there is some error in the compensation. Then if the transfer matrix from outside the compensation region, through to the IP is

$$T = \begin{pmatrix} M & m \\ n & N \end{pmatrix} \tag{57}$$

with m and n, 2×2 nonzero matrices, the vertical beam size at the interaction point is proportional to the horizontal beam size outside of the compensation region, and the luminosity is diluted. If there is a horizontal crossing angle, the closed orbits of the two beams have equal but opposite displacement and angle as they enter the compensation region. And the displacement and angle can not both be zero or there will be no crossing angle. In particular

$$\begin{pmatrix} x \\ x' \end{pmatrix}_{start} \sim M^{-1} \begin{pmatrix} 0 \\ x' \end{pmatrix}_{IP} \tag{58}$$

and $x'_{IP} = \theta_c^*$. It follows that the difference in the *vertical* orbits at the IP is related to the difference in *horizontal* orbits at the start of the compensation region:

$$\begin{pmatrix} \Delta y \\ \Delta y' \end{pmatrix}_{IP} = n \begin{pmatrix} \Delta x \\ \Delta x' \end{pmatrix}_{start} \tag{59}$$

In order to get some idea as to the size of such effects, suppose that the overall coupling of horizontal into vertical emittance is 1%. Now consider a 1% compensation error. That is, the matrix n has some combination of values such that 1% of the horizontal emittance is added to the vertical emittance at the IP, effectively doubling it. The beam size at the IP grows by a factor of $\sqrt{2}$ and the luminosity is similarly degraded. Meanwhile, the horizontal closed orbits of the two beams in the arcs are displaced an equivalent of $\sim 10\sigma_x$ in order that they be sufficiently well separated at the parasitic crossing points. The area in phase space bounded by the closed orbits of the two beams corresponds to an emittance about 100 times the horizontal emittance of a single beam since $\sigma_x \sim \sqrt{\epsilon_x\beta_x}$. The compensation error propogates 1% of that into the vertical at the IP leading to a vertical separation of $\Delta y \sim \sqrt{\beta_y\epsilon_x}$. The vertical beam size is already established to be $\sigma_y \sim \sqrt{0.01 \times 2 \times \epsilon_x\beta_y}$. The beams are thus separated by $\Delta y \sim 5\sqrt{2}\sigma_y$. We find that a compensation error that increases the beam size by 40% results in a vertical separation in excess of $5\sigma_y$. The beams effectively miss each other. Evidently a compensation error that has a moderate effect on luminosity in terms of increased beam size can have a devastating effect in terms of closed-orbit errors.

4 Conclusion

The symmetry of Maxwell's equations with respect to time reversal guarantees that, in machines in which guide fields are purely magnetic, orbits and twiss parameters be identical for both beams. But the basic e^+e^- collider has evolved to a multibunch machine with many more crossing points than collision points, and the optics must be specifically constrained to preserve common beam properties. The successful operation of ever more complicated configurations depends critically on an ability to diagnose machine errors and the flexibility to correct them. This is especially true in the regime of very small β^* and totally distinct closed orbits.

References

[1] Siemann, R., Meller, R.: Coherent Normal Modes of Colliding Beams, IEEE Trans. Nucl. Sci. NS-28, No. 3, 2431 (1981)

[2] Courant, E.D., Snyder, H.S.: Theory of the Alternating Gradient Synchrotron, Ann. of Phys. **3**,1(1985)

[3] Chao, A.W., Bambade, P., Weng, W.T.: Nonlinear Beam-Beam Resonances. Nonlinear Dynamics Aspects of Particle Accelerators, Proceedings, Sardinia 1985, ed. J.Jowett, M.Month, S.Turner, 93-94

[4] Seeman, J.: Observations of the Beam-Beam Interaction, Nonlinear Dynamics Aspects of Particle Accelerators, Proceedings, Sardinia 1985, ed. J.Jowett, M.Month, S.Turner, 126-137

[5] Piwinski,A.: Simulations of Crab Crossing in Storage Rings, SLAC-PUB-5430, February 1991

[6] Sands, M.: The Physics of Electron Storage Rings, SLAC-121, 1970, 100-102

[7] Servranckx, R., Brown, K.: SLAC Report 270 UC-28 (A), March 1984

[8] Rubin, D.: Doubling the Damping Decrement in CESR, CON 87-12

[9] For a more complete treatment along these lines see Ruth, R., Weng W.: Physics of High Energy Particle Accelerators, AIP Conference Proceedings, No. 87 (1981), ed. R.Carrigan, F.Huson, M.Month, 4-16

[10] Rubin, D. et al.: Beam-beam interaction with a horizontal crossing angle, to be published in Nuclear Instruments and Methods A.

How to Reach High Luminosity?

Y. Baconnier

CERN, Geneva, Switzerland

1 Introduction

How to reach high luminosity? is a question that successively the particle physicist, the accelerator designer and the machine experimenter have to face. The particle physicist and the machine designer have an essential job to do in common, namely to set 'reasonably optimistic' performance targets. This initial estimate will be the future yardstick to judge the performance of the machine after it has been built. Once the collider is built, the accelerator experimenter—often the same as the one who designed it—is supposed to achieve at least the design performance. At the same time the particle physicist is eagerly asking for the beam.

The aim of this lecture is more to give a survey of the problems to be solved, than to present a detailed scientific account of each of the effects which limit the performance. The main arguments leading to the basic choice of parameters are presented with insistence on the interrelation between the different effects.

2 Luminosity Definition

By definition the luminosity is the number of events produced by the collisions, per second, for events with a cross section of one square centimeter. Since a typical cross section unit is one nanobarn (1 nb $= 10^{-33}$ cm^2), a luminosity $\mathcal{L} = 10^{33}$ cm^{-2} s^{-1} only produces one such event per second, in which case the luminosity is said to be one inverse nanobarn per second. The figure that one quotes as luminosity is in general the peak luminosity of the machine, expressed in cm^{-2} s^{-1} which mostly interests machine designers.

Luminosity integrated over a week (see Fig. 1), or at least several runs is what physicists are interested in; it is often measured in inverse picobarn. Note that one inverse picobarn is one thousand times larger than one inverse nanobarn. In MKS unit: 1 pb^{-1} $= 10^{40}$ m^{-2}.

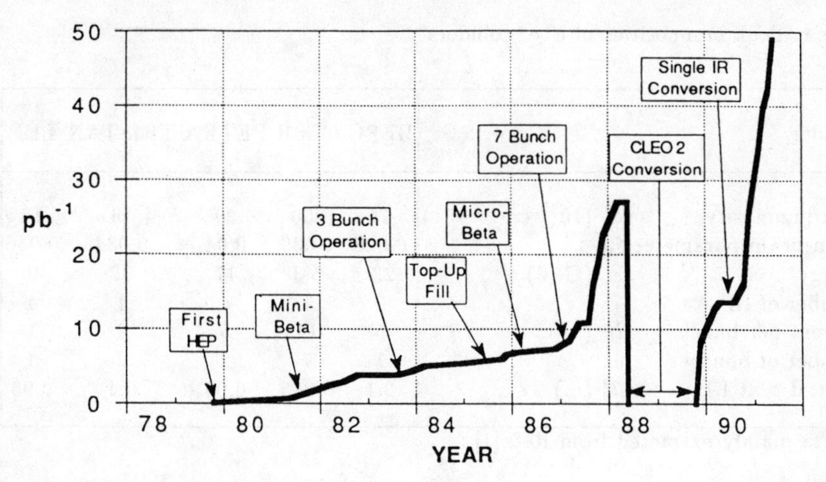

Fig. 1. CESR integrated luminosity per week

3 Performance

The performance of some of the most modern e^+e^- colliders is presented in Table 1. The world luminosity record ($\mathcal{L} = 2.10^{32}$ cm^{-2} s^{-1}) is presently—and presumably for a number of years—the property of the CESR machine at Cornell. It is interesting to see how this has been achieved. Figure 1 extracted from the presentation of D. Rice at the recent BEPC upgrade workshop [2] shows that with 12 years of constant effort the luminosity of CESR has been improved by a factor 100. The recipes are given in the figure: smaller beta at the Interaction Point, more bunches and only one Interaction Point.

The top-up fill mentioned in Fig. 1 corresponds to the injection of only the missing current at the end of the run rather than a complete refill. In fact the filling—or topping-up—time is an essential ingredient of the performance of a collider, it not only allows an increase of the average luminosity, but—more important— it allows the machine development sessions to be very efficient and consequently speeds up the rate of improvements. The dead time between two physics runs is obviously shorter if the injectors are powerful enough, that is if the injection is made at full energy and if the positron source can provide the several ampere required by factories in a short time. All factory designs include a full energy injector.

One more element cited by D. Rice [2] explains the performance of CESR: one quarter of the machine operating time is reserved for machine studies. This implies an extremely powerful team of machine experimenters, working in shifts. In spite of this large fraction of machine study time one can see from Fig. 1 that several months of hard work are required between the commissioning of an upgrade and the corresponding luminosity increase. In the case of the CLEO II conversion, the graph indicates that six months after the stop scheduled for the modification of the detector, the machine performance was still reduced to

Table 1. Peak luminosities of e^+e^-colliders*

Collider		BEPC	CESR	PETRA	TRISTAN	LEP
Peak luminosity	$\mathcal{L}$ (10^{30} cm^{-2} s^{-1})	7.	200	20	14	13
Beam-beam parameter	ξ_v	0.028	0.039	0.04	0.034	0.036
Energy	E (GeV)	2.2	5.3	17	32	46
Number of I.P.		2	1	4	1	4
Current per bunch	I_b (mA)	5	14	5	4	1
Number of bunches	n_b	1	7	2	2	4
Vertical β at I.P.	β_v^* (m)	0.1	0.015	0.06	0.1	0.05

* Data mainly extracted from Ref. [1]

half the performance achieved before the stop; life has not always been easy at CESR.

4 Design Process

The choice of the main parameters is made with the permanent preoccupation of leaving enough flexibility in the machine for progressive luminosity improvement by the accelerator experimenter.

4.1 Simulations

The search for the optimum values of beam dynamics parameters is made using simulation programs and a number of workshops have been dedicated to these. The major problem to simulate large numbers of turns without loosing the properties of conservation attached to Hamiltonian motion (Liouville theorem or more generally simplecticity) has been solved. Considerable progress has been made in introducing all possible effects in the simulation, including hardware imperfections. The simulation programs are now an essential tool of the designer. Nevertheless the predictive power of simulation programs in the search for the optimum working point (Q_h, Q_v) for example is not very good. Several machines have had to change their working point to reach their maximum luminosity.

The reason, or the consequence, is that the difference in the performance of the various colliders is not clearly understood. Experience proves that the luminosity is limited by various effects taking place in the bunch crossing and measured (in first approximation) by the tune shift induced in beam one by the space-charge effect of beam two. The beam-beam tune shift parameter (ξ) of most e^+e^- colliders is given in Table 2. A number of attempts have been made to find scaling laws applying to the variation of this parameter with energy, number of interactions or damping decrement. There may be a trend towards higher beam-beam parameter values with energy, and towards lower values with increasing number of interaction points, but even those apparently reasonable trends are not obviously confirmed by the performance of existing colliders. The maximum value of the beam-beam parameter has a remarkably constant value, varying between 0.025 and 0.06 in most cases, for a range of energy of two orders of magnitude, with the corresponding differences in damping time, synchrotron frequency and natural emittances; this over the 20 years which separate ADONE and LEP. It is nevertheless difficult to predict the beam-beam parameter and therefore the luminosity of a new machine to better than 30–40%.

4.2 Equations

The detail design of a collider is complex involving the simultaneous solution of a large number of equations. At the end of this school you will have a case full of them. The performance of a collider can nevertheless be described using a small number of key equations.

The first one is the luminosity definition, simplified to the case where the colliding bunches have the same number of particles N, and the same rms horizontal and vertical beam size at the interaction point, σ_h and σ_v:

$$\mathcal{L} = \frac{N^2 f}{4\pi\sigma_h\sigma_v} n_b \tag{1}$$

where f is the revolution frequency, and n_b the number of bunches. The fraction is sometimes called the single-bunch luminosity.

The second equation concerns the beam-beam parameter. This is denoted by ξ and its value is

$$\xi_v = \frac{r_e}{2\pi}\frac{\beta_v^*}{\gamma}\frac{N}{\sigma_v(\sigma_h+\sigma_v)} \tag{2}$$

where r_e is the classical electron radius. In e^+e^- accelerators the vertical beam size is negligible compared to the horizontal one so that the last fraction can be simplified. For the same reason the beam-beam parameter in the horizontal plane is less constrained.

The third equation is a result of the combination of the two above equations where the number of particles per bunch has been replaced by the single-bunch current I_b, using: $I_b = eNf$, where e is the electron charge. Then:

$$\mathcal{L} = \frac{1}{2er_e}\frac{\gamma\xi_v}{\beta_v^*} n_b I_b \tag{3}$$

Most of the high-luminosity constraints are present in this equation. Comparing with the successive CESR upgrades we see the beneficial effect of mini-beta's: lowering β_v^*, of many-bunch operation: increasing n_b and, as already mentioned, the effect on the ξ parameter of working with a single interaction point. The

Table 2. Beam-beam tune shift of e^+e^- colliders*

Collider	Energy (GeV)	ξ_v	Nb of IP
VEPP-2M	0.5	0.050	2
DCI	0.8	0.041	2
ADONE	1.5	0.070	6
SPEAR	1.2	0.018	2
	1.9	0.056	2
	2.1	0.055	2
BEPC	1.6	0.035	2
DORIS-2	5.3	0.026	2
VEPP-4	5.0	0.050	1
KEK-AR	5.0	0.030	2
	5.0	0.045	1
CESR	4.7	0.018	2
	5.0	0.022	2
	5.3	0.026	2
	5.5	0.028	2
	5.4	0.020	2
	5.4	0.035	1
PEP	14.5	0.045	6
	14.5	0.065	2
	14.0	0.050	1
PETRA	7.0	0.014	4
	11.0	0.024	4
	17.0	0.040	4
TRISTAN	30.4	0.034	4
LEP	45.6	0.035	4

* Data mainly extracted from Ref. [4]

place where the optimization of all these parameters raises the most delicate problems is the interaction area, but before discussing this we must consider the idea of the double-ring collider, which opens the way for large luminosity improvements.

5 Bunch Spacing

Of all the quantities in equation (3) the only one susceptible of increasing the luminosity by an order of magnitude, as requested by the new factory specifications, is the number of bunches (or its equivalent in this case, the bunch spacing around the ring). This is possible in two ways, either put more bunches in a single ring—The CESR solution—or have the two beams in two different rings, solution selected for future modern factories.

5.1 The Pretzel Scheme

The seven bunches of CESR circulating in the same ring, cross in 14 places around the ring. If nothing was done, the tune shift induced by the beam-beam effect in each of the crossings being of the order of 0.04, the total beam-beam tune spread would be $\delta Q = 14 \times 0.04 = 0.56$ and most of the beam would probably be lost in a few turns. The solution adopted by CESR—the so-called Pretzel scheme—is to have the two beams circulate on different orbits so that at the crossing points not used for experiments they are separated. At these parasitic crossings the beam-beam effect is considerably reduced by this separation, but it is still present so that one cannot increase the number of bunches at will. Moreover, each of the beams have to be accommodated in a smaller part of the vacuum chamber and the nonlinear optics are severely complicated by the requirement to 'comfortably' install two beams on two different trajectories inside the same vacuum chamber. Still, if you can, with time and effort, accommodate the two beams, you have a chance to win the world race for luminosity, which CESR did.

5.2 The Double-ring Collider

With the two beams each installed in their own vacuum chamber, the beams only see each other in the interaction area, the limitation to the number of bunches is now in the separation scheme in the interaction area. The obvious disadvantage is the cost of the installation of two rings instead of one. Also some specific problems have to be solved: the mechanical and magnetic stability of the two rings have to be carefully checked in order to avoid that the two beams move or vibrate at the collision point where the beam sizes are only a few microns.

The double ring makes it possible to avoid collisions in the arcs. However, in factories the bunch spacing considered is a few meters only as indicated in Table 3. The separation of the beams must be made in the interaction area. tab2

6 Low-beta Section

To achieve high luminosity low beta values are required at the interaction point. The assembly of elements used to achieve this, starting from the regular lattice, is called the low-beta section, or the interaction area optics. It usually includes, starting from the interaction point: a quadrupole doublet, a matching section, a dispersion suppressor, and a set of skew quadrupoles in order to compensate the effect of the detector solenoïd. In the case of double rings a set of beam separators is required. When the separation is made in the vertical plane a vertical dispersion matching is required. This in itself is an interesting exercise with the requirement to simultaneously match ten parameters. In the case of

Table 3. e^+e^- factories

Collider		ϕ-factory[1]	τ-c factory[2]	B-factory[3]	Z-factory[4]
Energy	(GeV)	.5	2.5	9	45
Circumference	(m)	98	360	2199	26659
Max. beam current	(A)	5.3	.5	1.5	0.027
Bending radius	(m)	1.5	12	165	3100
Energy loss per turn	(MeV)	0.01	0.2	3.6	260
Radiation losses/ring	(MW)	0.01	0.1	5.3	7
Rms bunch length	(cm)	3	.7	1	2
β-function at IP	(cm)	4.5	1	3	5
Beam-beam parameter		0.04	0.04	0.03	0.03
Bunch spacing	(m)	4.2	12	1.26	740

[1] Daphne[5]
[2] τ-charm project[6]
[3] High energy ring of SLAC design[7]
[4] High luminosity LEP with 36 bunches[8]

the B-factory this must be done separately for two different energies, and with elements common to the two beams close to the interaction point. The solutions proposed should be transparent enough that the experimenter can understand, measure, and correct possible imperfections.

Several considerations influence the design of the linear optics in the low-beta section:

- *Chromaticity*
  The very strong quadrupoles required in low-betas will focus more low energy particles of the beam than high-energy ones. This effect (the chromaticity induced by the low-beta optics) must be corrected to prevent head-tail instabilities. This correction is made using sextupoles, which reduce the dynamic aperture.
- *Quadrupole strength*
  The scaling laws for the quadrupole doublet of the low-beta optics (see e.g. Möhl [3]) indicate that in order to decrease the beta value by a factor a, and keep the chromaticity constant, the distance of the first lens to the interaction point must be reduced by the same factor a. In this scaling the quadrupole length is reduced by a, the k-value is multiplied by a^2, the aperture by $a^{1/2}$, so that the pole field is multiplied by $a^{3/2}$. This explains the fact that most designs select the superconducting technology for the low-beta quadrupoles.

– *Detector forward acceptance*
By pushing the quadrupoles closer and closer to the interaction point one increases the solid angle where the detector is blind, which makes the interpretation of events more uncertain. Quadrupoles with minimum transverse dimensions are therefore favored. This explains the choice of permanent magnet quadrupoles in some designs.

– *Detector masking*
The detector must be protected from stray radiation to avoid excessive background. Two sources of background must be considered: the synchrotron radiation and the circulating beam interaction with the residual gas or the vacuum chamber. Their effect is analyzed using tracking programs which include routines to describe the secondary particle production when the incident particle hits an obstacle. It is only recently that the comparison between simulations and actual measurements at LEP or DESY have proved satisfactory. The result of the study is a set of masks placed at convenient positions, close to the interaction point, to stop the incident particles. The situation is particularly delicate in the case of B-factories where the vertex detection imposes a small radius of the vacuum pipe at the interaction point. Moreover the vertex detector is itself extremely sensitive to radiation damage.

– *Mechanical layout*
It must be possible for the low-beta quadrupoles installed inside the detector to be removed or reinstalled when the detector is being repaired or replaced. This, together with the requirement to install in the minimum possible space the cooling of superconducting quadrupoles, and the required pumping speed close to the masks where the desorption is important, imposes severe constraints on the mechanical design of the machine parts inserted in the detector.

– *Separation*
The aim of separation is not only to send the two beams into their respective rings, but also to separate the beams as close as possible to the interaction point. This is to allow the injection of as many bunches as possible with a minimum number of crossings at places where the beams are separated but not yet isolated in their own vacuum chamber; these are called distant crossings. How distant these should be is still a matter of debate. For separation in the horizontal plane it seems that a distance d equal to 7 to 9 rms of the horizontal distribution at the point of crossing is sufficient, for vertical separation figures ranging from 3 to 7 times the rms of the horizontal beam size have been quoted. In any case the question must be solved for each particular collider because the parasitic crossing effect on luminosity depends very much on a number of other parameters such as the main beam-beam tune shift, the working point selected, the number of useful and of parasitic crossings etc. This is one of the many applications of simulation programs.

The separation is made using electrostatic deflectors in equal energy machines, and magnetic deflectors in B-factories where the two beams have different energies. This is the reason why B-factories have smaller bunch longitudinal distances: the separation is easier.

Optimization of the luminosity can only be attacked once these various effects have been understood and their consequences introduced in the design. This in general requires several rounds of successive versions of design. The final design and construction will be made keeping in mind these various possibilities so that they can be tested in the machine during the experimental search for maximum luminosity.

7 Radio-frequency Systems

A number of considerations enter into the design of factory RF systems. The latter are one of the key elements in the design of Z- and B-factories because of the enormous amount of synchrotron radiation losses.

- *Synchrotron radiation*
  The energy loss per turn due to synchrotron radiation scales as E^4/ρ where ρ is the average bending radius in the machine. The optimization of the machine gives a bending radius approximately proportional to the square of the energy so that the energy loss per turn is also approximately proportional to the square of the energy. Two single-cell accelerating cavities are required for DAPHNE whereas LEP requires a total of 640 cells in phase one.
  Not only the voltage required from the RF system can be very high, but the circulating current of these colliders is around one ampere, with the consequence that the RF power to be delivered to the cavities is very large indeed (Table 3). For this reason a double-ring Z-factory is not proposed, it would consume too much RF power.
- *Bunch length*
  In order that all particles cross at the waist of the low beta, the bunch length must be short compared to the value of β^*. How short the bunch should be is not clear. Experiments at CESR demonstrated that, when the bunch length is limited by the RF voltage available, the optimum value of β^* is equal to the rms bunch length. For the designer the question is rather: how short should be the bunch to maximize the luminosity at a given β^*? There are some indications that the bunch length should be shorter than β^* by a factor between 1.5 and 2. Since the RF voltage required varies with the inverse of the square of the rms bunch length, the RF specification depends very much on the answer given to this question.
- *Transient beam loading*
  The beam ionizes molecules of the residual gas, the ions collect in the potential well of the electron beam and induce blow-up by the space-charge effect. In order to eliminate the ions one suppresses a number of bunches in the circulating beam, the ions are destabilized by this 'hole' in the potential well and no longer accumulate. This hole, in turn, induces a considerable transient beam loading in the RF system at the revolution frequency. The

net effect is to modulate the RF voltage, the RF phase and therefore the longitudinal position of the bunch crossing in the interaction area. This is not acceptable. The solution is first to introduce a hole in the positron beam to equilibrate the transient so that the bunches cross at the interaction point, and second to arrange a low level RF control which can live with these powerful transients. The Z-factory bunches are too distant to allow ion collection but in the ϕ-factory the circumference is too short to install a hole so that the ions will have to be cleared using clearing electrodes.

– *Beam loading*
Apart from this transient beam loading effect, the low-level-RF engineer of the large current colliders is faced with a new problem linked to the high circulating beam current. Due to the beam loading at the fundamental frequency, which is not in phase with the required voltage, the RF cavity must be detuned so as to present a matched load to the RF source. This detuning can be as large as one revolution frequency in a machine with a large radius (small revolution frequency) and large current. The large impedance of the detuned cavities will excite the first mode (m = 1, n = 1) of longitudinal multibunch instabilities. The corresponding dipole oscillation of the bunch must be controlled by a feedback system within the RF low-level control.
– *RF technology*
The debate between normal and superconducting cavities is still open for these machines with the exception of the ϕ-factory were the normal conducting cavities are sufficient and the Z-factory were the conversion to superconducting cavities is underway. Normal conducting cavities consume more power for less accelerating voltage and present a higher impedance to the beam. They are reliable, simple to operate and not sensitive to synchrotron radiation. Recent experience with superconducting cavities in TRISTAN and LEP show a large sensitivity of superconducting cavities to the circulating current. The plans are, in general, to start with normal conducting cavities at lower luminosity and to test progressively superconducting cavities. This was the procedure for LEP and could be the procedure for future τ-charm and B-factories.

The RF specialist must combine the competence of the accelerator physicist, of the low-level-electronics engineer and of the high-power specialist. In this domain, as in all others, the designer should prepare the hardware for the adjustments or modifications he will have to make when his work changes from designer to experimenter during the commissioning stage, and later development.

8 Instabilities

Several types of instabilities can reduce the beam current in operational conditions. The catalogue of these instabilities, and of the correction technique has been made in detail. In his lecture J. Gareyte [9] provides some advice: 'The most important phase (of the correction of instabilities) is the process of understanding, which requires good observation skills coupled to a thorough knowledge of the theoretical models'.

Instabilities can affect the longitudinal or transverse motion, they come in two types: single bunch or multibunch. They are induced by the wake fields left after the passage of the bunch through the various sections of the vacuum envelope. A quantity called impedance has been introduced which measures the ratio between the volts applied to the beam due to these wake fields and the beam current. In the longitudinal case this ratio is measured in ohms. In the transverse case it is the beam transverse displacement which is both the source and the result of the instability so that the transverse impedance is measured in ohms per meter of transverse displacement.

- **Single bunch**
  Single-bunch effects require that the action is made in a short time, during the bunch passage. In the frequency domain this is equivalent to a broad-band effect. The instability is therefore characterized by the broad-band part of the impedance. This comes from the vacuum chamber resistive wall itself but is often dominated by broad-band resonant structures like bellows, kickers or instrumentation. Longitudinal instabilities induce longitudinal blow-ups and therefore bunch lengthening. The threshold can be increased by increasing the RF voltage. Transverse instabilities of the simple head-tail modes are stabilized by using positive chromaticity. The transverse, head-tail, mode-coupling instability, where two head-tail modes couple, is usually destructive. No cure has been found.

- **Multibunch**
  These instabilities are generated when a coherent pattern of transverse or longitudinal oscillations of all bunches around the ring enters into resonance with a parasitic mode of oscillation of the RF cavities (or other similar resonant equipment). A single mode of oscillation could be easily damped by damping the resonant mode in the cavity, or by a feedback system measuring the bunch position and feeding back with the proper phase a deflecting or accelerating force onto the beam, or by both actions simultaneously. The difficulty is that with a large number of bunches there is a large number of such resonant patterns covering a large frequency band. Similarly there are many parasitic Higher-order Modes (HOM) of oscillations of the cavities capable of entering into resonance with the rotating patterns. The damping of HOM in the accelerating cavities is indispensable because the strength of the resonance can be so big that blow up occurs before the bunch phase has changed enough for the feedback system to be efficient. The bandwith and power requirements for the feedback system are at the limit of present technologies. Prototypes are being built using digital feedback techniques [10].

– **Beam current**
The highest beam current ever reached in an e^+e^-storage ring is less than one ampere, this in extreme conditions with large beam blow-up. The circulating current in factories is of the order of one ampere. [1] The various effects at the origin of this blow-up (mainly instabilities) are now well understood and accelerator physicists have estimated that the risks in entering this new working condition are acceptable, providing the conditions stated by J. Gareyte are fulfilled.

9 Non-conventional Schemes

Two proposals do not follow the traditional scheme of head-on collisions at zero dispersion: the Cornell CLNS laboratory has decided to investigate and propose a crab-crossing scheme, at Novosibirsk studies of monochromators open the way for a proposal where the classical regime of beam-beam interaction is modified. In Cornell Yuri Orlov is investigating the possibility to have mm long bunches cross in mm low beta to reach still higher luminosity.

9.1 Crab-crossing

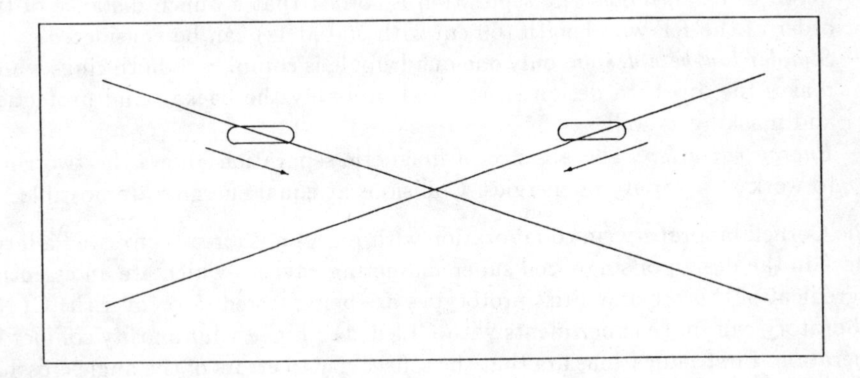

Fig. 2. Crab-crossing scheme

This scheme [12] was initially proposed [11] for linear colliders. The aim is to have the bunches cross at an angle in the interaction point. This provides the most efficient separation. In order to avoid the corresponding problem of the excitation of synchro-betatron resonances when the bunches do not perfectly overlap during the collision, the colliding bunches are tilted on their orbit (Fig.

[1] The Z-factory current must be limited to lower values to avoid excessive radiation losses with the corresponding problems of power consumption and cooling. The ϕ-factory will have 5 A of circulating current.

2). This way, the collision in the center of mass is made head-on even though the bunches cross at an angle.

The bunches are tilted by an RF cavity working in a deflecting mode, with a phase adjusted to zero at the center of the bunch so that the head and the tail are deflected in opposite directions. A second cavity placed at the other side of the interaction region, at a betatron phase advance of π with respect to the first deflection, restores the bunch to its normal trajectory. The scheme presents several difficulties:

- *Tolerances:* The deflection phase and amplitude of the two deflecting cavities must be adjusted with precision in order to avoid a residual angle of crossing, or a residual oscillation inducing synchro-betatron resonances.
- *Betatron phase advance:* A betatron phase advance error between the two compensated cavities has the same effect as an amplitude difference between them: synchro-betatron resonances are excited.
- *Parasitic modes:* The deflecting mode in a normal cavity is not (unlike the accelerating one) the mode of oscillation of the cavity with the lowest frequency. Therefore the corresponding lower-order modes are more difficult to damp and the cavities add to the general ring impedance which is already at the limit in these machines.

The advantages are also numerous:

- *Small bunch spacing:* The separation is so fast that a bunch distance of the order of the RF wavelength (60 cm with 500 MHz) can be considered.
- *Simpler low-beta design:* only one quadrupole is common to both rings which makes the low-beta design easier, and probably the background protection and masking as well.
- *Energy variation:* The absence of magnetic separation allows the two rings to work at a variety of energies. Collisions at equal energies are possible.

The Cornell laboratory (in collaboration with KEK) has already invested a large effort in the design of single-cell superconducting cavities which are an essential ingredient of this scheme. First prototypes are being tested. Moreover the CLNS laboratory can make experiments with CESR, the highest luminosity collider in operation. First indications are that the sensitivity to errors of the angle crossing scheme is not so severe as was feared. Experiments are continuing. A number of studies and experiments have been performed in various laboratories on this scheme, which is proposed for the high-luminosity version of the ϕ-factory under construction at Frascati.

9.2 Monochromators

The Novosibirsk laboratory has published [13] a study on a scheme which, at the origin, was invented [14] to insure a better energy resolution of the observed collisions. It turns out however that this scheme modifies the beam-beam interaction so that the current equations of luminosity limitation are no longer valid.

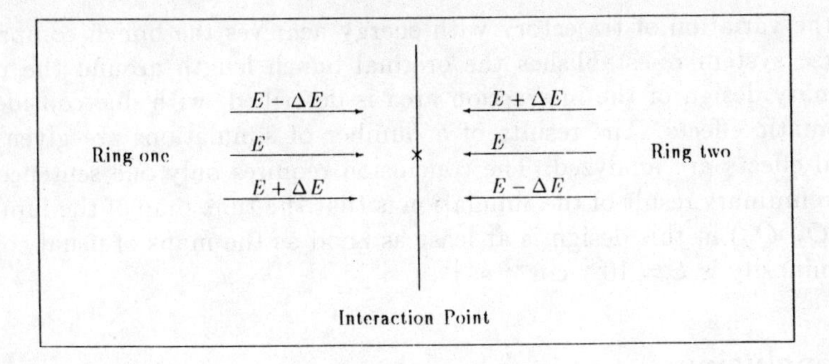

Fig. 3. Monochromator scheme

The idea is to take advantage of the two rings to install at the interaction point energy dispersions of opposite sign. In this way (see Fig. 3), particles with opposite energy difference collide so that the energy in the collision is always $2E$. The horizontal beam size at the interaction point is now dominated by the energy dispersion and no longer by the emittance.

The risk is obviously the excitation of synchro-betatron resonances, reason for which the dispersion function is usually made precisely equal to zero at the interaction point. The authors however make the point that to excite synchro-betatron resonances one needs nonlinearities in the betatron phase space and that if the emittance is reduced by a large enough factor, a monoenergetic part of the beam will only sample a very small horizontal fraction of the beam-beam force, so that these nonlinear terms will be considerably reduced. The linear part of the force (the beam-beam tune shift) is no longer a measure of the nonlinear terms so that this parameter does not have the same physical signification.

The scheme [15] has been applied to the tau-charm factory and combined with a proposal to achieve polarization in the same rings.

Obviously one should examine in detail this new scheme, and its consequences, before reaching conclusions. The scheme does not benefit from the several decades of experience accumulated in e^+e^- colliders. Other obstacles are no doubt on the way, but it is still one of the very few options left for trying to improve the luminosity of colliders. If a two-ring collider (Tau-charm or B-factory) is built one day with only one interaction area used, a monochromator could be tested using the second crossing point.

9.3 Micro-betas and Bunch Compression

Working at Cornell, Yuri Orlof has proposed [16] to further reduce, by an order of magnitude, the value of beta at the Interaction Point: β^*. Simultaneously a reduction of the bunch length is required in order to make good use of the low-beta value all over the bunch. In the proposed scheme this is achieved by using a classical bunch compression technique combining an RF cavity, which provides an energy modulation from the head to the tail of the bunch, and a magnet where the variation of trajectory with energy achieves the bunch compression. A reverse system re-establishes the original bunch length around the ring. A preliminary design of the interaction area is described, with due consideration to chromatic effects. The results of a number of simulations are given where essential effects are analyzed. The conclusion requires only one sentence: 'The main preliminary result of the simulation is that the tune map of the luminosity $\mathcal{L} = \mathcal{L}(Q_x, Q_y)$ in this design is at least as good as the maps of usual colliders. The luminosity is $\mathcal{L} = 10^{35}$ cm^{-2} s^{-1}.

10 Conclusion

In order to be competitive, there is a tendency for proposals for new colliders to announce rather high values for the expected luminosity. Past experience with electron colliders has proved that it is extremely long and painful to achieve the design peak luminosity once the machine is built. This is even more true with the large size of high energy machines. Still one should design for a high luminosity in order to avoid that the luminosity is limited by under-design rather than by more fundamental problems. The recipe used in all proposals is to keep the design flexible enough that the experimenter can try a variety of parameter sets. The hope is that in the future the simulation programs will make it possible to do a large part of this search on the computer.

In other words it is far easier to give a lecture on how to achieve high luminosity than to actually work on it for days and nights in a control room.

References

1. Zhang Chuang ed., Workshop on BEPC luminosity upgrades, Institute of High Energy Physics, Chinese Academy of Science, Beijing, China, June 1991
2. D. Rice, Luminosity upgrades at CESR, in Ref. [1].
3. D. Möhl, On the low-beta optics for round and flat beams, CERN PS/AR/Note 90-15
4. D.H. Rice, Beam-beam interaction: Experimental, AIP conf. Proc. 214,(1990)
5. INFN, Proposal for a ϕ-factory, LNF-90/031(R),(1990)
6. J. Jowett, Machine summary, Meeting on the tau-charm factory detector and machine, Universidad de Sevilla, Andalucia, Spain, (1991)
7. An asymmetric B factory based on PEP, SLAC-372, (1991)
8. E. Blucher et al. ed, Report of the working group on high luminosities at LEP, CERN 91-02, (1991)
9. J. Gareyte, Observation and correction of instabilities in circular accelerators, Proc. joint US-CERN school, Hilton Head Island, Springer- Verlag, Lecture Notes in Physics, No. 400, and CERN SL/91-09 (AP). (1991)
10. F. Pedersen, Feedback systems, Contribution to the workshop: B Factories: The state of the art in accelerators, detectors and physics, Stanford April 6-10 1992, to be published.
11. R. Palmer, SLAC-PUB-4707 (1988)
12. K. Oide, K. Yokoya, SLAC-PUB-4832 (1989)

13. I. Protopopov, A. Skrinsky, A. Zholents, Energy monochromatisation in storage rings, Preprint INP 79-6, Novosibirsk, (1979)
14. M. Bassetti, Frascati ADONE Internal Report E-18 (1974)
15. A. Zholents, Polarized J/ψ mesons at a tau-charm factory with a monochromator scheme, CERN SL/92-27 (AP) (1992)
16. Y. Orlov et al., B-factory optics and beam-beam interaction for millimeter β^* and locally shortened bunches, CLNS 91/1092 CBN 91-9, (1992)

Φ-Factory Design

C. Biscari
INFN, Italy

Introduction

The search for fundamental interaction laws continuously stimulates the accelerator community to increase the energy of accelerators. The present need for high-accuracy measurements in already explored energy regions requires a new generation of colliders, the $e^+ e^-$ factories. These should work at the energies of hadronic resonances and produce a very high rate of events with a luminosity increased by at least two orders of magnitude with respect to the present values. Such high luminosity poses interesting new problems to both the detector and accelerator physicists.

The lowest-energy factory is expected to work at the Φ resonance (~ 1 GeV) and is mainly dedicated to the investigation of CP violation. The construction of a collider in the Φ energy range requires a relatively small investment, in terms of budget and manpower, and can therefore be afforded by medium-size laboratories. A strong advantage of this kind of project is the small size of the groups concerned so allowing them to participate in all the problems connected with the detector and the accelerator.

The basic requirements of a Φ factory are:

- *Luminosity* = 10^{32} - 10^{33} cm^{-2} s^{-1} at the c.m. energy of 1.02 GeV. This is a challenging request: the maximum luminosity in this energy range has been obtained by VEPP-2M [1] and is two orders of magnitude lower ($\mathcal{L} \sim 5\ 10^{30}$ cm^{-2} s^{-1}). The maximum luminosity ever obtained is of the same order but has been obtained at an energy one order of magnitude larger ($\mathcal{L} \sim 1.8\ 10^{32}$ cm^{-2} s^{-1} at the energy of 5.3 GeV in CESR [2]).
- *High average luminosity*. The peak cross-section at the Φ resonance is:

$$\sigma_{peak} = 4.4\ 10^{-30}\ cm^2$$

  The Φ production required for the precision experiments on CP violation is at least 10^{10} events per year, which corresponds to $\mathcal{L} = 2.3\ 10^{32}$ cm^{-2} s^{-1} assuming 10^7 seconds operating time in a year; to obtain such a high average luminosity a very reliable storage ring is needed.
- *Large free solid angle* in the detector: the dream of the experimentalists is a storage ring with a high luminosity and no machine components near the luminosity source, which of

course are conflicting requirements. Coordinated designs of detector and accelerator are necessary.

Most experimental proposals require symmetric energies: two beams colliding at 510 MeV, with equal parameters (beam current, emittance, dimensions, energy spread). The possibility of using a Linac against a ring will not be treated in this lecture. A general overview will be given on the Φ-factory design strategy, on the main new techniques and on the existing projects.

1 Basic design parameters

The single-bunch luminosity for an electron-positron storage-ring collider is given by the well known formula:

$$\mathcal{L} = f \frac{N^2}{4\pi\, \sigma_x\, \sigma_y} \tag{1}$$

where f is the collision frequency, N is the number of electrons and positrons per bunch (assumed to be the same) and σ_x and σ_y are the horizontal and vertical rms beam sizes at the interaction point (IP).

The luminosity is limited by the beam-beam interaction. When two beams collide, one sees the other as a focusing lens, whose strength is given by a dimensionless parameter, the so called beam-beam tune shift

$$\xi_x = \frac{r_e\, N\, \beta_x}{2\pi\gamma\sigma_x\, (\sigma_x + \sigma_y)} \qquad \xi_y = \frac{r_e\, N\, \beta_y}{2\pi\gamma\sigma_y\, (\sigma_x + \sigma_y)} \tag{2}$$

r_e being the classical electron radius, $\beta_{x,y}$ the betatron functions at the IP, and γ the particle energy in units of its rest mass. The value of ξ has an empirical limit that will be discussed later. In order to maximize the luminosity $\mathcal{L}$ it is convenient to arrange for the beam-beam tune shifts to be a maximum and equal in the two planes; this is possible if the ratio between the emittances, the coupling factor κ, is equal to the ratio of the beam rms sizes at the IP. It follows that κ is also equal to the ratio of the betatron functions at the IP:

$$\kappa = \frac{\varepsilon_x}{\varepsilon_y} = \frac{\sigma_x}{\sigma_y} = \frac{\beta_x}{\beta_y} \tag{3}$$

With $\varepsilon = \varepsilon_x + \varepsilon_y$, the beam-beam tune shift can be rewritten as:

$$\xi = \frac{r_e\ N}{2\pi\gamma\,\varepsilon} \tag{4}$$

By introducing ξ in the luminosity expression, we obtain:

$$\mathcal{L} = \pi f \frac{\gamma^2}{r_e^2}\ \frac{\xi^2\ \varepsilon\ (1 + \kappa)}{\beta_y} = \pi f \frac{\gamma^2}{r_e^2}\ \frac{\xi^2\ \varepsilon\ (1 + \kappa)}{\kappa\beta_x} \tag{5}$$

This formula shows the relevant accelerator parameters that can be adjusted in order to maximize the luminosity; a discussion follows on each of them to illustrate the choices that can be made.

1.1 Linear Tune-shift Parameter ξ

When two beams collide the electromagnetic interaction between particles acts as a focusing lens producing a tune shift. The interaction force is linear only up to an extent of about one rms beam size and has its maximum at 1.6 σ (see Fig. 1). The beam-beam tune shift linear part, given by Eq. (2), is proportional to the beam density (strength of the interaction), to the betatron function β (sensitivity to interaction) and inversely to γ (rigidity of the beam). The focusing strength on a particle, and therefore its tune shift depend on the particle position inside the bunch: due to beam-beam interaction a tune spread arises whose width is well represented by ξ, because the maximum tune shift corresponds to particles which are inside the core of the bunch distribution and therefore see the linear part of the interaction force.

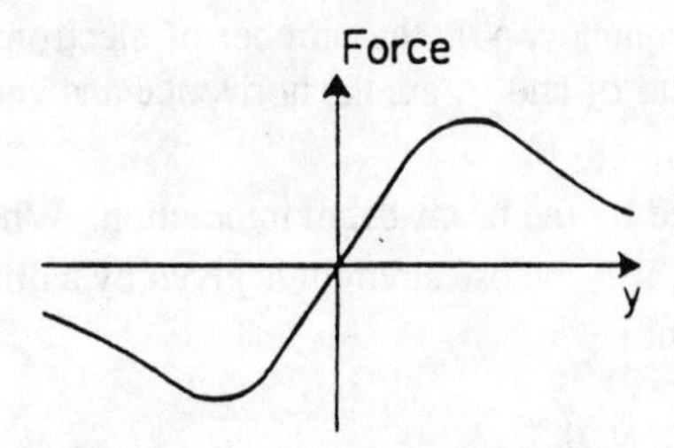

Fig. 1. Beam-beam force

It is clear from (5) that a luminosity increase can be obtained by increasing ξ. Unfortunately it seems not to be possible to go to high values of ξ: the beam-beam tune shift increases linearly with the beam current up to a maximum value. Further increase of the beam current results in beam blow-up or even in beam loss. Experimental evidence shows that the ξ limit is:

$$\xi = 0.038 \pm 0.013$$

averaged over most existing colliders. This number seems, in principle, independent of the energy, and of the age of the accelerators.

Fits of the experimental data have been performed by several authors, e.g. Seeman in 1985 [3] and more recently by Bassetti [4], analyzing the best performance of a group of e^+e^- colliders. It is not easy to find a general law from the collider measurements, because not all the machines were optimized to the same extent, and their performance can be limited by different problems such as single-bunch current threshold, rf, alignment, beam separations, etc. Seeman concluded with a possible dependence of ξ_y on the energy E, on the characteristic ring bending radius ρ and on the number of crossings per turn N_i:

$$\xi_y \propto \frac{E}{\sqrt{N_i\, \rho}}$$

The latest news from CESR [2] says that with the removal of one of the two IPs the luminosity and the value of ξ_y have fortunately almost doubled, which seems in disagreement with the above fit, even if one has to consider that some ring characteristics were changed between the two configurations: different beam separation schemes, different solenoid compensation, elimination of dispersion at the IP.

Bassetti has considered a larger number of colliders including old and low-energy accelerators, and new and high-energy ones like LEP, and found that the best fit for ξ_y is the following:

$$\xi_y \propto \frac{1}{(\rho \sqrt{N_i}\,)^{.4}} \left(E \sqrt{\frac{\beta_x}{\varepsilon_x}}\right)^{.5} \kappa^{-1.7}$$

It is interesting to note that the fit does not depend on the energy E, since even if E appears in the formula, ε_x being proportional to E^2, the E dependence cancels out. The fit favours flat beams rather than round ones. In any case, when designing a new factory, a more or less conservative value for ξ must be assumed, so determining N/ε. ξ may change with the tune in a difficulty predictable way: the ring lattice should therefore be very flexible, in order to compensate for unexpected effects in the beam-beam interaction.

1.2 Low Beta

From the luminosity formula it is evident that the smaller is β at the IP the higher the luminosity: the beam-beam interaction changes the divergence of particles; where the divergence is large the unwanted deflection is less troublesome. Anyway β cannot be squeezed at will, mainly because of the geometrical limitation of luminosity [3]. The betatron function, starting from a symmetry point, shows a parabolic dependence on the distance s:

$$\beta_y(s) = \beta_y^* \left(1 + \left(\frac{s}{\beta_y^*}\right)^2\right)$$

The betatron function within the bunch increases rapidly, so that to keep the advantage of having a low beta, β_y^* must be larger than the half bunch length σ_z which is determined by the ring rf system, impedances and single-bunch instability thresholds. Moreover low β_y^* implies high chromaticity with a corresponding reduction of dynamic aperture and also strong focusing near the IP, where normally there is a lack of free space because of detector requirements. Reasonable values for both σ_z and β_{oy} are in the order of a few centimeters.

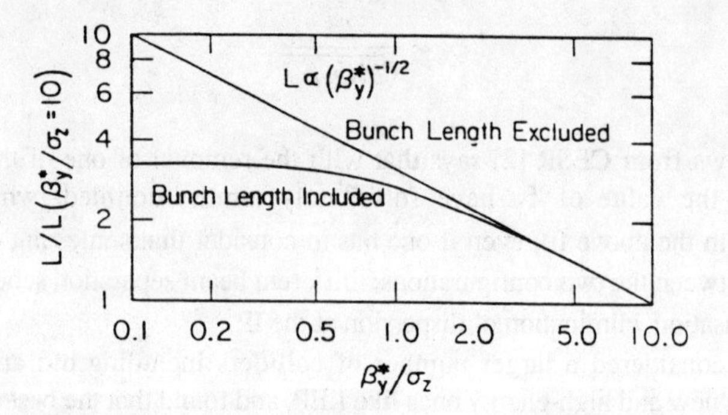

Fig. 2. Luminosity dependence on vertical betatron function [3]

1.3 Emittance

The equilibrium emittance is determined by the radiation process which depends mostly on the characteristic radius of the bending field, and on the properties of the lattice. To increase the luminosity one would like to have large emittance, but ε cannot be made arbitrarily large because of the machine physical aperture necessary for a reasonable beam lifetime.

The dynamic aperture is also sensitive to ε: large emittance means that most particles perform betatron oscillations in the magnetic field region where multipole components are more dangerous. Furthermore, ξ being proportional to N/ε, increasing ε means increasing the bunch current, to which beam instabilities are related.

The relatively large value of the emittance (~ 1 mm mrad) can be obtained either with small bending radius dipoles or with wigglers which also allows emittance tuning.

1.4 Coupling Factor κ

The coupling factor can be chosen to be between 0 and 1, corresponding respectively to a very flat, or to a round, beam.

$\kappa = 1$ implies *round beam* at the IP: $\beta_x = \beta_y$, $\varepsilon_x = \varepsilon_y$, $\sigma_x = \sigma_y$.

For the Φ factory, due to the relatively low energy of the beam, it is possible, at least in principle, to focus a round beam to the low-beta values with a solenoidal field, while, in higher-energy rings, quadrupoles are necessary to obtain strong focusing and it is critical to focus in both planes at the same point in the lattice.

With $\kappa = 1$ a factor 2 is directly gained on the luminosity (see formula (5)). Some beam-beam simulations claim that a higher ξ can be obtained [5,6] but this has not yet been tested. There exists a very interesting proposal to make round beams in VEPP-2M [7], to check the physics of beam-beam interaction in this configuration.

Strong focusing in both planes tends to increase the maximum value of the betatron functions near the IP. As a consequence, the chromaticity increases and the aperture requirements become more demanding in the horizontal plane. Also the vertical aperture becomes critical due to the vertical emittance, with the obvious drawback on vacuum and complexity of element designs.

A low value of κ implies a *flat beam*, with a vertical emittance much smaller than the horizontal one. Such a choice is convenient from the point of view of apertures and chromaticity, because only one of the planes is critical. It also implies reduction of ion trapping, which is very important in multibunch operation. Accurate orbit correction is necessary to ensure the design value of coupling.

1.5 Collision Frequency

There are two basic strategies to increase the collision frequency: to store the beams in two rings in multibunch operation and to intersect them at one, or maximum two, IPs, or alternatively to use very compact rings and only one bunch per beam.

The compact-ring design maximum collision frequency is a few tens of MHz and therefore the single-bunch luminosity

$$\mathcal{L}_o = \mathcal{L} / n_b$$

where n_b is the number of bunches in each beam, must be pushed well beyond the limit of existing machines. Conversely, in double-ring multibunch operation, the value of $\mathcal{L}_o$ is of the order of that achieved in presently operating machines and it is the number of bunches that must be increased.

Both choices have their advantages and their drawbacks, and these will be discussed later.

2 Projects and Proposals

There are several projects around the world for Φ factories; two of them (Novosibirsk and UCLA) follow the compact-ring approach. The other three are two-ring configurations. In the following table their main parameters are listed, together with those of VEPP-2M for comparison.

A general description of compact-ring characteristics and some words on each proposal will now be presented. Then follows a discussion of the two-ring configurations with special emphasis, of course, on the Frascati project.

3 Compact Ring

The size of a Φ detector is of the order of 3 to 5 m; to fit it inside a ring a minimum circumference of ~ 20 m is needed, corresponding to a maximum collision frequency of the order of 15 MHz. This relatively low value of f implies that to obtain the desired luminosity one or more of the following parameters must be pushed to a *critical value:* low beta, emittance, bunch density, or beam-beam tune shift.

The small bending radius can be obtained with superconducting dipoles, whose *high nonlinearities* can limit the dynamic aperture, already critical due to the large emittance.

The large *synchrotron radiation power density* on the vacuum chamber walls requires a special design of the vacuum chamber (for example the antichamber at the UCLA ring).

A fundamental problem of the compact ring is the high rate of particle loss due to single beam-beam bremsstrahlung which shortens the lifetime to a few minutes. Continuous filling of the ring is necessary.

Table 1. Main parameters of Φ-factory projects

	Novosibirsk[5] *Russia*	UCLA[8] *USA*	DAFNE[9] *Italy*	KEK[10] *Japan*	Mainz[6] *Germany*	VEPP2M[1] *Russia*
C (m)	35	17	98	120	29	18
β_y^* (cm)	1.0	3.9(.3)[a]	4.5	1.0	0.12	4.5
σ_z (cm)	~1	3(.3)	3	.5	1	
ε_x(mm mrad)	.47	3.2(1)	1.0	1.1	.22	.46
κ	1	0.2	0.01	0.01	1	0.01
N (10^{10})	20	40	9	6	8	3.7
I_{tot} (A)	.5	1.2(.5)	5	7	.8	.1
n_b	1	1	120	300	6	1
f (MHz)	17	17	368	750	62	17
θ (mrad)	0	0	12.5	20	0	0
ξ_x	.10	.05	.04	.03	.08	.05
ξ_y	.10	.05	.04	.03	.08	.02
α	.03 -.06	.11(~0.)	.005	.007	.018	
τ (min)	< 5	40	300	15	60[b]	
$\mathcal{L}_o$(10^{32})	10	2 (10)	0.04	0.1	1.6	0.07
$\mathcal{L}$(10^{32})	10	2 (10)	5	30	10	0.07

[a] Numbers in parenthesis correspond to QIR
[b] Only τ_{bb}

3.1 Novosibirsk Φ Factory

The Novosibirsk proposal is a very innovative idea. The ring is a bone-shaped machine (see Fig. 3) where the two beams, in the single-bunch mode, circulate in opposite directions and cross at one point thanks to the introduction of negative-curvature sections in the arcs.

Round beams are used: a SC solenoid (11 T) around the IP rotates the transverse planes by $\pi/2$: the betatron oscillation normal modes are vertical in one arc and horizontal in the other so that the emittances in the two planes are the same. Beams are focused to very small beta values in both planes by the same solenoid. In this was a very high value of ξ (> 0.1) is claimed to be obtainable.

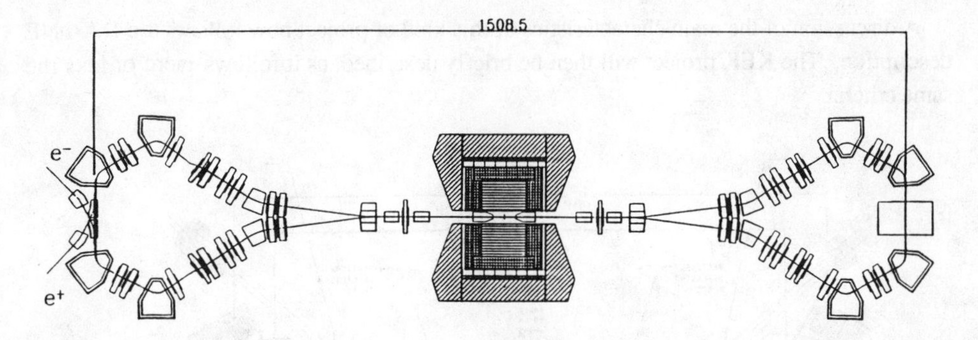

Fig. 3. Novosibirsk Φ factory layout

The working point is placed on the main coupling resonance $\nu_x - \nu_y = 0$. The tunes are shifted along the coupling resonance and do not trespass on the 2-dimensional coupling sidebands. SC dipoles (6.5 T) are used to obtain the small bending radius.

The beam lifetime of a few minutes is considered satisfactory and manageable with the injection chain, consisting of a full-energy linac and a positron cooling ring. A later stage with three bunches per beam is envisaged, applying electrostatic beam separation at the IP sides to reach luminosities higher than 10^{33} cm^{-2} s^{-1}.

3.2 UCLA Φ Factory

The UCLA Φ factory design consists of a small machine of ~ 20 m circumference, equipped with SC dipoles. A sketch of the proposed layout is given in Fig. 4. Three successive phases are envisaged, with increasing luminosity:

- The Compact Superconducting Ring (SMC) with $\mathcal{L} = 2\ 10^{32}$ cm^{-2} s^{-1} working in single-bunch mode, with low coupling ($\kappa = 0.2$) and high emittance.
- The Quasi Isochronous Ring (QIR) with $\mathcal{L} = 10^{33}$ cm^{-2} s^{-1}. The idea is to shorten the bunch to the millimeter range by making the first order momentum compaction vanish. This permits very small values of the vertical betatron function at the IP, so that, even with a lower current with respect to the phase I, a higher luminosity can be achieved.

The Linac-against-Ring option, finally, should lead to a luminosity larger than 10^{33} cm^{-2} s^{-1}.

4 Double-Ring Multibunch Operation

The choice of increasing the collision frequency by storing many bunches in two rings leads to a design based on conventional technology, as far as the lattice is concerned. The short distance between bunches obliges the beams to cross at an angle in order to suppress parasitic crossings. However, the main problem is multibunch instabilities which must be countered with specially designed feedback systems, rf cavity design with HOM suppression, and low impedance in all vacuum-chamber components.

A discussion of the main characteristics of this kind of project now follows the DAΦNE description. The KEK project will then be briefly described, as it follows more or less the same criteria.

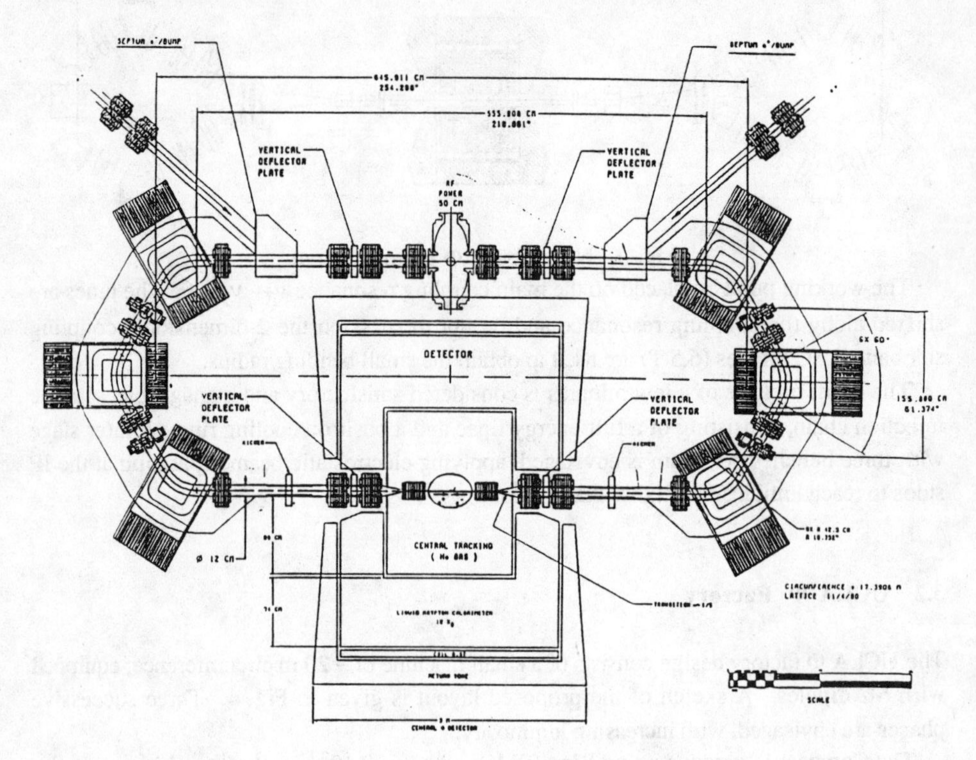

Fig. 4. UCLA Φ factory layout

4.1 DAΦNE

DAΦNE, the Frascati Φ factory, is so far the only funded project for factories anywhere in the world.

It consist of two rings, of about 100 m circumference, intersecting at an angle in the horizontal plane in two points, and will be installed in the old ADONE hall. It will accommodate two experiments, KLOE and FI.NU.DA.; the first specially designed for the CP-violation investigation, the second for nuclear physics. A layout of the rings is shown in Fig. 5. The ring is not symmetric with respect to the two IPs: looking at the figure it is clear that because of the crossing angle the arcs joining the two IPs are of different lengths.

Flat beams ($\kappa = 0.01$), multibunch operation and crossing angle are the key features of the project. The emittance is controlled by means of four wigglers in each ring, located in the dispersive zone.

Since the two physics detectors are equipped with strong solenoids (0.6 T x 2.1 m for

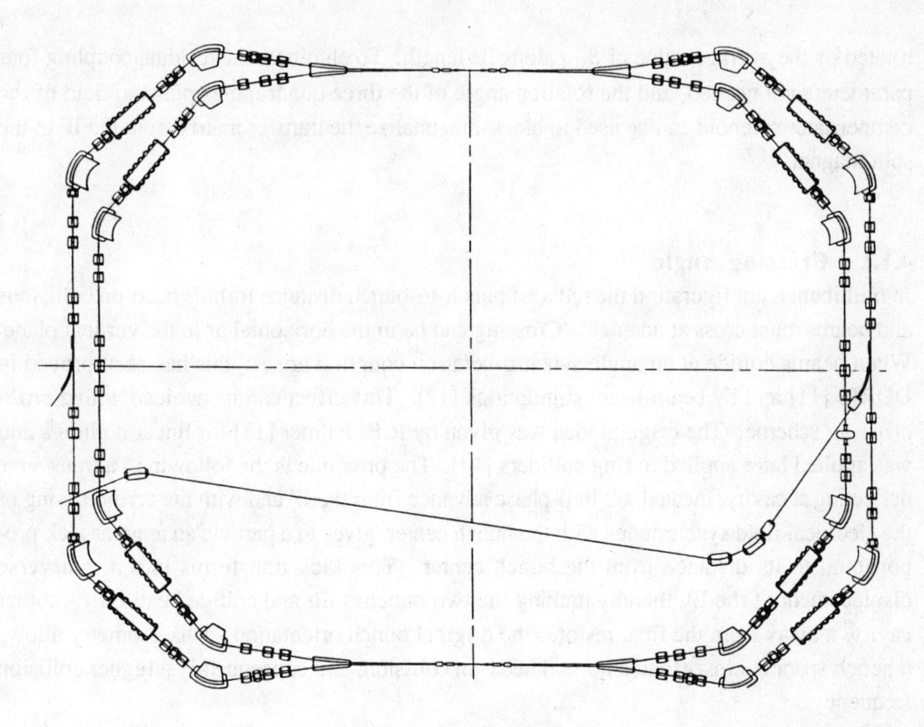

Fig. 5. DAΦNE layout

KLOE and 1.5 T x 2 m for FI.NU.DA.), the ensuing perturbation of the machine optics necessitates a coordinated design of the interaction region and the detector.

The interaction regions inbetween the two splitting magnets, which separate the two beam trajectories, are shared by the two rings. The magnetic axis of their elements is the bisectrice of the nominal closed orbits of the two rings. Both interaction regions are equivalent from the optics point of view, i.e. with the same first-order transfer matrix, so that the optical functions and the beam separation are the same at the ends of the interaction region. This solution simplifies injection, tuning and chromaticity correction in the different modes of operation.

In the interaction regions the transverse coupling due to the detector solenoid is corrected by two compensating solenoids at the detector ends whose total field integral is opposite to that of the detector. In both experiments all or some of the low-beta quadrupoles are immersed in the solenoidal field. To eliminate the coupling effect quadrupoles should be rotated following the angle given by:

$$\theta_{sol} = \int \frac{B_Z \, dz}{2B\rho}$$

This continuous rotation is not feasible with a real quadrupole, so each quadrupole will be rotated by the average value of θ_{sol} along its length. To eliminate the residual coupling four parameters are needed, and the rotation angle of the three quadrupoles plus the field of the compensator solenoid can be used to block-diagonalize the transfer matrix from the IP to the split magnet.

4.1.1 Crossing angle

In multibunch configuration the reduced bunch-to-bunch distance forbids head-on collisions and beams must cross at an angle. Crossing can be in the horizontal or in the vertical plane. When beams collide at an angle synchro-betatron coupling arises: this has been proved in DORIS [11] and by beam-beam simulations [12]. This effect can be avoided with a *crab-crossing* scheme. The original idea was given by R.B. Palmer [13] for linear colliders and was applied later applied to ring colliders [14]. The principle is the following: a transverse deflecting rf cavity, located $\pi/2$ in β-phase advance from the IP and with the zero-crossing of the electrical field synchronous with the bunch center, gives to a particle an angular kick proportional to its distance from the bunch center. This kick transforms into a transverse displacement at the IP, thereby making the two bunches tilt and collide head-on. Another cavity, π away from the first, restores the original bunch orientation. This geometry allows a bunch spacing closer than a normal head-on collision, and consequently a higher collision frequency.

The required cavity voltage, assuming that the rms bunch length σ_z is small compared to $\lambda_{rf}/4$, where λ_{rf} is the rf wavelength of the cavity, is:

$$V_{rf} = \frac{E.}{4\pi e} \frac{\theta \lambda_{rf}}{\sqrt{\beta^* \ \beta_C}}$$

θ being half the crossing angle, β^* the betatron function at the IP, β_C at the cavity location, both in the crossing plane.

This scheme requires rf phase φ and amplitude stability:

$$\Delta\varphi << \frac{4\pi\sigma}{\theta\lambda_{rf}} \qquad \frac{\Delta V_{rf}}{V_{rf}} << \frac{1}{\sqrt{N_\beta}} \frac{\sigma}{\sigma_z}$$

σ being the rms bunch size at the IP in the crossing plane and N_β the number of turns in a betatron damping time. With a flat beam, since β_x is much larger than β_y, the cavity amplitude V_{rf} is much lower and the tolerances much larger for the crossing in the horizontal plane. In fact tolerances are so large that for parameters characteristic of DAΦNE a cavity is not necessary.

The key parameter describing the effect of the crossing angle on beam dynamics is what is called the *badness factor a*:

$$a = \theta \frac{\sigma_L}{\sigma_x}$$

and is a measure of the coupling between the radial and the longitudinal phase spaces generated by the crossing angle. This coupling, experimentally observed in DORIS I [11], limits the maximum achievable tune shift and, consequently the luminosity.

The simplest hypothesis one can make on the relation between a and ξ is:

$$\xi a = \text{const}$$

Suggestions on this value come both from observations at DORIS I and computer simulations. At DORIS I the maximum tune shift value ever obtained and the value of a were:

$$\xi = 0.01 \quad , \quad a = 0.5$$

In DAΦNE $a \sim 14\ \theta$, so that an angle of ~ 10 mrad seems safe from this point of view. Simulations [15] of particle tracking, including beam-beam interaction with crossing angle, confirm that no dangerous synchro-betatron coupling occurs up to an angle of ~ 100 mrad.

The effect of *parasitic crossings* must also be considered: when two beams collide at an angle 2θ the first parasitic crossing (p.c.) occurs at distance $l/2$ (l is the spacing between bunches) from the IP if the bunch centers are at a distance given by $d = l\ \theta$.

The tune shift due to the parasitic crossing is:

$$\xi_x = \frac{r_o N \beta_x}{2\pi\gamma d^2} \qquad\qquad \xi_y = \frac{r_o N \beta_y}{2\pi\gamma d^2}$$

where β_x, β_y are the betatron functions at the p.c. From the consideration that one beam could act as a scraper for the opposite beam's particles passing inside its core, a criteria can be defined [16]: the full separation should be at least 7σ so that particles one wishes to keep in the beam stay out of the opposite beam's center.

To check this criteria simulations have been made with DAΦNE parameters [17], using the approximation of weak-strong interactions. Two different effects were discovered:

- Blow-up in the vertical emittance for particles inside the core so affecting the *luminosity* (σ_y increase).
- Long vertical tails for particles which pass near the core of the opposite bunch and which can affect the *beam lifetime*.

Both the effects depend on the current, on the beam separation, and on the *betatron functions* at crossing, but we obtained no evidence of the scraper effect. In fact a particle passing inside the opposite beam gets a tune shift which affects its betatron motion, but not necessarily cuts it out. In any case a particle which sees the other bunch at one turn will probably not pass inside it again in the successive turns during a betatron damping time.

With DAΦNE design parameters, 7σ at the p.c. is a safe number, but with a higher current a separation of only 7σ would reduce the beam lifetime.

4.1.2 Rf and feedback system

The rf system must overcome one of the most harmful problems in high current storage rings working in multibunch operation: the high-order modes (HOMs) of the rf cavities induce coupling of the relative motion of the n_b bunches through the wake fields left behind in the cavity by the bunches themselves, so giving rise to multibunch instabilities. The instability growth rate must be kept below the radiation damping rate by an appropriately designed cavity design and feedback system.

The rf voltage necessary to compensate the energy losses due to synchrotron radiation and to HOMs is not very high (<< 1 MV): room temperature cavities are used. The choice of the frequency is usually related to the availability of rf sources together with the bunch-length requirements: 350 or 500 MHz are most commonly used.

A large amount of R & D effort (Frascati, SLAC) [18] is in progress to develop an rf cavity having the lowest possible interaction with the beam spectrum. The research is essentially dedicated to two items:

- Cavity-shape optimization to reduce the number of "trapped" HOMs by means of long tapered tubes (R/Q is reduced by an order of magnitude). Figure 6 shows a sketch of the cavity shape proposed for DAΦNE.
- HOM damping by coupling the more dangerous parasitic modes with loops or with waveguides propagating at their frequencies.

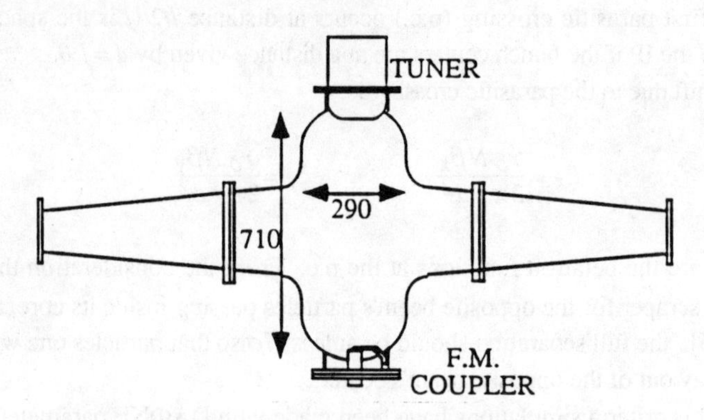

Fig. 6. DAΦNE cavity shape

Even with HOM damping or frequency shift the rise time of coupled-bunch (cb) modes of oscillation can be much faster than the natural radiation and Landau damping time; moreover the probability for a damped HOM to cross a cb mode frequency is larger, due to the wider bandwidth. An all-mode feedback system (f.s.) capable of damping all the cb modes and the injection transients is then necessary.

In most of the presently operating rings the usual approach consists in detecting the bunch synchrotron-phase error, rotating the signal in the longitudinal phase plane with a filter then amplifying it and giving an energy kick with a kicker. This method requires that the number of filters be equal to the number of bunches, which of course becomes very demanding in multibunch operation.

Alternatively, present-day electronic technology makes it possible to use a bunch-to-bunch, time-domain system [19]: in a mixed analog/digital f.s. the Digital Signal Processors (DSP) filters can serve several different channels, reducing the overall complexity.

Moreover the digital system can be programmed to maintain the correction signal just below the saturation limit of the power amplifier even in the presence of large phase excursions. A modular approach allows the system to be implemented for only a reduced number of bunches and then to be increased when a higher n_b is needed for the luminosity upgrade.

4.2 KEK Φ Factory

The two beams are stored in two rings superposed and horizontally crossing with an angle. The lattice provides flat beams with high currents and large emittance.

The rf frequency is 1.5 GHz. Every two buckets are filled to avoid very large values of ε_x.

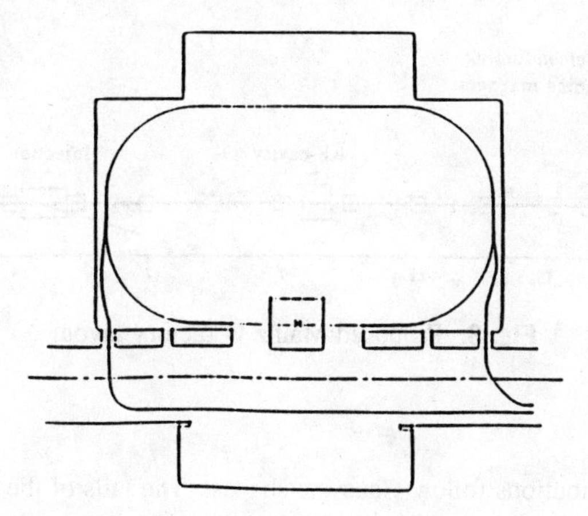

Fig. 7. KEK Φ factory layout

4.3 Φ Factory Design Studies at Mainz

Studies for designing a Φ factory are being carried out at Mainz. The proposal is a compromise between the compact ring and the double ring operation. The lattice consists of two rings intersecting at one point (see Fig. 8). Beams are round and following the Novosibirsk idea, are both focused and rotated by a superconducting solenoid; few bunches per beam (n_b = 6) give a collision frequency of 50 MHz. The short bunch length (σ_z = 1 cm) should be obtained with a special insertion made of dielectric material which should focus the beam longitudinally. A high value of ξ is proposed (0.08) and has been checked with beam-beam simulations.

5 Beam Lifetime

Beam lifetime in an accelerator is a measure of how 'peaceful' the accelerator is: almost any single-bunch instability and interaction between the beam and the surroundings reflects on the stability of the beam inside the vacuum chamber. A summary of the main considerations for beam lifetime follows:

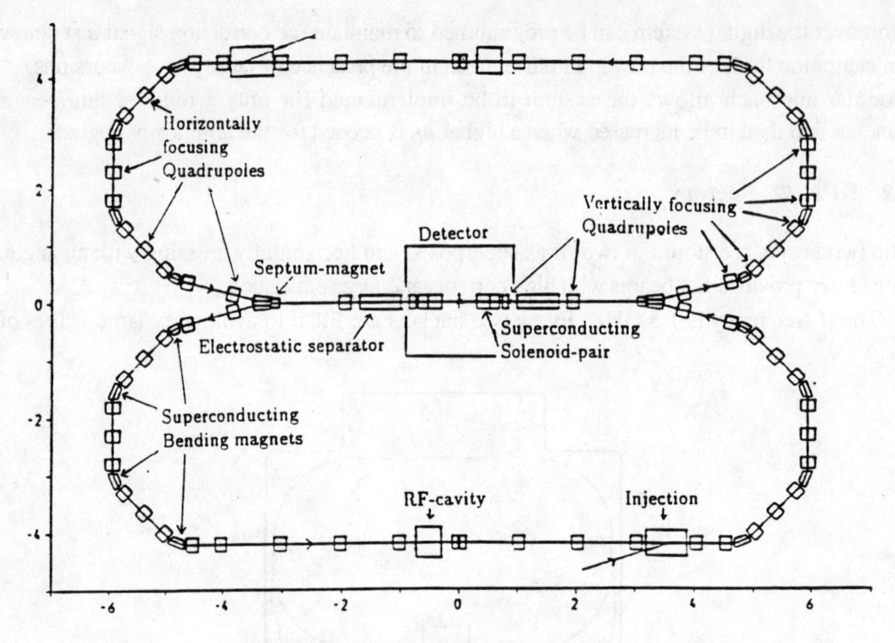

Fig. 8. Proposed Mainz Φ factory layout

a) Apertures

The bunch distributions follow Gaussian shapes. The tails of the distribution are lost due to hitting the vacuum chamber or the dynamic aperture; the quantum beam lifetime τ is given by:

$$\tau = \frac{\tau_\beta}{2\, r_\beta} e^{r_\beta} \qquad r_\beta = \frac{1}{2}\left(\frac{A}{\sigma}\right)^2$$

τ_β being the betatron damping time and A the aperture. For typical values of τ_β (tens of milliseconds) $A = 6\,\sigma$ gives a lifetime of several hours, while $5\,\sigma$ reduces the lifetime to a few minutes. Small rings with round beams and high emittances are of course critical from this point of view.

b) Single-bunch effects

Touschek effect: when scattered particles inside the bunch have a longitudinal momentum variation they can exceed the momentum acceptance of the ring, or if the scattering occurs in a dispersive zone the induced horizontal oscillation can exceed the transverse acceptance of the ring. This effect is very strong for low-energy rings, especially for very flat beams because of its dependence on the bunch density. It is in fact the main limitation for multibunch-operation rings working with flat beams.

Multiple Touschek effect: scattering between particles inside the bunch transfers momentum between the planes to which can be associated an increase in the transverse emittance which could exceed the transverse acceptance, either physical or dynamic. In

Φ factories this effect is not of much concern because the emittance is already large and its increment is only a few percent, well contained in the ring acceptance.

c) *Beam-gas interaction*

The scattering between beam particles and residual gas molecules can be elastic, producing betatron oscillations that can exceed the transverse acceptance, or inelastic (bremsstrahlung), producing energy oscillations exceeding the longitudinal acceptance. It depends on the vacuum pressure and on the apertures: it is more important in small rings because of the higher synchrotron radiation pressure.

d) *Beam-beam bremsstrahlung*

The beam-beam decay rate is proportional to $L/(n_b N)$. In single-bunch modes and at high luminosity the beam lifetime is reduced to a few minutes.

6 Injection

The total number of positrons required to obtain the quoted luminosity is very high: 10^{11} - 10^{12} for compact rings and ~ 10^{13} for double rings. Furthermore reliable operation requires a short injection time, typically a few minutes for the double-ring configuration, 1 minute for the compact ring, to store the full current of both beams. It is clearly convenient to inject at full energy.

An accumulator, at the same energy as the main ring, is usually proposed to accumulate at lower f_{rf} to accept the Linac pulses. It is used also to damp the e^+ emittance and energy spread to values acceptable to the main ring system, and to equalize the two beam characteristics.

References

[1] P.M. Ivanov et al, Luminosity and the Beam-Beam Effects on the Electron-Positron Storage Ring VEPP-2M with Superconducting Wiggler Magnet, Proceedings of the 3rd Advanced ICFA Beam Dynamics Workshop, Novosibirsk, 1989, p. 26.

[2] D.L. Rubin and L.A. SChick, Single Interaction Point of Operation of CESR, 1991 Particle Accelerator Conference, San FranciSCo, USA, 1991, p.144.

[3] J.T. Seeman, Observations of the Beam-beam Interaction, Nonlinear Dynamics Aspects of Particle Accelerators, S. Margherita di Pula, Italy, 1985, Springer Verlga, Lecture Notes in Physics 247, p.121.

[4] M. Bassetti, What are the Beam-beam limits in DAΦNE?, 1992, Unpublished.

[5] L. Barkov et al., Novosibirsk Project of Φ-Meson Factory, Proceedings of Workshop on Physics and Detectors for DAΦNE - Frascati, April 9-12 1991, p.67.

[6] A. Streun, Φ-Factory Studies at Mainz, in Ref. [5], p.99.

[7] A.N. Filippov et al., Proposal of the Round Beam Lattice for VEPP-2M Collider, Presented at the XVth International Conference on H.E.A., Hamburg, 1992.

[8] C. Pellegrini, The UCLA Φ-Factory Project, in Ref. [5], p.83.

[9] DAΦNE Project Team, DAΦNE Status Report, Proceedings of 3rd EPAC, Berlin, 1992.

[10] K. Hirata and K. Ohmi, Feasibility of a Φ-Factory in KEK, 1991 Particle Accelerator Conference, San Francisco, USA, 1991, p. 2847.

[11] A. Piwinski, IEEE Trans. on Nucl. Sci. NS-24, 1408 (1977).

[12] A. Piwinski, Synchro-betatron resonances, in Ref. [3].

[13] R.B. Palmer, SLAC - PUB - 4707 (1988).

[14] K. Oide and Y. Yokoya, Phys. Rev. D40, 315 (1989).

[15] M. Bassetti and M.E. Biagini, private communication.

[16] D.H. Rice, Beam-Beam Interaction: Experimental, AIP Conf. Proceedings 214, Berkeley 219 (1990).

[17] M. Bassetti and C. Biscari, Beam-beam simulations with pc, unpublished.

[18] P. Baldini et al., DAΦNE Accelerating Cavity: R&D, in Ref. [9].

[19] M. Bassetti et al., DAΦNE Longitudinal Feedback, in Ref. [9].

GENERAL B FACTORY DESIGN CONSIDERATIONS†

Michael S. Zisman

Exploratory Studies Group, Accelerator & Fusion Research Division,
Lawrence Berkeley Laboratory, Berkeley, CA 94720 USA

Abstract. We describe the general considerations that go into the design of an asymmetric B factory collider. Justification is given for the typical parameters of such a facility, and the physics and technology challenges that arise from these parameter choices are discussed. Cost and schedule issues for a B factory are discussed briefly. A summary of existing proposals is presented, noting their similarities and differences.

1 Introduction

For the past few years, there has been intense interest [1–8] in the design of a high-luminosity electron-positron collider to serve as a "B factory." The primary physics motivation for such a facility is to carry out a detailed and systematic study of the origins of CP violation, a phenomenon that is thought to be responsible for the dominance of matter over antimatter in the Universe. It is anticipated that this effect will be large in the $B\overline{B}$ system, making a B factory the ideal platform from which to launch such a study.

It was pointed out some years ago by Oddone [9] that the key to studying CP violation in the $B\overline{B}$ system was to create a moving center-of-mass. This could be achieved by building a so-called "asymmetric" collider, in which the electron and positron energies are different. With such an approach, the B and $\overline{B}$ decays can be separated spatially with modern silicon vertex detectors, permitting the easy reconstruction of their time difference. To enhance the cross section for producing the particles, we will focus here on colliders designed to operate at the Υ(4S) resonance, requiring a center-of-mass energy of 10.6 GeV. In terms of machine parameters, this requirement means that the product of the two beam energies must be $E_{+}E_{-} = 28$ GeV^2. As is obvious, the requirement of asymmetric beam energies dictates a two-ring collider. We will see later that other B factory parameters also lead to the requirement for two independent rings.

The required energy asymmetry for a B factory is a trade-off among competing

†This work was supported by the Director, Office of Energy Research, Office of High Energy and Nuclear Physics, High Energy Physics Division, U.S. Department of Energy, under Contract No. DE-AC03-76SF00098.

factors. A large energy asymmetry gives a higher boost to the decaying particles and thus makes it easier to separate the decays in the vertex detector. However, the solid angle for detection decreases as the decaying particles are kinematically focused into the forward direction (where it is difficult to detect them due to interference with the collider optical elements and beam pipe). There are equivalent trade-offs to be made in terms of accelerator design. In general, a larger energy asymmetry simplifies the beam separation optics in the interaction region (IR), as discussed in Ref. [10], but makes the beam current requirement more difficult to meet and tends to make the difference in synchrotron radiation damping times between the two beams quite extreme. We will return to these points later when we discuss the parameter choices for a *B* factory. For now, we simply note that the energy asymmetry studied by various designers has ranged from 6.5/4.3 to 12/2.3.

The other parameter we must consider for a *B* factory is the luminosity, $\mathcal{L}$. Based on a number of studies carried out in the past few years [11], it is generally agreed that a peak luminosity of $\mathcal{L} = 3 \times 10^{33}$ cm^{-2} s^{-1} is required for the study of *CP* violation. In reality, however, it is the ***integrated*** luminosity that is the proper figure-of-merit for a *B* factory. This is because the *CP* violation studies require an abundant sample of $B\overline{B}$ pairs. We consider the luminosity requirement to correspond to an integrated luminosity (over a "standard" year of 10^7 seconds) of

$$\int_{\text{year}} \mathcal{L}\cdot dt = 3 \times 10^{40}\ \text{cm}^{-2} = 30\ \text{fb}^{-1} \tag{1}$$

This is the meaning—and the challenge—of the "factory" aspect of a *B* factory.

In Section 2 of this paper we discuss the typical parameters of a *B* factory collider and the motivations for them. Section 3 will indicate the design approach that should be followed to achieve high integrated luminosity. In Section 4 we discuss the physics and technology challenges that result from the parameter choices arrived at in Section 2. Section 5 covers the typical construction scenario and cost issues for a *B* factory project. In Section 6 we briefly summarize the present status of existing *B* factory proposals and comment on their similarities and differences. Summary remarks are given in Section 7.

2 Typical Parameters

The luminosity of a *B* factory collider can be expressed in terms of the beam intensities of the electron and positron beams (N_- and N_+, respectively), the collision frequency (f_c), and the cross-sectional area of the beams at the interaction point ($4\pi\sigma_x^*\sigma_y^*$) as

$$\mathcal{L} = \frac{N_+ N_- f_c}{4\pi\sigma_x^*\sigma_y^*} \tag{2}$$

However, for machine design purposes, we write the luminosity in a different way that calls out explicitly the dependence on machine parameters (the +,– subscripts refer to the e^+ and e^- rings, respectively) with which the accelerator physicist deals:

$$\mathcal{L} = 2.17 \times 10^{34}\,\xi(1+r)\left(\frac{I \cdot E}{\beta_y^*}\right)_{+,-} \quad [\mathrm{cm^{-2}\,s^{-1}}] \tag{3}$$

where I is the total beam current in one of the rings, β_y^* is the vertical beta-function at the interaction point (IP), $r = \sigma_y^*/\sigma_x^*$ is the beam aspect ratio at the IP, E is the beam energy in GeV, and ξ is the design value for the beam-beam tune shift.

An important justification for writing the luminosity expression as in Eq. (3) is to call out the dependence on the beam-beam parameter, ξ, which has proved empirically to be a limit in all colliders [12]. In terms of the other beam parameters, the actual expressions for the beam-beam tune shift are

$$\xi_{y,+} = \frac{r_e N_- \beta_{y,+}^*}{2\pi\gamma_+ \sigma_{y,-}^*\left(\sigma_{x,-}^* + \sigma_{y,-}^*\right)}, \tag{4a}$$

$$\xi_{x,+} = \frac{r_e N_- \beta_{x,+}^*}{2\pi\gamma_+ \sigma_{x,-}^*\left(\sigma_{x,-}^* + \sigma_{y,-}^*\right)}. \tag{4b}$$

Although we indicated that the various parameters in Eq. (3) are available to be adjusted, in reality a number of them are constrained by other considerations. We have already commented on the accelerator physics limit associated with ξ. The beam aspect ratio, r, is also constrained, in this case by detector background considerations. The potential benefits of using a round beam are twofold. First, it appears by inspection of Eq. (3) that this choice would reduce by a factor of two the beam current required for producing a given luminosity. A second benefit may be an increase in the allowable beam-beam tune shift for round beams. This was predicted by Krishnagopal and Siemann [13], who estimated that $\xi \approx 0.1$ may be reachable. This aspect remains to be confirmed experimentally. Despite the possible advantages, early studies based on a round-beam ($r = 1$) design [14] showed that the required focusing and beam separation optics produced a prodigious amount of synchrotron radiation power (about 0.75 MW) in the region within a few meters of the IP. A masking solution to deal with this problem would be, at best, quite difficult. Furthermore, the assumption of a factor of two reduction in beam current for the same luminosity is predicated on being able to keep the same value for the vertical beta-function in the round-beam case. As has been observed by Willeke [15], the total chromaticity that can be handled in the ring is roughly fixed, so the minimum beta that can be tolerated in the standard flat-beam ($r \approx 0$) case will in general be about half that for a round-beam case, eliminating the perceived advantage. For now, it does not seem that the benefits of round beams are sufficient to compensate for the practical difficulties of creating them, and this option has not been followed by any of the B factory design teams.

Finally, the beam energies themselves are constrained by the need to operate at the ϒ(4S) resonance, as discussed in Section 1. Therefore, the product of the two beam energies is fixed and there is relatively little adjustment possible. Thus, we see that only the beam currents and beta-functions are really free parameters in the optimization of luminosity.

At the present time, the beam-beam interaction has not been studied experimentally for the case of asymmetric electron-positron energies. Therefore, lacking data on such collisions, we take our guidance from the observations in symmetric energy colliders [12]. For design purposes, most groups have adopted a beam-beam tune shift value of $\xi = 0.03$ for both beams in both transverse planes. The choice of equal tune shifts for both beams in both planes reduces the number of free parameters but is not otherwise justified on theoretical grounds. It is worth noting, however, that the adopted ξ value should be thought of, at this stage, as a *design value* rather than an actual beam-beam limit.

Most designers have also adopted a head-on collision configuration. Here too, the primary motivation is that this is the configuration we are familiar with from symmetric colliders. Recently, there have been theoretical arguments suggesting that a small crossing angle will not lead to excessive excitation of synchrobetatron resonances and therefore will not reduce the obtainable beam-beam tune shift. Experiments under way at CESR [16] are consistent with only a small decrease in ξ for an uncompensated crossing angle of 2.5 mrad.

A new possible scheme to collide two beams at a non-zero crossing angle without exciting synchrobetatron resonances has been proposed by Palmer [17] for linear colliders and subsequently extended to the circular collider case by Oide and Yokoya [18]. The technique is referred to as "crab crossing." The concept is to apply a transverse kick to each bunch by means of an RF cavity located an odd multiple of $\pi/4$ in phase advance from the IP. If the kick tilts the bunch by $\theta/2$, where θ is the full crossing angle, then the bunches pass through each other head-on in their center-of-mass system and excitation of synchrobetatron resonances is avoided. For a storage ring one uses a pair of transversely deflecting cavities ("crab cavities") located symmetrically about the IP, the first to tilt the bunch and the second to cancel the tilt. Studies of such a system show that the tolerances appear achievable [19] and designs for such cavities are under study at Cornell [20]. Nevertheless, the fact that there is no experience with the technique has convinced most *B* factory design groups to treat crab crossing as a "phase 2" option.

For now, all that can be done to justify the choice of the beam-beam tune shift parameter is to carry out simulations to show that the adopted choice is reasonable. Typical beam-beam simulation results for the PEP-II collider design, taken from Ref. [21], are shown in Fig. 1. It is fair to summarize the present situation by saying that there is no indication of any new physics issues associated with the asymmetric beam energies. It is also noteworthy in this regard that the HERA electron-proton collider, which is exceedingly asymmetric in terms of both energy and damping time, has not shown any surprises in terms of the single-bunch beam-beam behavior [22].

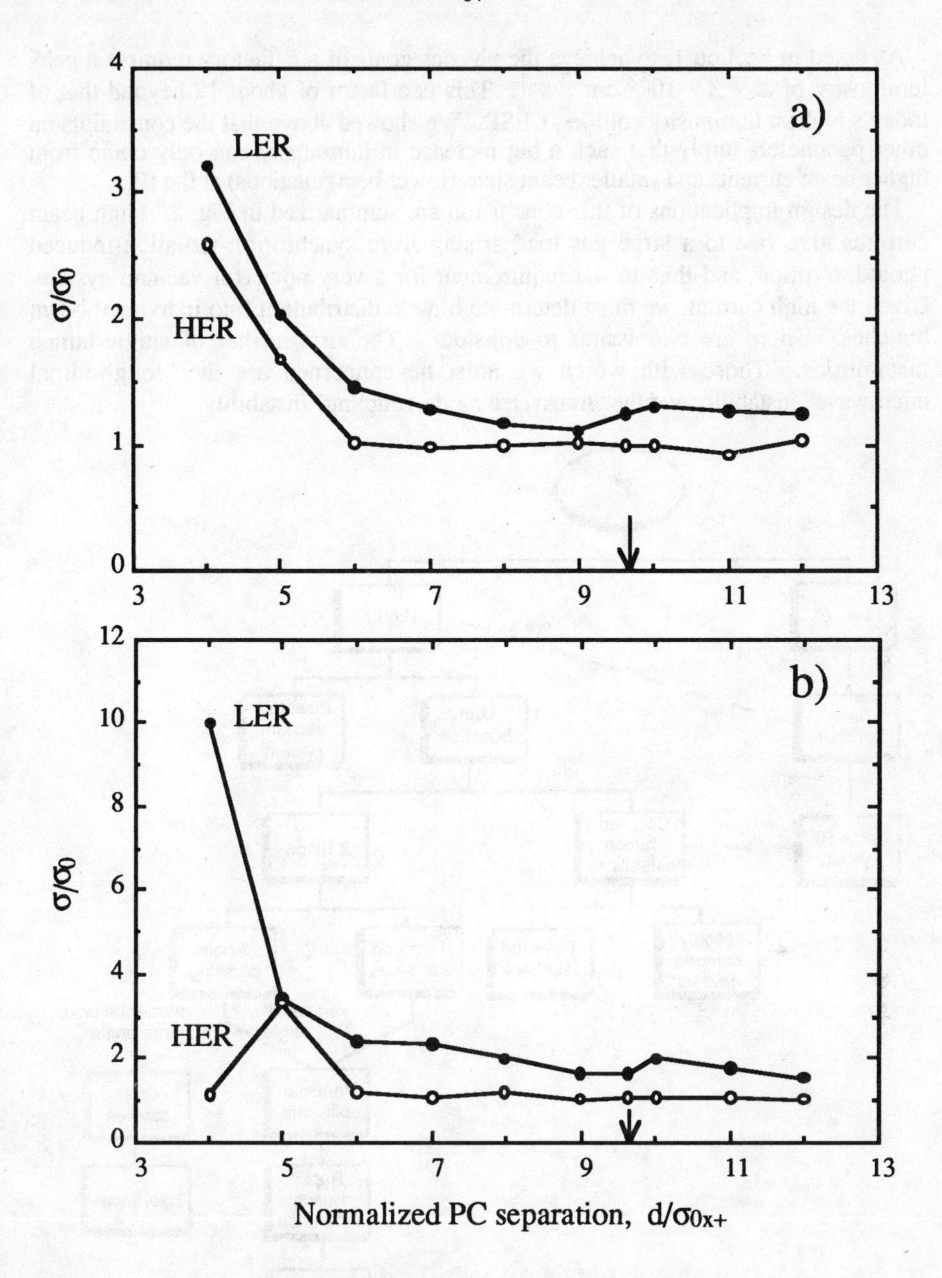

Fig. 1. Beam-beam simulation results for the PEP-II collider, showing vertical beam blowup (horizontal blowup is negligible) as a function of separation distance at the parasitic crossing (PC) points. The nominal separation distance is shown by the arrow. Case a) is for a beam-beam parameter of 0.03 and results in a luminosity of 2.9×10^{33} cm^{-2} s^{-1}. Case b) is for a beam-beam parameter of 0.05 and results in a luminosity of 5.7×10^{33} cm^{-2} s^{-1}.

As noted in Section 1, to achieve the physics goals of a B factory requires a peak luminosity of $\mathcal{L} = 3 \times 10^{33}$ cm^{-2} s^{-1}. This is a factor of about 12 beyond that of today's highest luminosity collider, CESR. We showed above that the constraints on other parameters imply that such a big increase in luminosity can only come from higher beam currents and smaller beam sizes (lower beta-functions) at the IP.

The design implications of this conclusion are summarized in Fig. 2. High beam currents give rise to a large gas load arising from synchrotron-radiation-induced photodesorption, and thus to the requirement for a very powerful vacuum system. Given the high current, we must determine how to distribute it into individual beam bunches. There are two issues to consider. The first is that of single-bunch instabilities. Those with which we must be concerned are the "longitudinal microwave" instability and the "transverse mode coupling" instability.

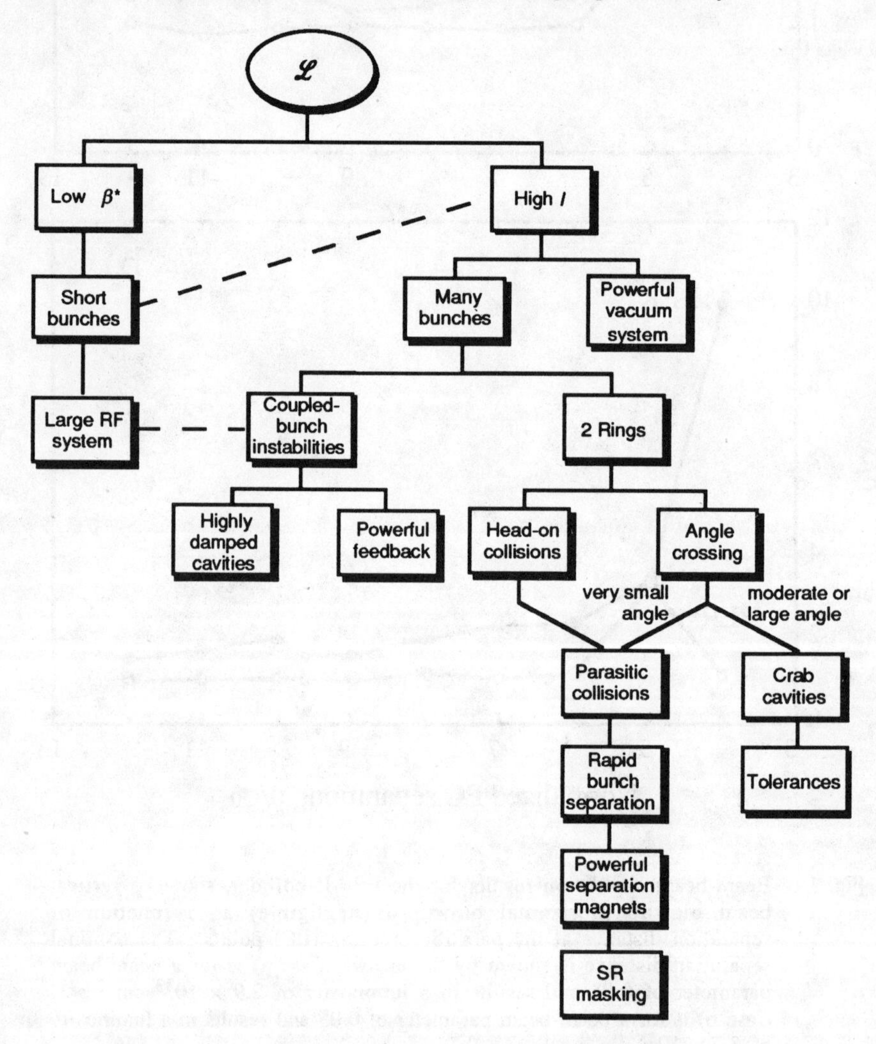

Fig. 2. Design implications for a B factory.

The longitudinal microwave instability does not lead to beam loss but to an increase in the bunch length and momentum spread when the single-bunch current exceeds a threshold value [23]. Insofar as we require short bunches to avoid luminosity loss from the hourglass effect [24] and a reasonably small momentum spread to take good advantage of the resonant cross section, it behooves us to keep the single-bunch current below the threshold value. The transverse mode coupling instability [25] is a more serious constraint, as it limits the current that can be stored in a single bunch.

Although a detailed description of these effects is beyond the scope of the present paper, we note that for either instability the only practical approach to increasing the threshold current is to decrease the broadband longitudinal or transverse impedance of the ring. Given that today's rings are already designed with quite low impedance, reducing the impedance sufficiently to accommodate a factor of 10 increase in single-bunch current is certainly easier said than done. Thus, the single-bunch instabilities in practice preclude a few-bunch, high-current operating mode for a *B* factory and we are led to a many-bunch configuration.

The limitation on the beam-beam tune shift discussed above also means that the luminosity increase must come from more—as opposed to higher intensity—bunches. The reason is that, for a given beam-beam tune shift, the increased current must correspond to an increase in the beam emittance (as implied by Eq. (4)). In today's colliders, the typical beam emittance (limited by reasonable magnet dimensions) is $\varepsilon_x \approx 100$ nm·rad. To increase the beam current by a factor of 10–20 in the same number of bunches would require magnet gaps 3–5 times larger than standard values—an expensive proposition.

It should be clear from the above arguments that a solution with many bunches is inevitable for a *B* factory. This puts us in the regime where coupled-bunch instabilities are of primary concern to the designers. It turns out, however, that the growth rates for coupled-bunch instabilities [26] are determined mainly by the *total* beam current, and depend only weakly on the distribution of that current into different bunch patterns. With this in mind, we are free to choose the number of bunches such that the single-bunch thresholds are not exceeded for reasonable values of broadband impedance. (Increasing the number of bunches is not totally without penalty in terms of coupled-bunch instabilities, because the bandwidth required for a multibunch feedback system increases as the bunch separation decreases, but this is not a significant limitation with modern technology, as will be discussed below.) There is nonetheless a lower limit to the acceptable bunch separation determined by consideration of the parasitic crossings. The general rule-of-thumb is that the beam separation at the parasitic crossing points should be at least 7σ [16]. In present designs, the bunch separation distance, s_B, falls in the range of 0.6–12 m.

The requirement for a low β^* value also implies the need for short bunches ($\sigma_\ell < \beta^*$) to avoid luminosity loss from the hourglass effect [24]. This, in turn, means that a high RF voltage is needed, and thus a large RF system. Because the impedance from the RF cavities is the main driving source for coupled-bunch instabilities, there is an intrinsic conflict between the need for high currents and for short bunches.

Based on the considerations discussed in this Section, we can summarize the typical *B* factory parameter choices as indicated in Table 1.

Table 1. Typical *B* factory parameters for $\mathcal{L} = 3 \times 10^{33}$ cm^{-2} s^{-1}.

ε_x [nm·rad]	100
I_b [mA]	1 – 5
σ_ℓ [cm]	0.5 – 1
I [A]	1 – 3
No. of bunches	100 – 2000

3 Design Approach

Before describing the technical details of a *B* factory, it is worth digressing briefly here to discuss the more "philosophical" issue of the proper way to approach the design of such a facility. The main message to give is that it is important to take the "factory" aspect seriously in order to provide the required integrated luminosity. To do so, it is a reasonable goal to minimize the extrapolations we must make in accelerator design. Clearly, however, this goal will be at odds with the desired extrapolation we wish to make in accelerator *performance*. The reality is that we cannot avoid doing *some* new things in order to provide a high-luminosity asymmetric collider, but we can—and should—pick our spots carefully. That is, we should strive in the design of a *B* factory to avoid problems we already know how to solve and save our strength for the inevitable surprises.

Another important idea is that it is crucial to design in reliability from the outset. This can be accomplished, for example, by making hardware designs modular to facilitate debugging, repair and replacement. Another method to improve reliability is to have in place—at the outset—good diagnostics hardware and a control system that can interpret it. Given a real-world commissioning scenario, it is also crucial to have a powerful injection system that permits quick recovery from the inevitable beam losses.

Lastly, it is important to have a design with sufficient flexibility to maintain "maneuvering room" in parameter space. As examples, it is important to design the vacuum system for beam currents that are higher than nominal, to design feedback systems to cope with expected (or unexpected) instabilities, to design the lattice to permit easy adjustment of the emittance and damping time, and to design the IR layout to accommodate either head-on collisions or a non-zero crossing angle.

In all of the designs discussed in this paper, parameters have been chosen to keep the challenges primarily in the realm of engineering rather than accelerator physics. The reason for this is not so much that the engineering aspects are easy, but that they are more amenable than the physics issues to being verified and optimized via suitable R&D. Put another way, thermal loading on the beam pipe is considerably more reliably estimated than the beam-beam effects.

4 Design Challenges

The design of a high-luminosity *B* factory gives rise to both physics and technology challenges [27]. In practice, the various design choices are interrelated. The proper optimization for a particular site is the "art" of the accelerator design.

4.1 Physics Challenges

We noted earlier that the *B* factory parameter regime of asymmetric energies, high beam currents, and many bunches forces the facility to be a two-ring collider. A two-ring collider leads to physics challenges in several areas.

In terms of lattice design, the issues to be dealt with include:

- providing low β^* values for both beams while maintaining adequate dynamic aperture

- providing rapid beam separation to avoid luminosity or lifetime degradation

- providing adequate masking both for synchrotron radiation and for lost particles due to beam-gas interactions or injection losses.

These issues are covered in detail in Refs. [10] and [28] and will not be discussed here.

Physics challenges also exist in the area of the beam-beam interaction. In a nutshell, the issue here is how to optimize the luminosity for an asymmetric, two-ring collider. Compared with single-ring colliders, a *B* factory has many more parameters that can be adjusted. In principle, the beam sizes, intensities, aspect ratios, beam-beam tune shift parameters, synchrotron tunes, betatron tunes, bunch lengths, etc., are all independently adjustable in the two-ring case. Optimization of a design in this larger parameter space is just beginning. As a practical matter, most designs have made use of some version of so-called "energy transparency" conditions [29, 30]. These are intended to make the two beams behave symmetrically with respect to the beam-beam interaction, and also serve to limit the parameter choices to a manageable number. In general, the beam-beam tune shifts, fractional betatron tunes, beam sizes at the IP, synchrotron tunes, and aspect ratios are all held equal for the two beams, and the damping decrements (damping rate between collisions) and β^* values are sometimes made equal as well.

There are two other aspects of *B* factory design associated with the necessity of a two-ring collider. The first is to keep the beams in collision—a condition that cannot be viewed as "automatic" in a case with two independent rings. Effects that can differentiate between the two rings include thermal motion, mechanical vibration, and power supply drifts. Moreover, there is no guarantee that the two relatively flat beams will not be tilted with respect to each other, as illustrated schematically in Fig. 3. As demonstrated in Fig. 4, for a given tilt angle the luminosity decreases rapidly

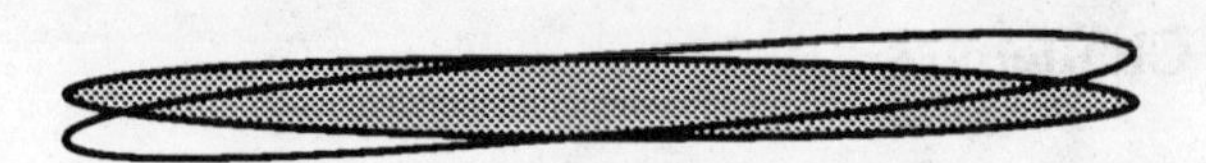

Fig. 3. Schematic illustration of bunches tilted at the IP.

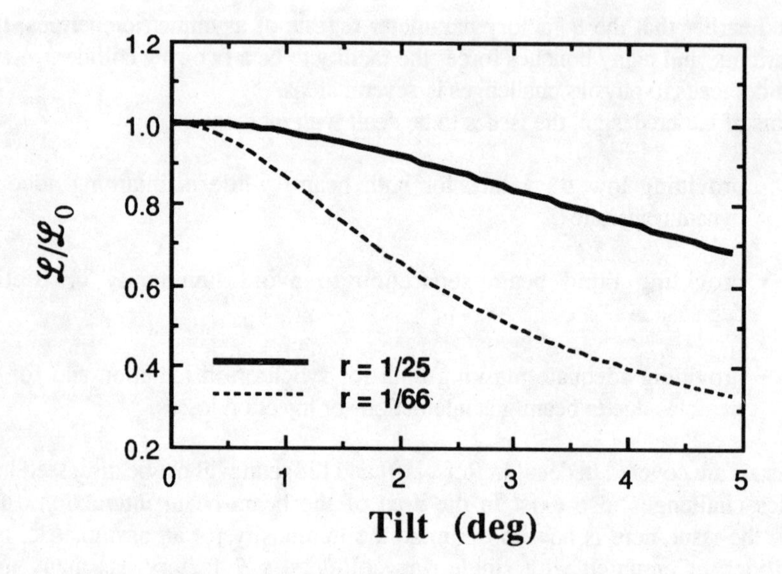

Fig. 4. Sensitivity of luminosity to relative beam tilt illustrated in Fig. 3.

as the aspect ratio of the two beams, r, gets smaller. Techniques for utilizing the beam-beam effect to diagnose such alignment problems are being developed [31]. Secondly, we must pay attention to minimizing the deleterious effects of the parasitic crossings. In addition to the obvious approach of increasing the separation distance, a prudent choice of betatron tune can also reduce the effects of the parasitic crossings.

As discussed earlier, B factory parameters are invariably chosen such that single-bunch instabilities are not a problem. That is, we use enough beam bunches to keep the single-bunch intensity below the instability thresholds. A broadband longitudinal impedance of $|Z/n| \approx 1\ \Omega$, which requires care—but not heroic effort—in vacuum chamber design, is sufficient. However, coupled-bunch instabilities in a B factory are expected to be very severe. The growth rates for coupled-bunch instabilities scale with the *total* beam current, and depend only weakly on the bunch pattern. The impedance that drives these instabilities is due to the trapped higher-order modes

(HOMs) of the RF cavities (of which many are needed to provide the short beam bunches). As will be discussed below, the problem is dealt with via a two-pronged attack wherein the HOMs are damped in the cavities and powerful broadband feedback systems are employed to deal with the remaining instability.

In the *B* factory parameter regime, the luminosity lifetime is limited by two effects: beam-gas bremsstrahlung (a single-beam effect) and radiative Bhabha scattering ($e^+e^- \rightarrow e^+e^-\gamma$, a beam-beam effect). The former effect has a very weak dependence on lattice parameters, and the only means of controlling it is to lower the pressure. The latter effect, which scales with the luminosity itself, will ultimately be dominant in a very high luminosity collider. Because the lifetime from the radiative Bhabha process increases with the ring circumference, a larger ring is an advantage in this regard.

4.2 Technology Challenges

The physics challenges discussed in Section 4.1 make certain implicit assumptions about *B* factory hardware. In a sense, one can summarize the technology challenges implied above by saying that the technology choices should not make liars out of the accelerator physicists. Examples of what is meant by this include:

- beam-lifetime and detector-background estimates assume a low background-gas pressure in the rings, despite the copious photodesorption
- luminosity estimates assume the required high beam currents can be supported without melting anything
- coupled-bunch instability growth rate estimates assume the HOMs of the RF cavities are heavily damped
- performance estimates (in the sense of integrated luminosity) assume the hardware components are sufficiently reliable that the machine does not "spend all of its time in the shop."

For all *B* factory designs, the main technological challenges lie in the areas of vacuum system, RF system, and feedback system design. In addition, the injection requirements for a *B* factory are nontrivial and are worthy of considerable attention.

4.2.1 Vacuum System Challenges

There are two main challenges for the vacuum system of a *B* factory collider:

- withstanding the high thermal flux from synchrotron radiation power
- maintaining a low pressure in the face of severe synchrotron-radiation-induced photodesorption.

The linear thermal power density is given by

$$P_L = \frac{P_{SR}}{2\pi\rho} \propto \frac{E^4 I}{\rho^2}. \tag{5}$$

This value can reach up to about 25 kW/m in some designs. Most designers choose a copper vacuum chamber, which is better able to dissipate the heat than is the more commonly used aluminum chamber and has other advantages mentioned below.

A *B* factory vacuum system must maintain a pressure range of 1–10 nTorr at the full beam currents in the rings. The photodesorption gas load is given by

$$Q_{gas} = 2.42 \times 10^{-2}\, E\, I\, \eta \qquad [\text{Torr}\cdot\text{L/s}] \tag{6}$$

where E is the beam energy in GeV and I is the beam current in amperes. The photodesorption coefficient, η, which gives the number of gas molecules desorbed per incident photon, depends on the vacuum chamber material, its history (in terms of manufacturing and handling), and the photon dose to which the chamber has been exposed. A typical design value for estimating the required pumping speed is $\eta = 2 \times 10^{-6}$ molecules per photon. This low value requires "scrubbing" the chamber walls with photons until this level of photodesorption is reached. The dependence of the desorption coefficient on photon dose is shown in Fig. 5, taken from Ref. [32]. Note that both the horizontal and vertical scales are logarithmic, so the required dose to reduce η takes progressively longer to reach.

In addition to the functional requirements just discussed, the chamber must be designed in such a way as to minimize the impedance seen by the beam. In essence, this means that the chamber profile should be very smooth. All unavoidable changes in cross section should be done in a gently tapered fashion and any discontinuities (gaps in flanges, bellows, masks, etc.) should be made as smooth as possible. The HOM power lost by the beam scales with the square of the beam current and can be quite high for a *B* factory.

There are two approaches to the vacuum chamber design that are being pursued by *B* factory design teams: the standard chamber design (similar to that used for a single-ring collider) and the "antechamber" design. In cases where the parameters are such that a pumping speed on the order of 100 L/s/m is required to achieve the design operating pressure, a standard chamber will suffice. In more extreme cases where much higher pumping speed is required, an antechamber design is called for. For example, both titanium sublimation pumps (TSPs) and non-evaporable getter (NEG) pumps are employed to deliver a pumping speed of 2,500 L/s/m in the hard-bend region of the CESR-B design [4].

As mentioned earlier, most designers favor a copper chamber, similar to that used in the HERA collider [33]. In addition to its excellent thermal properties, copper has a low desorption coefficient, as has been verified in R&D studies (see Fig. 5). Furthermore, in the *B* factory energy range, copper provides the additional benefit of being self-shielding for the synchrotron radiation emitted by the beam, avoiding the complications of a lead shield on the chamber.

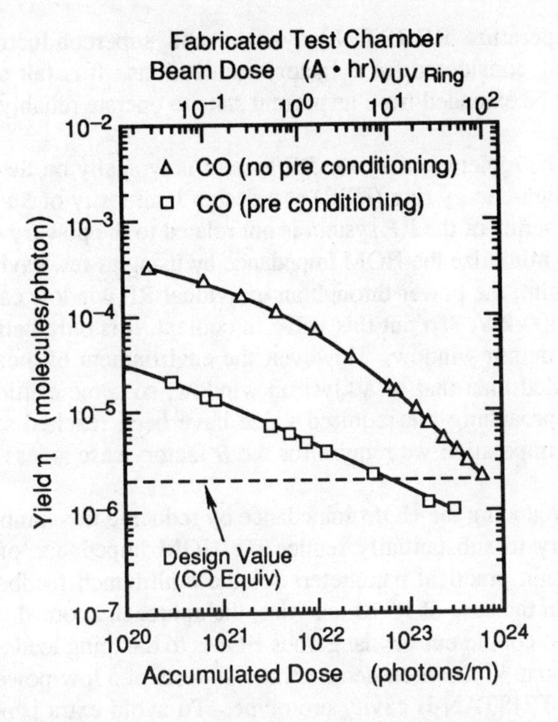

Fig. 5. Photodesorption coefficient for copper vacuum chamber measured at the BNL VUV ring. The PEP-II design goal of $\eta \leq 2 \times 10^{-6}$ is reached at relatively low photon dose. An argon glow-discharge pre-conditioning further reduces the dose required.

It is worth noting here that the IR is an especially difficult design challenge. On the one hand, there is a special need for low pressure (to minimize lost-particle backgrounds) while, on the other hand, there is a substantial gas source (the masks to protect the detector) combined with little room to install pumps. Innovative solutions are needed here, and all design teams are hard at work to deal with this issue.

4.2.2 RF System Challenges

The main challenges for a *B* factory RF system include:

- providing the large synchrotron radiation power loss
- minimizing the HOM impedance seen by the beam
- managing the transients associated with the gap in the bunch train (used for ion-clearing purposes)
- dealing with very heavy beam loading.

Both room-temperature RF (RTRF) systems and superconducting RF (SCRF) systems are being considered for *B* factory applications. It is fair to say that either technology must be extended from its present state to operate reliably in the *B* factory parameter regime.

The power to be replenished by the RF system is typically on the order of 5 MW for a *B* factory high-energy ring (HER) at a design luminosity of 3×10^{33} cm^{-2} s^{-1}. The difficulty in terms of the RF system is not related to the power *per se*, but results from the need to minimize the HOM impedance by using as few cavities as practical. With this constraint, the power through an individual RF window can be quite high, on the order of 500 kW. To put this value in context, it is only half the power of a typical klystron output window. However, the environment of a cavity window is less well controlled than that of a klystron window, so some caution is warranted. Power levels approaching the required value have been reached at both PEP and CESR, so the extrapolation we require for the *B* factory case is less than a factor of two.

In addition to reducing the HOM impedance by reducing the number of cavities, it is also mandatory to substantially reduce the HOM impedance of the individual cavities to maintain practical parameters for the multibunch feedback system (see Section 4.2.3). In the case of RTRF cavities, the approach adopted (see Fig. 6) is to use waveguides to couple out the dangerous HOMs to damping loads [34]. Damping to Q-values of about 30 has been demonstrated in tests of a low-power PEP-II cavity and also for the TRISTAN-II cavity prototype. To avoid extra penetrations of the cavity body in the case of SCRF, the approach used (Fig. 7) is to provide a large-aperture beam pipe to permit HOMs to propagate to a room-temperature load on the inner surface of the tube. Low-power tests of this scheme have demonstrated roughly the same performance as the RTRF scheme, with Q-values reduced to about 50.

Those *B* factory designs with a non-zero crossing angle will also need to develop transversely deflecting crab cavities. The design of crab cavities is generally based on SCRF, likely to be a good choice for this application as the requirements are for high voltage but low power. Note that the crab cavities must operate not at their fundamental mode but at the TM110 deflecting HOM.

Especially for RTRF cavities, the transient response of the heavily beam-loaded cavities to the ion-clearing gap in the bunch train is of concern. The issue is not so much that there is a transient, but that the transients must be well matched in the two rings in order to ensure that there is not excessive longitudinal displacement of the IP. As shown by Pedersen [35], it is possible to match the transient responses of the two *B* factory rings quite well by properly tailoring the gap in the positron ring.

Another concern that affects primarily the RTRF system is that of the heavy beam loading. To avoid excessive reflected power, the RF cavity must be substantially detuned in frequency. For a ring with large circumference and high beam current, the required detuning can exceed the rotation frequency. In this circumstance, the cavity fundamental mode strongly drives certain coupled-bunch modes unstable. Because the driving impedance is the fundamental mode itself, we cannot apply the usual trick of simply "de-Qing" the mode. The solution developed [35] involves a special RF

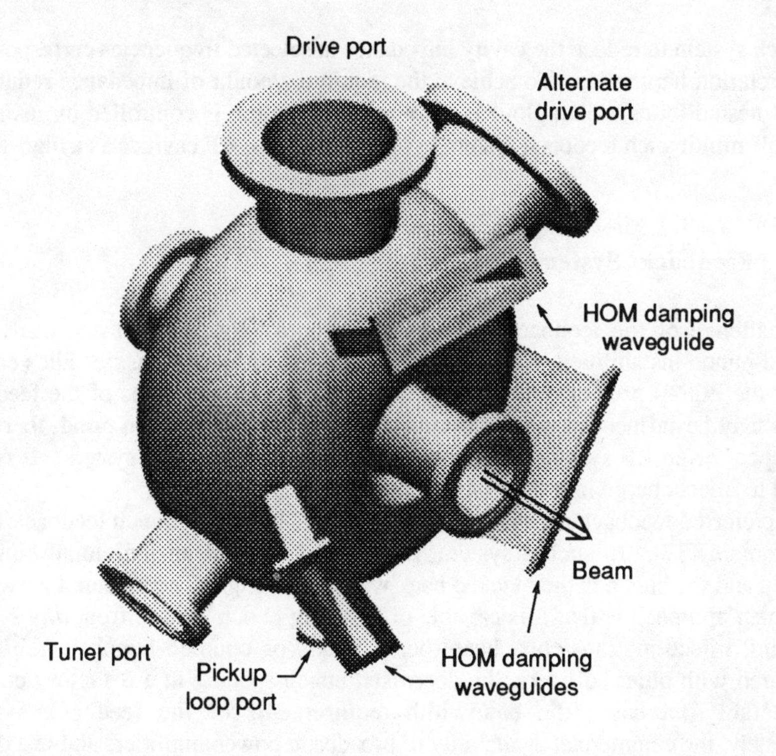

Fig. 6. RTRF cavity design from Ref. [6]. HOM damping loads are placed in waveguides that penetrate the cavity body.

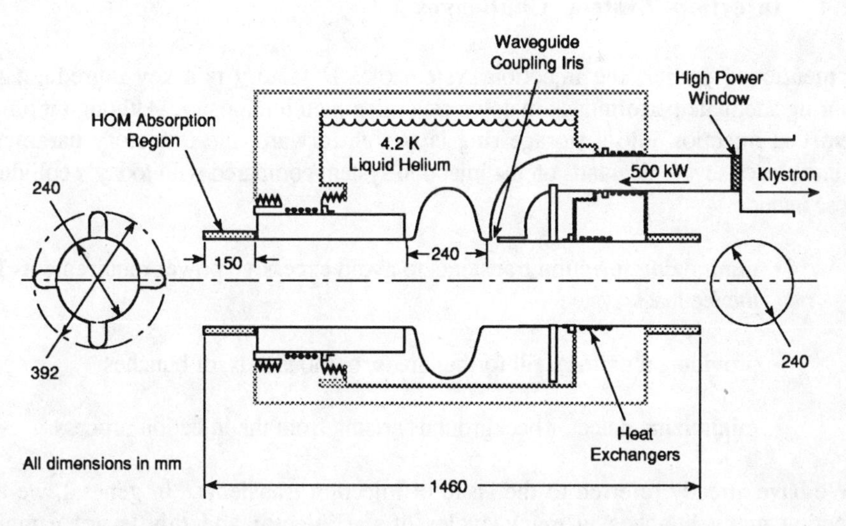

Fig. 7 SCRF cavity design from Ref. [4]. Coaxial HOM damping loads are located on the beam pipe inner surfaces.

feedback system to reduce the cavity impedance at selected frequencies corresponding to the rotation harmonics. To achieve the required amount of impedance reduction, several nested loops are employed. Any residual motion is controlled by using the "normal" multibunch feedback system to operate with the RF cavities as a high-power kicker.

4.2.3 Feedback System Challenges

The challenge of the feedback systems is to control longitudinal and transverse coupled-bunch instabilities with growth times on the order of 1 ms. Studies carried out for the PEP-II project [36] suggest that the power requirements of the feedback system may be influenced mainly by injection errors. With this in mind, there is a preference for an RF system that is phase-locked to the injection system. It is also helpful to inject charge into the rings in small increments.

The preferred feedback approach for a *B* factory is bunch-by-bunch feedback in the time domain [37]. In such a system, the displacement of an individual bunch is detected and the bunch is then kicked back where it belongs. The potential advantage of such an approach is that it is capable of damping bunch motion from *any* source, including injection transients, beam-beam kicks, or coupled-bunch instabilities. Compared with other colliders, the decreased bunch spacing in a *B* factory collider considerably increases the bandwidth requirements of the feedback system. Fortunately, the commercial availability of broadband power amplifiers and fast digital signal processing (DSP) chips makes this approach entirely feasible and tractable for a *B* factory.

4.2.4 Injection System Challenges

As mentioned earlier, the injection system of a *B* factory is a key ingredient for ensuring adequate performance in delivering integrated luminosity. Although it might seem that injection into a storage ring is straightforward, the *B* factory parameter regime places severe demands on the injection system compared with today's colliders. These include:

- minimizing injection transients to avoid excessive power requirements for the feedback system
- providing a uniform fill for hundreds, or thousands, of bunches
- minimizing detector backgrounds arising from the injection process.

We have already referred to the issue of injection transients. In general, we are injecting many bunches in many cycles of the injector and this is not a major problem. With a linac, the bunches are typically injected one at a time, with a fraction of the bunch current in each cycle. Nonetheless, it is important to keep the phase errors to a reasonable value. The uniform filling of the many bunches implies the ability to monitor the charge in each individual bunch at the level of a few percent of the nominal charge. This capability has not been developed in any of the *B* factory designs to date.

The ability to inject the two rings without excessive beam losses in the detector area is of paramount importance. One technique that has been adopted [6] is to use a so-called "graded aperture" in the ring. In this approach, the IR has the largest aperture in the ring (in units of the rms beam size), with progressively smaller apertures farther from the IR.

To optimize the luminosity performance of a *B* factory, it is crucial to be able to inject beam into the rings quickly. It is also beneficial to maintain the highest possible average luminosity, which implies relatively frequent injection (top-off mode). It remains to be seen whether it will be practical to reach the limiting case of continuous injection ("trickle charging"). This decision will require a careful study of detector backgrounds, an analysis of failure modes during injection, and an assessment of whether the injection process can be carried out without any readjustment of beam parameters, that is, with the beams in collision.

5 Cost and Schedule Issues

Clearly, the construction of a *B* factory depends on many considerations, such as:

- site details
- reusability of existing components
- manpower constraints
- funding profile constraints.

It is also important to be aware of a difference in cost accounting for U.S. projects compared with those elsewhere: all labor costs are explicitly included in U.S. cost estimates. (The significance of this accounting difference depends greatly on the amount of the project carried out with in-house personnel. In Japan, for example, many of the technical components are purchased from industry, in which case the labor costs are implicitly included in the overall project budget. In-house personnel, however, are not accounted for in the project cost estimate.) For these reasons, it is difficult to make any sweeping generalizations. Here, we will confine ourselves to those issues that are generally common or relevant to any *B* factory project.

5.1 Schedule

In the U.S., it is becoming more and more difficult to follow an optimum ("technology-limited") schedule. There are several reasons for this, including fluctuations in annual funding levels and the administrative overhead associated with environmental safety and health (ES&H) considerations. Still, it is the technology-limited schedule that is invariably presented at the design stage of a project, as this is the only thing that can be discussed sensibly. Clearly, it is the funding-limited schedule that is actually followed, and the stretch-out in this case can be one or two years.

Most *B* factory projects involve an upgrade of an existing facility. In this case there is little or no conventional construction (tunnels, buildings) required. In addition, there are many technical components (magnets, pumps, injector,...) available for reuse. It is worth commenting here that the equipment to be reused should be carefully inspected—and refurbished as necessary—to ensure its long-term reliability in *B* factory service.

In general, any accelerator project schedule is paced by a few key items. For a *B* factory, the main long-lead items are the vacuum system hardware and the RF system hardware (cavities, klystrons). These are the key to a timely project completion. In addition, there are certain labor-intensive operations that can pace a project schedule, such as installation, survey and alignment, and magnetic measurements, so it is also necessary to ensure that there are adequate resources to achieve the proposed schedule.

For a *B* factory, there are various R&D activities needed to verify design choices and optimize the design. It is important that these activities be fully integrated into the project schedule so that the construction team can decide what to buy or build in a timely way, especially for the long-lead-time items.

Although not a necessity, there are considerable benefits to scheduling the project such that some of the commissioning activities can begin early. One recommendation is to get one of the two *B* factory rings up and running. This will permit important tests of the vacuum system, the RF system, the feedback system, and the injection system. (The disadvantage of this approach is that it almost inevitably causes conflicts between the installation and commissioning teams, but this author is convinced that the advantages far outweigh this inconvenience.) An important point to note is that the detector is an integral part of the collider itself, so its schedule must also be integrated into the overall project schedule. In general, one envisions initial commissioning activities occurring prior to installing the radiation-sensitive portions of the detector. It is an open question whether the detector solenoid, which affects the beam orbit, is required for the initial commissioning.

6 Existing *B* Factory Proposals

In the past few years there have been six *B* factory design initiatives worldwide. A list of these projects is given in Table 2.

A summary of the key parameters for the various *B* factory designs is given in Table 3. It is difficult to make detailed comparisons among the projects, because the design choices are strongly influenced by site considerations. Nonetheless, it is striking in Table 3 that the range of parameters adopted by the design teams is actually rather narrow.

Table 2. List of existing *B* factory proposals.

Country	Laboratory	Project	Status
Germany	DESY	HELENA	Inactive
Japan	KEK	TRISTAN-II	Active (seeking funding)
Russia	INP-Novosibirsk	VEPP-5	Active at low level
Switzerland	CERN/PSI	BFI	Inactive
United States	Cornell	CESR-B	Active (incremental approach)
United States	SLAC	PEP-II	Active (seeking funding)

Table 3. Key parameters for *B* Factory designs.[a]

	PEP-II	CESR-B	BFI	HELENA	TRISTAN-II	VEPP-5
$\mathcal{L}$ (10^{33} cm^{-2}s^{-1})	3	3	1	3	2	5
E (GeV)	9.0,3.1	8.0,3.5	8.0,3.5	9.3,3.0	8.0,3.5	6.5,4.3
C (m)	2200	765	963	2304	3018	714
I (A)	1.5,2.1	0.9,2.0	0.6,1.3	0.7,1.1	0.2,0.5	0.7,1.0
No. of bunches	1658	164	80	640	1024	170
s_B (m)	1.3	3.0	12.0	3.6	3.0	4.2
Separation	magnetic	angle + crab	magnetic	magnetic	magnetic	magnetic
σ_ℓ (cm)	1	1	2	1	0.5	0.8
V_{RF} (MV)	18.0,9.5	33.0,12.0	13.0,2.0	17.0,4.5	47.0,20.0	7.0,4.5
RF technology	RT	SC	RT	RT	RT	RT→SC[b]
Vacuum technol.	DIP	NEG+TSP	NEG	DIP+LIP	DIP+NEG	LIP+TSP
P_L, max. (kW/m)	5.1	46.0	7.6	2.1	1.6	7.3
Pumping (L/s/m)	125	2500	500	670	100	500
$\varepsilon x/\varepsilon y$	25	36	33	20	100	20,16
β_y^* (cm)	3.0,1.5	1.5,1.5	3.0,3.0	2.0,1.0	1.0,1.0	1.0,1.0
D^* (m)	0.0	0.0	0.0	0.0	0.0	0.4
ξ_y	0.03	0.03	0.03	0.04	0.05	0.05
ξ_x	0.03	0.04	0.03	0.04	0.05	0.012

[a]Where two entries are given, they refer to the high- and low-energy ring, respectively. Information summarized here was collected mainly from Siemann, SLAC-PUB-5637, and references contained therein.

[b]Initial operation with RT system; final luminosity makes use of SC system as an upgrade.

In terms of energy asymmetry, the designs based upon a large ring circumference generally favor a higher value. This is because the beam separation is eased in this way and, for a large ring, the countervailing increase in synchrotron radiation power is not an overriding concern. The bunch spacing is usually selected to be a good match to the natural emittance of the HER. As is clear from Table 3, the range of values is not large. Dispersion at the IP is absent in most designs, as conventional wisdom says it should be avoided. The Novosibirsk "monochromatization" scheme takes the opposite tack, making use of a large dispersion and a small beam emittance to decouple the synchrotron and betatron motion. The choice of RF technology (RT or SC) depends mainly on whether the limiting factor is the requirement for high voltage or high power. In the former case SCRF is desirable, whereas the benefits of SCRF are not very significant in the latter case, especially if the voltage requirements are moderate. As mentioned in Section 4.2.2, the HOM damping requirements in either case (to $Q \approx 50$) are essentially the same.

7 Summary

As should be clear from the discussion in this paper, major progress is being made in the various *B* factory R&D programs. The groups are focusing on the proper issues and have successfully eliminated technical uncertainties. To date, no significant new technical issues have arisen. Physics and technology constraints tend to force the various groups to make similar parameter choices (though implementation details often differ), so the ongoing R&D programs are of interest, and have application, to all projects. Indeed, the issues being studied in *B* factory design are of interest to the entire new generation of colliders and storage rings, including Φ and τ-charm factories, hadron colliders (SSC and LHC) and synchrotron light sources.

The excellent combination of physics and accelerator physics makes the *B* factory an important and exciting project. The good mix of accelerator physics and engineering challenges will also make it a fun one! We are all looking forward to the construction start of the first project, hopefully in the next few years.

In closing, it is appropriate to remind the reader that making such a large jump in luminosity will not be an easy undertaking. The key ingredient in ensuring our success will be to remember to treat the challenges with proper respect.

8 Acknowledgments

It is a pleasure to thank the many designers of *B* factories for numerous stimulating discussions. I am especially grateful to my colleagues in the PEP-II project from LBL, SLAC, and LLNL for the information they have given me over the years. I would also like to thank Miguel Furman, Swapan Chattopadhyay, and Jonathan Dorfan for a careful reading of this paper, and Andrea Chan for help with preparation of the figures.

References

1) *CP* Violation and Beauty Factories and Related Issues in Physics, ed. by D. B. Cline and A. Fridman, Ann. of N.Y. Acad. of Sci., Vol. 619, February, 1991.
2) K. Wacker *et al.*, "Proposal for an Electron-Positron Collider for Heavy Flavour Particle Physics and Synchrotron Radiation," PSI PR-88-09, July, 1988.
3) "Feasibility Study for a *B* Meson Factory in the CERN ISR Tunnel," ed. T. Nakada, CERN 90-02, PSI PR-90-08, March, 1990.
4) K. Berkelman, *et al.*, "CESR-B, Conceptual Design for a *B* Factory Based on CESR," CLNS 91-1050, February, 1991.
5) A. N. Dubrovin and A. A. Zholents, in Proc. of 1991 IEEE Particle Accelerator Conference, San Francisco, May 6–9, 1991, p. 2835.
6) "An Asymmetric *B* Factory Based on PEP, Conceptual Design Report," LBL PUB-5303, SLAC-372, CALT-68-1715, UCRL-ID-106426, UC-IIRPA-91-01, February, 1991.
7) "Accelerator Design of the KEK *B* Factory," S. Kurokawa, K. Satoh, and E. Kikutani, eds., KEK Report 90-24, March, 1991.
8) H. Albrecht, *et al.*, "HELENA: A Beauty Factory in Hamburg," DESY 92-041, March, 1992.
9) P. Oddone, in Proc. of UCLA Workshop on Linear Collider $B\bar{B}$ Factory Conceptual Design, 1987, p. 243.
10) A. A. Garren, "Lattice and Interaction Design for *B* Factories," this school.
11) Workshop on Physics and Detector Issues for a High-Luminosity Asymmetric *B* Factory at SLAC, D. Hitlin, ed., SLAC-373, LBL-30097, CALT-68-1697, March, 1991.
12) J. Seeman, Lecture Notes in Physics, Vol. 247, p. 121, 1985.
13) S. Krishnagopal and R. Siemann, in Proc. of 1989 IEEE Particle Accelerator Conference, Chicago, March 20–23, 1989, p. 836.
14) "Feasibility Study for an Asymmetric *B* Factory Based on PEP," S. Chattopadhyay and M. Zisman, eds., LBL PUB-5244, SLAC-352, CALT-68-1589, October, 1989.
15) F. Willeke, private communication.
16) D. Rice, *et al.*, in Proc. of Conference on *B* Factories: The State of the Art in Accelerators, Detectors, and Physics, Stanford, April 6–10, 1992, SLAC-400, D. Hitlin, ed., December, 1992, p. 118.
17) R. Palmer, SLAC-PUB-4707, 1988.
18) K. Oide and K. Yokoya, SLAC-PUB-4832, 1989.
19) A. Piwinski, DESY HERA Report 90-04, 1990.
20) K. Akai, *et al.*, in Proc. of Conference on *B* Factories: The State of the Art in Accelerators, Detectors, and Physics, Stanford, April 6–10, 1992, SLAC-400, D. Hitlin, ed., December, 1992, p. 181.
21) J. Eden and M. Furman, "Further Assessments of the Beam-Beam Effect for PEP-II Designs APIARY 6.3D and APIARY 7.5," PEP-II AP Note 2-92, October, 1992; J. Eden and M. Furman, "Assessment of the Beam-Beam Effect for Various Operating Scenarios in APIARY 6.3D," PEP-II ABC Note 62, May, 1992.
22) F. Zimmermann, *et al.*, in Proc. of Conference on *B* Factories: The State of the Art in Accelerators, Detectors, and Physics, Stanford, April 6–10, 1992, SLAC-400, D. Hitlin, ed., December, 1992, p. 125.

23) D. Boussard, CERN Report 11/RF/Int. 75-2, 1975.
24) M. Furman, in Proc. of 1991 IEEE Particle Accelerator Conference, San Francisco, May 6–9, 1991, p. 422.
25) Y.-H. Chin, CERN/SPS-85-2, 1985; R. Ruth, in Accelerator Physics Issues for a Superconducting Super Collider, Ann Arbor, December 12–17, 1983, M. Tigner, ed., Univ. of Michigan Report UM HE 84-1, p. 151, 1984.
26) M. Zisman, *et al.*, in Research Directions for the Decade, Snowmass 1990, E. Berger, ed., World Scientific, Singapore, 1992, p.683.
27) M. Zisman, in Proc. of 1991 IEEE Particle Accelerator Conference, San Francisco, May 6–9, 1991, p. 1.
28) R. Ehrlich, "Backgrounds at e^+e^- Factories," this school.
29) Y.-H. Chin, in Proc. of Conf. on Beam Dynamics Issues of High-Luminosity Asymmetric Collider Rings, Berkeley, February 12–16, 1990, AIP Conference Proceedings 214, A. Sessler, ed., 1990, p. 424.
30) S. Krishnagopal and R. Siemann, Phys. Rev. D 41, 1741 (1990).
31) M. Furman, *et al.*, "Closed-Orbit Distortion and the Beam-Beam Interaction," LBL-32435, DAPNIA/SPP 9203, SLAC ABC Note 49, June, 1992.
32) W. Barletta, *et al.*, "Photodesorption from Copper Alloys," to be published in Proc. of 1992 High Energy Accelerator Conference, Hamburg, July 25–29, 1992; C. Foerster, *et al.*, "Desorption Measurements of Copper and Copper Alloys for PEP-II," to be published in Proc. of 12th Intl. Vacuum Congress 8th Intl. Conf. on Solid Surfaces, October 12–16, 1992, The Hague.
33) R. Ballion, *et al.*, Part. Accel. 29, 145 (1990).
34) R. Rimmer, "RF Cavity Development for the PEP-II *B* Factory," LBL-33360, to be published in Proc. of Intl. Workshop on *B* Factories: Accelerators and Experiments, Tsukuba, November 17–20, 1992.
35) F. Pedersen, in Proc. of Conference on *B* Factories: The State of the Art in Accelerators, Detectors, and Physics, Stanford, April 6–10, 1992, SLAC-400, D. Hitlin, ed., December, 1992, p. 192; see also F. Pedersen, "Multibunch Feedback," this school.
36) G. Lambertson, "Bunch-Motion Feedback for *B* Factories," LBL-33359, October, 1992.
37) J. Fox, *et al.*, in Proc. of Conference on *B* Factories: The State of the Art in Accelerators, Detectors, and Physics, Stanford, April 6–10, 1992, SLAC-400, D. Hitlin, ed., December, 1992, p. 208.

Lattice and Interaction Region Design for Tau-Charm Factories

John M. Jowett

CERN
CH-1211 Geneva 23
Switzerland

1 Introduction

A Tau-Charm Factory (τcF) is a high luminosity e^+e^- circular collider working in the energy range of the J/ψ resonance and τ-lepton production threshold. A τcF would continue mining the rich seam of physics opened up by SPEAR at SLAC (Stanford) and further explored by BEPC (Beijing). Since the first proposal in 1987 [1, 2], a number of designs for such a collider have been produced. As the requirements of the physics community have evolved, these designs have become more complex and now make essential use of some of the most interesting concepts proposed in the 30-year history of e^+e^- storage rings. While some of these were first mooted many years ago, a τcF is almost certainly going to be the first opportunity to put them into practice. Ideas such as monochromatization and longitudinal polarization were previously somewhat compromised by being conceived as modifications of existing storage rings. From the point of view of an accelerator physicist, a τcF presents not only a particularly demanding challenge in the performance required but also an exciting opportunity to work these elegant, yet untried, ideas into the design from the outset.

In the past, frontier e^+e^- colliders have, on the whole, been designed to meet the same goals, essentially those of the highest possible luminosity and low background at their chosen energies. Colliders from 1 to 200 GeV centre-of-mass energies are broadly similar in design (if the appropriate scaling is applied). In the "factory era" this has changed. The experimental conditions required of ϕ-, τ-c- and B-factories are so specialised and diverse that the machine designs differ much more than one might expect, given that their centre-of-mass energies span just one order of magnitude.

This lecture discusses the principles of τ-Charm Factory design, referring to several examples to be found in the literature, rather than laying down a single path to the ideal τcF. Each of these has something to teach us. A final design, incorporating all the best features in the happiest synthesis, is yet to be made and certainly requires further preliminary research on certain aspects: we can do no more than take a snapshot of a rapidly developing field. The list of references is intended as a starting point for further study. The emphasis here is on optics, particularly the less familiar aspects. Other critical aspects such as collective instabilities, the beam-beam effect, vacuum, the RF system, feedback and other hardware topics are treated by other lecturers.

2 Requirements For τcF Physics

Among present proponents of a τcF, one can distinguish a broad consensus, driven by the needs of specific measurements and moderated by realism, on what the performance goals should be. The machine would have a two- or three-phase programme as follows.

Initial high luminosity phase The first designs of a τcF aimed to provide an extremely high luminosity with maximum reliability and flexibility. The collider should work at centre-of-mass energies in the range

$$w = \sqrt{s} = 2E_0 = 3\text{–}5\,\text{GeV} \tag{1}$$

and attain its highest luminosity

$$L \simeq 1 \times 10^{33}\,\text{cm}^{-2}\text{s}^{-1} \tag{2}$$

at $\sqrt{s} = 4\,\text{GeV}$, a little above the τ-pair production threshold. As we shall see below, meeting this objective with as conventional a design as possible leads to a design for a high emittance, high current, multi-bunch collider. Comparing with the $L = 8.2 \times 10^{30}\,\text{cm}^{-2}\text{s}^{-1}$ which is the best so far realised by BEPC [3], shows that even this first stage of the τcF is ambitious.

Monochromatic phase At the energy of the J/ψ resonance

$$\sqrt{s} = 3.097\,\text{GeV} \tag{3}$$

or $E = 1.548\,\text{GeV}$, some 10^{10} J/ψ decays/year are needed for CP-violation studies in the decays of $\Lambda^0\bar{\Lambda}^0$ and $\Xi\bar{\Xi}$ pairs [4]. However, the full width $\Gamma(\text{J}/\psi) = 0.086\,\text{MeV}$ is very narrow. With a typical collider's centre-of-mass energy resolution, $\sigma_\epsilon = \langle (E - E_0)^2/E_0^2 \rangle \simeq 10^{-3}$ only a small fraction of the collisions would occur with centre-of-mass energy within $\Gamma(\text{J}/\psi)$. One would need a luminosity of

$$L = 5 \times 10^{33}\,\text{cm}^{-2}\text{s}^{-1} \tag{4}$$

to create enough J/ψ s. Otherwise the only way is to reduce the energy spread in collision by means of a "monochromatic" collision scheme (a special insertion design which has never been tested). We shall see that this requires a low-emittance optics.

Ultimate phase Monochromatization of collisions would be even better if combined with longitudinally polarized beams, another long-standing idea which has never been realised at any e^+e^- collider. A polarization scheme which does not use the usual Sokolov-Ternov radiative mechanism to polarize the beams was recently proposed [5] for a τ-Charm Factory.

We shall see that the requirements of these different phases pull the parameters of the machine in different directions: from high to low emittance. It is obvious that they cannot all be met by a single configuration. Present thinking is that we should try to design a machine which can be *adapted* relatively easily. Ideas for achieving the necessary flexibility within the same basic structure have been proposed recently by several authors [5, 6, 7, 8, 9]. Within a given configuration, further flexibility in the accessible range of parameters is achieved by means of various wiggler magnets.

For as long as no-one implements a monochromator scheme, doubts can always be cast on the feasibility of the second phase. Studies would be done during the operation of the first phase which must, for the sake of safety, produce a substantial output of physics. It will do so if it attains luminosities of a few times $10^{32}\,\mathrm{cm}^{-2}\mathrm{s}^{-1}$ which must, therefore, be the primary initial goal of the project.

3 Basic Parameter Choices

The design of a collider starts with the dynamics of the collision process itself (essentially the beam-beam effect). Working outwards from the interaction point (IP) this determines the size, shape and intensity of the beams which in turn determine the focusing structure, size and other gross features of the storage ring (or rings).

In this section we shall relate the luminosity to the phase-space distribution of the beam in order to see how to choose the parameters which determine this distribution. The formulation will be rather more general than usual since we want to cover cases (such as monochromator schemes) with non-vanishing dispersion at the IP. Although the treatment is meant to be as self-contained as possible, some basic notions from the statistical mechanics of electron storage rings [10, 11] will be found helpful.

3.1 Beam Distribution

In a discussion of luminosity, it is convenient to work with the basic set of phase space coordinates of particles in each beam denoted by $\mathbf{X} = (x, p_x, y, p_y, z, \varepsilon)$ where x and y are the radial and vertical *total displacements* from the closed orbit, $\varepsilon = (E - E_0)/E_0$ is the energy deviation and p_x, p_y and z are appropriate conjugate variables.

Note that these are *not the normal modes* of linearised motion around the closed orbit. In a further stage of the analysis, it is customary to split the transverse displacements into components due to betatron and synchrotron motion by introducing the *dispersion functions*, $D_x(s)$ and $D_y(s)$ so that

$$x = x_\beta + D_x \varepsilon, \qquad y = y_\beta + D_y \varepsilon. \tag{5}$$

This canonical transformation decouples the energy and transverse oscillations up to quadratic terms in the Hamiltonian. In almost all cases of interest, it is further arranged that the betatron coupling effects (terms $\propto x_\beta y_\beta$) are weak and that two of the true normal modes coincide with x_β and y_β at the IP. Then it can be shown [11] that the equilibrium between radiation damping and quantum excitation results in gaussian phase space distributions of particles *at the IPs*:

$$f^{\pm}(\mathrm{X}) = \frac{\tilde{f}^{\pm}(p_x, p_y, z)}{\sqrt{8\pi^3 \beta_x^* \epsilon_{xc} \beta_y^* \epsilon_{yc} \sigma_\varepsilon^2}} \exp\left\{ -\frac{(x - D_x^{*\pm}\varepsilon)^2}{2\beta_x^* \epsilon_{xc}} - \frac{(y - D_y^{*\pm}\varepsilon)^2}{2\beta_y^* \epsilon_{yc}} - \frac{\varepsilon^2}{2\sigma_\varepsilon^2} \right\}. \tag{6}$$

Quantities referring to the IP are "starred" and positron and electron bunches are distinguished with superscripts + or −. Apart from noting that $\tilde{f}^{\pm}$, like $f^{\pm}$ itself, is normalised to unity, the details of $\tilde{f}^{\pm}$ are of no concern here. The distribution is characterised by just three parameters: ϵ_{xc} and ϵ_{yc} are the natural emittances, determined by the magnetic lattice of the ring(s) and some weak ($\kappa \ll 1$) betatron coupling, and σ_ε is the fractional energy spread as we shall see below.

The gaussian distribution (6) is generally valid over most of the beam in the limit of low beam intensities. The "tails" of the distribution at large amplitudes may be altered by non-linear single-particle effects. At high intensities, both the core and the tails may be modified, e.g., by the beam-beam interaction [12, 13]. Even then, however, one can often approximate the distribution by (6) provided different values of ϵ_{xc} and ϵ_{yc} are used to account for "beam-beam blow-up". We shall implicitly do so several times in the following sections; it should be clear which formulae depend on this assumption; the advantage of the gaussian distribution is that all averages discussed below can be calculated analytically using some straightforward, if tedious, transformations.

3.2 Beam Sizes and Correlation Functions

For any function $\mathcal{A}(\mathrm{X}^+, \mathrm{X}^-)$ (possibly depending on variables of particles from both bunches), let us define averages over the distribution functions of the bunches at the IPs as

$$\langle \mathcal{A} \rangle^{\pm} = \int f^{\pm}(\mathrm{X}^{\pm}) \mathcal{A}(\mathrm{X}^+, \mathrm{X}^-)\, d\mathrm{X}^{\pm}, \qquad \langle \mathcal{A} \rangle^* = \left\langle \langle \mathcal{A} \rangle^+ \right\rangle^- . \tag{7}$$

Then we can evaluate the beam sizes as, e.g.,

$$\sigma_x^* = \sqrt{\langle x^2 \rangle^{\pm}} = \sqrt{\beta_x^* \epsilon_{xc} + D_x^{*2} \sigma_\varepsilon^2}, \tag{8}$$

$$\sigma_y^{*2} = \sqrt{\langle y^2 \rangle^{\pm}} = \sqrt{\beta_y^* \epsilon_{yc} + D_y^{*2} \sigma_\varepsilon^2} = \sqrt{\beta_y^* \kappa^2 \epsilon_{xc} + D_y^{*2} \sigma_\varepsilon^2} = \kappa \sigma_x^* \sqrt{\beta_y^* / \beta_x^*}. \tag{9}$$

The fractional energy spread is

$$\left\langle \varepsilon^2 \right\rangle^{\pm} = \sigma_\varepsilon^2. \tag{10}$$

It is also easy to calculate *correlation functions* between position and energy such as deviation

$$\langle x\varepsilon \rangle^{\pm} = D_x^{*\pm} \sigma_\varepsilon^2, \qquad \langle y\varepsilon \rangle^{\pm} = D_y^{*\pm} \sigma_\varepsilon^2. \tag{11}$$

This shows the dispersion functions in a new light. Rather than thinking of them as "derivatives of the closed orbit with respect to momentum", i.e., as related to shifts of the centre of the beam distribution with RF frequency, we can also consider them as normalised correlation functions of position and energy *within* the beam distribution.

3.3 Luminosity and Differential Luminosity

In terms of the numbers of particles in each bunch, $N^{\pm}$, the number of bunches, k_b, and the revolution frequency f_0, the general formula for *luminosity* is

$$L = k_b f_0 N^+ N^- \left\langle \delta(x^+ - x^-)\delta(y^+ - y^-)\right\rangle^* . \tag{12}$$

The Dirac delta-functions express the fact that luminosity is related to *collisions* in an intuitively appealing way. If we assume that the beams pass through each other head-on, only the transverse coordinates enter in (12). When a positron with energy deviation ε^+ collides with an electron of energy deviation ε^-, the centre-of-mass energy is

$$w = \hat{w}(\varepsilon^+, \varepsilon^-) \stackrel{\text{def}}{=} 2E_0\sqrt{1+\varepsilon^+}\sqrt{1+\varepsilon^-} \simeq E_0(2+\varepsilon^+ + \varepsilon^-) \tag{13}$$

and the *differential luminosity* [14, 15, 16] or luminosity per unit centre-of-mass energy is:

$$\Lambda(w) = k_b f_0 N^+ N^- \left\langle \delta(x^+ - x^-)\delta(y^+ - y^-)\delta\left(w - \hat{w}(\varepsilon^+, \varepsilon^-)\right)\right\rangle^* . \tag{14}$$

Clearly,

$$L = \int_0^\infty \Lambda(w)\, dw. \tag{15}$$

In the special case of the distribution (6), w also has a gaussian distribution about the mean $2E_0$ with standard deviation

$$\sigma_w^2 = L^{-1}\int_0^\infty w^2 \Lambda(w)\, dw - 4E_0^2. \tag{16}$$

Standard High Luminosity Optics The first phase of the τcF will use a conventional flat-beam collision optics with no dispersion at the IPs:

$$D^*_{x\pm} = D^*_{y\pm} = 0. \tag{17}$$

There is therefore *no correlation* between position and energy in the colliding bunches (the functions in (11) vanish). Evaluating the luminosity using (6) we get the familiar result

$$L = L_0 = \frac{k_b f_0 N^+ N^-}{4\pi\sigma_x^*\sigma_y^*}, \tag{18}$$

where $\sigma_x^* = \sqrt{\beta_x^* \epsilon_{xc}}$, $\sigma_y^* = \sqrt{\beta_y^* \epsilon_{yc}}$.

The differential luminosity is simply

$$\Lambda(w) = \Lambda_0(w) = \frac{L_0}{\sqrt{2\pi}\sigma_w} \exp\left\{\frac{-(w-2E_0)^2}{2\sigma_w^2}\right\}, \qquad \sigma_w = \sqrt{2}\,\sigma_\varepsilon E_0 \tag{19}$$

where the root-mean-square (RMS) spread in centre-of-mass energies σ_w is typically $\simeq 1\,\mathrm{MeV}$.

Monochromator Optics The second phase of the τcF will use a monochromator optics [14] whose purpose is to focus the luminosity onto a narrower region of particle spectrum than given by (19). This is achieved by means of *opposite correlations* between (vertical) position and energy in the beam distribution as indicated in Figure 1. According to (11) such correlations require opposite vertical dispersions for the two beams.

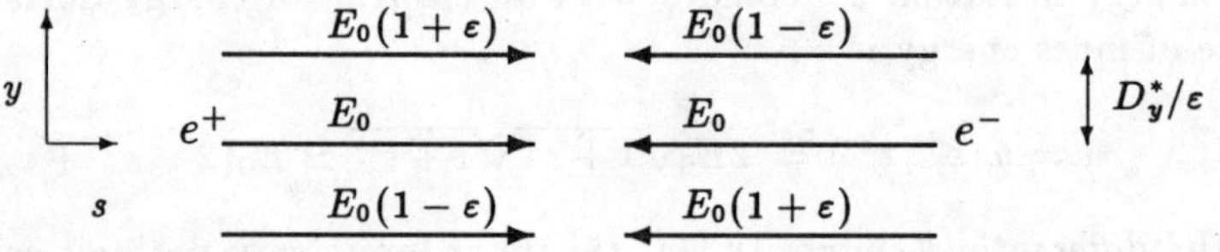

Fig. 1. Principle of a monochromator scheme: positrons with higher-than-average energy tend to meet electrons with lower-than-average energy and vice-versa. Positrons with energy $E_0(1+\varepsilon)$ will, on average, have the vertical position $y = D_y^*/\varepsilon$ for the collision optics specified in (20).

To illustrate the difference between equal and inverted dispersions, let us assume that there is in addition some horizontal dispersion D_x^* (possibly due to errors) which has the same sign for the two beams:

$$D_{x+}^* = +D_{x-}^* = D_x^*, \qquad D_{y+}^* = -D_{y-}^* = D_y^*. \tag{20}$$

Other parameters being equal, the total luminosity is then reduced by the dispersions

$$L = \frac{L_0}{\sqrt{1+\frac{D_x^{*2}\sigma_\varepsilon^2}{\sigma_x^{*2}}}\sqrt{1+\frac{D_y^{*2}\sigma_\varepsilon^2}{\sigma_y^{*2}}}}. \tag{21}$$

The differential luminosity is

$$\Lambda(w) = \frac{\Lambda_0(2E_0)}{\sqrt{1+\frac{D_x^{*2}\sigma_\varepsilon^2}{\sigma_x^{*2}}}} \exp\left\{-\left(\frac{D_y^{*2}}{\sigma_y^{*2}} + \frac{1}{\sigma_\varepsilon^2}\right)\frac{(w-2E_0)^2}{2(\sqrt{2}\,E_0)^2}\right\}. \tag{22}$$

It is simpler to look at its value at the centre of the distribution

$$\Lambda(2E_0) = \frac{\Lambda_0(2E_0)}{\sqrt{1 + \frac{D_x^{*2}\sigma_\epsilon^2}{\sigma_x^{*2}}}}, \tag{23}$$

which is undesirably reduced by the D_x^*. On the other hand, the energy resolution is improved by D_y^*

$$\sigma_w = \frac{\sqrt{2}\,E_0\sigma_\epsilon}{\sqrt{1 + \frac{D_y^{*2}\sigma_\epsilon^2}{\sigma_y^{*2}}}}, \tag{24}$$

i.e., σ_w has been reduced (in comparison to (19)) without reducing σ_ϵ. It is customary to define the "monochromatization factor" as the enhancement of energy resolution:

$$\lambda = \frac{\sqrt{2}\,\sigma_\epsilon E_0}{\sigma_w} = \sqrt{1 + \frac{D_y^{*2}\sigma_\epsilon^2}{\sigma_y^{*2}}}. \tag{25}$$

It is easy to show from (13) and (11) that $\langle wy \rangle^\pm = 0$: there is no correlation between w and the vertical position of the interaction vertex. The equal horizontal dispersion does correlate w with the position of the interaction vertex: $\langle wx \rangle^\pm = 2D_x^*\sigma_\epsilon^2$, a fact which may be of some use experimentally. From here on, however, we shall assume that the horizontal dispersion is designed to vanish, $D_x^* = 0$, in the monochromator optics. The total luminosity L is apparently reduced by the factor λ. Formally, from (19) and (23), we also have $\Lambda(2E_0) = \Lambda_0(2E_0)$.

For large values of $\lambda \gg 1$, as aimed for in present monochromator designs, the luminosity formula becomes

$$L \simeq \frac{k_b f_0 N^+ N_-}{4\pi\sigma_x^* D_y^*\sigma_\epsilon} = \frac{k_b f_0 N^+ N_-}{4\pi\sqrt{\beta_x^*\epsilon_{xc}}\,D_y^*\sigma_\epsilon} \tag{26}$$

implying that the experimental insertion design should aim for small β_x^* and larger β_y^*.

3.4 Luminosity at High Intensity

Standard High Luminosity Optics If the emittances and beam sizes have their "natural" values, i.e., those calculated in the usual way [10, 11] from the radiation effects in the lattice, then luminosity formulae like (18) are valid only at sufficiently low intensity that there is no blow-up of the beam cores [12]. With the beam-beam effects thus "switched off", we can re-write (18) as

$$L_0 = \frac{k_b I_b^2}{4\pi e^2 f_0 \sigma_x^* \sigma_y^*} = \frac{k_b I_b^2}{4\pi e^2 f_0 \epsilon_x \sqrt{\beta_x^* \beta_y^*}}\left(\kappa + \frac{1}{\kappa}\right), \tag{27}$$

where $I_b = eN_{\pm}f_0 = I/k_b$ is the current per bunch (now assumed to be the same for both beams).

The "unperturbed" vertical beam-beam tune-shift parameter (calculated on the basis of the "natural" σs) is

$$\xi_{yo} = \frac{(I_b/ef_0)r_e\beta_y^*}{2\pi(E_0/m_ec^2)(\sigma_x^* + \sigma_y^*)\sigma_y^*} \tag{28}$$

with an analogous formula for ξ_{xo}. The formulae are simpler if we assume conventional flat beams and adjust the betatron coupling to the "optimal" value $\kappa^2 = \epsilon_{yc}/\epsilon_{xc} = \beta_y^*/\beta_x^*$ which maximises L_0 and makes $\xi_{yo} = \xi_{xo}$. (In practice it is better to try to minimise σ_y^* by all means available!)

To reach the beam-beam limit, corresponding to a specified ξ_{yo} and a given emittance ϵ_x, we need a bunch current

$$I_b = \frac{2\pi ef_0(E_0/m_ec^2)\epsilon_x\xi_{yo}}{r_e} \tag{29}$$

and (27) becomes

$$L_0 = \frac{\pi k_b(1+\kappa^2)f_0(E/m_ec^2)^2}{r_e^2\beta_y^*}\epsilon_x\xi_{yo}^2. \tag{30}$$

If I_b is limited (likely in the initial year or two of operation while single- or multi-bunch instabilities are being mastered), then (29) has to be re-written as an equation determining the emittance. Provided it is possible to stay at the beam-beam limit in such a case, $L_0 \propto I_b$, and it does not pay to have a large ϵ_x. For this reason it is important, even in the initial high-luminosity phase, to arrange for the emittance to be variable over a given range. The large values needed once design current is achieved should be accessible with the help of wigglers, starting from a lower value corresponding to the "bare" lattice. In LEP, for example, we now use a relatively low-emittance lattice in conjunction with emittance wigglers [17, 18] to squeeze out the maximum luminosity for all the beam current values which occur throughout a fill. Concocting a parameter list for "design luminosity" and designing the optics to achieve the corresponding emittance is *the first path to a machine with too large an emittance.*

If we now consider the *true* luminosity, experiment and simulation show us that, for large enough ξ_{yo}, ϵ_{yc} is blown up somewhat while ϵ_{xc} normally does not change significantly[1]. The formula (30) for L_0 is an overestimate of the luminosity and is *the second path to a design with too large an emittance.*

Since the complexity of beam-beam phenomena [12, 13] cannot be meaningfully taken into account at this stage, we shall use a simple parametrised phenomenological model (as for LEP luminosity estimates in the design phase). According to this, the beam-beam effect reduces an unperturbed $\xi_{yo} \simeq 0.06$ to an effective saturated value $\bar{\xi} \simeq 0.04$ which may depend on other parameters

[1] In most rings, LEP being an exception.

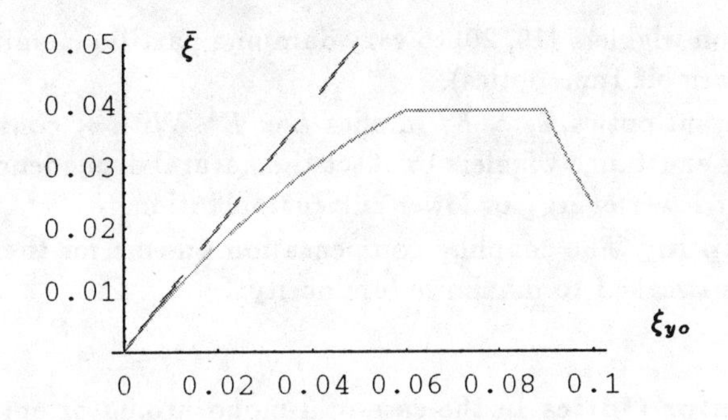

Fig. 2. **Typical relationship between the unperturbed beam-beam tune-shift ξ_{yo} and the effective beam-beam parameter $\bar{\xi}$. The dashed line, $\bar{\xi} = \xi_y 0$, corresponds to no beam-beam effects. Note that $\bar{\xi}/\xi_{yo} \to 1$ as $\xi_{yo} \to 0$.**

such as the damping time (c.f. $\bar{\xi} = 0.04$ at BEPC [3]) giving a more realistic estimate

$$L = L_0 \frac{\bar{\xi}(\xi_{yo}, \tau_y, \ldots)}{\xi_{yo}} = \frac{\pi k_b(1+\kappa^2) f_0 (E/m_e c^2)^2}{r_e^2 \beta_y^*} \epsilon_x \xi_{yo} \bar{\xi}(\xi_{yo}, \tau_y, \ldots). \qquad (31)$$

Defining the bunch separation $S_b = c/f_0 k_b$, this can be expressed numerically as

$$[L/\mathrm{cm}^{-2}\mathrm{sec}^{-1}] = 1.09 \times 10^{38} \frac{(1+\kappa^2)[E/\mathrm{GeV}]^2[\epsilon_x/\mathrm{m}]}{[S_b/\mathrm{m}][\beta_y^*/\mathrm{m}]}. \qquad (32)$$

Again if it turns out that $\bar{\xi} > 0.04$ then a lower ϵ_x or higher I_b may be needed to reach the beam-beam limit and maximise L.

From (32), we can see that, as usual, high luminosity will be obtained with the help of:

Low β_y^* A tightly-focusing micro-β insertion can attain values $\beta_y^* \simeq 1\,\mathrm{cm}$ using the gradients available (typically $30\,\mathrm{T\,m^{-1}}$) from superconducting quadrupoles. However this generates a lot of chromaticity which has to be corrected and places constraints on the bunch length (see Section 3.5).

Minimum bunch spacing S_b For a given circumference, this is equivalent to storing as many bunches as possible. The smallest bunch spacing is determined by separation requirements connected with parasitic beam-beam interactions at the next bunch encounter at $s = S_b/2$ from the IP. All τ-Charm Factory designs have opted for a double ring with some kind of separation scheme.

Here again it is desirable to provide a range of possible k_b (machine commissioning will start with a single bunch and work up).

Large emittance ... but only once the necessary I_b has been achieved! The lattice should be designed with fairly large ϵ_x but include means to increase

it (Robinson wigglers [19, 20] to vary damping partition, emittance wigglers, possibly variable tune optics).
For a constant optics, $\epsilon_x \propto E^2$ implies $L \propto E^4$. With ϵ_x constant (Robinson wigglers or emittance wigglers to offset the natural dependence), we can keep $L \propto E^2$ for lower energy or lower current operation.

Variable coupling The coupling compensation scheme for the detector solenoid will be tweaked to maximise luminosity.

Monochromator Optics In the case of a monochromator optics, the parameter choice is somewhat different. Here we follow the treatment of [5] to which the reader is referred for further details. We can still define the beam-beam strengths as in (28) except that the contribution of the energy spread dominates the vertical beam size in (9) and the beam is flat in the vertical plane:

$$\sigma_y^* \simeq D_y^* \sigma_\epsilon \gg \sigma_x^* = \sqrt{\epsilon_{xc}\beta_x^*}. \tag{33}$$

Not surprisingly, the beam-beam dynamics is considerably modified. According to pioneering studies [13, 21, 22] this results in a condition on the vertical beam-beam parameter:

$$\xi_{yo} \simeq \frac{(I_b/ef_0)r_e\beta_y^*}{2\pi(E_0/m_ec^2){D_y^*}^2\sigma_\epsilon^2} = \frac{(I_b/ef_0)r_e}{2\pi(E_0/m_ec^2)\mathcal{H}_y^*\sigma_\epsilon^2} \lesssim 0.015, \tag{34}$$

where $\mathcal{H}_y^* = {D_y^*}^2/\beta_y^*$ is the value[2] of the vertical dispersion "invariant" which is constant between the IP and the first vertically bending element. According to the same studies, there is also a condition on the horizontal beam-beam parameter:

$$\xi_{xo} \simeq \frac{\beta_x^*}{\beta_y^*}\frac{\sigma_x^*}{\sigma_y^*}\xi_{yo} \lesssim 0.05. \tag{35}$$

Therefore the emittance is determined by

$$\epsilon_{xc} \simeq \left(\frac{\xi_{yo}}{\xi_{xo}}\right)^2 \frac{\mathcal{H}_y^*}{\beta_y^*}\beta_x^*\sigma_\epsilon^2 \tag{36}$$

and tends to be small. Taking these conditions literally constrains the parameters rather tightly. However it is clear that even with a monochromator optics it is important to preserve as much flexibility as possible to cover the inevitable uncertainties in the analysis and estimation of parameters.

[2] It is assumed here that the optics is symmetric about the IP so that $\alpha_y = D_y^{*\prime} = 0$ at the IP itself.

3.5 Bunch Length

Equations (32) and (34) show that small values of the β-functions at the IP are needed for high luminosity. This is arranged by means of strong final focusing of the beam in a low-β insertion. The stronger the focusing, however, the sharper the waist of the β-functions. And it is well known [12, 13] that the value of the smaller β should not be significantly less than the bunch length σ_z. So we must also satisfy

$$\sigma_z = \frac{c\alpha_c}{\omega_s E_0} \lesssim \begin{cases} \beta_y^* \text{ (standard optics)} \\ \beta_x^* \text{ (monochromator optics)} \end{cases} . \tag{37}$$

Since (for conventional lattices) the momentum compaction α_c is more or less determined ($\alpha_c \simeq 1/Q_x^2$) [10], the only way to get a sufficiently short bunch is to increase the synchrotron tune $Q_s = \omega_s/\omega_0 \propto \sqrt{V_{\mathrm{RF}}}$ by applying a large RF voltage. A low-emittance monochromatic optics is at an advantage here over a standard high-emittance optics with its larger α_c.

Since various experiments and simulations suggest a range of values for the ratio, it is important to provide enough RF voltage to shorten the bunch to perhaps $\sigma_z/\beta_y^* \simeq 0.6$ although it is to be hoped that the collider can run closer to $\sigma_z/\beta_y^* \simeq 1$ to alleviate the problem of bunch-lengthening at high currents.

4 Designing the Collider

4.1 General Configuration

We have seen that a τ-Charm Factory should be a double-ring collider with many bunches. The question of how many interaction points it should have is often brought up. Although this depends partly on whether a single detector is adequate for the anticipated physics programme, there are strong arguments for a single interaction point on the basis of performance. Experience has shown that the total tune-spread from beam-beam interactions comes into the determination of $\bar{\xi}$. Moreover, with a single IP, the spectrum of coherent beam-beam modes is as simple as can be since each bunch is only aware of the presence of a single bunch in the other beam.

Most designers of a τ-Charm Factory [2, 23, 24, 25, 26, 27, 28, 6, 29, 30, 8, 7, 9, 31] have chosen a head-on collision geometry with an electrostatic separation scheme in the vertical plane. Notable exceptions include [32, 5] which use a combination of horizontal and vertical separation and [33, 34] which have crossing angles in the horizontal plane.

A layout of the accelerator complex from [6] is shown in Figure 3. It includes a synchrotron light source which could be built to take advantage of the powerful positron injector system. Filling a light source storage ring (of energy around 2 GeV) would take only a small part of the duty cycle of the injectors [26, 23, 35].

In contrast to the earlier generations of e^+e^- rings, about half the circumference is devoted to the two straight sections. With many bunches in two rings, there is no longer any need to maximise the revolution frequency *per se* and

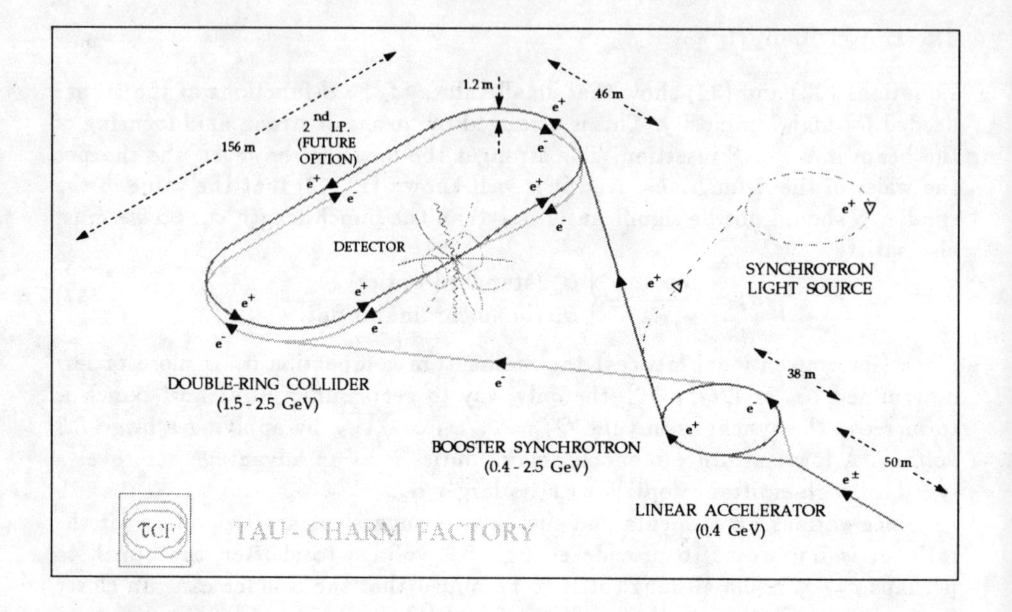

Fig. 3. **Schematic layout of a proposed τcF [6, 23]. Note the addition of a storage ring for the production of synchrotron light which could comfortably share the powerful injector system needed by the collider.**

we shall see that the long experimental straight section is necessary to properly match its optics. On the other side of the ring, the space provided by the long utility straight section is useful for a number of purposes.

In this layout, newly-injected electrons and positrons must make almost a full turn before reaching the interaction region, thereby minimizing detector backgrounds during injection. This leaves open the option of injection into synchrotron phase space at a dispersive location to eliminate transverse oscillations of the injected beam in the experimental straight section. If the backgrounds can be kept low enough, then this may open the possibility of topping up the stored beam during physics data-taking.

An additional benefit of having two rings, is that at the price of some asymmetry between them, radiation-sensitive elements such as the superconducting RF cavities and the detector can always be placed *upstream* of the principal sources of synchrotron radiation (wigglers, arc dipoles).

4.2 Arc Optics

Most τcF designs have used conventional FODO cells in the main arcs. These are most economical in space and can be equipped with sextupoles to correct chromaticity in the well-known way. Correction with just two sextupole families can provide adequate dynamic aperture [23, 26] and can be tuned to a variety of different phase advances per cell in order to vary ϵ_x: $(\mu_x, \mu_y) = (60°, 60°)$,

(60°,90°), (90°,90°) .... Alternatives to this traditional scheme will be discussed in Section 5.3. More sextupole families may improve dynamic aperture but would have to powered differently for different (μ_x, μ_y). Since the basic ideas are rather well-known, we shall not discuss FODO-based schemes in much more detail.

Injection insertions with fast kickers and non-zero dispersion (see above) should be incorporated into the arcs.

4.3 Dispersion Suppressors

The arcs must be connected to the dispersion-free straight sections by suitable dispersion suppressors. The simplest scheme with half the normal bend angle per cell and quadrupoles in series with arc only works for $\mu_x = 60°$. A more flexible scheme with several independent power supplies (c.f. LEP's [36]) is needed if (μ_x, μ_y) are to be variable.

Emittance (pure dipole) and Robinson (combined-function dipole-quadrupole) wigglers, which both need large values of D_x, can be included in these insertions. Such devices produce a vertical tune-shift $\Delta Q_y \propto B^2/E_0^2$ from edge-focusing which must be compensated with the nearby quadrupoles.

4.4 Utility Insertions

The dispersion-free utility insertion has a number of purposes. It can house the RF cavities and damping wigglers to increase injection efficiency or reduce emittance. This linear optics of this part of the ring will also be used to adjust the tunes and make other compensations. One attractive and flexible idea is to construct it from a set of quadrupole triplets as in [5]. This can be used to provide low β-functions at the locations of RF cavities or wigglers. Other designs use a string of FODO cells.

4.5 Interaction Region

The most complicated part of the lattice is the experimental insertion consisting of a micro-β insertion and separation scheme. Unlike the other parts of the ring discussed above, it must change substantially in the transition from standard to monochromator optics.

Standard Optics Most designs use doublet or triplet focusing to achieve $\beta_y^* \simeq$ 1 cm. Engineering solutions for this could be based on iron-free superconducting quadrupoles in a common cryostat protruding into the detector or on a hybrid consisting of small permanent magnet quadrupoles as close as possible to the IP, backed up with superconducting coils [37]. The idea of the latter solution is to prevent excessive growth of $\beta_y(s) = \beta_y^* \left(1 + s^2/\beta_y^{*2}\right)$ by starting the focusing some 15 cm closer to the IP than the end of the cryostat would be in a pure superconducting scheme. However it raises questions of field quality and engineering construction which are beyond the scope of this lecture. The detector design allows the closest quadrupole to approach to $L_1 = 0.8\,\mathrm{m}$ with an outer radius not greater than 20 cm.

The micro-β insertion must also incorporate a scheme to compensate the betatron coupling generated by the detector solenoid. There are a variety of possibilities for this:

- Compensating solenoid coils, wound around the quadrupoles, to directly cancel the solenoid effects. This is the most attractive scheme but it remains to be seen if it can be made compact enough for the solid-angle requirements of the detector.
- Dedicated skew quadrupoles interspersed with the focusing quadrupoles have the disadvantage of increasing the distance to the electrostatic separator.
- Rotatable micro-β quadrupoles as used with success in CESR. However the construction of separately rotatable *superconducting* quadrupoles inside their common cryostat would be another technical challenge [37].

As soon as possible after the micro-β quadrupoles, the beams must be separated into their two rings. The separation can be initiated with electrostatic (or, perhaps, RF-magnetic) separators. The amount of this initial separation is limited by the integral of the vertical electric field over the separator. The peak value (at the top energy of the collider) of this should be kept below about $2\,\mathrm{MV\,m^{-1}}$ to minimise the risk of sparking, especially if there is a significant quantity of synchrotron radiation[3]. Once the separation is sufficient, these can be followed with a suitable vertically bending septum magnet. Further vertical bends finish the separation and bring the two beams back onto horizontal orbits in their respective rings. The vertical dispersion thus generated has to be matched to zero before the beams enter the horizontal dispersion suppressors. This requires a vertical phase advance shift $\Delta\mu_y \simeq 2\pi$ between the separator and the final vertical dipole [38, 26, 5]. Several similar interaction region optics have been worked out. Figure 4 is typical.

The overall length of the common part of the rings determines the minimum totally safe[4] bunch spacing. In practice, encounters are allowed at distances where the beams are separated by $\gtrsim 10\sigma_x$ which is enough to reduce the parasitic beam-beam effects to an acceptable level. For separation at the top energy (say, 2.2 GeV) the minimum bunch spacing works out to be $S_b \simeq 12\,\mathrm{m}$ in typical cases [2, 23, 25, 26, 31]. This means that the first encounter takes place some 6 m from the IP, a point which may be inside the electrostatic separator.

Monochromator Optics In the τcF, the necessary vertical dispersion can be generated in the vertical separation scheme. Since the initial separation is elec-

[3] This may have several sources: the vertical bends, micro-β quadrupoles, etc. The design of the interaction region must be carried out in parallel with a detailed study of such sources to estimate detector backgrounds and other unwanted effects like separator sparking.

[4] In the sense that each bunch only ever experiences fields from bunches of the opposing beam in the collision process at the IP.

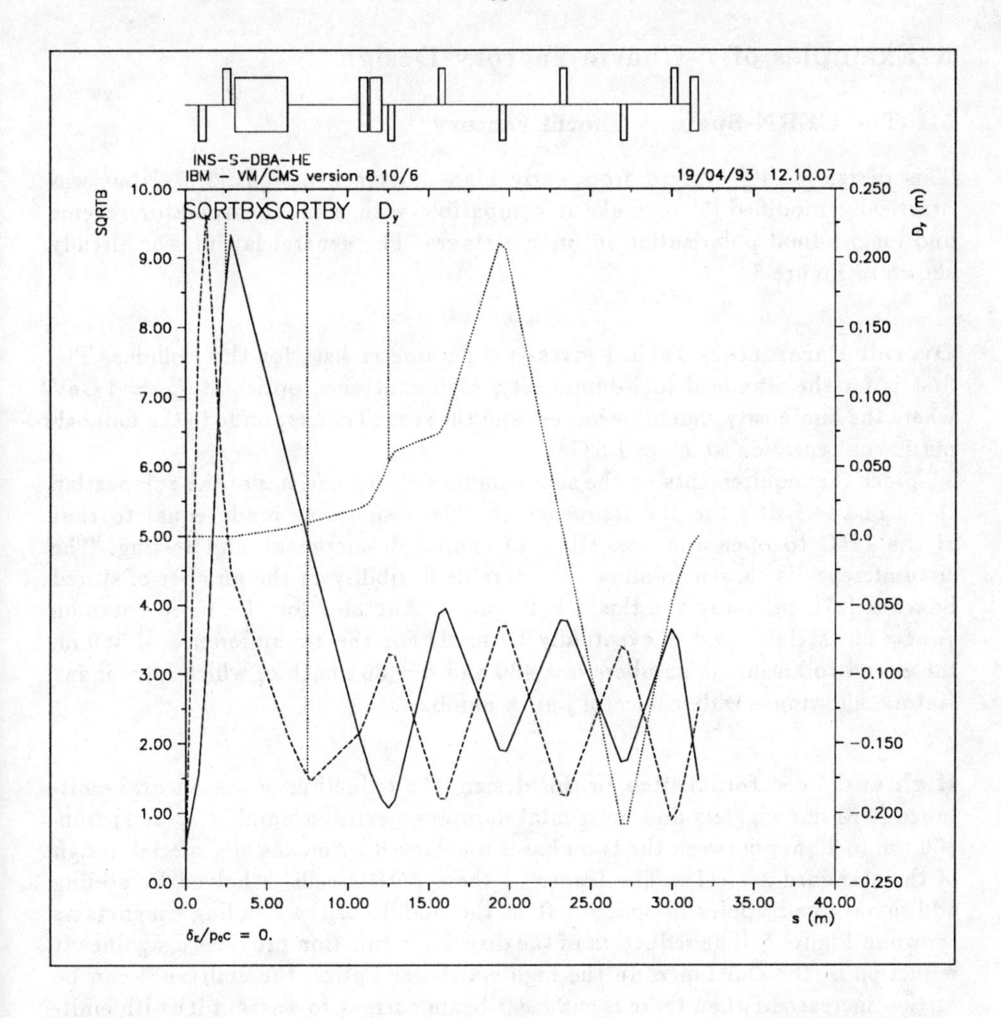

Fig. 4. **Interaction region optics for a standard high-emittance optics, showing $\sqrt{\beta_x}$, $\sqrt{\beta_y}$ and the vertical dispersion D_y. In this and similar plots (made by the optics program MAD [39]) horizontally (vertically) focusing quadrupoles are shown as boxes above (below) the beam axis and bending magnets or electrostatic separators are shown as boxes straddling the beam axis. In this case, the first such box is the electrostatic separator and the others are the vertical bending magnets.**

trostatic, this dispersion naturally has opposite signs for the two beams. As seen from (25) and (34), the performance of the scheme hinges on being able to generate a large value of $\mathcal{H}_y^*$ at the IP. A large β_y in the vertical bends helps to achieve this.

5 Examples of τ-Charm Factory Design

5.1 The CERN-Spain τ-Charm Factory

This design [6, 40] evolved from early high-emittance designs [2, 24] but was drastically modified [5] to make it compatible with a monochromator scheme and longitudinal polarization in further stages. The general layout was already shown in Figure 3.

Overall Parameters Table 1 gives two parameter lists for this collider. The first is for the standard high-luminosity, high-emittance optics at $E_0 = 1\,\text{GeV}$ where the luminosity should be largest and the second corresponds to the monochromatic configuration at $E_0 = 1.5\,\text{GeV}$.

Since the requirements on the superconducting RF cavities of the τcF overlap those of the LHC, the RF frequency in this design was made equal to that of the LHC to open the possibility of shared development and testing. The circumference is chosen to allow considerable flexibility in the number of stored bunches [27], not only for this RF frequency but also for the more common $f_{\text{RF}} = 500\,\text{MHz}$ should it eventually be used. For the circumference of 360 m, these lead to harmonic numbers $h = 480$ and $h = 504$ both of which have many factors, allowing a wide choice of bunch numbers.

High and Low Emittance In this design, the reduction of the natural emittance (with no wigglers and horizontal damping partition number $J_x = 1$) from 100 nm to 20 nm between the two phases is achieved by means of a special design of the standard arc cells. The length of these FODO cells is halved by adding additional quadrupoles in spaces left in the middle of the bending magnets as shown in Figure 5. The reduction of the dispersion function provides a significant reduction in the emittance. In the high-emittance optics, the emittance can be further increased (when there is sufficient beam current to warrant it) with emittance and Robinson wigglers to reduce J_x. In the monochromatic case, damping wigglers (in locations with zero dispersion) and Robinson wigglers (now working in the opposite direction to increase J_x) are used to reduce the emittance still further.

A disadvantage of this scheme of halving the cell length is that it becomes very difficult to find space for additional elements such as sextupoles, corrector magnets, beam-position monitors (BPMs), vacuum pumps, etc. Since some of the lattice options are to have $\mu_x = 90°$, it is important to have more than one BPM per cell to ensure adequate sampling of the orbit. The only way to pursue this approach seems to be to make the cells longer and accept a larger circumference. An alternative is to reduce the emittance simply by varying the phase advance in a fixed FODO cell layout. This approach has been followed in [7, 31] where a factor of 3 reduction in emittance is obtained by changing from $\mu_x = \mu_y = 90°$ to $\mu_x = \mu_y = 60°$ (this is similar to what has been done in practice in LEP [18]).

Table 1. Parameter lists for the CERN-Spain τcF [6, 5]. Parameters for the standard high luminosity phase are quoted at the energy just above the τ-pair production threshold where maximum luminosity is required. For the monochromatic phase the relevant energy is that of the J/ψ resonance.

Quantity		Standard	Monochr.	Units
Nominal Beam Energy	E	2	1.5	GeV
Circumference	C	360		m
Revolution frequency	f_0	832.76		kHz
Bending radius	ρ	11.8		m
β-function at IP	β_x^*	0.2	0.01	m
	β_y^*	0.01	0.15	m
Betatron tunes	Q_x	9.3	17.28	
	Q_y	9.2	10.18	
Momentum compaction	α	0.020	0.0036	
Natural horizontal emittance ($J_x = 1$))	ϵ_x	108	20	nm
Emittance with wigglers etc.		300	10	nm
Vertical emittance		7	2	nm
Fractional energy spread	σ_ϵ	4.5×10^{-4}	8×10^{-4}	
Radiative energy loss per turn	U_0	130	70	keV
Radiation damping times	τ_x	70	26	msec
	τ_y	35	52	msec
	τ_ϵ	14	26	msec
RF frequency	f_{RF}	399.72339		MHz
RF voltage	V_{RF}	8.0	2.3	MV
Synchrotron tune	Q_s	0.077	.02	
Number of bunches	k_b	30		
Bunch spacing	S_b	12		m
Total current per beam	I	573	315	mA
Particles per bunch	N_b	1.37×10^{11}	0.8×10^{11}	
Radiated power per beam	P_{rad}	75	22	kW
RMS bunch length	σ_z	6.6	8	mm
Longitudinal impedance ($\omega \to 0$)	$\|Z/n\|_0$	$\simeq 1.0$		Ω
Longitudinal impedance (effective)	$\|Z/n\|_{eff}$	$\simeq 0.13$	0.17	Ω
Beam-beam parameters	ξ_x	0.04	0.04	
	ξ_y	0.039	0.014	
Lifetime	τ	$\simeq 4$	1	hours
Luminosity	L	1×10^{33}	4×10^{32}	cm^{-2}sec^{-1}
CM energy spread	σ_W	1.3	0.10	MeV

The dispersion suppressor optics, which is similar to the LEP design [36] and shown in Figure 6 is sufficiently flexible to deal with dispersions in the different arc optics.

Interaction Region For the Standard case, the optics in the straight section is similar to that shown in Figure 4. For the monochromatic case, the interaction region scheme (Figure 7) is based on that presented (for more extreme conditions

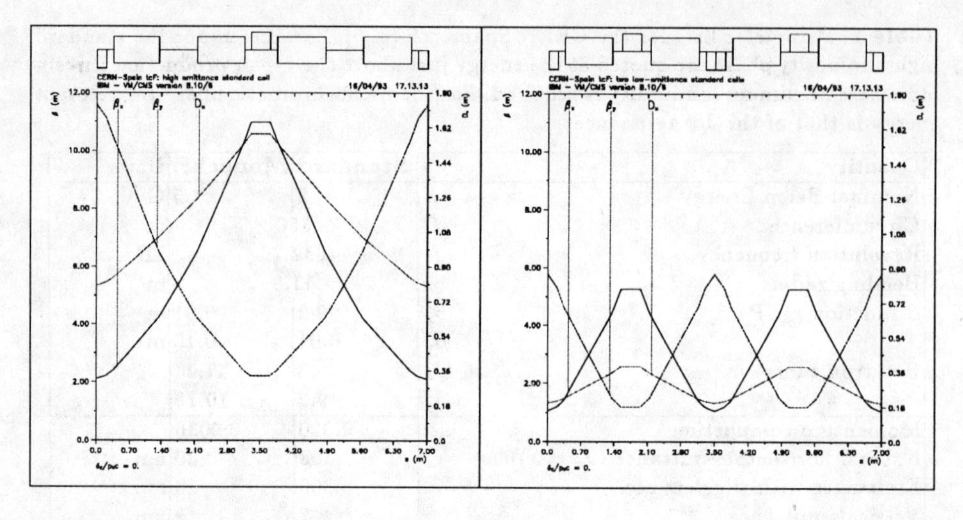

Fig. 5. Example of emittance reduction by adding "missing quadrupoles" to make two FODO cells out of one (following [5]). In both cases, the horizontal and vertical phase advances are 90° although $\mu_y = 60°$ was used in [5] as it is advantageous for dynamic aperture.

of emittance) in [32]. This design incorporates further special features:

- The initial electrostatic separation is *horizontal*. Once they are sufficiently separated the beams are separated vertically by bending magnets.
- The bulk of the correction of the chromaticity of the micro-β insertion is done locally using *electrostatic sextupoles* to take advantage of the opposite sign dispersion which exists near the IP.

Although there are some technical challenges associated with this concept, the 3D separation scheme offers advantages for background screening and the design of the first vertical bending magnet. Other monochromatic insertion designs separate only in the vertical plane [7, 8, 9, 30]

Beam Lifetime The beam lifetime in the two cases is determined by different effects: In the Standard case, beam-beam bremsstrahlung is the dominant loss mechanism. In the Monochromatic case the high beam density, particularly in the interaction region, causes the Touschek effect to dominate [41, 5].

Longitudinal Polarization The final stage of the collider proposed in [5] further extends the physics reach of the monochromator scheme with longitudinally polarized beams at the IP. This scheme is based on ideas in [42, 43] and does not involve the usual radiative self-polarization in the collider. Instead the beams are pre-polarized in special rapidly-polarizing (i.e., high magnetic field) accumulation rings (an upgrade of the injection system ...) and then injected through

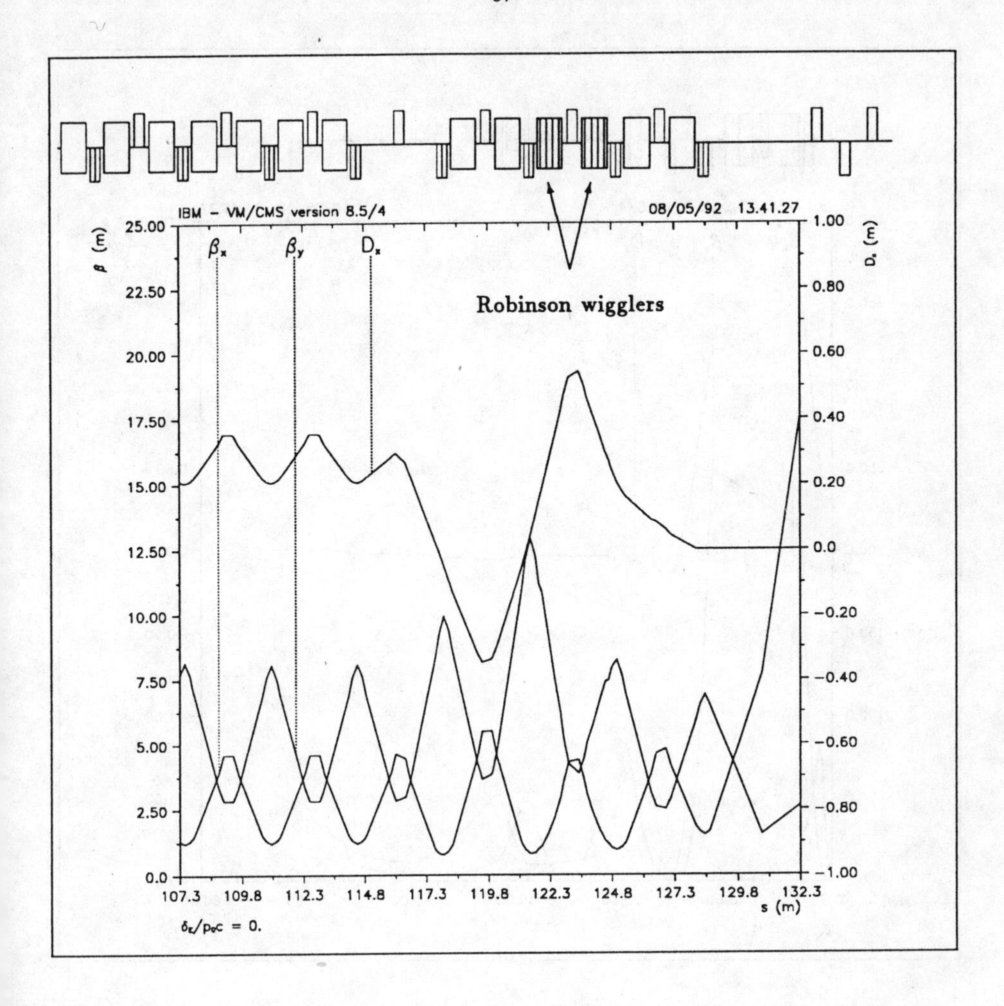

Fig. 6. Dispersion suppressor and an arc cell in the monochromatic optics of [5]. The Robinson wigglers to vary J_x are installed in a high dispersion location.

spin-rotators in the transfer lines so that their polarization is in the horizontal plane of the collider. The initial angle with respect to the orbit is chosen so that the precession around the vertical axis brings the spins into the longitudinal direction at the IP. A compact superconducting solenoid in the utility insertion further rotates the spin around the longitudinal direction so that the total spin-tune (number of precessions per turn) is exactly 1/2. This ensures that the polarization will again be longitudinal on the next turn. An analysis of the depolarizing effects [5] shows that a decent level of longitudinal polarization can be maintained for the duration of a fill. For further details, we refer the reader to [5].

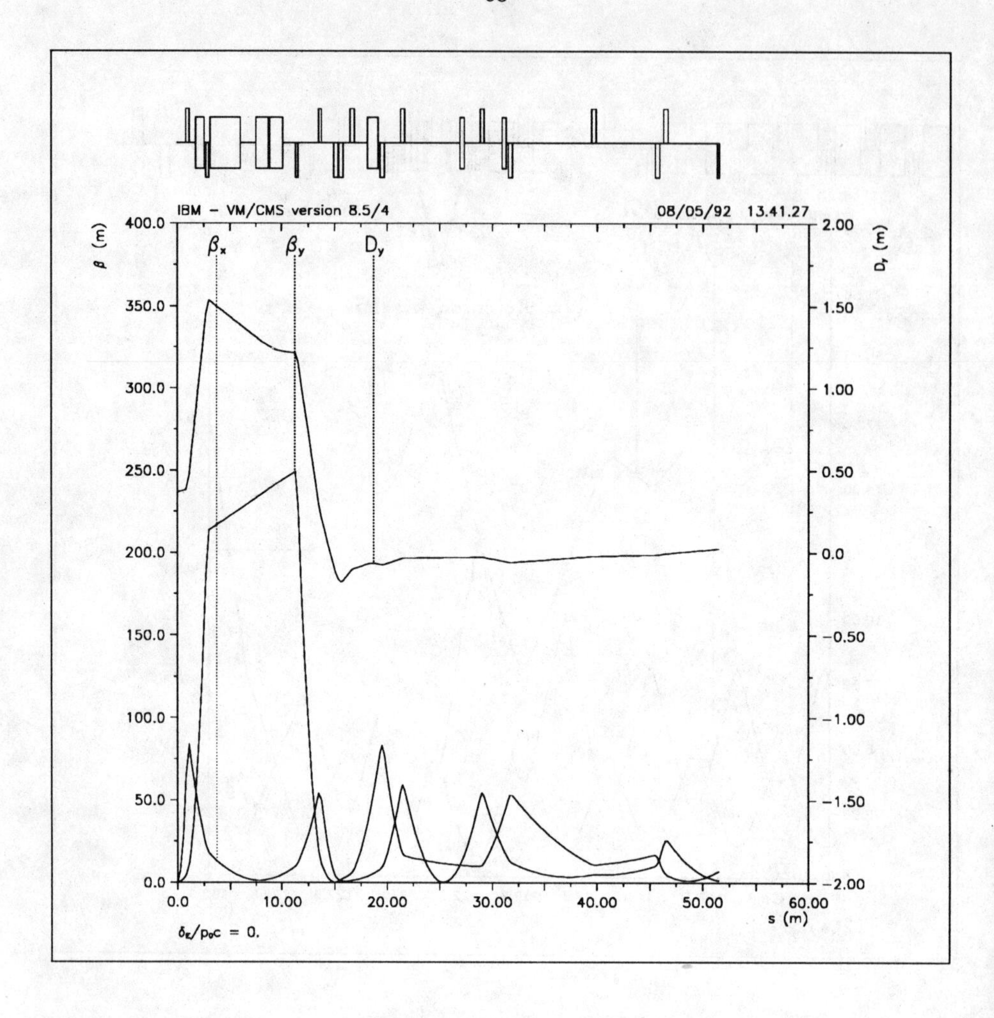

Fig. 7. Optics of the experimental insertion from [5]; the small horizontal dispersion generated by the initial separation in that plane is not shown.

5.2 Monochromator with Flat Beams

Early ideas for monochromatic colliders [14, 16] considered the introduction of vertical dispersion as a modification of the usual flat beam collision scheme with $\beta_y^* \ll \beta_x^*$. It may seem attractive to try to implement such a scheme as a relatively minor modification of the Standard scheme. However it has been shown [44] that this procedure leads to a number of difficulties including an awkward geometry and poorer performance than can be expected with schemes using beams flattened in the vertical plane. Most authors are therefore developing schemes of the latter type.

5.3 Non-FODO Arc Cells

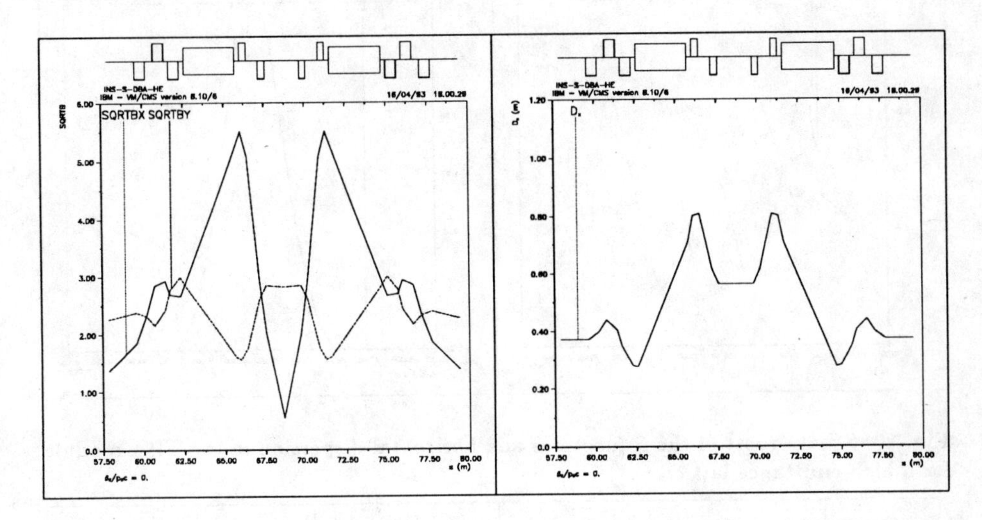

Fig. 8. Square roots of the β-functions and horizontal dispersion in the mis-matched DBA module for a high-emittance lattice.

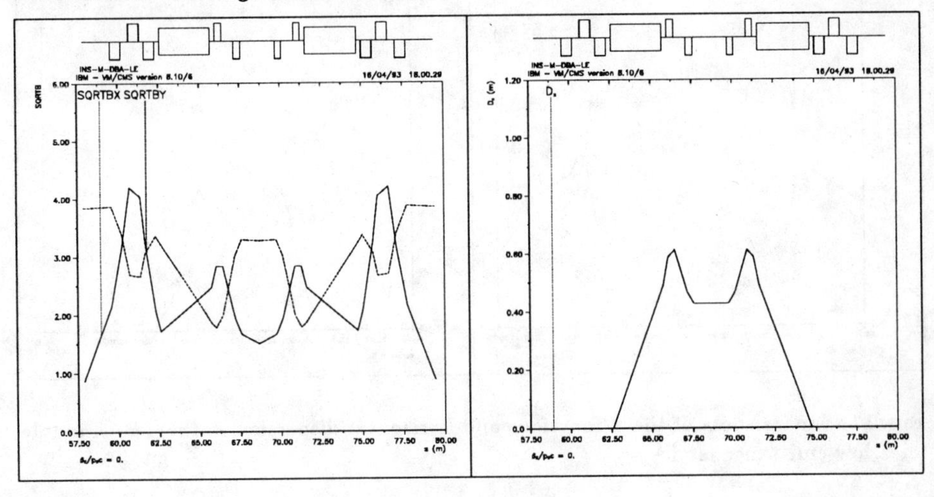

Fig. 9. Square roots of the β-functions and horizontal dispersion in the DBA module for a low-emittance lattice.

Storage rings designed for the production of synchrotron light from insertion devices such as wigglers and undulators require low emittances. Their designs have departed from the FODO cell structure used for nearly all the high-energy physics colliders (see, e.g., [45] for a review). Instead, a variety of achromatic modules can be used as the basic lattice units. In this context, achromatic means that the dispersion function is zero at the entrance and exit of the module.

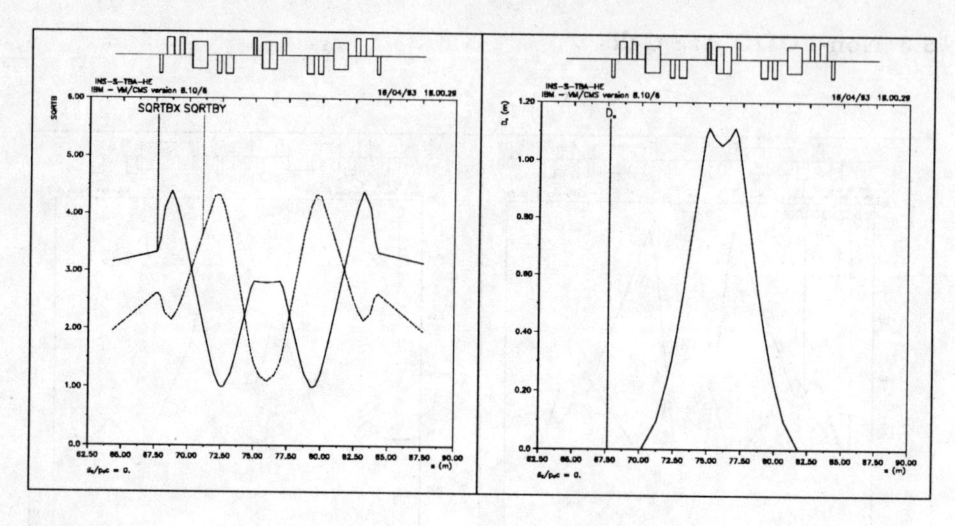

Fig. 10. Square roots of the β-functions and horizontal dispersion in the TBA module for a high-emittance lattice.

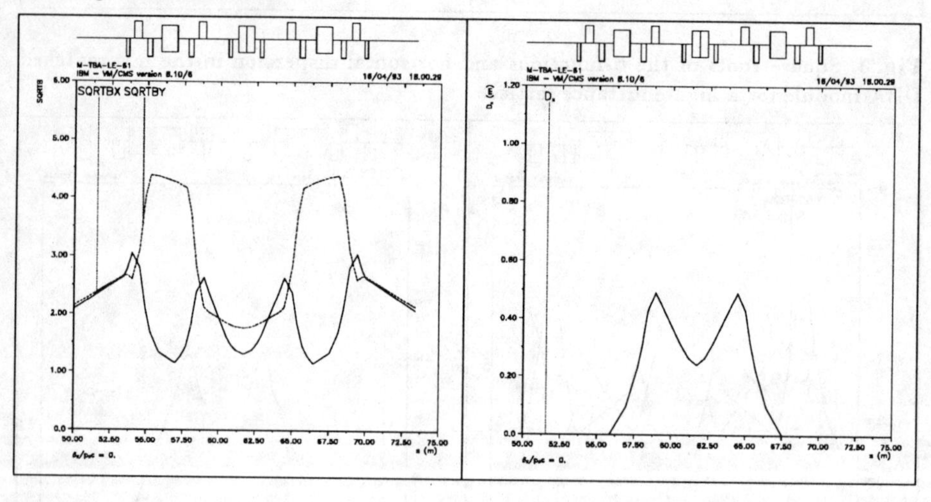

Fig. 11. Square roots of the β-functions and horizontal dispersion in the TBA module for a low-emittance lattice.

A recent study [9], from which the examples in this section are taken has shown how some of these, namely the Double-Bend Achromat (DBA) and Triple-Bend Achromat (TBA) can be used to construct flexible optics for a two-phase τcF. The optics of high- and low-emittance versions of these basic modules is shown in Figures 8–11.

Achromatic arc modules eliminate the need for a dispersion suppressor since the dispersion is automatically matched to zero at the ends of each module.

An exception to this among the four cases shown in Figures 8–11 is the high-emittance DBA lattice where it is necessary to allow the dispersion to be non-zero at the ends of the module to achieve a large contribution to the emittance. This allows $\mathcal{H}_x$ to increase in the bending magnets. The dispersion is finally suppressed by a special matching of the last module in each arc as shown in Figure 12.

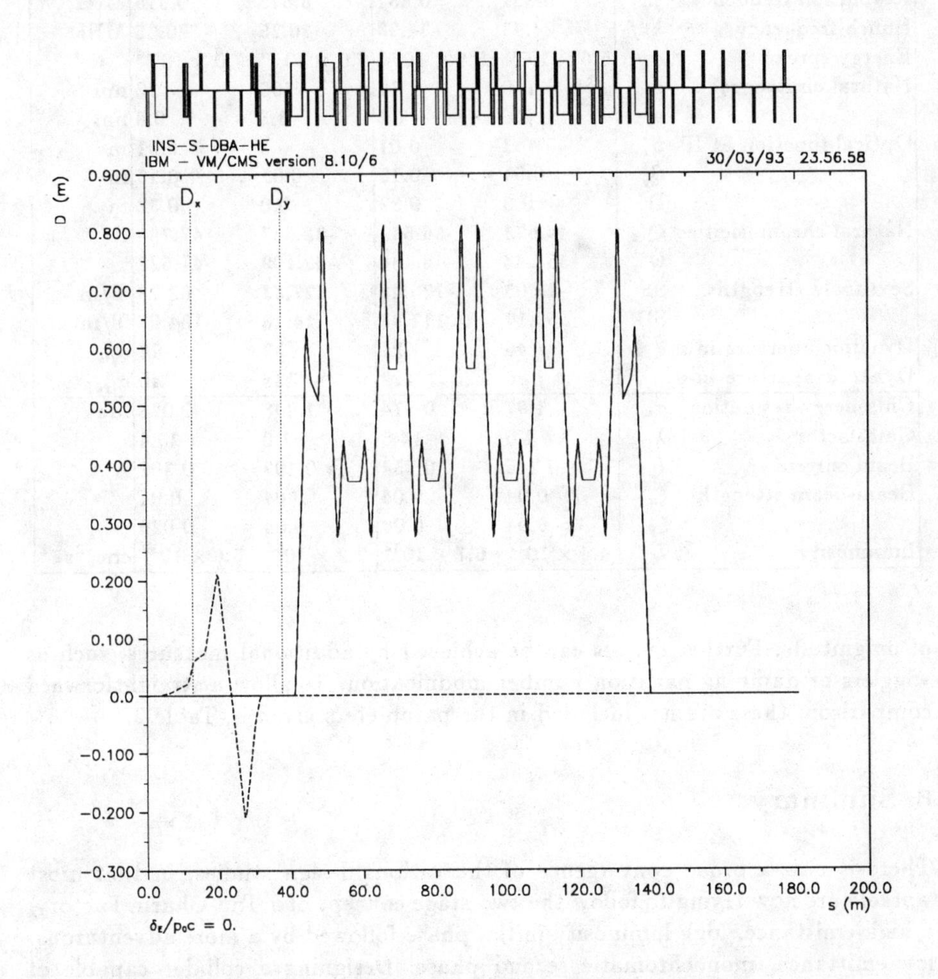

Fig. 12. Dispersion functions in half the ring for the high-emittance DBA lattice.

In the TBA lattice, the change in emittance is achieved by a simple rematching with the quadrupoles which changes $\mathcal{H}$ dramatically in the bending magnet in the centre of the achromat.

These lattices use a straightforward FODO structure for the utility insertion.

The parameters obtained by this set of lattices are given in Table 2. In each case, just rematching the basic lattice module changes the emittance by an order

Table 2. Performance of the Double- and Triple-Bend Achromat lattices in their two modes of operation.

Quantity		DBA		TBA		Units
		Standard	Monochr.	Standard	Monochr.	
Nominal beam energy	E_0	2.0	2.0	2.0	2.0	GeV
Circumference	C	360.9	360.9	444.2	444.2	m
Revolution frequency	f_0	0.831	0.831	0.675	0.675	MHz
Bunch frequency	$k_b f_0$	24.93	24.93	20.25	20.25	MHz
Energy spread	σ_ϵ	5.6×10^{-4}	5.6×10^{-4}	7.0×10^{-4}	7.0×10^{-4}	
Natural emittance	ϵ_{x0}	170	17	100	12	nm
	ϵ_{y0}	5.6	1.2	3.3	0.8	nm
Optical function at IP	β_x^*	0.3	0.01	0.3	0.01	m
	β_y^*	0.01	0.15	0.01	0.15	m
	D_y^*	0.0	0.32	0.0	0.32	m
Natural chromaticities	Q'_x	-44.973	-56.531	-32.667	-47.756	
	Q'_y	-35.548	-44.134	-41.109	-47.627	
Sextupole strengths	SF	-18.03	-126.48	-27.83	62.78	T/m
	SD	52.17	117.83	14.18	-104.07	T/m
Dynamic aperture in x		46	21	33	17	$\sigma_{\beta x}^*$
Dynamic aperture in y		20	22	139	40	$\sigma_{\beta y}^*$
CM energy resolution	σ_w	1.67	0.114	1.998	0.098	MeV
Gain factor	λ	1.0	14.5	1.0	20.4	
Beam current	I_b	0.245	0.354	0.102	0.302	A
Beam-beam strength	ξ_x	0.04	0.04	0.04	0.04	
	ξ_y	0.04	0.04	0.04	0.029	
Luminosity	L	4.4×10^{32}	6.7×10^{32}	2.2×10^{32}	5.6×10^{32}	$cm^{-2}s^{-1}$

of magnitude. Further factors can be achieved by additional measures, such as wigglers or damping partition number modification. To allow a straightforward comparison, these are not included in the parameters given in Table 2.

6 Summary

There is now a broad convergence of the various design studies, in that most workers are now trying to follow the two-stage concept of a Tau-Charm Factory: a high-emittance, high luminosity initial phase followed by a more adventurous, low-emittance, monochromatic second phase. Designing a collider capable of both of these is highly demanding and will require a good deal of work yet. However a number of promising lattices are in the air, among them some which depart from the FODO arcs traditional for colliders. In this flow of ideas we see the synchrotron light sources beginning to pay back their debt to the high-energy colliders. We also see that the intellectual challenges and scope for innovation in the field of e^+e^- storage rings are far from being exhausted.

Anyone embarking on a project must never forget that, since the feasibility of a monochromatic scheme will not be demonstrated until someone actually builds one, the initial high-luminosity phase, which is difficult enough already, should not be compromised.

Acknowledgements

I would like to thank the many colleagues with whom I have worked over the years on the design of the τ-Charm Factory. Special mention must be given to Joel Le Duff, Jasper Kirkby, Marc Muñoz, Juan-Antonio Rubio, Carlos Willmott and Alexander Zholents. The material for Section 5.3 was very kindly supplied by Angeles Faus-Golfe.

References

1. J. Kirkby, "A τ-Charm Factory at CERN", CERN-EP/87-210 (1987), and Proc. Int. School of Physics with Low-Energy Antiprotons on Spectroscopy of Light and Heavy Quarks, Erice, Sicily, 1987 (Plenum Press, New York, 1989) 401.
2. J.M. Jowett, "Initial Design of a τ-Charm Factory at CERN" CERN LEP-TH/87-56 (1987).
3. C. Zhang, "BEPC: Status and Performance", Proc. XVth International Conf. on High Energy Accelerators, Hamburg, 20-24 July 1992, Int. J. Mod. Phys. A (Proc. Suppl.) 2A (1993) 439.
4. E.M. Gonzalez Romero and J.I. Illiana, "CP Violation Study in the Processes $e^+e^- \to J/\psi \to \Lambda\bar{\Lambda}$ and $J/\psi \to \Xi^-\Xi^+$", in [27].
5. A. Zholents, "Polarized J/ψ Mesons at a Tau-Charm Factory with a Monochromator Scheme", CERN SL/92-27 (AP) (1992).
6. J.M. Jowett, A. Zholents, C. Fernandez-Figueroa, M. Munoz, J.-M. Quesada, C. Willmott, "The Tau-Charm Factory", same volume as [3] 439.
7. P. Beloshitsky, "A Magnet Lattice for a τ-Charm Factory Suitable for both Standard Scheme and Monochromatization Scheme", Orsay Report, LAL/RT 92–09 (1992).
8. M.V. Danilov *et al*, "Conceptual Design of Tau-Charm Factory in ITEP", same volume as [3] 455.
9. A. Faus-Golfe and J. Le Duff, "A Versatile Lattice for a τ-Charm Factory that includes a Monochromatization Scheme (Low Emittance) and a Standard Scheme (High Emittance)", Proc. Particle Accelerator Conf., Washington May 1993 (to appear).
10. M. Sands, "The Physics of Electron Storage Rings", SLAC-PUB 121 (1970).
11. J.M. Jowett, *in* M. Month and M. Dienes (Eds.), "Physics of Particle Accelerators", AIP Conf. Proc. 153 (1987) 864.
12. R. Siemann, this school.
13. A.A. Zholents *in* M. Dienes, M. Month, S. Turner (Eds.), Frontiers of Particle Beams: Intensity Limitations, Springer-Verlag, Berlin 1992.
14. A. Renieri, Frascati Preprint /INF-75/ 6 (R), (1975).
15. I.Ya. Protopopov, A.N. Skrinsky and A.A. Zholents, "Monochromatization of Colliding Beam Interaction Energy in Storage Rings", Proc. 5th All-Union Conf. on Charged Particle Accelerators, Dubna, 1978, (Dubna, 1979) 132.
16. A.A. Avdienko et al., Proc. 12th International Conf. on High Energy Accelerators, Batavia, 1983.

17. J.M. Jowett and T.M. Taylor, "Wigglers for Control of Beam Characteristics in LEP", IEEE Trans. Nucl. Sci. **30** (1983) 2581.
18. J. Poole (Ed.), "Proc. Third Workshop on LEP Performance", Chamonix, January 11–16, 1993, CERN SL Report, to appear 1993.
19. K.W. Robinson, "Radiation Effects in Circular Accelerators", Phys. Rev. **111**, 373 (1958).
20. Y. Baconnier, R. Cappi, J.-P. Riunaud, H.H. Umstätter, M.P. Level, M. Sommer, H. Zyngier, Nucl. Instr. and Methods **A234** (1985) 244.
21. A.L. Gerasimov and A.A. Zholents, "Beam-Beam Effects in Storage Rings with Monochromator Scheme", Proc. 13th International Conf. on High Energy Accelerators, Novosibirsk, 1986, (Novosibirsk, 1987) 82.
22. A. Gerasimov, D. Shatilov and A. Zholents, "Beam-Beam Effects with a Large Dispersion at the Interaction Point", Nucl. Instr. and Meth., A305, No.1, (1991)25.
23. W.T. Kirk and M.L. Perl (Eds.), "Proc. Tau Charm Factory Workshop", SLAC, California, 23–27 May 1989, SLAC Report 343 (1989).
24. J.M. Jowett, "The τ-Charm Factory Storage Ring", Proc. 1st European Particle Accelerator Conf., Rome, 7–11 June, 1988, World Scientific, Singapore, 1989, p. 368.
25. B. Barish *et al*, "Tau-Charm Factory Design", SLAC-PUB 5180 (1990).
26. J. Gonichon, J. Le Duff, B. Mouton, C. Travier, "Preliminary Study of a High Luminosity e^+e^- Storage Ring at a C.M. Energy of 5 GeV", Orsay Report, LAL/RT 90-02 (1990).
27. J.M. Jowett *in* J. Kirkby and J.-M. Quesada (Eds.) Proc. Meeting on the τ-Charm Factory Detector and Machine, Univ. of Seville, 29 April–2 May 1991.
28. V.S. Alexandrov et al., "JINR Tau-Charm Factory Design Considerations", Proc. IEEE Particle Accelerator Conf., San Francisco, 6-9 May 1991, 195.
29. V.A. Bednyakov, G.A. Shelkov (Eds.), "Proc. of the Workshop on JINR C-Tau Factory", 29-31 May 1991, JINR, Dubna 1992.
30. E.A. Perelstein *et al*, "JINR Tau-Charm Factory Study", same volume as [3] 448.
31. M. Munoz, private communications, to be published.
32. Yu.I. Alexahin, A.N. Dubrovin and A.A. Zholents, "Proposal on a Tau-Charm Factory with Monochromatization", Proc. 2nd European Particle Accelerator Conf., Nice, 1990 (Éditions Frontiéres, 1990) 398.
33. K. Oide, in [23].
34. Y. Alexahin, "On the Scheme of Monochromatic τ-Charm Factory with Finite Crossing Angle", private communication, to be published.
35. J. Le Duff, this school.
36. A. Hutton, M. Placidi, T.M. Taylor, "Improvements to the LEP Lattice Design", same volume as [16] 164.
37. T.M. Taylor, private communications.
38. M. Donald and A. Garren, in [23].
39. H. Grote, F.C. Iselin, "The MAD Program, Version 8.10, User's Reference Manual", CERN/SL 90–13 Rev. 3 (1992).
40. J. Kirkby, "Physics and Design of the Tau-Charm Factory in Spain", CERN-PPE/92-30; Int. Workshop on Electroweak Physics Beyond the Standard Model, Valencia, Spain, 2-5 October 1991.
41. J. Le Duff, "Single and Multiple Touschek Effects", CERN Accelerator School, Berlin, 14-25 September 1987, CERN 89-01 (1989).
42. Ya.S. Derbenev and A.M. Kondratenko, "Acceleration of Polarized Particles", Soviet Physics Doklady, **20** (1975/1976) 562.

43. Ya.S. Derbenev et al., "Project of Obtaining Longitudinal Polarized Colliding Beams in Storage Ring VEPP4 at 2 GeV", Proc. 12th International Conf. on High Energy Accelerators, Fermilab (1983) 410.
44. A. Faus-Golfe and J. Le Duff, "A Versatile Lattice for a τ-Charm Factory that includes a Monochromatization Scheme", LAL/RT 92-01 (1992).
45. A. Ropert *in* CERN Accelerator School on "Synchrotron Radiation and Free Electron Lasers", CERN Report 90-03 (1990).

Lattice and Interaction Region Design for Z-Factories

Eberhard Keil

CERN, CH-1211 Geneva 23

Methods for increasing the luminosity of LEP by about an order of magnitude or more are discussed when LEP is operated at the energy of the Z. The notation and scaling laws for luminosity, beam-beam tune shifts, and synchrotron radiation are presented first. Limitations of the bunch current and of the total beam current are discussed. Methods for the design of the lattice and the interaction regions are presented. Two schemes for raising the total circulating current and hence the luminosity – pretzels and bunch trains – are discussed in detail. The paper ends with a discussion of combining pretzels with bunch trains, and with lessons for future Z factories.

1 Introduction

This paper discusses methods for increasing the luminosity L of LEP by about an order of magnitude or more when it is operated at the energy of the Z. The standard formulae and notation are introduced in Sect. 1.1. Effects of synchrotron radiation are summarized in Sect. 1.3. Bunch and beam current limitations are discussed in Sect. 1.4 and 1.5, respectively. Since these topics are also covered by other lectures at this school, the results are just quoted, and no derivation is shown. The lattice and interaction region design is discussed in Sect. 2. Two methods of increasing the total circulating current and hence the luminosity – pretzels and bunch trains – are presented in Sect. 3 and 4, respectively. Section 5 contains a discussion of combining pretzels with bunch trains, and lessons for the future.

1.1 Standard Formulae and Notation

When the vertical amplitude function at the interaction points β_y is much larger than the bunch length σ_s, i.e. for $\beta_y \gg \sigma_s$, the luminosity L and the beam-beam tune shift parameter ξ_z in the z-plane (z is x=horizontal or y=vertical) are given by:

$$L = \frac{N^2 f}{4\pi\sigma_x\sigma_y} = \frac{N^2 f_{\text{rev}} k}{4\pi\sigma_x\sigma_y} \qquad \xi_z = \frac{N r_e \beta_z}{2\pi\gamma(\sigma_x+\sigma_y)\sigma_z} \tag{1}$$

The meaning of frequently used symbols and their values are explained in Tab. 1. I assume throughout that the e^+ and e^- parameters are the same. Correction factors for L and ξ_y with finite β_y/σ_s are known [1, 2].

Table 1. Explanation and Standard Values of Frequently Used Parameters

$C_q = 3.84 \times 10^{-13}$ m	$55\hbar/32\sqrt{3}mc \approx$ Compton wavelength
$f = f_{rev}k$	Bunch collision frequency
$f_{rev} = 11.245$ kHz	Revolution frequency
$I = Nef = 0.5$ mA	Bunch current
$J_s = 2$	Damping partition number for synchrotron oscillations
$J_x = 1$	Damping partition number for horizontal betatron oscillations
$J_y = 1$	Damping partition number for vertical betatron oscillations
$dJ_s/d\Delta p/p = 394$	Slope of damping partition number J_s
k	Number of bunches in one beam
$\ell_B = 70$ m	Total length of dipoles in an arc cell
$\ell_Q = 3.2$ m	Total length of quadrupoles in an arc cell
$N = 2.775 \times 10^{11}$	Bunch population
$Q = \rho\mu/\ell_B = 69$	Contribution of arcs to tune
$Q_s = 0.08$	Synchrotron tune
$R = 3494$ m	Average radius of arcs
$r_e = 2.818 \times 10^{-15}$ m	Classical electron radius
$U_s = 137.3$ MeV	Synchrotron radiation loss per turn
$\alpha = 1$ mrad	Full horizontal crossing angle
$\beta_x = 1.25$ m	Horizontal amplitude function at IP
$\beta_y = 0.05$ m	Vertical amplitude function at IP
$\gamma = 89237$	Particle energy E in units of its rest energy E_e
$\epsilon_x = \sigma_x^2/\beta_x = 30$ nm	Horizontal emittance
$\epsilon_y = \sigma_y^2/\beta_y = 1.2$ nm	Vertical emittance
$\mu = \pi/2$	Phase advance in an arc cell
$\rho = 3076$ m	Bending radius
$\sigma_e = 1.0654 \times 10^{-3}$	Rms momentum spread
$\sigma_s = 10.7$ mm	Bunchlength
$\sigma_x = 0.194$ mm	Horizontal rms beam radius at the collision points IP
$\sigma_y = 7.75$ μm	Vertical rms beam radius at the collision points IP
$\tau_s = 29.5$ ms	Damping time for synchrotron oscillations
$\tau_x = 59.0$ ms	Damping time for horizontal betatron oscillations
$\tau_y = 59.0$ ms	Damping time for vertical betatron oscillations

1.2 Manipulations of the Standard Luminosity Formula

To gain further insight we use ξ_y to eliminate one power of N from the L equation:

$$L \approx \frac{N f \gamma \xi_y}{2 r_e \beta_y} = \frac{I k \gamma \xi_y}{2 e r_e \beta_y} \tag{2}$$

These equations tell us that the products Nf or Ik have to be increased by the same factor as the desired increase in L, because γ is fixed, $\xi_x \approx 0.03$, $\xi_y \approx 0.03$ are fairly fixed, and β_y cannot vary much as will be discussed in Sect. 2.2. These equations also imply that ϵ_x is adjusted such that the design value of ξ_x is reached at the design current I:

$$\epsilon_x = \frac{N r_e}{2\pi\gamma\xi_x} \tag{3}$$

The lattice design must ensure that this emittance is reached at the design energy and at the design current, and that the lattice is flexible enough to reach the beam-beam limit for a range of energies and bunch currents around the design values. The emittance ϵ_x increases in proportion to N or I. To have $\xi_x = \xi_y$, we need $\epsilon_y/\epsilon_x = \beta_y/\beta_x$. In a typical low-$\beta$ insertion $\beta_y \ll \beta_x$, hence beams are flat with $\epsilon_y \ll \epsilon_x$ and $\sigma_y \ll \sigma_x$. There are two extreme ways for increasing Nf: One may either increase N and keep f fixed, or one may keep N fixed and increase f. Intermediate solutions don't contribute anything fundamentally new. The ultimate limit on L is determined by the total beam current Ik.

1.3 Effects of Synchrotron Radiation

Synchrotron radiation causes several effects in e^+e^- storage rings. The radiation loss per turn U_s in terms of voltage which must be compensated by the acceleration in the RF system, and the radiation damping rates τ_z^{-1} for z-th mode are given by:

$$U_s = \frac{4\pi r_e E_e \beta^3 \gamma^4}{3\rho} \qquad \tau_z^{-1} = \frac{U_s f_{\mathrm{rev}} J_z}{2E} \tag{4}$$

Note that U_s and τ_z are both proportional to ρ^{-1}. Hence the smaller the storage ring the higher the synchrotron radiation losses and synchrotron radiation power, and the more expensive the RF system. This dependence on ρ differs from that of the length of the arcs which varies in proportion to ρ. The cost of the arc tunnel and of the equipment inside is also likely to be about proportional to ρ. One way of arriving at a choice for ρ is balancing the cost of the tunnel against that of the RF system [3]. The damping partition numbers J_z always satisfy the relation $J_x + J_y + J_s = 4$. If the machine has a median plane like LEP, the vertical damping partition number satisfies $J_y = 1$. If in addition the bending magnets have no gradient, then also $J_x \approx 1$, and hence $J_s \approx 2$. Errors in the alignment and excitation of the magnets in LEP cause errors in the closed orbit and in the J's. The damping partition numbers also depend on the relative momentum error $\Delta p/p$ since the closed orbit for an off-momentum particle passes through the quadrupoles off-axis. Hence, they act like combined-function magnets. The slopes of J_s and J_x are given by:

$$\frac{\mathrm{d}J_s}{\mathrm{d}\Delta p/p} = \frac{2\ell_{\mathrm{B}}}{\ell_{\mathrm{Q}}}\frac{4+\sin^2\mu/2}{\sin^2\mu/2} = -\frac{\mathrm{d}J_x}{\mathrm{d}\Delta p/p} \tag{5}$$

Since both J_s and J_x must be positive, this slope determines the damping aperture, bounded by: $-2(\mathrm{d}J_s/\mathrm{d}\Delta p/p)^{-1} \leq \Delta p/p \leq (\mathrm{d}J_s/\mathrm{d}\Delta p/p)^{-1}$. The balance between quantum excitation and radiation damping determines the relative energy spread σ_{e} and the horizontal emittance ϵ_x:

$$\sigma_{\mathrm{e}} = \frac{C_{\mathrm{q}}\gamma^2}{\rho J_s} \qquad\qquad \epsilon_x \approx \frac{C_{\mathrm{q}}\gamma^2}{J_x Q^3}\frac{R}{\rho} \tag{6}$$

Both equations apply to a machine in which all dipoles have the same bending radius ρ. The equation for ϵ_x assumes that the bending angle and the phase advance μ in a cell are both small. The correction factor for finite bending angle and finite μ is known [4]. The desired value of ϵ_x can be achieved by adjusting the contribution Q of the arcs to the tune. Small variations in ϵ_x can be obtained by varying J_x in principle, by small variations of the momentum error $\Delta p/p$ on the closed orbit that are achieved by changes of the RF frequency. However, in practice such changes are not desirable in LEP because they interfere with the energy calibration.

In LEP with a rather low field in the dipoles in the arcs it is easy to increase the emittance and the energy spread with wiggler magnets, while it is difficult to reduce the emittance, and impossible to reduce the energy spread. Hence, it is advantageous to make the focusing in the arcs stronger than needed, i.e. to choose Q higher than needed from the equation above, and to increase the emittance with wiggler magnets. The wigglers in LEP have just three poles, a central pole with a field up to 1 T in the same direction as the main dipoles, and outer poles with a field up to 0.4 T in the opposite direction to limit adverse effects on the polarization [5]. The total deflection in a wiggler vanishes. The damping wigglers are installed in the straight sections with $D_x = 0$. The emittance wigglers are positioned in the dispersion suppressors with $D_x \neq 0$. They increase the emittance from 12.26 nm to 30.25 nm.

1.4 Bunch Current Limitations

The transverse mode coupling instability limits the bunch current I at injection where the beam energy E is lowest. In the broad-band resonator model its threshold is approximately given by:

$$I \approx \frac{2\pi(E/\mathrm{e})Q_s f_\perp(\sigma_s)}{\sum_i \bar{\beta}_i (hR/Q_\perp)_i} \tag{7}$$

Here, $Q_s \approx 0.08$ is the synchrotron tune, $h = f_{\mathrm{r}}/f_{\mathrm{rev}}$ is the harmonic number of the resonator resonating at frequency f_{r}, $R/Q_{\perp i}$ is the transverse impedance of the i-th group of equipment, $\bar{\beta}_i$ is the average value of the β-function there, and

$f_\perp(\sigma_s)$ is a form factor that has a minimum of about unity when the bunchlength σ_s is about half the vacuum chamber radius, and increases roughly proportionally to σ_s for higher values. Therefore, the bunches are made longer with wigglers in routine operation. In LEP, about 2/3 of the sum in the denominator is due to the Cu RF cavities. Hence, it is attractive in the long run to replace them by s.c. cavities which by design have a larger beam hole and a lower transverse impedance. The remaining 1/3 of the sum is due to bellows and difficult to change. There is excellent quantitative agreement between observations of the threshold in LEP and the theory underlying the scaling law above [6]. All other single-beam collective effects do not limit the bunch current in LEP.

The beam size is proportional to the bunch current $I^{1/2...4/3}$ and will eventually reach the edge of the – physical or dynamic – aperture when I is increased. The value of the power of β_y depends on the scaling procedure. This limitation arises first in the low-β insertions, and will be discussed in Sect. 2.2.

Collimators protect the experiments from synchrotron radiation background. Synchrotron radiation background from the arcs is reduced by distance, i.e. by the length of the straight section, by weak dipoles, experiments not being in direct line of sight, and by the synchrotron radiation being at least scattered twice on collimators. Synchrotron radiation background from the quadrupoles in the straight section increases in proportion to the beam size, and must be shielded by collimators with an aperture proportional to the beam size. Observations in LEP show that there is a practical upper limit for the horizontal emittance at $\epsilon_x = 50$ nm.

All three reasons prevent increasing the bunch current by an order of magnitude.

1.5 Limitation of Total Beam Current

All schemes for increasing the LEP luminosity by an order of magnitude or more require an increasing of the total circulating current by about the same factor. Several phenomena may limit the total beam current. The voltage delivered by the RF system must be larger than synchrotron radiation losses $U_s \approx 125$ MV. This figure is much smaller than voltage required at 90 GeV. The power P from the RF system

$$P = 2U_s Ik \approx 2 \cdot 125\ \text{MV} \cdot 1\ \text{mA} \cdot 36 \approx 9\ \text{MW} \tag{8}$$

must be at least equal to the synchrotron radiation power. In this equation the factor of two takes care of the two beams, and the factor 36 is the maximum number of bunches which we will arrive at in Sect. 3. The power above is much smaller than the power installed for LEP-200. The power absorbed by the vacuum chamber per running metre $P/2\pi\rho \approx 0.5$ kW/m is within the capabilities of the present cooling system. The power P_{hom} of the higher-mode losses in the superconducting cavities is given by:

$$P_{\text{hom}} = 2k_\parallel k I^2 f_{\text{rev}}^{-1} \tag{9}$$

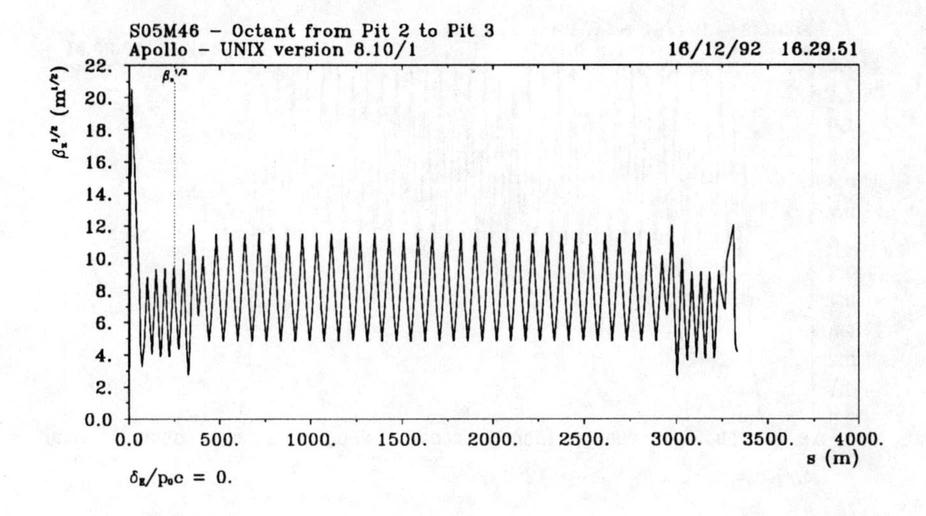

Fig. 1. Horizontal Orbit Function $\beta_x^{1/2}$ in a LEP Octant

With a loss factor $k_{\|} \approx 0.25$ V/pC [7] and a revolution frequency $f_{\text{rev}} = 11245.5$ Hz, the specifications of the present higher-mode couplers [8] $P_{\text{hom}} = 200$ W impose a limit $kI^2 \approx 4.5\ \text{mA}^2$. The maximum number of bunches k decreases quadratically with the bunch current from $k = 18$ at $I = 0.5$ mA, through $k = 8$ at $I = 0.75$ mA to $k = 4$ at $I = 1$ mA.

2 Lattice Design

The LEP lattice consists of eight – almost – identical octants. The octant boundaries coincide with the eight interaction points. The octants in turn consist of three modules each, half a straight section, an arc, and half a straight section. Figures 1 and 2 show the orbit function β_x, and the dispersion D_x, respectively. The behaviour of β_y is similar to that of β_x. All figures in this lecture were made with MAD [9]. When the distance s along the orbit is used as abscissa, a schematic layout diagram often appears on top of the graph. Blocks above and below the central line represent dipoles, blocks above (below) the central line represent horizontally focusing (vertically focusing) quadrupoles, etc. The three orbit functions are periodic in the arcs which consist of FODO cells with a phase advance $\mu = \pi/2$. The dispersion D_x vanishes in the straight sections. This is achieved by dispersion suppressors at both ends of the arcs. A weak dipole at the entrance of the dispersion suppressor prevents synchrotron radiation from the arcs falling directly into the detectors at the interaction points.

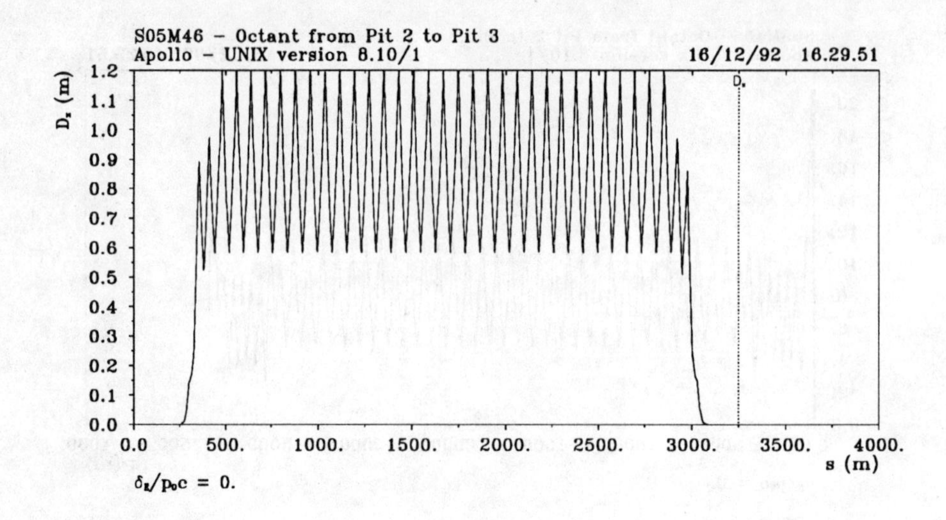

Fig. 2. Horizontal Dispersion D_x in a LEP Octant

2.1 Straight Section Design

Figure 3 shows β_x, β_y and D_x in a smaller region of LEP, starting at the interaction point in Pit 2, including the low-β insertion, the RF section with a periodic cell lattice for minimum $\bar{\beta}_y$, and the dispersion suppressor. Each of these three lattice modules has a specific purpose: The low-β insertion matches the optical functions α and β between the values in the RF section and the low values for the β-functions wanted at the interaction point in Pit 2. The RF section consists of 5/2 cells of FODO lattice with the space between the quadrupoles chosen such that there is space for eight Cu RF cavities and associated equipment. The phase advance there is adjusted such that the average value $\bar{\beta}$ is at a minimum in order to raise the threshold of the transverse mode coupling instability. The dispersion suppressor matches the horizontal dispersion D_x from zero in the straight section to the non-zero value in the arc, and the optical functions α and β between the values in the RF section and those in the arc, with prescribed values of the phase advances. This is achieved by adjusting the strengths of eight individually powered quadrupoles in each dispersion suppressor. In this manner, the dispersion suppressor can be matched to arc cells with a range of phase advances. Much simpler dispersion suppressors are possible if the phase advance in the arc cells is fixed [10]. I will show below that a high gradient in the first superconducting quadrupole is desirable, and that the minimum value of β_y which can be achieved in practice increases with ϵ_x.

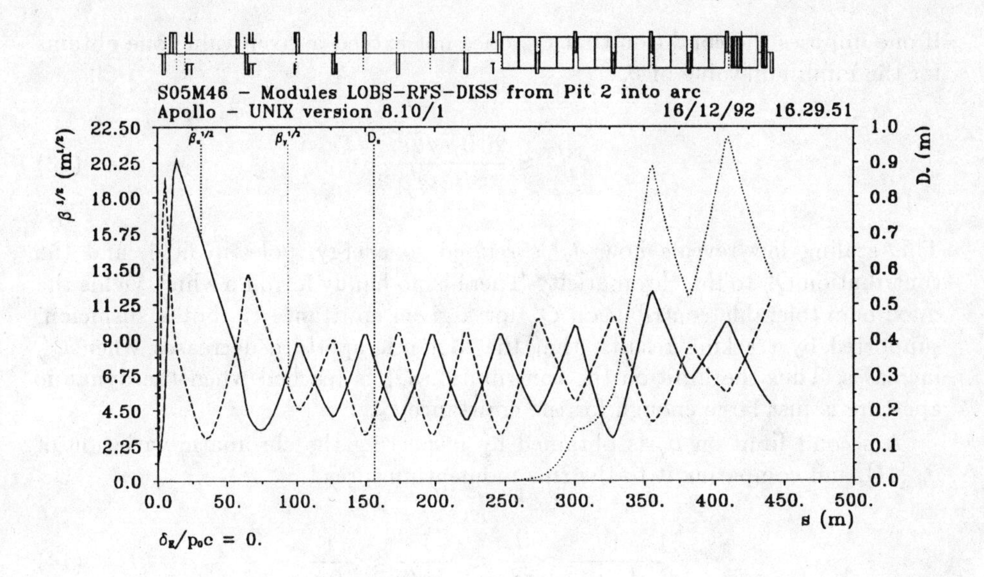

Fig. 3. Orbit Functions $\beta_x^{1/2}$, $\beta_y^{1/2}$ and D_x from Pit 2 into Arc

2.2 Interaction Region Design

The low-β insertion in LEP is a classical low-β insertion with $\beta_y \ll \beta_x$. It is matched by using a computer program. Approximate parameters and scaling laws can be obtained analytically. There are relations between the focal length d and the distance l_x of the centre of the first quadrupole from the interaction point, between d, the aperture a, the "pole-tip field" B_Q, and the length of quadrupole l_Q, and between l_Q and l_x:

$$d \approx l_x/2 \qquad d \approx \frac{B\rho a}{l_Q B_Q} \qquad l_Q \approx l_x/2 \tag{10}$$

The aperture a must exceed the beam size by a factor which is typically 10. With the optimum emittance ratio $\epsilon_y/\epsilon_x = \beta_y/\beta_x$, the beam is approximately round on the front face of the first quadrupole. Allowing for "full coupling" with $\epsilon_y/\epsilon_x = 1/2$ the vertical beam size becomes larger than the horizontal one, and determines the aperture a. We therefore write for a, where $\beta_Q \approx \ell_x^2/\beta_y$ is taken at the quadrupole:

$$a \approx 10\sqrt{\epsilon_x \beta_Q/2} \approx 10\sqrt{\epsilon_x l_x^2/2\beta_y} \tag{11}$$

The two vertically focusing quadrupoles contribute Q'_y to the vertical chromaticity:

$$Q'_y \approx \beta_Q/2\pi d \approx l_x/\pi\beta_y \tag{12}$$

If one imposes the condition that Q'_y does not exceed a given value one obtains for the minimum value of β_y:

$$\beta_y^{3/2} \approx \frac{20B\rho\sqrt{2\epsilon_x}}{\pi B_Q Q'_y} \tag{13}$$

This scaling law reveals how β_y is related to energy, pole-tip field, and the contribution Q'_y to the chromaticity. There is no handy formula which yields the maximum tolerable contribution Q'_y for a given emittance ϵ_x, but a suspicion, supported by tracking results, that the dynamic aperture decreases when Q'_y increases. Thus the limit on the contribution Q'_y is reached when the dynamic aperture is just large enough for the emittance ϵ_x.

A second limit on β_y is obtained by inspecting the chromatic variation of β_y [11], and comparing it to the rms momentum spread σ_e:

$$\frac{1}{\beta_y}\frac{d\beta_y}{d\Delta p/p} \approx \frac{l_x}{\beta_y} \approx \frac{1}{\Delta p/p} \approx \frac{1}{10\sigma_e} \tag{14}$$

This leads to the following minimum value of β_y from the momentum aperture $\Delta p/p$:

$$\beta_y^{3/2} \approx \frac{20B\rho(\Delta p/p)\sqrt{2\epsilon_x}}{B_Q} \tag{15}$$

With the standard parameters and $B_Q = 2$ T, one finds $\beta_y = 24$ mm. Hence, there is room for reducing β_y. The other parameters follow from β_y by back substitution.

3 Pretzels

In a pretzel scheme the e^+ and e^- beams are separated over most of the circumference by exciting forced, closed-orbit oscillations with $n\pi$ phase difference by electrostatic separators. It follows that a pretzel scheme increases the aperture needed by the amplitude of the orbit oscillation, and that it can only be installed in machines which have an aperture which is larger than that needed without pretzels. This is typically the case when machines are operated below their design energy. Such a scheme was first installed and used in the CESR storage ring at Cornell University, Ithaca NY, USA, and the name pretzel was invented there [12, 13]. The bunch spacing is arranged such that beam-beam collisions in the arcs occur only where the orbits are well separated. A pretzel scheme for LEP was proposed by Rubbia [14]. It led to a feasibility study [15], and eventually to a report of the Working Group on High Luminosity at LEP [16]. Figure 4 shows a typical e^+ orbit and the horizontal dispersion D_x in an octant of LEP. The e^- orbit and dispersion are almost mirror images.

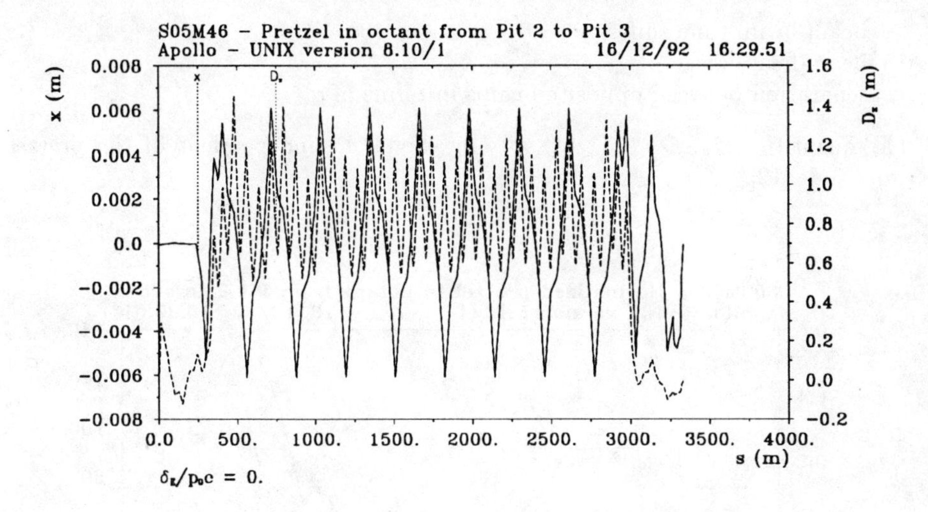

Fig. 4. Horizontal Orbit Offset x and Dispersion D_x for a Pretzel from Pit 2 to Pit 3

3.1 Pretzel Design Criteria

A pretzel scheme is designed such that the collisions at the interaction points still happen. However, the e^+ and e^- orbits are separated over most of the circumference. The separation is done in the horizontal plane for the following reasons:

- In a typical e^+e^- storage ring the horizontal aperture is larger than the vertical aperture.
- A vertical orbit offset in the sextupoles, caused by a vertical pretzel, would excite a skew gradient which in turn would cause coupling between the horizontal and vertical betatron oscillations.
- A vertical orbit offset in the sextupoles, caused by a vertical pretzel, would excite the vertical dispersion D_y which in turn would cause an increase of the vertical emittance ϵ_y by quantum excitation beyond the desirable value $\epsilon_y = \epsilon_x \beta_y / \beta_x$.

The pretzel scheme is also designed such that there is no offset in the RF cavities to avoid exciting synchro-betatron resonances. Symmetries in the ideal machine are used to compensate the consequences of the differences between the e^+ and e^- orbits. In particular, the pretzels are made anti-symmetric about the interaction points to avoid accumulating effects from the orbit differences. In a real machine, these symmetries are broken by alignment and RF and magnet excitation errors. Their consequences must be corrected. The choice of the pretzel amplitude is determined by three criteria that will be discussed in Sect. 3.3:

- beam-beam tune shifts
- beam-beam kicks
- separation between opposite beams in terms of σ_x

The quantities Q_x, Q_y, Q'_x, Q'_y, β_x, β_y should be independent of the pretzel amplitude [12].

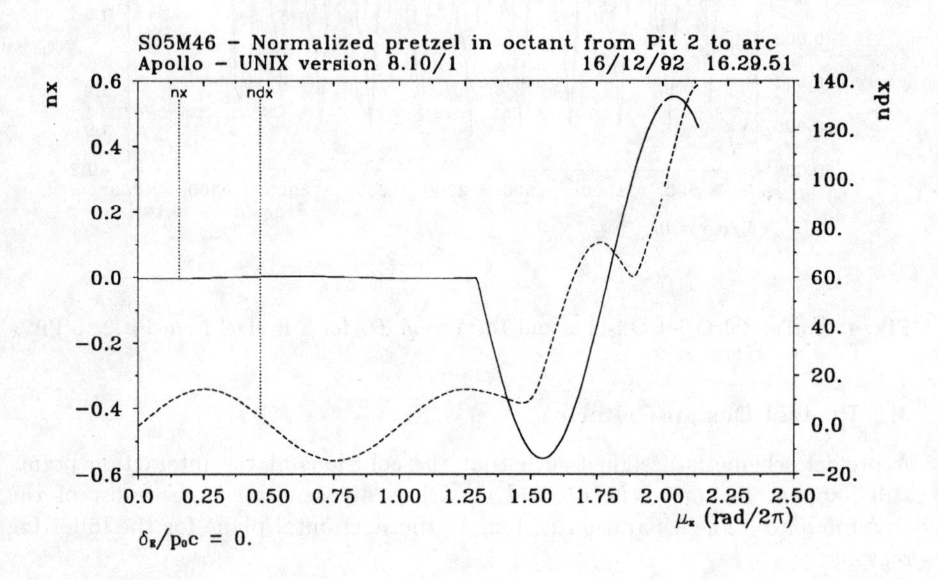

Fig. 5. Scaled Horizontal Offset $x/\beta_x^{1/2}$ and Dispersion $D_x/\beta_x^{1/2}$ in $\mu m^{1/2}$ for a Pretzel in Octant from Pit 2 to Arc

3.2 Implementation of Pretzels in LEP

Clearer pictures are obtained by plotting the normalized pretzel $x/\beta_x^{1/2}$ as a function of the horizontal phase advance μ_x. Figure 5 shows the normalized pretzel starting at Pit 2. The orbit separation starts at the end of the straight section beyond the RF system. Therefore, the two beams pass through the RF cavities without offset. The two beams collide head-on in Pit 2, where the dispersion D_x vanishes, i.e. $D_x = 0$ by symmetry, but its slope D'_x does not, i.e. $D'_x \neq 0$. Hence, D_x does not vanish in the straight sections, as shown in Fig. 4. In the straight sections, D_x behaves like a betatron oscillation. Therefore, the normalized dispersion $D_x/\beta_x^{1/2}$ behaves like a sine wave when plotted against μ_x, as shown in Fig. 5. A $D_x \neq 0$ in the RF system near Pit 2 drives synchro-betatron resonances [18]. The pretzel on the other side of Pit 2 is symmetrical. The pretzels in the other even pits are very similar.

Figure 6 shows the normalized pretzel ending at Pit 3. By symmetry, the orbit offset x and dispersion D_x vanish at Pit 3, i.e. $x = 0$ and $D_x = 0$, while the orbit and dispersion slopes, x' and D'_x do not vanish, i.e. $x' \neq 0$ and $D'_x \neq 0$. The pretzel on the other side of Pit 3 is symmetrical. The pretzels in the other odd pits are very similar. In LEP, we adjust the horizontal phase advance μ_x in the straight sections next to the odd pits to close the pretzel. It should be noted that the partition numbers change on pretzel orbits for the same reason as on off-momentum orbits. The change of the J's is quadratic in the pretzel amplitude. For the example shown in Fig. 4 the changes are $\Delta J_x \approx -\Delta J_s \approx 0.17$.

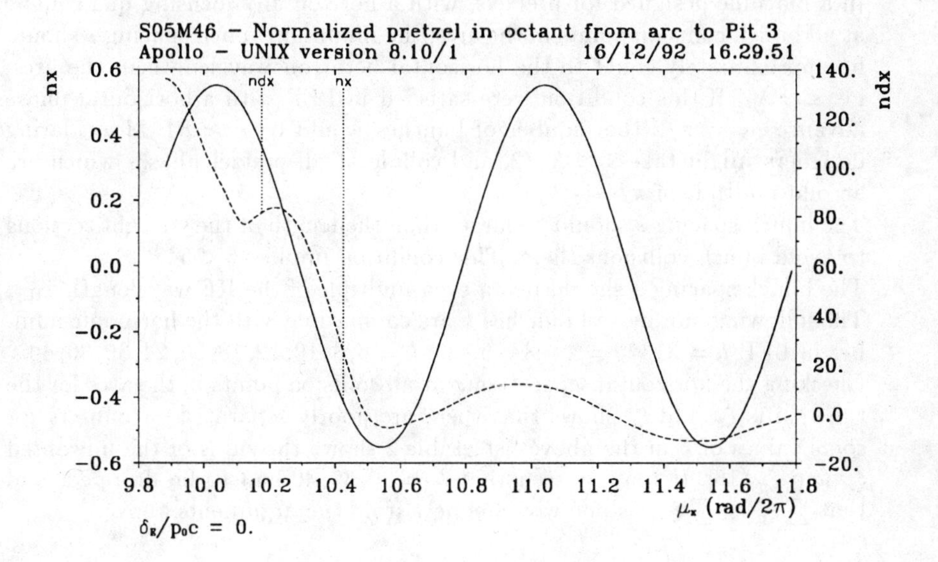

Fig. 6. Scaled Horizontal Offset $x/\beta_x^{1/2}$ and Dispersion $D_x/\beta_x^{1/2}$ in $\mu m^{1/2}$ for a Pretzel in Octant from Arc to Pit 3

3.3 Accelerator Physics of Pretzels

The beam-beam tune shifts ξ_x and ξ_y of beams with full horizontal separation $x \gg \sigma_x$ may not be negligible. They are given by:

$$\xi_x = -\frac{N r_e \beta_x}{2\pi\gamma x^2} \qquad \xi_y = \frac{N r_e \beta_y}{2\pi\gamma x^2} \tag{16}$$

The beam-beam tune shifts for separated beams have opposite sign, contrary to those for head-on collisions. Expressions for ξ_x and ξ_y for arbitrary horizontal separation are known [16]. For vanishing separation x, ξ_x and ξ_y remain finite, and Eq. (1) applies. Inspection of these equations shows that it is advantageous to arrange for the separated beam-beam collisions with horizontal separation to occur close to a peak of the sine wave because ξ_x depends only on the horizontal phase, since the orbit offset is given by $x = \beta_x^{1/2}\sin(\mu_x + \phi_x)$, where ϕ_x is an arbitrary phase. Hence, the expression β_x/x^2 appearing on the right hand side of the equation for ξ_x is an invariant. The separated beam-beam collisions should also be arranged close to a horizontally focusing quadrupole where β_y is small and x is large and ξ_y is small.

The following arguments lead to a choice of the bunch spacing s and hence of the number of evenly spaced bunches k:

- In a machine designed for pretzels, with a horizontally focusing quadrupole at all bunch collision points in the arcs, the minimum bunch spacing s should be approximately equal to the horizontal betatron wavelength in the arcs, i.e. $s \approx \lambda_\beta$. If this condition were satisfied in LEP with a horizontal phase advance $\mu_x = \pi/2$ the number of bunches would be $k \approx 84$. More daring designers might take $s \approx \lambda_\beta/2$, and collide at all pretzel phases which are an odd multiple of $\pi/4$.
- The bunch spacing s should be larger than the length of the straight sections to avoid bunch collisions there. This condition implies $k < 54$.
- The bunch spacing s should be an even multiple of the RF wavelength λ_{RF}. The following numbers of bunches k are compatible with the harmonic number in LEP $h = 31320 = 2^3 \cdot 3^3 \cdot 5 \cdot 29$: $k = 6, 8, 10, 12, 18, 20, 24, 30, 36, 40$.
- Checking the horizontal separation x at all collision points in the arcs for the tune shifts ξ_x and ξ_y shows that there are poorly separated encounters for some values of k in the above list. Table 2 shows the sums of the unwanted ξ_x and ξ_y [16]. It may be seen that $k = 20, 30, 40$ had to be dropped, and that $\sum \xi_y \ll \sum \xi_x$, as one would expect from the arguments above.

Table 2. Sums of the unwanted beam-beam tune shifts ξ_x and ξ_y for good values of the number of bunches k in the LEP pretzel scheme with a bunch current $I = 0.75$ mA

k	6	8	10	12	18	24	36
$\sum \xi_x$	.0187	.0018	.0079	.0386	.0253	.0437	.0571
$\sum \xi_y$	.0024	.0001	.0044	.0046	.0057	.0049	.0113

The separated beam-beam collisions also cause a horizontal beam-beam kick x'_{bb} which with full horizontal separation $x \gg \sigma_x$ and $N = 2.775 \times 10^{11}$, at 45.6 GeV, $x = 12$ mm is given by:

$$x'_{\mathrm{bb}} = -\frac{2Nr_e}{\gamma x} \approx 1.5\ \mu\mathrm{rad} \tag{17}$$

These beam-beam kicks have the following consequences:

- Beam-beam kicks cause a horizontal orbit distortion that is proportional to the population of the encountered bunch N.
- All bunches travel along their private closed orbit, given by the sequence and populations of the bunches in the other beam that they meet.
- To get an idea about the magnitude of these kicks they may be compared to the horizontal divergence which with $\epsilon_x = 30$ nm, $\beta_x = 132$ m is given by $\sigma'_x = (\epsilon_x \gamma_x)^{1/2} \approx 15$ μrad where $\gamma_x = (1+\alpha_x^2)/\beta_x$.
- If the beam-beam kick is not small compared to the divergence the two bunches are likely to miss each other at the desired interaction points. Hence, we obtain a tolerance on the inequality of the bunch populations N which is easily satisfied:

$$\frac{\Delta N}{N} \ll \frac{\gamma x}{2 N r_e} \sqrt{\epsilon_x \gamma_x} \tag{18}$$

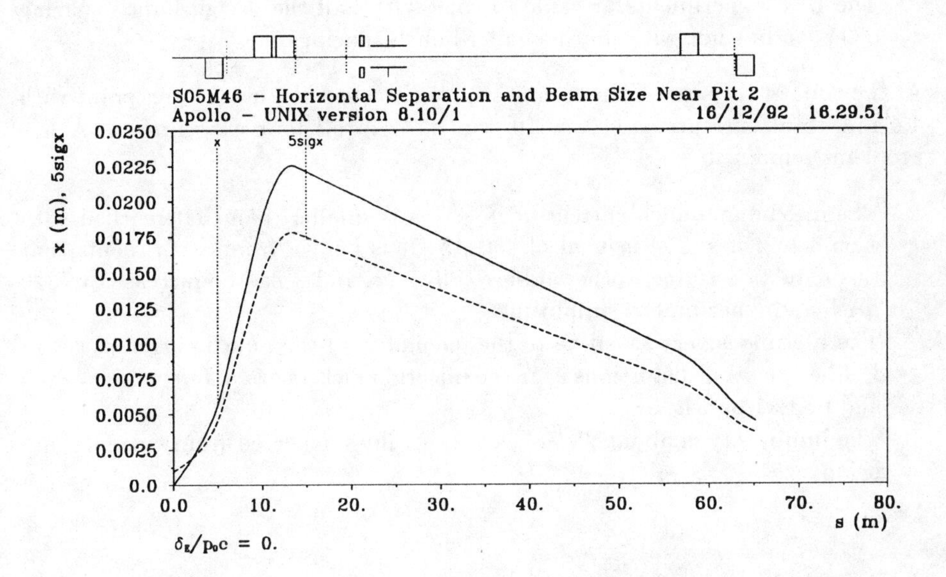

Fig. 7. Full Horizontal Separation x and $5\sigma_x$ in Units of metres for a Bunch Train Near Pit 2

A further criterion on the pretzel amplitude follows from a comparison between the separation $2x$ and the horizontal beam radius σ_x. A separation of $5\ldots6\sigma_x$ is needed to ensure a good lifetime of the two beams [17]. This observation is often discussed in a way analogous to the quantum lifetime in an RF system [19]. The probability that a particle reaches, by quantum excitation, a horizontal betatron amplitude large enough to make it pass through the core of the opposite beam is a steeply decreasing function of that amplitude, just as the probability for a particle to get its energy changed enough to jump out of the RF bucket. The separation will be discussed further in Sect. 5.1.

3.4 Status of Pretzels in LEP

Pretzels were added to LEP as an afterthought [14]. Half the electrostatic separators required were installed by spring 1991. This was enough to study the behaviour of two regular bunches with a pretzel bunch between them. The remaining separators were installed by May 1992, and machine development with up to eight bunches in each beam was started. Physics runs with pretzels took place from 19 October 1992. The number of bunches is $k = 8$ for two reasons:

1. In the LEP RF system [20] the Cu RF cavities are coupled to storage cavities, and the coupled system operates with two frequencies. The beat frequency between them corresponds to $k = 8$.
2. The LEP experiments are able to cope with half the design bunch spacing of 11 μs, but not with much smaller bunch spacings.

At the time of the school, the luminosity had reached the break-even point with 4.4 mA total current, and $L \approx 10^{31}\text{cm}^{-2}\text{s}^{-1}$ in all four experiments. A few problems remained:

- The maximum bunch current at injection is smaller than that reached with 4 on 4 bunches, 250 instead of 450 μA. It is known from experiments that this reduction is due to the mid-arc collisions, and it can be increased to 320 μA by a higher pretzel amplitude.
- The lifetime is very sensitive to the inequality of the bunch currents.
- Deliberate orbit distortions near the injection kickers are different for normal and pretzel bunches.
- The luminosity is about 25 % below the values expected from currents and beam sizes.

4 Bunch Trains

In LEP, four trains of closely spaced bunches might be an alternative to pretzels. In this case, the two beams have to be separated only over a distance equal to the length of the bunch train, near the interaction points. Since the two beams should still collide at the interaction point, an asymmetric separation at a crossing angle is appropriate. A crossing with a horizontal angle α makes it possible to satisfy one of the criteria for the crossing angle which is related to synchro-betatron resonances and requires that $\alpha \ll \sigma_x/\sigma_s$ [21]. The design of a horizontal separation scheme with a full crossing angle $\alpha = 1$ mrad is shown in Fig. 7. Also shown there is $5\sigma_x$, assuming a horizontal emittance $\epsilon_x = 30$ nm to demonstrate that the criterion $x > 5\sigma_x$ is satisfied.

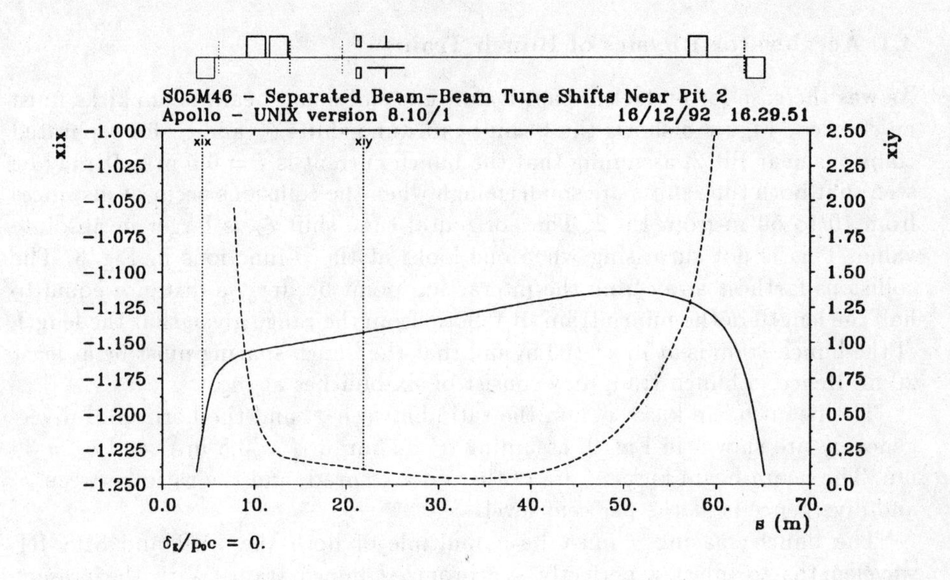

Fig. 8. Beam-Beam Tune Shifts ξ_x and ξ_y in Units of 0.001 for a Bunch Train Near Pit 2

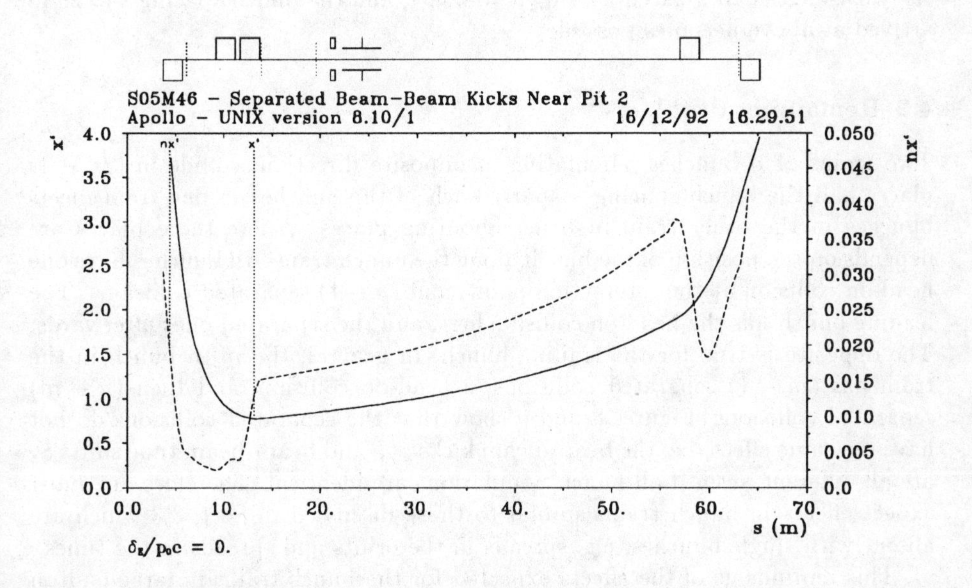

Fig. 9. Beam-Beam Kicks x' in μrad and Ratio nx' between x' and Divergence σ'_x for a Bunch Train Near Pit 2

4.1 Accelerator Physics of Bunch Trains

As was the case for pretzels, beam-beam tune shifts and beam-beam kicks must be checked. Figure 8 shows the beam-beam tune shifts ξ_x and ξ_y for separated collisions near Pit 2, assuming that the bunch current is $I = 0.5$ mA. It may be seen that both tune shifts are small enough when the collisions occur at distances from 10 to 50 m from Pit 2. The horizontal tune shift ξ_x is larger in absolute value. This is not surprising when one looks at the β-functions in Fig. 3. The collisions farthest away from the interaction point occur at a distance equal to half the length of the bunch train. It follows from the range given that the length of the bunch train is at most 100 m and that the bunch spacing must be at least 20 m. Hence, a bunch train may consist of six bunches at most.

The beam-beam kicks x' and the ratio between x' and the horizontal divergence σ'_x are shown in Fig. 9, assuming $\alpha = 1$ mrad, $I = 0.5$ mA, and $\epsilon_x = 30$ nm. The beam-beam kicks x' are of the order of μrad, and the ratio between x' and divergence is at the per cent level.

The bunch spacing s must be a multiple of both the LEP and SPS RF wavelengths to inject a perfectly synchronized bunch train. With the present RF harmonic numbers in LEP and SPS – $h_{\mathrm{LEP}} = 31320$ and $h_{\mathrm{SPS}} = 4620$ – the RF wavelengths are related by $58\lambda_{\mathrm{LEP}} = 33\lambda_{\mathrm{SPS}}$. Hence, the minimum bunch spacing is $s \approx 50$ m, thus allowing just three bunches in a train of 100 m length. If the harmonic number in the SPS is increased by 20 units to $h_{\mathrm{SPS}} = 4640$, the RF wavelengths are related by $7\lambda_{\mathrm{LEP}} = 4\lambda_{\mathrm{SPS}}$, and the bunch spacing $s \approx 20$ m arrived at above becomes possible.

4.2 Remaining Problems

Two trains of n bunches, circulating in opposite direction, collide in $(2n - 1)$ places half the bunch spacing s apart. Each of the bunches in one train meets bunches in the other train in n neighbouring places. Where these places are depends on the position of the bunch along the bunch train. All bunches have one head-on collision at the interaction point, and $(n - 1)$ separated collisions. The leading bunch has the head-on collision first, and the separated ones afterwards. The opposite is true for the trailing bunch. In general, the m-th bunch in the train has $(m - 1)$ separated collisions, a head-on collision, and then $(n - m)$ separated collisions. Figures 8 and 9 show that the separated collisions do not have the same effect, i.e. the beam-beam kicks x'_{bb} and beam-beam tune shifts ξ_z are all different, even if all bunch populations are identical. Therefore one must expect effects in bunch trains similar to those discussed in Sect. 3.3 which are absent with single bunches, e.g. spreads in the orbits and spreads in the tunes.

This reminds us of the effects expected for the bunch trains in large hadron colliders like the LHC and SSC [22]. However, there is a difference between these bunch trains and those considered here. In the LHC or SSC, the bunch trains are much longer than the distance over which they circulate in the same ring and affect each other by their electromagnetic fields. Therefore, the number of collisions is different for bunches at or near the head or tail of the train from

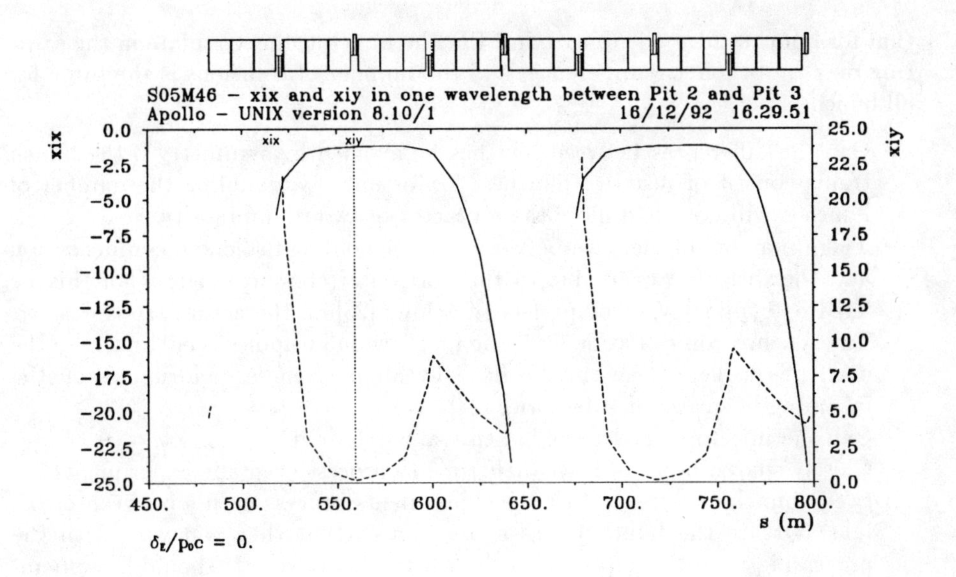

Fig. 10. Beam-Beam Tune Shifts ξ_x and ξ_y in Units of 0.001 for a Betatron Wavelength in the Arc from Pit 2 to Pit 3

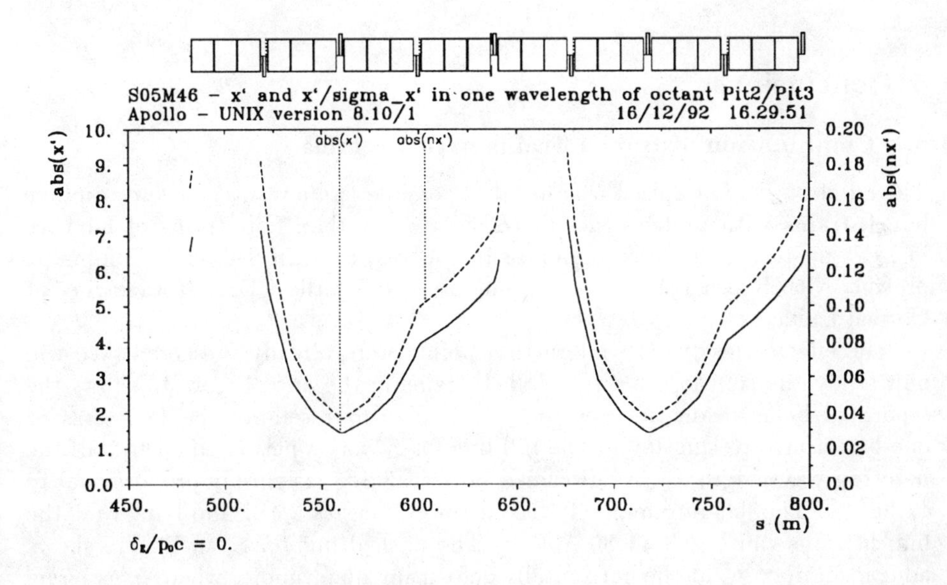

Fig. 11. Beam-Beam Kick $|x'|$ in μrad and Ratio $|nx'|$ between $|x'|$ and Divergence σ'_x for a Betatron Wavelength in the Arc from Pit 2 to Pit 3

that for bunches near the centre. In LEP, the bunch trains circulate in the same ring over the whole circumference, and the number of collisions is the same for all bunches.

- Can the differences between bunches be avoided by symmetry if the bunch trains consist of just two bunches? Unfortunately, doubling the number of bunches will not yield more than a factor of two in luminosity.
- There may be interference between the solenoid and solenoid compensation and the distorted orbit due to the horizontal crossing angle. Can this be fixed by symmetry, or by perfect matching [23] on the actual orbit?
- The synchrotron radiation from the first few quadrupoles is enhanced by the orbit offset there. Can this be fixed by aligning these quadrupoles on the incoming beam, as in B factories [24]?
- Can the injectors deliver the bunch trains required?
- Closely spaced bunches leave little time to process events in experiments between bunch crossings. Must the experiments process events between crossings? Or are the bunch trains short enough that they can wait with the processing until the gap between bunch trains arrives? It should be remembered that most collisions do not have an event. Therefore, it appears that by accumulating data over a bunch train the experiment accumulates background, but events do not pile up. This is different from the situation in the LHC where the bunch spacing is similar, but every collision contains one or more events.

5 Conclusions

5.1 Combination of Bunch Trains with Pretzels

The question may be raised whether it is possible and advantageous to combine bunch trains with pretzels, i.e. to consider more than four trains of bunches in LEP, up to 36, the maximum possible, as discussed in Sect. 3.3. Combining pretzels with bunch trains is also being studied for the CESR storage ring at Cornell University.

The effects close to the interaction points were already presented. We will now study the collisions between bunch trains in the arcs. Figure 10 shows the separated beam-beam tune shifts ξ_x and ξ_y in four cells of the LEP arcs or one betatron wavelength λ_β. The full line for ξ_x has a period equal to half the betatron wavelength λ_β. As discussed in Sect. 3.3, ξ_x is simply proportional to $1/\sin^2 \mu_x$. The flat minimum of ξ_x is about as long as a cell, implying that the bunch trains can be about 80 m long. The dashed line for ξ_y in Fig. 10 shows additional maxima at the horizontally defocusing quadrupoles where β_y is large. The length over which ξ_y is small enough is only about 50 m. Hence, the bunch trains can at most have about that length.

Figure 11 shows the separated beam-beam kicks x' and the ratio between x' and the horizontal divergence σ_x' in four cells of the LEP arcs. The kicks x' are a few μrad and the ratio x'/σ_x' is a few per cent.

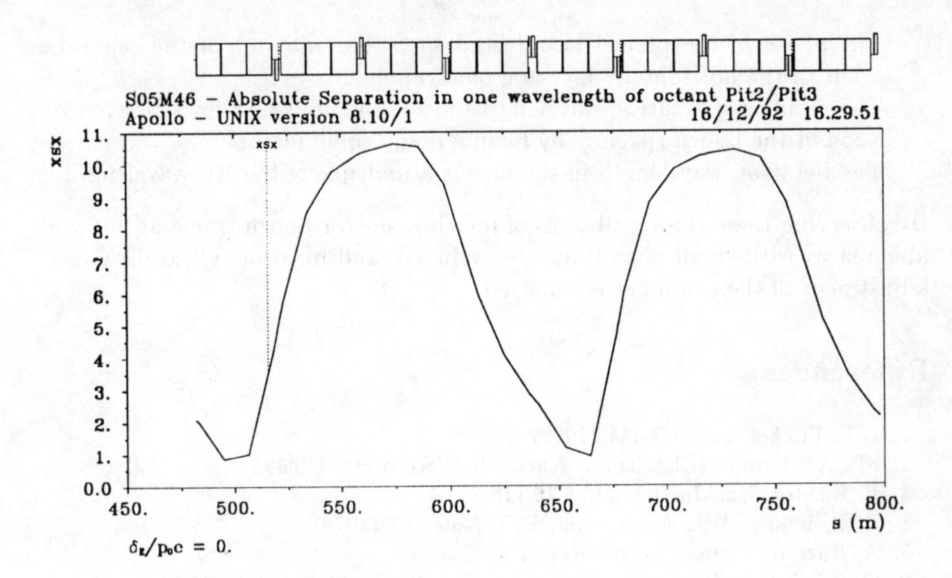

Fig. 12. Separation $|x|$ in Units of σ_x for a Betatron Wavelength in the Arc from Pit 2 to Pit 3

Figure 12 shows the ratio between the absolute value of the separation $|x|$ and the horizontal rms beam radius σ_x for a betatron wavelength in an octant of LEP, and demonstrates that the criterion $|x| > 5\sigma_x$ is well satisfied.

Figures 10, 11, and 12 show that there are regions in the arcs where bunch trains of up to 50 m length can collide without ξ_x, ξ_y, x', x'/σ'_x and $|x|/5\sigma'_x$ becoming too large. It remains to be demonstrated that this applies to all collision points foreseen in a specific implementation of a pretzel and bunch train scheme in LEP.

5.2 Lessons for the Future

What can be learned from this lecture for a hypothetical Z factory which includes pretzels and/or bunch trains as a design feature, not as an afterthought? In such a machine, collisions between bunches and/or bunch trains occur at regular intervals around the circumference. Therefore, its lattice should be designed such that they systematically occur in the vicinity of maxima of the pretzel separation. This is achieved if the lattice design includes the following features:

- Horizontally focusing quadrupoles are installed at the bunch collision points, and therefore are more regularly spaced through the whole factory, i.e. arcs, dispersion suppressors, insertions, than they are in LEP now.
- The horizontal phase advance is a simple rational fraction like 1/6 or 1/4 through the whole range of the pretzels.

- The phase of the pretzel is adjusted such that maxima of the separation occur in the horizontally focusing quadrupoles.
- The number of betatron wavelengths inside the pretzels allows an easy variation of the bunch spacing, by having many small factors.
- The betatron wavelength in the arcs is a multiple of the RF wavelength.

By observing these rules, collisions of bunches and/or bunch trains at unfavourable places with small separation are excluded, and maximum flexibility in the adjustment of their number is achieved.

References

1. G.E. Fischer, SPEAR-154 (1972)
2. SPEAR Group, IEEE Trans. Nucl. Sci. **NS-20** (1973) 838
3. B. Richter, Nucl.Instr.Meth. **136** (1976) 47
4. R.H. Helm and H. Wiedemann, PEP Note 303 (1979)
5. A. Hutton, Particle Accelerators **7** (1976) 177
6. G. Besnier, D. Brandt, B. Zotter, Particle Accelerators **17** (1985) 51
7. B. Zotter, CERN SL/92-29 (DI) (1992) 457
8. E. Haebel, this school.
9. H. Grote and F.C. Iselin, CERN SL/90-13 (AP) Rev. 3 (1992)
10. E. Keil, CERN 77-13 (1977) 22
11. B.W. Montague, LEP note 165 (1979)
12. R. Littauer, IEEE Trans. Nucl. Sci. **NS-32** (1985) 1610
13. D.L. Rubin, Particle Accelerators **26** (1990) 63
14. C. Rubbia, Proc. European Particle Accelerator Conf. (Rome 1988) 290
15. J.M. Jowett, Proc. IEEE Particle Accelerator Conf. (Chicago 1989) 1806
16. E. Blucher et al. (eds.), CERN 91-02 (1991)
17. D.H. Rice, CBN 92-9 (1992)
18. A. Piwinski and A. Wrulich, DESY 76-07 (1976)
19. M. Sands, SLAC-121 (1970)
20. LEP Design Report, Vol. II, CERN-LEP/84-01 (1984)
21. A. Piwinski, IEEE Trans. Nucl. Sci. **NS-24** (1977) 1408
22. J.L. Tennyson, SSC-155 (1988)
23. S. Kamada, Particles & Fields '91 (Vancouver 1991) 1055
24. An Asymmetric B Factory, LBL PUB-5303 (1991)

Lattice and Interaction Region Design for B Factories

Al Garren
Lawrence Berkeley Laboratory and SSC Laboratory

1 INTRODUCTION

A B factory consists of two beams, one of electrons and the other of positrons, colliding at an interaction point (IP). Each beam is a train of equally spaced bunches. The energies of the beams, E^+, E^- correspond to the $Y(4S)$ resonance at $E_{cm} = 10.58$ GeV, where $E_{cm} = \sqrt{4E^+E^-}$. If the energies are unequal, it is called an asymmetric B factory. The asymmetry enhances detection of the B-meson decays. A typical energy ratio is about 3:1 and the luminosity L required is in the range 10^{33} to 10^{34} cm^{-2} s^{-1}.

The beams of a B factory may be provided by combinations of rings and/or linacs:

- Linac + linac
- Linac + storage ring
- One storage ring with two counter-rotating beams (equal energies only)
- Two storage rings

The topic of this paper is asymmetric, two-ring B factories. Lattice problems are illustrated by PEP II design choices.[1] These are not unique, but they illustrate the decisions affecting the lattice that must be made. For further discussion of PEP II design considerations, see the paper by Zisman.[2]

Work on the lattice and interaction region (IR) design was done mainly by Martin Donald, Mike Sullivan, and the author.

2 BEAM PARAMETERS

The lattice design of a B factory begins with the choice of beam parameters, which critically affects the interaction region (IR) design. Therefore we begin with one method for selecting a self-consistant set of parameters. The notation is defined below:

subscript	n	horizontal x or vertical y direction
superscript	j	+ or − beam
superscript	k	− or + beam = $-j$

β_n^j	beta-function value at the IP	N^k	number of particles per bunch
ε_n^j	emittance	I^k	current (amperes)
σ_n^j	rms beam size at IP for j-th beam	s_B	bunch spacing
σ_n	rms beam size at IP—equal ± sizes	f_c	bunch collision frequency
ξ_n^j	beam-beam tune shift parameter	r_e	classical electron radius
ξ	common tune shift parameter	L	luminosity
γ^j	relativistic energy	E^j	energy (GeV)

2.1 Interaction Point Relations

The lattice design is strongly coupled to the values of energies, currents, bunch spacing, emittances, and IP beta values. Selection of these values is much simplified by adoption of the following two conditions:[3,4,5]

i. The beams should overlap exactly at the IP.

ii. All four beam - beam tune shifts should be equal.

Overlap condition:

$$\sigma_n^k = \sqrt{\beta_n^k \varepsilon_n^k} = \sigma_n, \quad \text{for} \begin{cases} k = + \text{ or } - \\ n = x \text{ or } y \end{cases} \tag{1}$$

Tuneshift condition:

$$\xi_n^j = \frac{r_e \beta_n^j N^k}{2\pi \gamma^j \sigma_n^k \left(\sigma_x^k + \sigma_y^k\right)} = \xi, \quad \text{for} \begin{cases} j = \pm, \; k = -j \\ n = x \text{ or } y \end{cases} \tag{2}$$

Combined overlap-tuneshift condition:

$$\xi = \frac{r_e \beta_n^j N^k}{2\pi \gamma^j \sigma_n \left(\sigma_x + \sigma_y\right)}, \quad \text{for} \begin{cases} j = \pm, \; k = -j \\ n = x \text{ or } y \end{cases} \tag{3}$$

Ratios between electron and positron beams: Equating the right sides of Eq. (3) for the same $n = x, y$ but different values of j, k gives

$$\xi = \frac{r_e \beta_n^- N^+}{2\pi \gamma^- \sigma_n \left(\sigma_x + \sigma_y\right)} = \frac{r_e \beta_n^+ N^-}{2\pi \gamma^+ \sigma_n \left(\sigma_x + \sigma_y\right)} \quad \Rightarrow \quad \frac{\beta_n^-}{\beta_n^+} = \frac{\gamma^-}{\gamma^+} \frac{N^-}{N^+} = b$$

where b is a design parameter to be chosen. By using Eq. (1) again we find

$$\sigma_n = \sqrt{\beta_n^- \varepsilon_n^-} = \sqrt{\beta_n^+ \varepsilon_n^+} \quad \Rightarrow \quad \frac{\beta_n^-}{\beta_n^+} = \frac{\varepsilon_n^+}{\varepsilon_n^-}.$$

Thus there are three equal ratios between e^+ and e^- beam quantities at the IP:

$$\frac{\beta_n^-}{\beta_n^+} = \frac{\varepsilon_n^+}{\varepsilon_n^-} = \frac{\gamma^-}{\gamma^+} \frac{N^-}{N^+} \equiv b \quad \text{for } n = x \text{ or } y \tag{4}$$

and both beams have the same value of the following ratio:

$$\frac{\gamma^+ N^+}{\beta_n^+} = \frac{\gamma^- N^-}{\beta_n^-}. \tag{5}$$

Vertical to horizontal aspect ratio: Equating the right sides of Eq. (3) for the same beam j but different $n = x, y$ values gives

$$\xi = \frac{r_e \beta_y^j N^k}{2\pi \gamma^j \sigma_y \left(\sigma_x + \sigma_y\right)} = \frac{r_e \beta_x^j N^k}{2\pi \gamma^j \sigma_x \left(\sigma_x + \sigma_y\right)} \quad \Rightarrow \quad \frac{\beta_y^j}{\beta_x^j} = \frac{\sigma_y}{\sigma_x} = r$$

where the aspect ratio r is another design parameter. Using (1) again, we have

$$r = \frac{\sigma_y}{\sigma_x} = \frac{\sqrt{\beta_y^j \varepsilon_y^j}}{\sqrt{\beta_x^j \varepsilon_x^j}} = \sqrt{r} \sqrt{\frac{\varepsilon_y^j}{\varepsilon_x^j}} \quad \Rightarrow \quad r = \frac{\varepsilon_y^j}{\varepsilon_x^j},$$

so that

$$\frac{\beta_y^j}{\beta_x^j} = \frac{\varepsilon_y^j}{\varepsilon_x^j} = \frac{\sigma_y}{\sigma_x} \equiv r, \quad \text{for } j = + \text{ or } - \cdot \tag{6}$$

Also,

$$\frac{\varepsilon_y^j}{\beta_y^j} = \frac{\varepsilon_x^j}{\beta_x^j} \quad \Rightarrow \sigma_y{}' = \sqrt{\frac{\varepsilon_y}{\beta_y}} = \sqrt{\frac{\varepsilon_x}{\beta_x}} = \sigma_x{}' = \sigma',$$

which shows that the horizontal and vertical beam divergences at the IP are equal!

Emittance: If one takes Eq. (3) with $n = x$, replaces $\sigma_y = r\,\sigma_x$ from Eq. (5) and $\sigma_x^2 = \beta_x^j \varepsilon_x^j$ from Eq. (1), then

$$\xi = \frac{r_e \beta_x^j N^k}{2\pi\gamma^j \sigma_x(\sigma_x + \sigma_y)} = \frac{r_e \beta_x^j N^k}{2\pi\gamma^j \sigma_x{}^2(1+r)} = \frac{r_e \beta_x^j N^k}{2\pi\gamma^j (\beta_x^j \varepsilon_x^j)(1+r)} = \frac{r_e N^k}{2\pi\gamma^j \varepsilon_x^j (1+r)}.$$

Solving for ε_x^j and using Eq. (5) again gives

$$\varepsilon_x^j = \frac{r_e N^k}{2\pi\gamma^j \xi(1+r)}, \quad \varepsilon_y^j = r\varepsilon_x^j, \quad \text{for } \begin{cases} j = + \text{ or } - \\ k = -j \end{cases} \tag{7}$$

and, in terms of the current $I = ecN/s_B$ in amperes and energy $E = mc^2\gamma$ in GeV,

$$\varepsilon_x^j = \frac{mc^2 r_e}{2\pi ec} \frac{s_B I^k}{\xi E^j (1+r)} = 4.77 \frac{s_B I^k}{\xi E^j (1+r)} \text{ (nm)} . \tag{8}$$

Thus, for a given current, the emittance varies directly with the bunch spacing. Note that there is no dependence on the beta-function values at the IP.

Luminosity: The luminosity is given by

$$L = \frac{f_c N^+ N^-}{2\pi\sqrt{\left(\sigma_x^{+2} + \sigma_x^{-2}\right)\left(\sigma_y^{+2} + \sigma_y^{-2}\right)}} = \frac{c N^+ N^-}{4\pi s_B \sigma_x \sigma_y}$$

where the collision frequency is $f_c = c/s_B$, and Eq.(1) has been used. From Eqs. (6) and (7),

$$\sigma_x \sigma_y = \sqrt{\beta_x^+ \varepsilon_x^+}\sqrt{\beta_y^+ \varepsilon_y^+} = \sqrt{\frac{\beta_y^+}{r}\varepsilon_x^+}\sqrt{\beta_y^+ \, r\varepsilon_x^+} = \beta_y^+ \varepsilon_x^+$$

$$= \beta_y^+ \frac{r_e N^-}{2\pi\gamma^+ \xi(1+r)} = \beta_y^- \frac{r_e N^+}{2\pi\gamma^- \xi(1+r)}.$$

Substituting into L and using Eq. (5) gives

$$L = \frac{c}{2r_e} \frac{\xi(1+r)}{s_B} \left(\frac{\gamma N}{\beta_y}\right)^{\pm} = \frac{\xi(1+r)}{2er_e mc^2} \left(\frac{EI}{\beta_y}\right)^{\pm},$$

or numerically

$$L = 2.17 \times 10^{34} \xi(1+r) \left(\frac{EI}{\beta_y}\right)^{\pm} \quad (\text{cm}^{-2}\ \text{s}^{-1}) \tag{9}$$

where E is in GeV, I in amperes, and β_y in cm. Thus, the overlap and equal tuneshift conditions imply a simple dependence of the luminosity on the three independent quantities ξ, r, and the ratio EI/β_y (of either beam).

Round *vs* flat beams: At first sight, the factor $1+r$ in the luminosity formula would seem to argue for round beams, but the gain turns out to be illusory. We compare two cases with equal energy, current, and luminosity:

1. A fully-coupled, round-beam case with $r=1$ and beta-function values at the IP of $\beta_y = \beta_x = \beta_c$.
2. A flat-beam case with $r \ll 1$, $\beta_y = \beta_{yf}$, and $\beta_x = \beta_{xf} = \beta_{yf}/r$.

Equating the luminosities, Eqs. (5) and (7) give

$$\beta_{xf}^{\pm}/\beta_c^{\pm} = (1+r)/2r, \qquad \beta_{yf}^{\pm}/\beta_c^{\pm} = (1+r)/2\,,$$

and, by comparing the emittances, we find from Eq. (6) that

$$\varepsilon_{xf}^{\pm}/\varepsilon_c^{\pm} = 2/(1+r), \qquad \varepsilon_{yf}^{\pm}/\varepsilon_c^{\pm} = 2r/(1+r),$$

so that the beam divergences at the IP are

$$\sigma_n' = \sqrt{\varepsilon_n/\beta_n} \quad \Rightarrow \quad \sigma_{nf}'/\sigma_c' = 2\sqrt{r}/(1+r), \qquad \text{for } n = x,y.$$

It was also noted previously that the x,y divergences at the IP are equal. Now, in addition, we see that they scale roughly with $\sqrt{r}$.

In the drift region between the IP and the first quadrupole the beta-functions vary as

$$\beta_n(s) = \beta_n\left[1+(s/\beta_n)^2\right] \underset{s/\beta_n \gg 1}{\approx} s^2/\beta_n,$$

so that, for $n = x$ or y,

$$\sigma_n(s) = \sqrt{\varepsilon_n\beta_n(s)} \underset{s\gg\beta_n}{\approx} \sqrt{\varepsilon_n s^2/\beta_n} = s\sigma_n' \Rightarrow \sigma_{nf}(s)/\sigma_c(s) \underset{s\gg\beta_n}{\approx} \left(2\sqrt{r}\right)/(1+r).$$

Thus, if the drift space is long enough, the x and y beam sizes become equal for any r value. Flat beams at the IP become smaller than round ones after a long drift by the ratio $(2\sqrt{r}/(1+r)$. For example, if $r = 1/25$, then, when $s \gg \beta_n$, $\sigma_{nf}(s)/\sigma_c(s) = 0.19$.

Summarizing, beams that are flat at the IP become smaller in the IR quadrupoles than beams that are round at the IP. Consequently, the quadrupoles have the smallest apertures, highest gradients, shortest lengths, lowest peak beta-function values, and lowest chromaticity from the IR region in the case of beams that are flat at the IP. These advantages of flat beams have been pointed out by Andrew Hutton.

Other possiblilities: Clearly, the two conditions at the beginning of this section could be replaced by others. For example, Eden and Furman[6] have carried out simulations exploring the effect of modifying the four-equal-tuneshifts condition in two different ways. In one case they kept the x-y tuneshifts equal but allowed differences between those for the e^+e^- beams, and in the other case they did the opposite. In these variations both the nominal luminosity and the product of the tuneshifts were kept constant. In both cases they found that the optimum tuneshift values were split. Although the corresponding dynamic luminosity values increased only slightly above those with equal tuneshifts, there may be other reasons to exploit the enlarged parameter space implied by these studies.

3 PEP II LATTICE CHOICES

In this section we set out, for completeness, the PEP II design assumptions relevant to the lattice design.

3.1 Parameters

The required E_{cm}, luminosity, and tuneshift enable one to choose a subset of the relevant parameters and obtain the others from the above formulas. Table I shows the PEP II parameters. The columns refer to the low-energy ring (LER), the high-energy ring (HER), and both rings. Chosen subset values are printed in boldface type, and those derived from the above conditions in normal type. Since this paper is primarily about lattices we give only brief comments about the choices.

Table I PEP II Parameters

		LER	HER	Both	
Energy	E	3.1	**9.0**		GeV
Current	I	2.14	1.48		A
Beta-V at IP	β_y	**1.5**	3.0		cm
Beta-H at IP	β_x	37.5	75.0		cm
Vertical emittance	ε_y	3.9	1.9		nm
Horizontal emittance	ε_x	96.5	48.2		nm
$-:+$ β^* ratio at IP	b			**2**	
$y:x$ β^* ratio at IP	r			**0.04**	
Bunch spacing	s_B			**1.26**	m
Beam-beam tuneshift	ξ			**0.03**	
Luminosity	L			$\mathbf{3 \times 10^{33}}$	$cm^{-2}\ s^{-1}$

Energy: The HER energy should be high enough to give good asymmetry, but not so high as to aggravate the radiation problem in the IR region.

Beta-function values at the IP: Because the vertical beta of the LER is the smallest of the four, it was chosen to have a value comparable to the bunch length.

Ratio between electron and positron beam quantities, b: The value chosen, two, leads to currents believed to be achievable. However there are 'transparancy' arguments in favor of the value unity.

Ratio between vertical and horizontal beam quantities, r: As argued above, small r values are favorable for the IR design, but it was decided not to rely on extremely good suppression of coupling to achieve high luminosity.

Bunch spacing: A small bunch spacing decreases emittances and beam widths, but the spacing must be large enough that parasitic crossings occur only after the beams are separated. It is quantized to be a multiple of the rf wavelength.

Beam-beam tuneshift: The value chosen, 0.03, is typical of present electron-positron colliders.

Luminosity: The value chosen is believed sufficient to reach the physics goals.

3.2 Interaction Region Design

Mode of collisions: Head-on collisions at the IP. A crab crossing would ease the initial separation and masking problems, but since it requires another rf system, it is reserved for a possible future upgrade.

Beam separation method: First separation is done by a common horizontal dipole and offset quadrupoles; second separation by a vertical step in the LER.

Symmetry about interaction point: Horizontal S-bend: horizontal bending reflects antisymmetrically; vertical bending and focusing reflect symmetrically.

3.3 Overall Configuration

Geometry: The LER is located directly above the HER. The arc quadrupoles of the HER and LER are horizontally coincident. The straight sections of the HER and LER are horizontally coincident at their centers. Both rings have approximately the sixfold periodicity of PEP.

Structure common to both rings: Geometrically, both rings have six nearly equal sextants, each containing a circular arc and a long straight section. Each arc contains twelve standard cells in the center and two long cells at each end (**Table II**).

Table II Dimensions Common to the HER and LER

Circumference	2199.318	m	Normal half-cell length	7.5625	m
Sextant length	366.553	m	Long half-cell length	8.006625	m
Arc length	245.553	m	Normal cells per arc	12	
Straight section length	121.0	m	Long cells per arc	4	
Vertical ring separation	0.895	m			

There are several advantages to making the rings nearly coincident, horizontally at least. One is the beam-beam effect: the same pairs of bunches always collide. Another is that the LER is a busy ring; it needs as many long straight sections for various functions as the HER does. Finally, it is more economical to have only one tunnel with shared facilites.

On the down side, the optics design of the LER arcs, particularily that of the dispersion suppressors, could be simplified if they did not have to follow the geometry of the HER.

4 HIGH-ENERGY RING

The HER has six geometrically identical arcs (except for a few special dipoles in the IR sextant), and six long straight sections. There are two types of arcs and four different types of long straight section (LSS):

Arc type	Number
Normal	2
Emittance control	4

LSS type	Number
IR	1
Normal (rf)	2
Phase control	2
Injection	1

Since all of the arcs and long straight sections have identical matching conditions at their interfaces they can be linked together in different ways. That chosen for the HER is shown in Fig. 1.

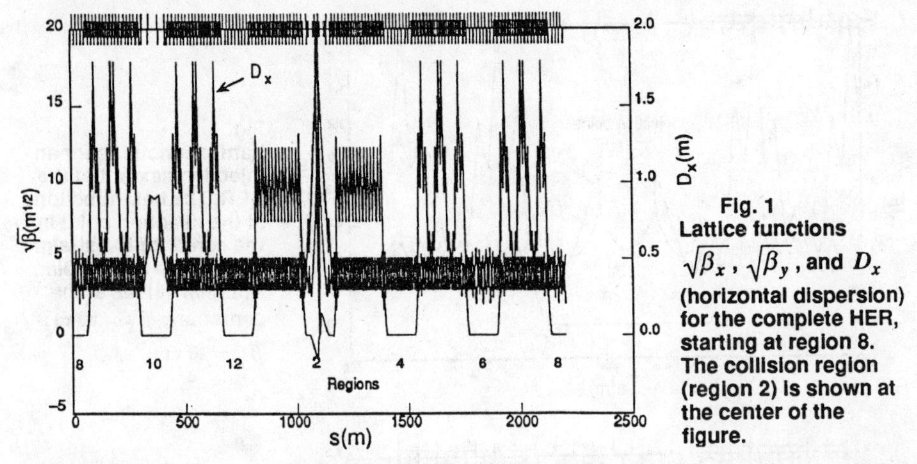

Fig. 1
Lattice functions $\sqrt{\beta_x}$, $\sqrt{\beta_y}$, and D_x (horizontal dispersion) for the complete HER, starting at region 8. The collision region (region 2) is shown at the center of the figure.

Figures 2 to 6 show the the different kinds of arcs and long straight sections assembled into sextants. Unlike the sextants adopted in the design, those in the figures have the same arc type on both sides of the straight section.

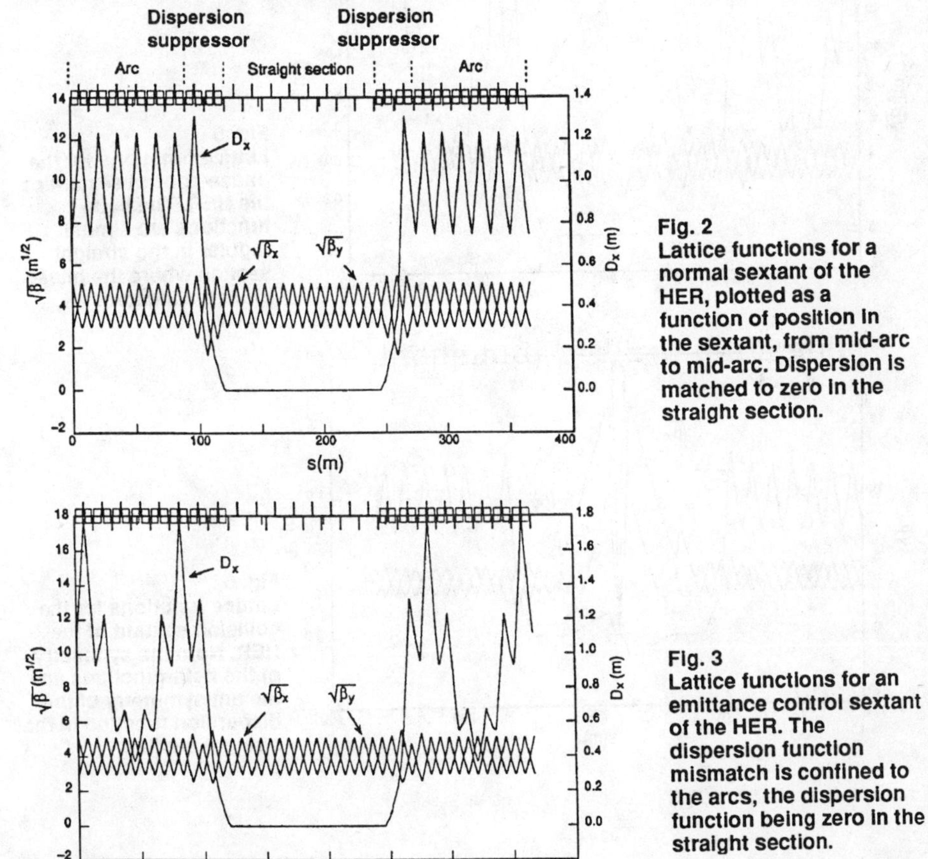

Fig. 2
Lattice functions for a normal sextant of the HER, plotted as a function of position in the sextant, from mid-arc to mid-arc. Dispersion is matched to zero in the straight section.

Fig. 3
Lattice functions for an emittance control sextant of the HER. The dispersion function mismatch is confined to the arcs, the dispersion function being zero in the straight section.

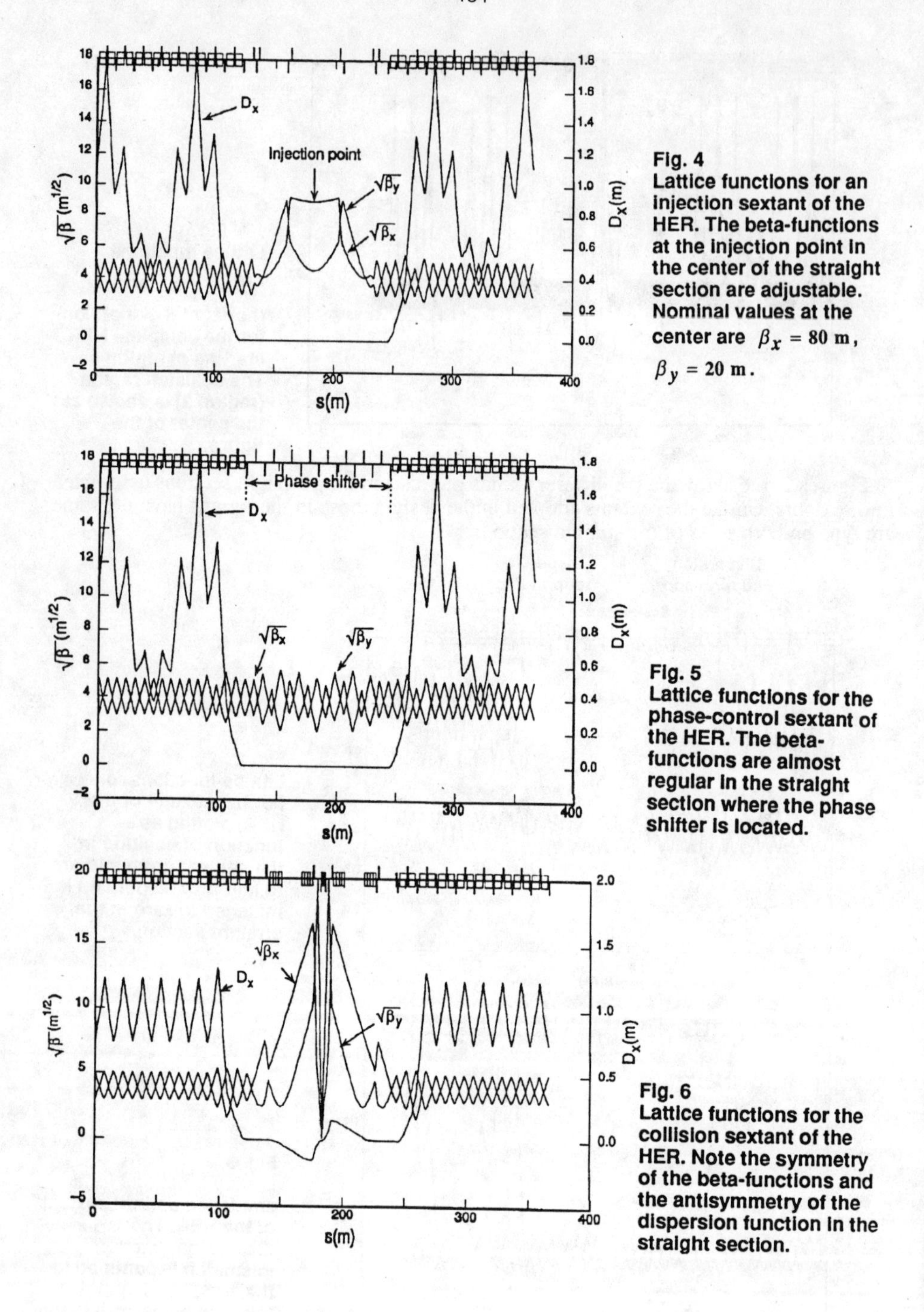

Fig. 4
Lattice functions for an injection sextant of the HER. The beta-functions at the injection point in the center of the straight section are adjustable. Nominal values at the center are $\beta_x = 80$ m, $\beta_y = 20$ m.

Fig. 5
Lattice functions for the phase-control sextant of the HER. The beta-functions are almost regular in the straight section where the phase shifter is located.

Fig. 6
Lattice functions for the collision sextant of the HER. Note the symmetry of the beta-functions and the antisymmetry of the dispersion function in the straight section.

4.1 HER Arcs

Each arc has twelve regular cells in the center and a two-cell dispersion suppressor at each end of the arc.

Regular cells: The twelve regular cells have 60° phase advance and thus together they comprise an achromat.

Dispersion suppressors: The suppressor cells have 90° phase advance and are longer than the regular ones. Their length is chosen so that their periodic dispersion is half that of the regular cells at the boundary point. The dispersion then makes a 180° sine wave from the cell value at one end to zero value at the other. Since the suppressor has a −1 transfer matrix, the beta-functions are mapped from one end to the other without change.

In an emittance-control sextant, the suppressors are perturbed to produce a dispersion wave through the arc, which does not leak into the adjacent straight sections.

5 LOW-ENERGY RING

The LER has six identical arcs (except for a few special dipoles in the IR sextant), and six long straight sections. There are four different types of long straight section, and for each type there is a corresponding type of sextant. Figure 7 shows the optics functions of the whole LER ring.

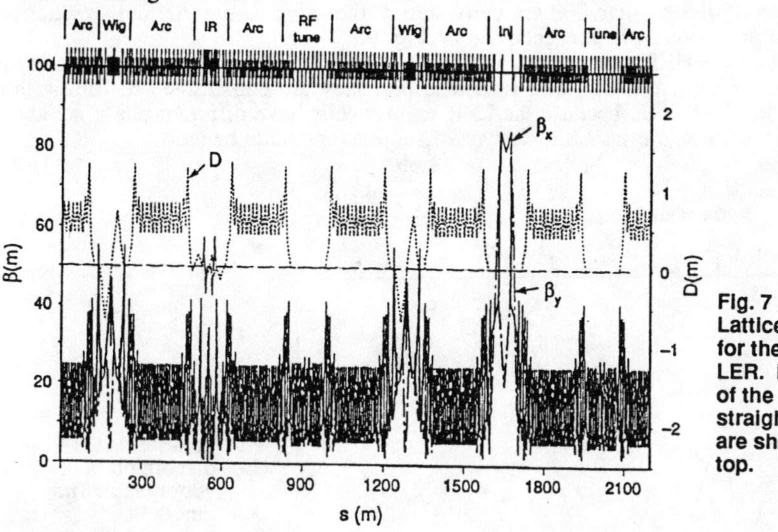

Fig. 7 Lattice functions for the complete LER. Functions of the six long straight sections are shown at the top.

5.1 LER Arcs

Regular cells: Each arc has nine regular cells in the center and a 3 1/2-cell dispersion suppressor at each end. The length of these cells is the same as that of the HER cells. The regular cells have 80° phase advance, thus the nine comprise an achromat.

The LER cells must have higher phase advance than those of the HER so that the momentum compaction will not be too large, which in turn would lead to a high rf voltage requirement.

Unlike the dipoles in the HER cells, those in the LER are shifted from the half-cell centers to permit synchrotron radiation from the beam to be absorbed in the straight sections between the dipoles and the quadrupoles. (See Fig. 8.)

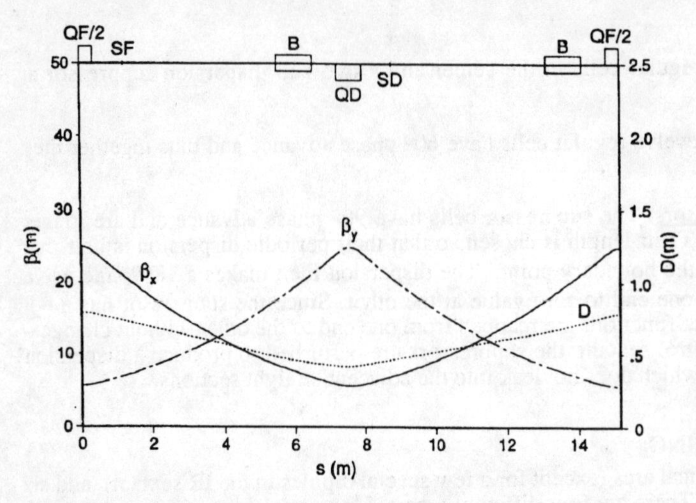

Fig. 8 Standard LER arc cell. Note the offset of the dipoles B from the half-cell centers.

Dispersion suppressors: Each dispersion suppressor has two long cells adjacent to the long straight section, and three normal-length half cells. It is matched at one end to the beta-functions of the regular 80° arc cells, and at the other to the 90° cells (without dipoles) of a normal long straight section. (See Fig. 9.)

In contrast to the HER suppressors, those in the LER require different excitations in each of their six quadrupoles. The reason is that they are constrained to follow the geometry of the HER, but, because the LER regular cells have different phase advance than those of the HER, the same simple type of suppressor cannot be used.

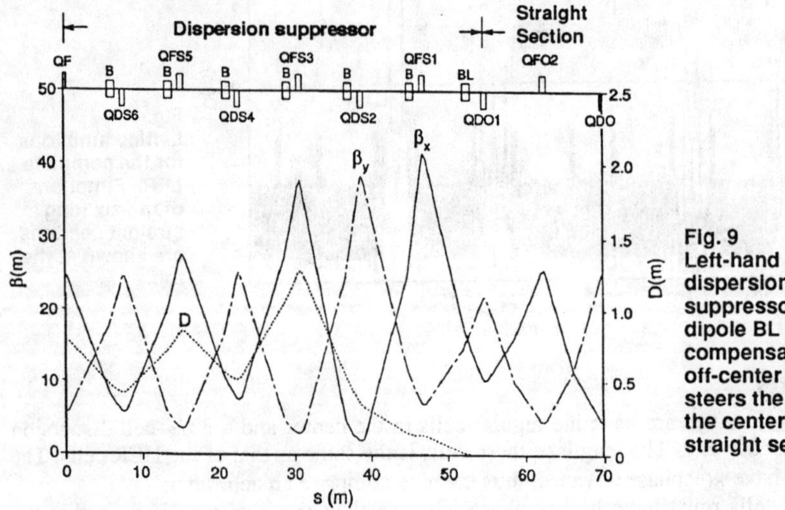

Fig. 9 Left-hand LER dispersion suppressor. The dipole BL compensates for the off-center dipoles; it steers the orbit to the center of the straight section.

Normal sextant: A complete normal sextant, from mid-arc to mid-arc, is shown in Fig. 10. One can see the $4\,^1/_2$ normal cells at each end, then the $3\,^1/_2$-cell dispersion suppressors, and then the long straight section in the center, which is made up of eight 90° FODO cells without dipoles.

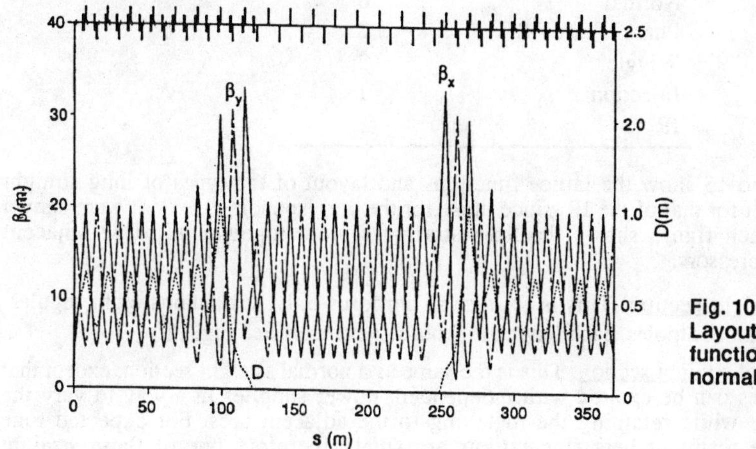

Fig. 10 Layout and lattice functions for a normal LER sextant.

IR sextant: The other sextants differ from the normal one only with respect to the long straight section (except that the IR sextant has some additional dipoles in its dispersion suppressors to complete the steering of the beamline from the IP to the arc). Figure 11 shows the IR sextant of the LER. Note that the scales are different from those of the normal sextant.

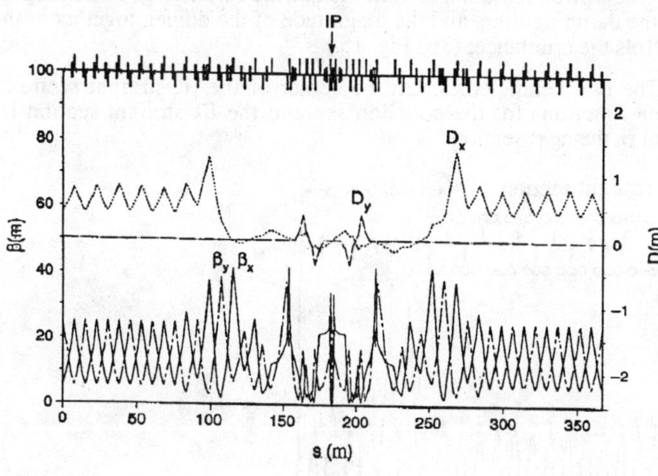

Fig. 11 Complete IR sextant of the LER. Note the symmetry of the vertical dispersion function and the antisymmetry of the horizontal dispersion function, which reflects the geometry.

5.2 Long straight sections

The LER has five types of long straight sections; one of which is used as a template.

LSS type	Number in ring
Normal	0
Phase control–RF	2
Wiggler	2
Injection	1
IR	1

Figures 12 to 15 show the lattice functions and layout of the types of long straight sections, except for that of the IR, since those for the corresponding sextant were shown in Fig. 11. Each figure shows the long straight section together with the adjacent dispersion suppressors.

Normal straight section: This is for design purposes only. It contains eight regular-length cells without dipoles, each having 90° phase advance. (See Fig. 12.)

Phase-control straight section: This is the same as a normal straight section, except that the quadrupoles can be excited with independent power supplies in a way to vary the phase advance while retaining the matching to the adjacent arcs. For expected tune variations, the resultant beta fluctuations are small; therefore one of these straight sections is used for rf cavities. (See Fig. 13.)

Injection straight section: This is nearly identical to that in the HER. It features a long drift space in the center with high beta values in the injection plane. (See Fig. 14.)

Wiggler straight section: Wiggler magnets are located in this straight section along a dogleg that directs their synchrotron radiation toward the outside of the ring. The strength of the wigglers controls the damping time, and the magnitude of the dogleg together with the wiggler strength controls the emittance. (See Fig. 15.)

IR straight section: The two beams collide in the center of the IR straight section. Figure 11 shows the beta-functions for the collision sextant; the IR straight section is discussed in greater detail in the next section.

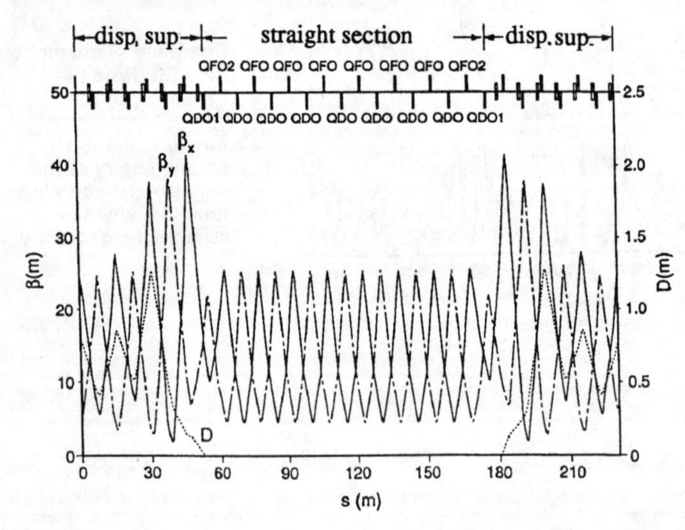

Fig. 12
Normal LER long straight section. It consists of eight 90° cells without bends.

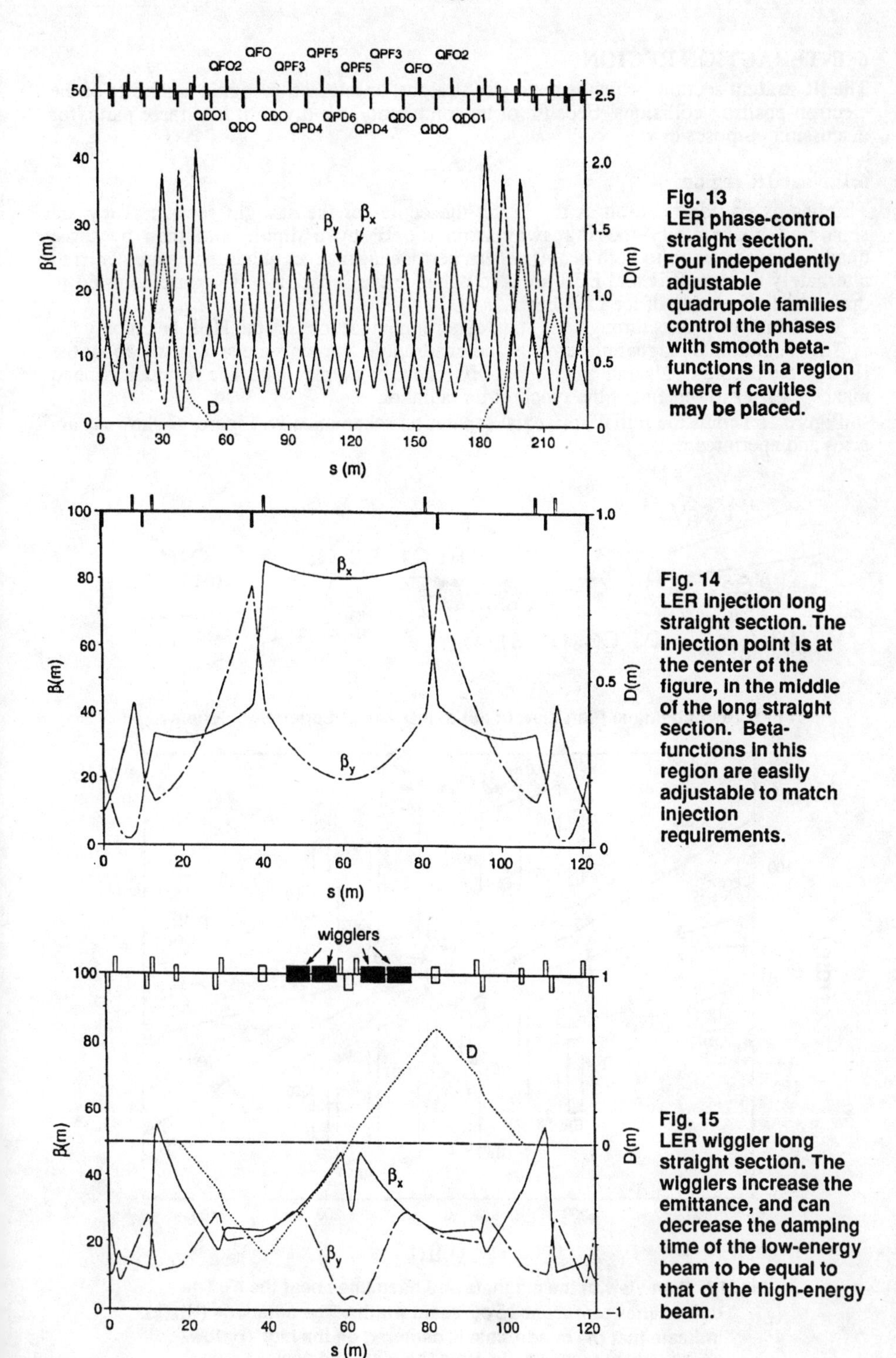

Fig. 13
LER phase-control straight section. Four independently adjustable quadrupole families control the phases with smooth beta-functions in a region where rf cavities may be placed.

Fig. 14
LER injection long straight section. The injection point is at the center of the figure, in the middle of the long straight section. Beta-functions in this region are easily adjustable to match injection requirements.

Fig. 15
LER wiggler long straight section. The wigglers increase the emittance, and can decrease the damping time of the low-energy beam to be equal to that of the high-energy beam.

6 INTERACTION REGION

The IR straight section (which is not really straight) or interaction region is the site of the electron-positron collisions. Because of its complexity, we divide it into three parts, for discussion purposes only.

6.1 Near IR region

The beams collide head-on at the IP, in the center of the straight section. They are separated horizontally–moving away from the IP–by a dipole and then by three quadrupoles common to both beams. These permanent-magnet quadrupoles are centered alternately on the HER and LER so as to enhance the separation. They are adjusted to optimize the focusing of the LER beam.

The beams then pass through a septum quadrupole that focuses the HER beam only.

The longitudinal magnet placements and quadrupole gradients are symmetric about the IP, and the dipole fields and quadrupole offsets are antisymmetric. The resultant S-bend makes it easier to screen out the synchrotron radiation.

Figure 16 shows the initial horizontal separation schematically. Figure 17 shows beam sizes and apertures.

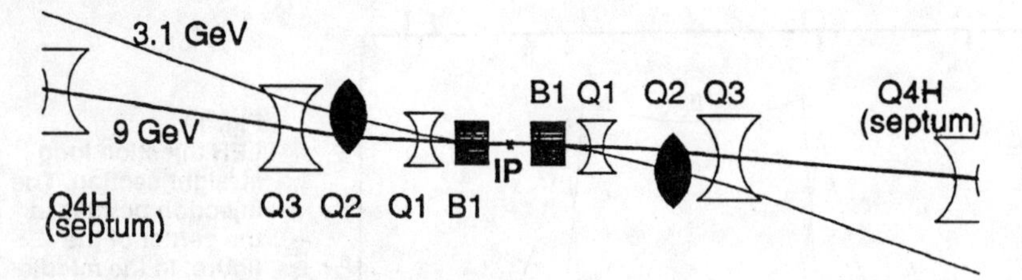

Fig. 16 Schematic Plan View of Initial-Horizontal-Separation Scheme.

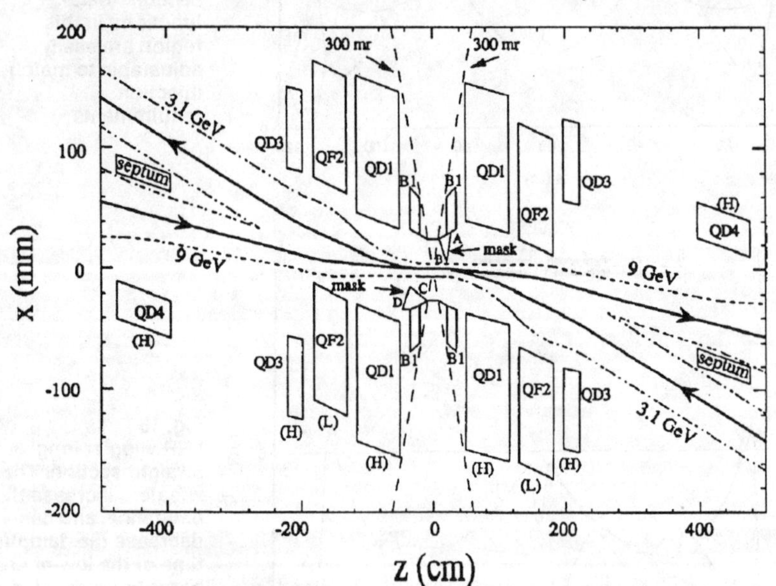

Fig. 17 Plan view of the magnets and beamlines near the IP. The dotted lines represent $15\sigma_x$ beam widths. The notations (H), (L) indicate that the quadrupole is centered on the high- or low-energy beam respectively. Note the distorted scale.

Lattice functions of the LER and HER through the initial separation region are shown in Figs. 18 and 19. Because of the symmetry of the focusing, it suffices to show only one side of the IR.

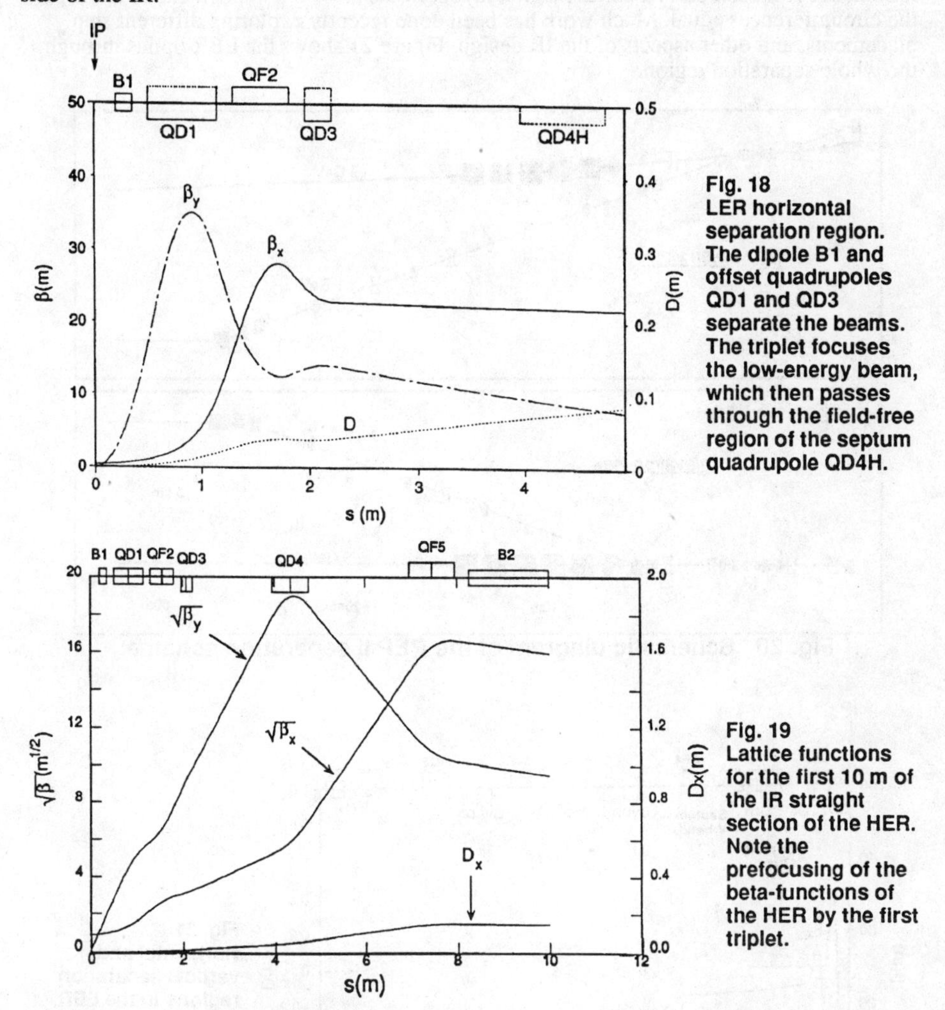

Fig. 18
LER horizontal separation region. The dipole B1 and offset quadrupoles QD1 and QD3 separate the beams. The triplet focuses the low-energy beam, which then passes through the field-free region of the septum quadrupole QD4H.

Fig. 19
Lattice functions for the first 10 m of the IR straight section of the HER. Note the prefocusing of the beta-functions of the HER by the first triplet.

6.2 Intermediate part of IR straight section

The HER beam is focused primarily by the QD4 septum quadrupole and by QF5, then it enters two strings of dipoles that bring its dispersion to zero. Meanwhile, the LER beam is deflected upwards by a vertical step to a plane 0.895 m above the HER. Quadrupoles along the step produce a 360° wave in the vertical dispersion and at the same time bring the horizontal dispersion to zero.

Figure 20, though not strictly accurate, shows the vertical-horizontal beam separation scheme qualitatively. The vertical step complicates the optics of the LER. The main reasons for it are to obtain a better magnet layout in the arcs, and make it easier to keep the circumferences equal. Much work has been done recently exploring different step placements, and other aspects of the IR design. Figure 21 shows the LER optics through the whole separation region.

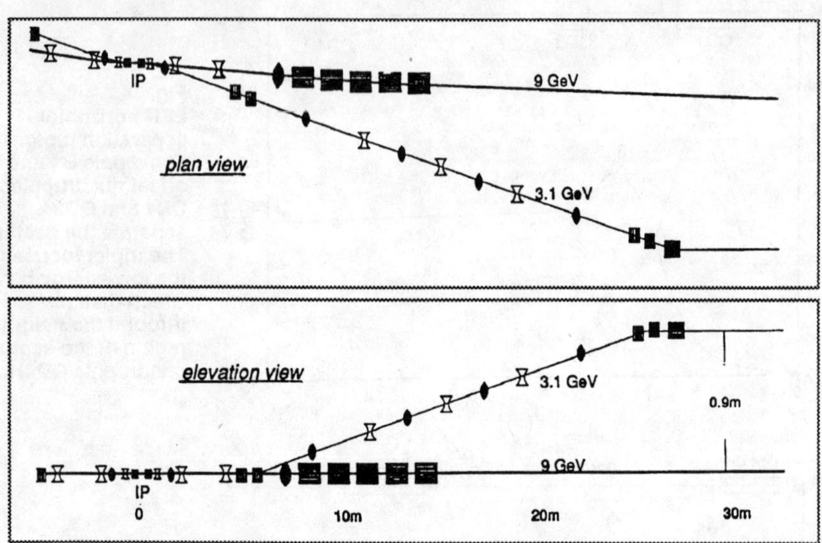

Fig. 20 Schematic diagram of the PEP II separation scheme.

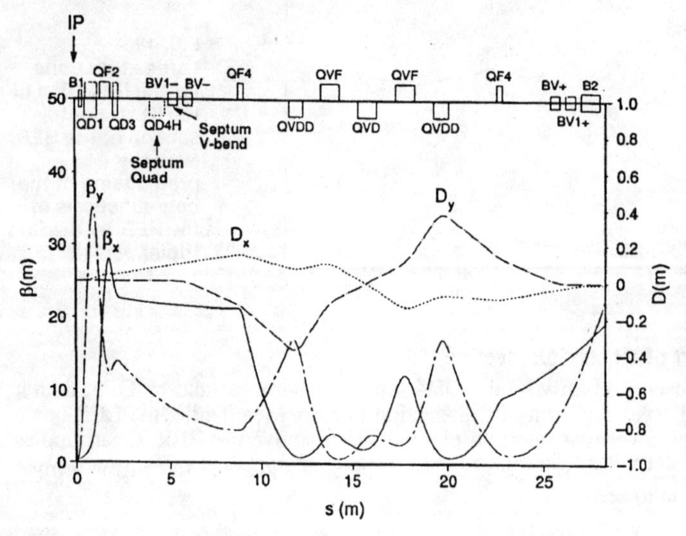

Fig. 21 Horizontal and vertical separation regions in the LER. To avoid horizontal-vertical coupling, vertical bending is confined to a region free of horizontal dipoles.

6.3 End part of IR straight section

The remainder of the IR straight section in each ring is devoted to matching of the beta-functions and dispersion into the arc, and steering the beams so that they enter the tunnel properly. (See Figs. 22 and 23.)

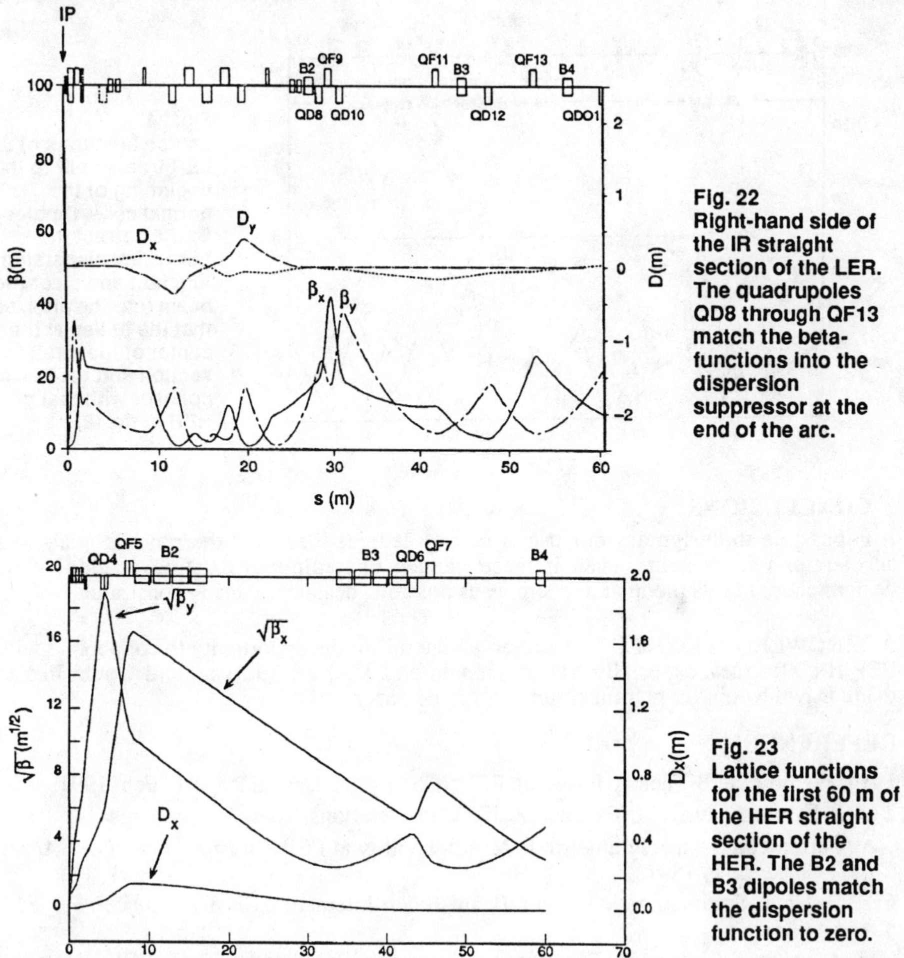

Fig. 22
Right-hand side of the IR straight section of the LER. The quadrupoles QD8 through QF13 match the beta-functions into the dispersion suppressor at the end of the arc.

Fig. 23
Lattice functions for the first 60 m of the HER straight section of the HER. The B2 and B3 dipoles match the dispersion function to zero.

Figure 24 shows the LER—from the IP, through the dispersion suppressor, to the normal cells. Horizontal steering magnets in the straight section and in the dispersion suppressor bring the orbit directly above that of the HER in the arc region.

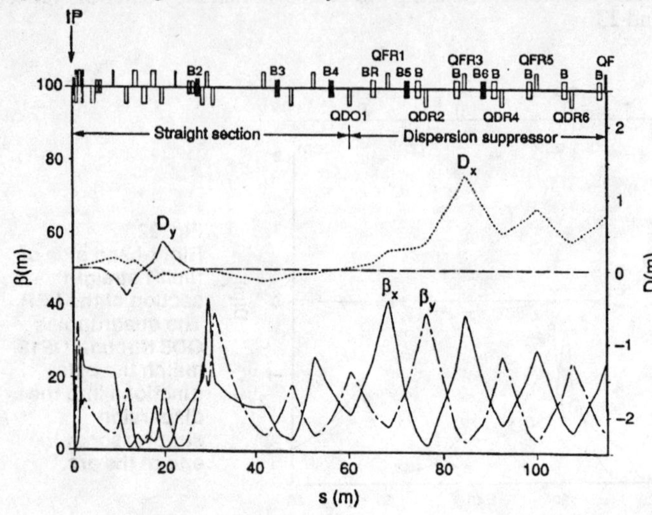

Fig. 24
Lattice functions of the LER from the IP to the beginning of the normal cells. Dipoles B2-B6 correct the horizontal dispersion function and steer the beam into the arcs, so that the IP lies at the center of the straight section and the beam is colinear with that of the HER at the IP.

7 CONCLUSIONS

It is possible to design asymmetric B factory lattices that meet the physics goals and accelerator requirements, even if these include upgrading an existing machine. The designs should be as modular and simple as possible, despite the many constraints.

ACKNOWLEDGEMENTS: The author is grateful for the opportunity to collaborate with PEP II colleagues, especially Martin Donald and Michael Sullivan, and would like to thank David Robin for helpful comments on the paper.

REFERENCES

1 An Asymmetric B-Factory Based on PEP, Conceptual Design Report, Feb. 1991.

2 M. S. Zisman, General B Factory Design Considerations, These Proceedings.

3 A. Garren et al., An Asymmetric B-Meson Facility at PEP, in *Proc. 1989 Part. Accel. Conf.,* Chicago, p. 1847.

4 Y. H. Chin, Symmetrization of the Beam-Beam Interaction, in *AIP Conf. Proc.* 214, p. 424, 1990.

5 M. A. Furman, Luminosity Formulas for Asymmetric Colliders with Beam Symmetries, PEP-II note ABC-25, 1991.

6 J. R. Eden and M. A. Furman, Simulation of the Beam-Beam Interaction for Pep-II with Unequal Beam-Beam Parameters, PEP-II/AP Note 6-92, Feb. 1993.

Linear Colliders as Factories for Z^0 and Heavier Particles

G. Guignard

Cern, Geneva, Switzerland

Abstract. After a brief review of the physics potential of linear colliders as Z^0 and heavier particles factories, the conditions required for the luminosity and the background to achieve good detection conditions are discussed. Considering an interval of RF frequency ranging from about 10 GHz to 30 GHz, the reasons for operating at such a high frequency when high accelerating gradient is required are recalled. Effects of strong wake fields on the linac energy-spread, transverse beam-stability and multibunch dynamics are summarized. The state of the art in building accelerating structures and keeping the trajectory within tight limits is given. Prospects for microwave klystrons as high-power RF sources for normal-conducting e^+e^- linacs are described briefly. Developments concerning a two-stage accelerator scheme proposed at CERN for generating the necessary power are presented. Finally, consistent sets of guessed parameters for linear-collider factories are proposed to estimate the performance that can be expected.

1 Perspectives and Main Issues

1.1 Introduction on Physics Potential

According to the physicists interested in the field of particle interactions [1], the open problems in the Standard Model at high energies can at present be divided schematically into three sectors: the fermion sector with the search for the top quark and its properties, the gauge boson sector including the $W^{\pm}$ and Z couplings as well as their static properties, and the Higgs sector with a quest either for their characteristic properties if they exist or for WW scattering if they do not. Beyond the Standard Model, possible extensions concern the gauge symmetry with the emergence of new gauge bosons $\left(W_R^{\pm},\ Z_R\right)$, heavy neutrinos together with exotic quarks and leptons on the one hand, and the supersymmetry implying Higgs spectrum, scalar partners of quarks and leptons and spin-1/2 partners of gauge bosons and Higgs particles on the other hand.

For this kind of research, both hadron colliders and the lepton colliders can be used and are complementary given their specific characteristics and the domains of energy they are likely to cover. Considering the first type, there are actually two proton–proton colliders in the design or project stage respectively, the Large Hadron Collider at CERN (LHC) and the Superconducting Super Collider (SSC) at the SSC Laboratory in Dallas. The LHC is designed for a centre-of-mass energy $\sqrt{s}$ of 15.4

TeV and the SSC is aiming at 40 TeV for the same quantity. These figures correspond to an effective part available for the reaction smaller than or equal to about 2 TeV in the first case and perhaps 5 TeV in the second. By comparison, the design or studies of lepton colliders are far less advanced; it is nevertheless clear that they will be linear colliders owing to the fact that cost and size of rings for energies beyond LEP energies become prohibitive because of the unavoidable strong synchrotron radiation emitted by electrons and positrons travelling along a curve. Nowadays, the energies that look reasonable and are most often considered by both the engineers and the physicists are lumped around two levels which could be associated with two phases of linear colliders; a phase I with $\sqrt{s}$ around 300 to 500 GeV and a phase II with $\sqrt{s}$ between 1 and 2 TeV say. Phase I, adequate for top and Higgs, would be complementary to proton–proton colliders and phase II equivalent and complementary to pp colliders, though very promising in the search for novel phenomena and particles.

Since we are talking about linear colliders as factories, the target luminosity should be at least an order of magnitude above that of LEP (Large Electron–Positron ring at CERN), say. The LEP luminosity is at present about 10^{31} cm^{-2}s^{-1} and will likely not exceed 10^{32} for a long time in the future, even with the multibunch scheme known as the "pretzel scheme".

Hence, the linear-collider factories should aim at a luminosity $\mathcal{L}$ of 10^{33} cm^{-2}s^{-1} or higher. Starting with this tentative figure, one will try to convince the reader all the way through this review of the conceptual studies currently being carried out on linear colliders, that the desired goal at the energies considered is not out of reach. Before doing this, however, let us briefly indicate what such a luminosity means for the physics research.

Let us come back to the three sectors of the Standard Model. The mass of the top quark, that is a scale of the electroweak symmetry breaking, plays a key role in the fermion sector of the model and is a gate to the physics at high energies beyond the model. This mass is large, above 91 GeV and in the interval 130 to 150 GeV (±30 GeV) according to recent experiments carried out at CERN and Fermilab for instance. With a luminosity of 10^{33} cm^{-2}s^{-1}, the expected production rate would be of the order of 10 000 top/year. Top production, production threshold and decay modes could be studied and, for the mass range quoted above, predictions say that the mass of the top could possibly be determined with an uncertainty smaller than 500 MeV, to be compared with the uncertainty of about 5 GeV expected in a proton–proton collider. Considering next the interactions of electroweak bosons, there are many independent coupling parameters and possible constraints on these parameters to be determined. Linear colliders are interesting in this case, for they would provide bounds for these parameters that are more stringent than at hadron colliders. The third sector concerns the Higgs in the Standard Model, i.e. with masses below 700 GeV approximately. Let us note that in hadron colliders the search for Higgs is easy for masses between 140 and 700 GeV, but difficult below 140 GeV (and very difficult above 700 GeV). As complementary instruments, linear colliders look ideal for the Higgs search in an intermediate mass range from below 140 GeV continuing up to 350 GeV if $\sqrt{s}$ = 500 GeV. Considering again $\mathcal{L} = 10^{33}$ cm^{-2}s^{-1} (corresponding to an integrated luminosity of about 10 fb^{-1}/year), the production rate of Higgs with masses near 140 GeV would be ~ 1 000 events per year.

The next field to mention concerns the extensions of the Standard Model. Within extended gauge theories, linear colliders seem interesting in the search for novel leptons in all cases, precision studies (masses, couplings) of all novel fermions, and as factories of the neutral Z′ (issued from the theory) if the centre-of-mass energy is sufficient. By comparison, pp colliders are ideal in this case for direct production of vector bosons up to about 5 TeV. Next comes the supersymmetry that aims at the unification of bosons with fermions and is a plausible step towards quantum theory of gravity. The associated particle spectrum includes five physical Higgs particles (three neutral and two charged, with a non-zero probability that all masses be below 250 GeV), sleptons and squarks (spin-0 partners to fermions), and gluinos and Higgsinos (spin-1/2 partners to gauge and Higgs bosons). Again, linear colliders look ideal for the search for these Higgs in the intermediate mass range (while pp colliders are appropriate for higher masses up to 800 GeV) and for measurements of spin or couplings.

In summary, linear colliders that could serve as Z^0 factories (though then in competition with existing ones) could also be very interesting as top and Higgs factories provided the appropriate performance can be reached, and the questions of the energy spread and of the number of underlying $\gamma\gamma$ events are solved. It indeed appears that precision measurements of the top's characteristics and coupling parameters are then possible as well as the search for and exploration of new phenomena in the lepton /Higgs sector.

1.2 Background and Interaction-Point Issues

During the collisions, beam–beam forces enhance the luminosity and the angular opening of the spent beam. But, at the same time, electrons and positrons radiate "beamstrahlung" photons in the bending field generated by the opposite bunch. These photons have basically two detrimental effects: they carry energy from the electrons and positrons, create parasitic e^+e^- pairs and generate a background of $\gamma\gamma$ processes. The beam parameters at the interaction point (I.P.) must therefore be adjusted to keep these effects at an acceptable level.

The first detrimental effect corresponds to a degradation of the resolution on the centre-of-mass (c.m.) energy of the collision, described by the luminosity distribution versus s (c.m. energy square) termed "differential luminosity". This degradation corresponds to a dilution of the luminosity towards low energies [2] that can be characterized by three most significant parameters:

i) the pinch enhancement factor H_D of the luminosity due to the bunch field-forces,
ii) the average fractional energy loss δ_E due to beam–beam radiation ("beamstrahlung"),
iii) two thirds of the fractional average critical photon energy in the classical regime (noted Y).

Flat beams with a high aspect ratio σ_x/σ_y make it possible to decrease all three parameters with respect to the values they take with round beams. The reduction of the enhancement factor H_D might seem unfavourable at first view, but it is associated with a beneficial reduction of the luminosity dilution and of the average energy loss δ_E (which remains large with respect to the acceptance of the final-focus system that precedes the I.P., but smaller than the one determined by the initial-state radiative corrections). This means that with flat beams there are more annihilation events with a total energy close to the c.m. energy available. For instance, with the CLIC parameters and an aspect ratio of ~ 11, the fraction of luminosity above 95% of the c.m. energy is about 70%.

The second important effect is the presence of background events in addition to the wanted e^+e^- annihilation processes [3]. It turns out that background is due to the two-photon process, since the incident electron and positron beams in a linear collider carry with them virtual photons, usually called "Weiszäcker–Williams" photons, as well as real "beamstrahlung" photons that can interact together in one of the following ways:

$$\gamma\gamma \to \text{hadrons} \quad \text{(real–real)}$$
$$\gamma e^{\pm} \to e^{\pm}\ \text{hadrons} \quad \text{(real–virtual)}$$
$$e^+e^- \to e^+e^-\ \text{hadrons} \quad \text{(virtual–virtual)}$$

The convolution of the virtual and real photon energy spectrum gives the effective differential luminosity due to two-photon events, which can then be compared with the annihilation differential luminosity. If the former is too high, the annihilation events' study can be difficult, either because the readout trigger is swamped by two-photon events that cannot easily be distinguished from the wanted events, or because too many valuable annihilation triggers would contain secondaries from simultaneous two-photon interaction. It is moreover not excluded that particles be produced at high transverse momenta by two-photon collisions in which photons transform virtually to vector mesons, and constituent quarks or gluons from the meson collide to produce high-transverse-momentum (p_T) "minijets". Nevertheless, the dominant process by which photons interact to produce hadrons is through a state in which both photons become virtual hadrons. Since the interaction volume is typical of hadron dimensions (~1 fm^3), the final products tend to have limited p_T. By contrast, e^+e^- annihilations produce quark–antiquark pairs in a pointlike process and the quarks generate hadrons in two jets with high transverse momenta. This difference and the fact that the $\gamma\gamma$ frame relative to the laboratory is boosted with final particles coming out at rather small angles suggest that event selection criteria can be found to suppress or reduce two-photon background relative to annihilations.

Possible detection criteria have recently been studied with the support of numerical simulations [3]. Any general-purpose detector is sensitive only to tracks emitted with angles greater than some limit θ_{min} optimized to obtain acceptable parasitic e^+e^- pairs (θ_{min} of the order of 150 mrad). Taking this into account, trigger requirements to keep $\gamma\gamma$ background low can be based on the total energy and transverse momentum in the final state, depending on what precedes. Total energy, close to $\sqrt{s}$ only in the absence of beamstrahlung degradation and the escape of secondaries, can for instance be set to be above a threshold of the order of 5% of nominal $\sqrt{s}$ to prevent $\gamma\gamma$ events from triggering without affecting too much the efficiency for annihilation events. To help in this challenge, triggers can also be based on p_T

requirements for the secondaries; one possibility is to define a minimum requirement on the "transverse energy" sum, while another way consists of requiring that at least one track in the event have a momentum p_T larger than some specified threshold of the order of 1% of $\sqrt{s}$. Monte Carlo results indicate that it should not be difficult to get clean annihilation measurements in linear colliders in this way, in particular for events at energies above 0.9 s, assuming one is prepared to use higher thresholds than those quoted and to lose the corresponding small fraction of good events taking place at lower energies.

Although $\gamma\gamma$ events are sufficiently different from e^+e^- events to make their contributions to the trigger rate insignificant, the probability of such an event occurring within the resolving time of the detector could be so high that every legitimate annihilation would be superposed on a two-photon event (in particular for colliders with high luminosity per bunch-crossing). This effect can be characterized by the mean number of two-photon events per bunch-crossing, also called occupancy

$$\Omega \sim \mathcal{L}_b\left(1+n_\gamma\right)^2 \sim \frac{RN^2}{f_{rep}(1+R)^2}$$

where $\mathcal{L}_b$ is the luminosity per bunch crossing and n_γ the average number of photons per incident particle. To reduce detector occupancy to the desired level, it is hence judicious to lower the number of particles N per bunch, increase the repetition rate f_{rep}, and, last but not least, use flatter beams (higher aspect ratio R that also limits the luminosity dilution, as mentioned above). Reducing N while keeping the same total luminosity might imply the presence of several bunches per RF pulse. This would, however, reduce the occupancy only if the detector can resolve events coming from the successive bunch-crossings. Once these few parameters have been selected, the experimenter can still use different strategies to minimize the effects of the underlying $\gamma\gamma$ events, like increasing the cut-off angle θ_{min} (background cuts are faster than signal cuts) and profiting from the jet structure of the annihilation events with balanced p_T.

If it is obviously essential to achieve workable experimetal conditions, the beam focusing at the I.P. also contains critical issues [2] briefly recalled hereafter. The required high-demagnification telescope induces transverse aberrations that necessitate the presence of a chromaticity correction section with sextupoles and subsidiary bending magnets to create dispersion. It also implies tight tolerances on field errors and misalignments to limit luminosity losses due to beam offset and coupling. In some cases, wake fields in small-aperture elements may generate a too large emittance growth. But designs of final-focus systems taking these effects into account have been worked out.

1.3 Accelerating Gradient and RF Frequency

The total average RF power that is required in one linac is given by

$$P_{RF} = \frac{P_b}{g^2\eta} \qquad (1)$$

where g^2 is the filling efficiency of the accelerating sections (i.e. the fraction of input energy left at the end of the fill time τ_0) and η the fractional energy extraction by the beam [4]. The beam power P_b is proportional to the final particle energy eU, the number N of particles in the beam and the repetition rate f_{rep},

$$P_b = eUNf_{rep} \tag{2}$$

while the fraction of stored energy extracted by a charge Ne is given by

$$\eta = \frac{NeZ_0\omega_0^2}{2\pi c\, E_0}. \tag{3}$$

The quantity Z_0 is frequency-independent and only depends on the shape of the accelerating structure; it is the shunt impedance over Q factor per RF wavelength and is related to the shunt impedance over Q factor per unit length r_0' via

$$\omega_0 Z_0 = 2\pi c\, r_0'. \tag{4}$$

The circular frequency ω_0 is equal to 2π times the RF frequency f_0, and E_0 stands for the accelerating gradient.

Since a high-energy extraction is desirable, the expression (3) advocates for high RF frequencies at a given beam charge. In particular, when aiming at very high gradients in order to limit the total length of the linac, very high frequencies f_0 are required. The highest possible value is eventually limited by the manufacturing and alignment tolerances as well as the wake fields, and this limit constrains the gradient E_0 actually attainable.

Other constraints on E_0 come from the difficulty in generating the required power. The peak power per unit length $\hat{P}/L_0$, in a structure of shunt impedance per unit length R_0' is indeed given by

$$\frac{\hat{P}}{L_0} = \frac{E_0^2}{g^2\alpha R_0'\eta_2} \tag{5}$$

where α is the power flow attenuation constant and η_2 is the efficiency of the energy transfer from the source to the linac structure. The power requirements (5) related to the desired E_0 may be large and it is a challenge to develop power sources delivering the necessary energy and working at high frequency.

The scaling of E_0 with ω_0^2 suggested above keeps constant the fraction η and the average power P_{RF}, by virtue of Eqs. (1)–(3). In these conditions the peak power per unit length $\hat{P}/L_0$ is proportional to

$$\frac{\hat{P}}{L_0} \sim \omega_0^{-1/2} E_0^2 \sim \omega_0^{7/2} \tag{6}$$

which includes the fact that R_0' varies with $\sqrt{\omega_0}$.

Increasing E_0 in this way can be considered until a limit either on the manufacturing tolerances or on the development of power sources is reached, as mentioned already.

A third limitation may arise from the wake fields whose peak values depend on ω_0, the iris aperture a and the loss factor k_0 as follows [4]:

$$\hat{W}_T^\delta \sim \frac{k_0}{\omega_0 a^2} \sim \frac{R_0}{Q_0 a^2} \sim \frac{\omega_0}{a^2} \sim \omega_0^3$$

$$\hat{W}_L^\delta \sim k_0 \sim \frac{\omega_0 R_0}{Q_0} \sim \omega_0^2 \qquad (7)$$

The rapid increase of the point-charge wakes (mainly transverse) with ω_0, and the concomitant beam blow-up are the sources of this limitation.

Since both the achievable gradient and the detrimental effects increase with ω_0, the choice of the RF frequency results from a compromise. The frequency range considered in this study report goes from about 10 GHz to 30 GHz, while the accelerating gradient is supposed to be between 50 and 100 MV/m. This corresponds to linacs suitable for either CLIC (CERN Linear Collider using a Drive Linac) or NLC (Next Linear Collider using pulsed RF generators) as envisaged at SLAC, KEK or IHEP. Table I gives considered values of the parameters discussed above, for the different proposals.

Table I. Examples of parameters

		E_0 (MV/m)	ω_0 (GHz)	$\hat{P}/L_0$ (MW/m)
CLIC		80–100	30.0	150
NLC:	SLAC (NLC)	50	11.4	60–240
	KEK (JLC)	100	11.4	
	IHEP (VLEPP)	100	14.0	

2 Beam Dynamics in the Linacs

2.1 Wake Fields and Single-Bunch Dynamics

As mentioned above, high-current beams induce in a high-gradient accelerator strong electromagnetic fields that increase rapidly with the RF frequency. These wake fields are responsible for energy loss, energy spread and transverse blow-up. Longitudinal wakes directly influence the distribution of energy inside the bunch, its contribution diminishing the accelerating field seen by the particles. Since these wakes are not uniform within a bunch, the energy loss is accompanied by an energy spread that induces variation of the focusing strength and dispersion of the trajectories inside the bunch in the presence of external magnetic quadrupoles. Transverse wakes deflect parts of the bunch and these deflections depend on the particle momentum and position with respect to the accelerator axis. Because of the energy spread and trajectory dispersion, dipole wakes produce kicks changing along the bunch and eventually producing an apparent beam blow-up (the "head" of the bunch deflecting the "tail"). Quadrupole wakes may also be present but they have been either considered as negligible or not yet studied in the different proposals.

Some compensation of the energy spread σ_E using wake potential versus RF sine wave is possible to high orders [4,5]. This is strongly desirable if the momentum acceptance at the exit of the linac is limited, as for instance in a final-focus system that typically accepts a $\Delta p/p$ of $\pm 2‰$ to $\pm 5‰$. The total accelerating gradient seen by a particle at position z in the bunch can be written:

$$G(z) = G_{RF} \cos(\omega_0 \frac{z}{c} - \phi_0) - W_L(z) \tag{8}$$

and the balance between the RF wave and the longitudinal wake will obviously depend on the phase ϕ_0, but also on the bunch charge N and bunch length σ_z that enter in W_L. Knowing G(z) it is possible to find ϕ_0 and σ_z values which minimize the spread σ_E for a given N. Furthermore, the energy distribution $\nu(E)$ can be calculated [5] in order to study its properties and dependence on G_{RF}, ϕ_0, σ_z and N, by using

$$\nu(E) = \frac{1}{N}\frac{dN}{dE} = \frac{1}{N}\frac{dN}{dz}\frac{dz}{dE} = \frac{\rho(z)}{dE/dz} = \frac{\rho(z)}{e\,dG(z)/dz} \tag{9}$$

where $\rho(z)$ is the charge distribution (Gaussian). Results concerning CLIC are given as examples [5]. Figure 1 shows G(z) for different charges N with $\phi_0 = 7–8.5°$ and $\sigma_z = 0.14–0.17$ mm. Figure 2 gives the corresponding energy distributions for $N = 5 \times 10^9$ and 6×10^9 per bunch. The flatter G(z), the smaller σ_E, and, in these two cases, the values obtained are about 4.6‰ and 7.7‰ respectively (in the absence of truncation of $\nu(E)$). Note also that this adjustment ends up with a minimal tail population and a somewhat narrow core size (Fig. 2).

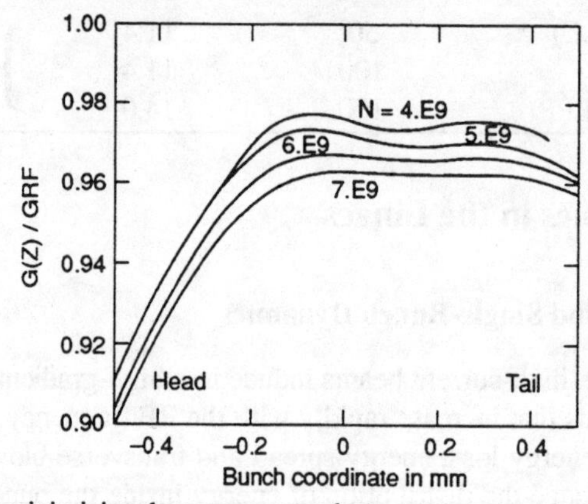

Fig. 1. Total gradient minimizing the energy spread σ_E

Transverse instabilities due to dipole wakes are all the more critical the smaller the emittance. In the linac of colliders, the vertical emittance may be very small since the beam is usually flat. Having a flat beam or a large aspect ratio $R = \sigma_x/\sigma_y$ at the collision point reduces the average energy loss due to synchroton radiation. This energy degradation due to beamstrahlung induces a large energy spread that should not, however, exceed the energy spreading related to background processes (imposing a limit of ~5% for tolerable σ_E from beamstrahlung). Furthermore, a large ratio R allows the avoidance of an excessive repetition rate that varies for constant luminosity and beam–beam radiation σ_E as follows:

$$f_{rep}(R) = f_{rep}(R=1)\frac{4}{RH_y} \tag{10}$$

where H_y is the vertical pinch enhancement factor (typically between 2 and 2.5). For these reasons, values of R and normalized vertical emittance $\gamma\varepsilon_y$ are respectively large and small in the proposals quoted (Table II).

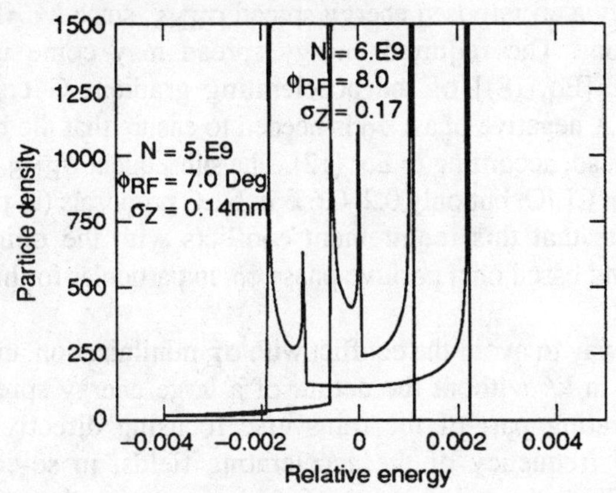

Fig. 2. Energy distribution minimizing σ_E for $N = 5{\cdot}10^9$ and $6{\cdot}10^9$

Table II. Beam aspect ratio at final focus and V-emittance

	R	$\gamma\varepsilon_y$ (rad m)
CLIC	10–20	$2{\cdot}10^{-7}$
NLC	100	$3{\cdot}10^{-8}$

The presence of strong dipole wakes (almost 20 times larger in CLIC than in NLC) implies large kicks originating from misalignments of the structure and off-centred trajectories. The corresponding equation of the transverse motion of single particles is given by

$$y'' + k_0^2 y = \left[k_0^2 - k^2(z,s)\right]y + \frac{r_0}{\gamma}\int_{-\infty}^{z} \rho W_T^{\delta}\left(z^* - z\right)y\left(z^*,s\right)\, dz^* \tag{11}$$

where k^2 characterizes the linac focusing (k_0^2 being its "average" value, independent of z) and the wake-field effect is integrated over the heading part of the bunch of charge distribution ρ. To counteract this effect, the idea consists of obtaining a coherent motion by imposing the same oscillation period to all particles. This condition, called autophasing by its author [6], comes directly from inspecting Eq. (11):

$$k^2(z,s) = k_0^2 + \frac{r_0}{\gamma}\int_{-\infty}^{z} \rho W_T^{\delta}\left(z^* - z\right) dz^* \tag{12}$$

All the proposals for high-gradient, high-frequency linear accelerators strive to satisfy condition (12) or its linearized version [7]

$$\left.\frac{\partial k^2}{\partial z}\right)_{z=0} = \frac{r_0}{\gamma} N \left.\frac{\partial W_T}{\partial z}\right)_{z=0} \tag{13}$$

to limit beam blow-up. There are basically two ways for achieving this variation of the focusing strength with the position z inside the bunch:

1) Using the external focusing of a magnetic FODO lattice, the change of k^2 with z can be obtained via an imposed energy spread σ_{BNS}, since $k^2 = k_0^2 / p$ if p is the particle momentum. The required energy spread may come in turn from the dependence on z [Eq. (8)] of the accelerating gradient G combined with an adjustment of ϕ_0. A negative phase ϕ_0 is needed to ensure that the bunch tail is more focused than the head, according to Eq. (12). The subsequent σ_{BNS} ensuring stability is as large as 5% in CLIC, but only 0.2–0.6% in NLC proposals (in proportion to their wake fields). Note that this requirement conflicts with the minimization of σ_E discussed above and based on a positive phase ϕ_0, in particular for high-RF frequency linacs.

2) An elegant way to avoid the conflict with σ_E minimization and simultaneously create the spread in k^2 without the detour of a large energy spread does exist. It consists of generating part of the transverse focusing directly from RF fields oscillating at the frequency of the accelerating fields, in so-called microwave quadrupoles [8]. Since the radial electric field in a narrow slit vanishes in the mid-plane, the effective magnetic gradient due to the axial electric field and deduced from Maxwell's equations is given by [8]

$$G_m(T/m) = \frac{\pi}{c\lambda_{RF}} E_0(MV/m)\sin\phi_1 \qquad (14)$$

where E_0 is the peak accelerating gradient, λ_{RF} the RF wavelength and ϕ_1 the RF phase angle measured from the top. In the direction perpendicular to the slit, the electric field is doubled (compared with circular aperture) and overcompensates the magnetic gradient by exactly a factor 2, thus forming a quadrupole. In practice an oval cavity with circular aperture (Fig. 3) is preferred to a circular cavity with slotted iris [8], in order to have the required radius and surface finish at the aperture.

Fig. 3. Microwave quadrupole cell with flat cavity (CLIC)

In CLIC, where transverse wakes are large, it is proposed to generate the spread of k^2 with microwave quadrupoles using a phase ϕ_1 close to the phase ϕ_0 that minimizes σ_E, i.e. running near the maximum accelerating voltage. In this solution [9], the main basic focusing k_0^2 is created by external magnetic quadrupoles, while the microwave quadrupoles are only responsible for the variation $k^2 - k_0^2$. Hence, transverse instabilities can be damped (Fig. 4) keeping the energy spread low.

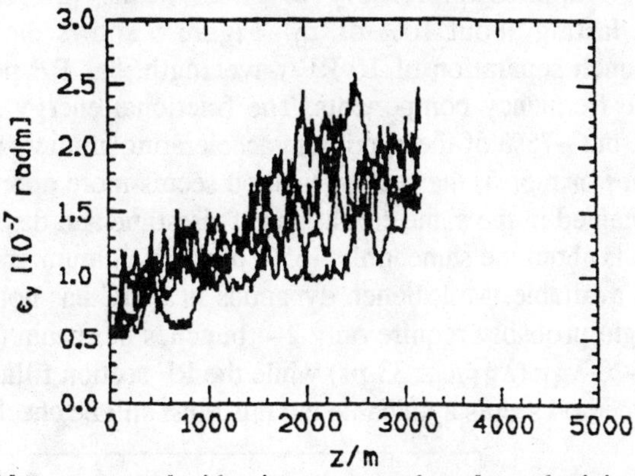

Fig. 4. Example of blow-up control with microwave quadrupoles and minimum σ_E

2.2 Wake Fields and Multibunch Dynamics

Accelerating long bunch trains that extend over a period comparable to the filling time τ_0 of a cavity section may provoke instabilities of the whole train along the linac, due to interactions with RF fields. The shape, timing and modulation of the RF pulse, as well as long-range longitudinal wake field, are responsible for bunch-to-bunch energy variations. The subsequent energy spread can cause filamentation and emittance growth beyond the acceptance at the exit of the linac. Interactions with resonant transverse electromagnetic fields in disk-loaded waveguides, in particular with the so-called HEM_{11} dipole modes, produce increasing deflections along the bunch train that drive a transverse instability, called beam break-up.

The bunch-to-bunch energy variation produces effects that are more severe with long trains. The fundamental mode, as well as high-order modes of longitudinal wakes, is at the origin of inter-bunch beam loading and the actual RF pulse influences the energy spectrum. There is consequently a need to control bunch-to-bunch energy spreads and some compensation schemes have been studied [10,11]:

1) Matched filling, i.e. adjustment of the injection timing of the bunch train with respect to the RF pulse and appropriate choice of the bunch spacing. The idea is to have sufficient extra energy in the RF section fill between bunches to cope for the energy lost in accelerating the preceding bunches.

2) Staggered timing, i.e. delay of a subset of klystrons so that some accelerating sections are only partially filled during build-up of the beam-loading voltage to its steady-state value. The number of delayed klystrons is selected to produce a voltage equal to about twice the steady-state beam-loading voltage in the linac.

3) Modulation of RF input, i.e. phase adjustments or small klystron variations during the time when the bunch train is passing through a cavity section. This makes use of the propagation out of the section of the leading edge of the pulse while the train is passing over.

The results of such compensation schemes depend on the bunch length with respect to the section filling-time and on the bunch separation. In the NLC for instance [10], the first method applies preferably to short trains (10 bunches of 10^{10} particles/bunch, lasting about 10% of τ_0). Figure 5 shows the energy deviation obtained with bunch separation of 16 RF wavelength, 5ns RF pulse rise-time and dispersion of RF frequency components. The fractional energy deviation remains below about 3‰ but ~75% of the maximum accelerating gradient is used. With long trains (70 bunches or more), the second method seems more appropriate and Fig. 6 shows results obtained in the same conditions of rise-time and dispersion. If the total energy deviation is about the same, only ~65% of the maximum gradient (assumed to be 50 MV/m) is available. Multibunch dynamics in CLIC has not yet been studied. Performance might probably require only 2–4 bunches per beam (perhaps up to 10) separated by 10–20 λ_{RF} ($\lambda_{RF}/c \cong 33$ ps) while the RF section filling-time is 11.1 ns. Hence, the first method seems applicable but this must still be checked.

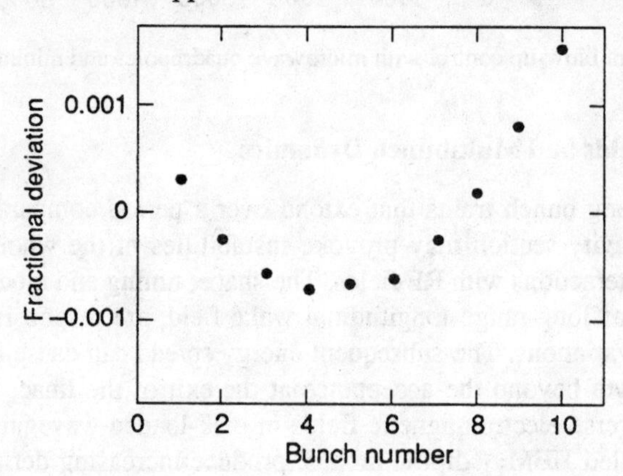

Fig. 5. Fractional energy deviation in a short train after matched filling [10]

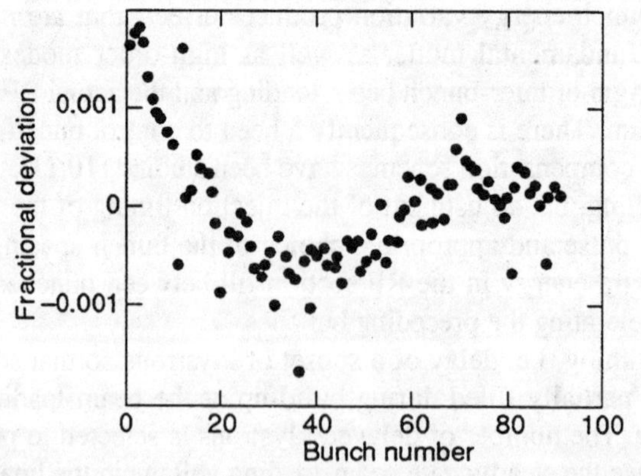

Fig. 6. Fractional energy deviation in a long train after staggered timing [10]

Cumulative beam break-up due to long-range dipole modes of transverse wake fields can be severe. The transverse beam modulation is carried along the linac from accelerating section to accelerating section, through the beam. Blow-up then occurs and manifests itself along the linac as an amplitude growth from the head to the tail of the bunch train. Two possible cures have been investigated [12,13,14]:

1) damped structures; modified disk-loaded waveguides in which the power of the wake field modes is coupled out to lossy regions through radial slots in the disks and/or azimuthal rectangular waveguides, whereby the external Q-factor of the undesirable HEM mode is lowered to values typically below 20. This was studied in SLAC and KEK, the latter requiring Q-values between 15 and 70 for the predominant TM_{11} modes [15] (in order to limit the emittance growth within a factor $\sqrt{2}$ and alignment tolerances within 80 μm). The limitation of this method might come from the low Q-values required and the large number of cells involved, the practical difficulties increasing with the RF frequency.

2) staggered tuning [12]; variation in the cell dimensions in each accelerating section resulting in a cell-to-cell spread (by a few per cent) of the dipole mode frequencies. These modes are split into N_f frequency-components, whose distribution can be varied. The best frequency distribution seems to be a truncated Gaussian, since the initial roll-off of the wake is strong, with low partial recoherence within the length of a (short) bunch train.

Whilst the fabrication of damped structures has been tested [14], the possibility of staggered tuning has been investigated by numerical simulations [16] and experimental measurements [17]. They concern a detuned 50-cavity disk-loaded structure with iris diameter ranging from 0.83 cm to 1.22 cm and Gaussian HEM frequency population centred at 14.45 GHz for a standard deviation of 1.07 GHz. The Advanced Accelerator Test Facility at Argonne makes it possible to measure the energy variation and the horizontal position of a witness bunch following the driving bunch in a time interval between 0 and 1 ns. The former yields the longitudinal wake potential and the latter gives the transverse potential as the structure is swept horizontally. Figure 7 compares calculation with experiment [17] and confirms roll-off expectation in this particular case, though it is not clear if recoherence takes place at larger distances z from the driving bunch (in NLC bunch separation is about 4 cm and a "short" train would cover ~ 3.5 m). No emittance blow-up simulations are known to the author for RF frequencies between 10 and 30 GHz, but the gain expected from staggered tuning can be illustrated by simulation results [18] obtained with 180 bunches, at lower frequency (3 GHz) and with $N_f = 25$ (Fig. 8). The effectiveness of staggered tuning is visible from the fact that without it about half of the bunches have too large amplitudes at the end of the linac, and from the difference in the scale used for the top picture (mm) and for the bottom one (μm).

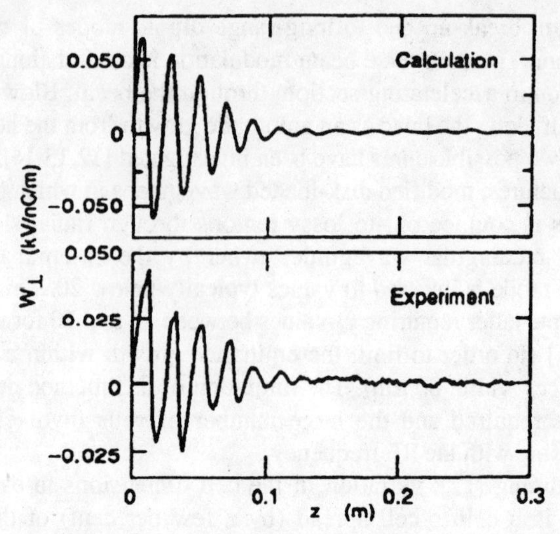

Fig. 7. Calculated and measured transverse wake potential for a detuned 50-cavity structure [17]

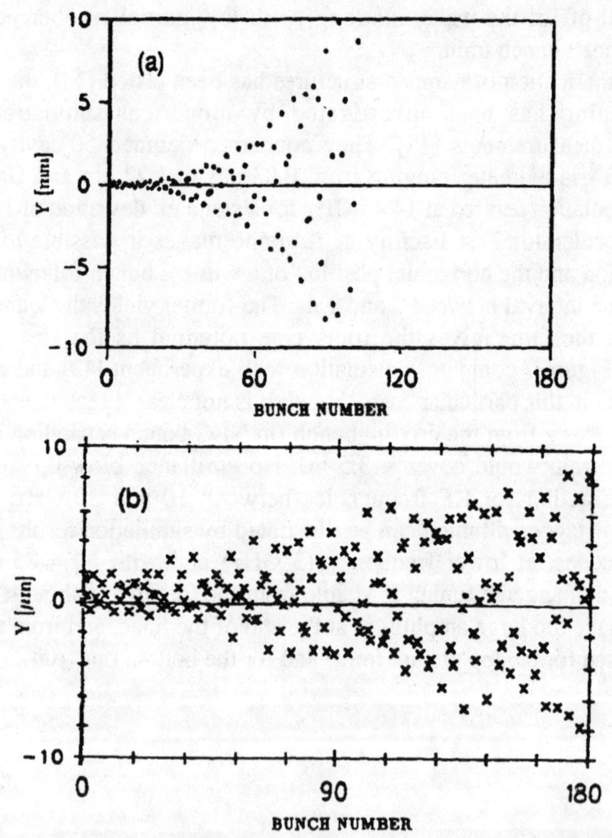

Fig. 8. Transverse bunch offsets at linac end, without (a) and with (b) staggered tuning [18]

3 Linac Hardware Challenges

3.1 Constraints on the Fabrication of Accelerating Sections

The requirement for high gradient at high frequency calls for tight tolerances in the fabrication of the accelerating cavities. Approximately the same principles have been adopted in the design of the accelerator sections proposed at CLIC and at the JLC or VLEPP versions of NLC. The most promising manufacturing method, tested at CLIC, is the brazing of machined copper cups, and the construction of prototypes of CLIC structures [19] proved that this technique could be successfully acquired by industry.

For the CLIC 30-GHz structure, the tolerances on the cell dimensions (iris diameter of 4 mm and cell diameter of 8.7 mm) are of the order of 2–5 μm (in order to limit the total phase error to 5° over a section) and on the surface finish of about 0.05 μm (to obtain 95% of the nominal Q-value). The high-precision copper cups were made on Pneumo diamond-tool lathes, with laser interferometric feedback with 25 nm resolution. The machining accuracy consistently achieved by manufacturers would eliminate the need for dimple tuning, since the phase shift error was approximately 0.1^o/cell. High-quality brazed joins were produced, with two diffusion-bonded annular surfaces at the inside and outer edges of the disks, to prevent braze leakage. Unacceptable frequency changes could be avoided during brazing operations. Complete structures (Fig. 9) with radial holes and vacuum manifolds for pumping and channels for cooling have been manufactured and measurements in the laboratory confirmed the expected parameter values [19]. In the JLC design, for example, damping slots could be incorporated in addition.

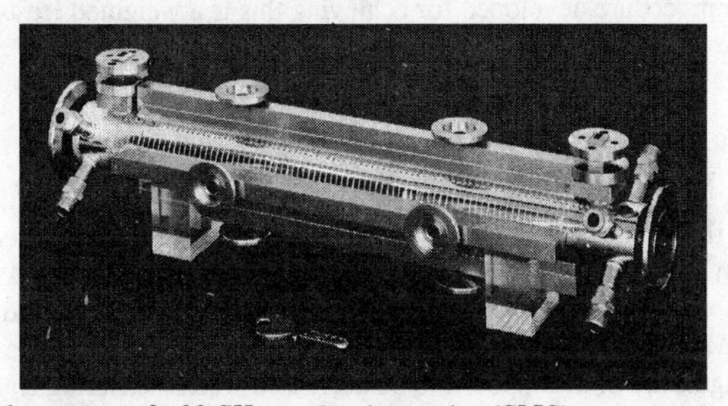

Fig. 9. Finished prototype of a 30 GHz accelerating section (CLIC)

Field-emitted electrons can create multipactoring resonance discharge at well defined field levels, and also dark currents due to the capture, bunching and acceleration of these electrons, eventually producing a parasitic beam. Recently, work was carried out on possible scaling laws [20] for these phenomena with the operating frequency. Basically, both phenomena scale with the frequency ω_0 for exact geometry scaling, but are influenced by the cavity shape near the axis for the same iris opening a. Working at high frequency with a large iris opening (or large a/λ_{RF} ratio) should be favourable in this respect. Numerical estimates [20] for CLIC cavities at 30 GHz indicate absence of electron capture and dark currents up to a gradient as high as 1400 MV/m. A possible limitation around 100 MV/m seems rather to come from second- or third-order multipactoring, in which emitted electrons lose energy by successive impacts, drift to the outer diameter, and eventually give rise to a breakdown of the structure. Further investigations are needed to better understand these phenomena that already look more critical at the lower end of the frequency interval considered.

3.2 Tolerance on the Alignment of Linac Elements

The dominant effect associated with misalignments is the blow-up of the projected emittance resulting from the incoherent motions due to transverse dipole wake fields (as already mentioned above). The acceptable growth of the normalized emittance $\gamma\varepsilon_y$ may vary from only 14% to a factor 3 or 4 depending on the proposal requirements. The subsequent alignment tolerances can be deduced from these numbers assuming first a simple one-to-one trajectory correction aiming at centring the beam in each position monitor (BPM) by moving the preceding lattice quadrupole. To give examples, the corresponding alignment tolerances for CLIC and NLC (SLAC version) are the following (r.m.s. values):

CLIC 3 μm on quadrupoles, 5 μm on structures,
NLC 7 μm on quadrupoles, 4 μm on structures.

In order to relax these tolerances while keeping the same acceptable emittance growth, a trajectory correction more efficient than the one-to-one method has to be applied. SLAC [21] proposed compensating for the dispersion while correcting the orbit and minimizing the wake-field dilutions caused by the corrected trajectory. The minimization procedure developed for achieving this is a weighted least-squares that minimizes the following sum:

$$\Phi = \sum_{j\in\{BPM\}} \frac{(\Delta y_{QF})_j^2}{2\sigma_{prec}^2} + \frac{(\Delta y_{QD})_j^2}{2\sigma_{prec}^2} + \frac{y_j^2}{\sigma_{al}^2 + \sigma_{prec}^2} = \Phi_{min} \qquad (15)$$

where y_j is the corrected trajectory amplitude, Δy_{QF} the difference trajectory resulting from both QF-variations and corrector adjustments (idem for Δy_{QD}, related to QD-variations however), σ_{prec} is the r.m.s. precision of the BPM readings and σ_{al} the r.m.s. BPM misalignments. Since the QF- and QD-fields are opposite and the QF- and QD-strengths are both supposed to be decreased when measuring Δy_j, the sum of these two variation terms mimics the effects of the dispersion (trajectory shift with momentum or quadrupole-strength deviation), while their difference mimics the effects of the wake field (sign depending only on the side where the trajectory is off-centred and not on the quadrupole field polarity). With this algorithm, the NLC (SLAC) alignment tolerances could be relaxed to 70 μm for both the quadrupoles and the accelerating structures. If all the quadrupoles are simultaneously detuned, only one difference Δy corresponding to the dispersion appears in the function Φ. Minimizing it was called dispersion-free correction while minimizing (15) was termed wake-free correction by its author [21]. Figure 10 shows beam distributions after one-to-one, dispersion-free and wake-free corrections in NLC. Plots on the left are projections onto the y-y′ phase plane, while right-hand plots are projections onto the y-z plane (z: longitudinal coordinate). One sees the strong dilution in (a), reduced to wake-field tail bend in (b) and minimized in (c).

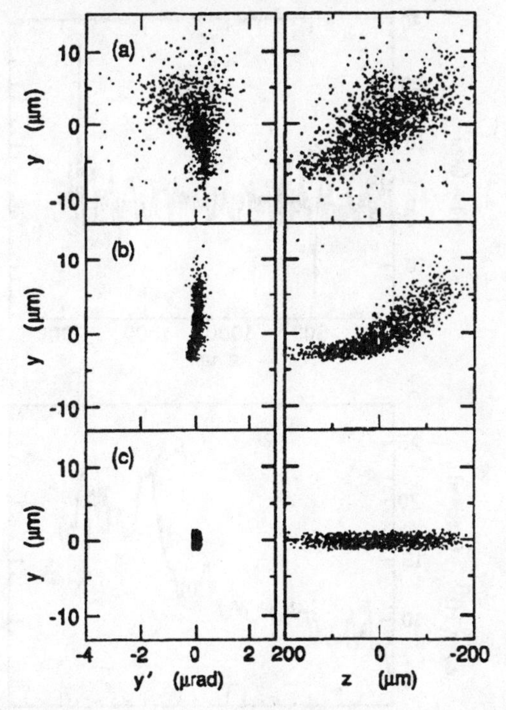

Fig. 10. Beam distributions after one-to-one (a), dispersion-free (b), and wake-free (c) correction in NLC [21]

CLIC started from this idea and focused on an achromatic trajectory correction [22] to higher order, developing the trajectory differences Δy_j (measured) and ΔY_j (due to corrections) in $\delta = \Delta p/p$,

$$\Delta y_j = \sum_n a_n^j \delta^n \ , \quad \Delta Y_j = \sum_n A_n^j \delta^n \tag{16}$$

Calling y_j the measured trajectory and Y_j the one due to corrections, the function to be minimized becomes,

$$\Phi = \sum_{j \in \{BPM\}_y} \left\{ w_0 \frac{y_j + Y_j}{\sigma_{al}^2 + \sigma_{prec}^2} + \sum_{n \geq 1} w_n \delta^{2n} \frac{a_n^j + A_n^j}{2\sigma_{prec}^2} \right\} = \Phi_{min} \tag{17}$$

where the sum applies to the quadrupoles focusing in the plane considered, in order to avoid too large a wake dilution. Each variation term of the second sum in Eq. (17) represents the dispersion order by order. In the applications, second-order corrections are actually implemented. The efficiency of this global correction is illustrated in Fig. 11 showing the residual trajectory and the emittance dilution in a sector of CLIC, for 5 μm alignment tolerances on the quadrupoles [23]. Combining this with a one-to-one algorithm realigning the quadrupoles with the beam and a peculiar lattice scaling, the CLIC alignment tolerances could probably be relaxed to about 50 μm for the quadrupoles and 10 μm for the accelerating structures.

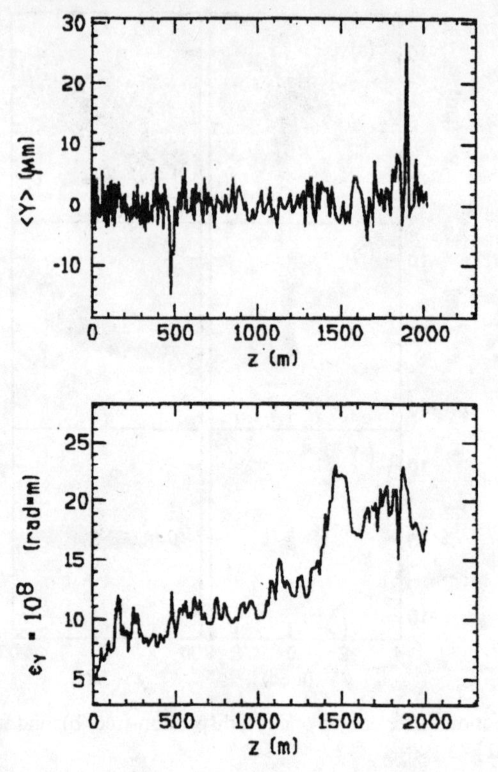

Fig. 11. Trajectory (top) and blow-up (bottom) after global correction in a 2 km-long CLIC sector

4 Power Generation

4.1 Microwave Klystrons as RF Power Sources

Conceptual designs of linacs for future colliders operating around 11 or 14 GHz call for microwave klystrons able to deliver as much as 100–200 MW in pulse lengths of the order of 1 μs. These requirements cannot be satisfied with existing microwave tubes and new klystron designs meet a certain number of challenges briefly recalled hereafter. The maximum power capability is limited by the area available to dissipate beam or RF losses and shrinking with the inverse of ω_0^2. Good power transfer efficiency from the beam to the output circuit and possible release of the intrabeam space charge forces favour high RF voltage V as well as low perveance defined by $I/V^{3/2}$; this implies a large RF gradient across the output gap. Permitting greater beam current I and power makes it necessary to achieve a higher beam convergence that involves better confinement and more precise beam optics. Finally, these high-current, high-voltage conditions increase the risk of failure mechanisms limiting the power; this concerns possible RF breakdowns mainly in the output circuit as well as intrapulse heating due to beam interception, the two mechanisms being interrelated. Typical figures considered are 40–50% for the power transfer efficiency, about 1 or 2 $\mu A/V^{3/2}$ for the perveance, beam convergence as high as 200, a gradient lower than 6 MV/cm in the output gap and a scraped beam fraction below 1%.

There are different means envisaged for trying to find solutions. Working at low perveance by increasing V is an important element for beam control, as empirically demonstrated for operational RF sources. Using excellent beam optics near the output circuit and reducing gap voltages by replacing the resonant cavity by multiple gaps or travelling-wave output are possible improvements. The klystron gun voltage can be divided by many intermediate anodes in order to provide the correct potential profile for beam formation and focusing, as in the VLEPP klystron design [24]. In this intricate design, the beam is composed of many separate "beamlets" produced by different regions of the cathode and is switched by a non-intercepting control electrode (offering the possibility of pulsing the klystron using a quasi-d.c., high-voltage supply and a low-voltage modulator in series with the grid). In order to achieve very low perveance, a "cluster klystron" is proposed by a BNL–SLAC collaboration [25]; it is a collection of 42 separate beams, each comprising a 40-MW klystron and all sharing a common superconducting solenoid. Finally, the use of lumped shavers in order to control beam interception might be necessary.

Some characteristics of high-power klystron development projects, incorporating (separately) the features mentioned, are given in Table III. Achievements so far, as quoted in a recent review paper [26], are briefly summarized below, together with the reasons for the limitation observed. The SLAC XC klystron achieved 40 MW, 0.8 μs or 72 MW, 0.1 μs pulses, performance being limited by RF breakdown in the double-gap output circuit stimulated by beam interception. In the KEK klystron, the output power was restricted to 22 MW by failure of the gun ceramic insulator. In the VLEPP klystron, 50 MW has been achieved, limited by a substantial beam loss that amounts to 70 out of 200 A. The cluster klystron is still at the design stage.

Table III. High-power klystron projects

Klystron	Wave-length (cm)	Power out (MW)	Pulse length (μs)	Voltage (kV)	Microper-veance ($\mu A/V^{3/2}$)	Frequency (GHz)
SLAC 5045	10.5	67	3.5	350	2.0	2.86
SLAC	10.5	150	1.0	450	2.0	2.86
SLAC XC	2.6	100	1.0	550	1.2	11.4
KEK	2.6	120	1.0	550	1.2	11.4
VLEPP	2.1	150	1.0	1000	0.3	14.2
Cluster	2.6	1680	1.0	400	0.4	11.4

4.2 Drive Linac as High-Frequency Power Source

Owing to difficulties met in the development of microwave klystrons operating in the frequency range of 11–14 GHz, it is unthinkable to use similar tubes to deliver the 30-GHz, ~150 MW/m peak power per unit length required in the CLIC linac structure. The CLIC scheme for generating the necessary power is based on a two-stage accelerator [27]; there is a drive beam that runs parallel to the main beam and ensures power flow to the main linac. The drive linac contains strings of travelling-wave transfer structures, in which short and intense bunches induce the required power that is then fed to the main linac, and sectors of superconducting (SC) cavities supplying energy to the beam when necessary. The use of SC cavities to reboost the drive beam as well as to accelerate it up to its initial energy is dictated by the concern of good extraction efficiency; LEP-type cavities operating at ~ 350 MHz with a gradient of ~ 6 MV/m are considered. The drive-beam energy should be of the order of a few GeV, which implies no longitudinal mixing inside the bunches and no phase slip with respect to the main beam. Owing to the unavoidable intermittent reacceleration, the drive beam has to be arranged in discrete trains of dense bunchlets that all contribute to the build-up of a decelerating field in the transfer structure and are separated by the 30-GHz wavelength. To generate a pulse of length equal to the main structure filling time τ_0 (11.1 ns), four such trains are needed, separated by the 350-MHz wavelength. The total charge needed per train is about 1.65 μC and the number of bunchlets per train depends on the matching of the decelerating field build-up to the SC cavity accelerating gradient. Simply minimizing the deviation of the linear build-up from the sinusoidal acceleration wave limits this number to 11, but the use of voltage harmonics schemes [28] (their sum giving almost a linear function as shown in Fig. 12) might allow for 43 bunchlets (with correspondingly lower charge per bunchlet).

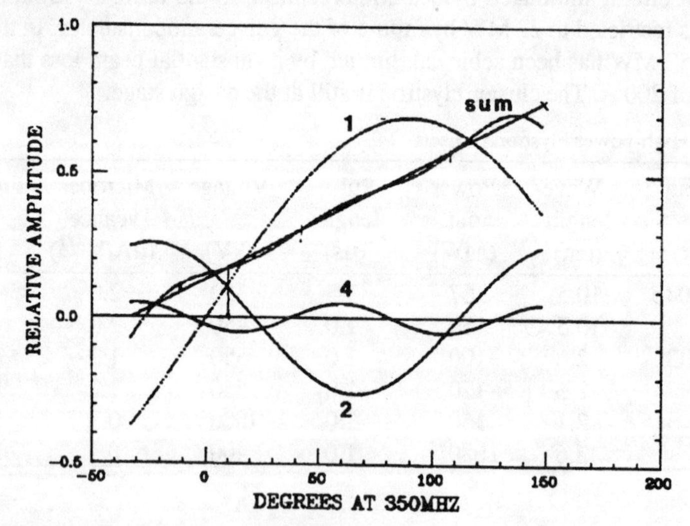

Fig. 12. Accelerating ramp obtained with 3 RF harmonics

This drive-linac conceptual design implies certain challenges which have been addressed, mainly the bunchlet generation, the transfer structure design and the beam dynamics control. Owing to the difficulty of generating short (1 mm r.m.s.) and dense (up to 10^{12} particles) bunchlets, a test facility (CTF) has been built [29]. It includes an RF gun, a beam line acting as magnetic spectrometer, acceleration to 60 MeV at 3 GHz and RF power generation at 30 GHz (using prototype structures). A charge up to 30 nC per beam should be obtained using a laser-driven photocathode, synchronized with the RF. During first tests using a prototype of the main linac structure instead of an actual transfer structure, a 2.7 MW–12 ns power pulse was extracted. The most recent transfer structure design proposed [30], is based on power-collecting rectangular waveguides that run along the outside of the beam pipe (either on each side or above and below) and are coupled to the inside via slits about 0.5 m long (Fig. 13). The phase velocity in the waveguide is adjusted by periodic indentations. Numerical calculations and model work are being carried out in parallel to check the possibility of generating the required flat power pulse and the amplitude of the wake fields. This design must indeed provide the low impedance needed (~ 4.5 Ω/m for R'/Q) and the required decelerating field per bunchlet (~ 65 kV/m for a population of 10^{12}), while minimizing the undesirable wake-field modes that could compromise beam stability. The dynamics of such a beam includes special features: the energy differences between bunchlets are unusually large owing to the increasing decelerating field, the energy spread within each bunchlet is wide since the bunch length is not short with respect to λ_{RF}, and the amplitudes of the synchronous wake fields may be disturbing. The impact of these features on the beam transport and dynamics, in the presence of alignment imperfections, trajectory correction, variable wakes and magnetic focusing, is being investigated by numerical simulations [31]. Figure 14 gives an example of initial- and final-energy distributions in a train of 11 bunches travelling over ~ 3.5 km. Phase plots of the emittances (Fig. 15) show that all the bunchlets remain within the beam-pipe acceptance (circle tangent to the frame), with the assumptions retained for this calculation. Further investigations are needed to check the feasibility of the scheme.

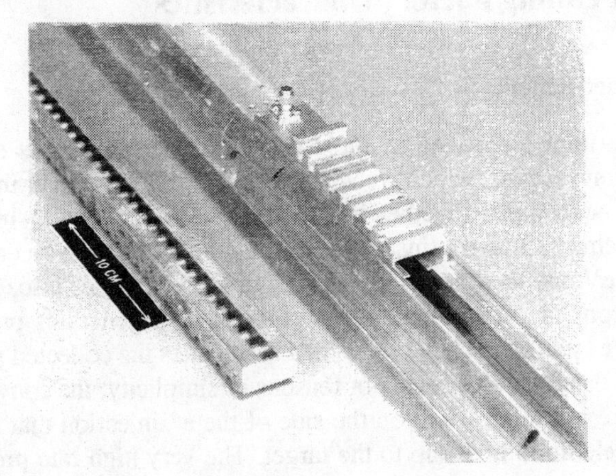

Fig. 13. Model of the most recent transfer structure design

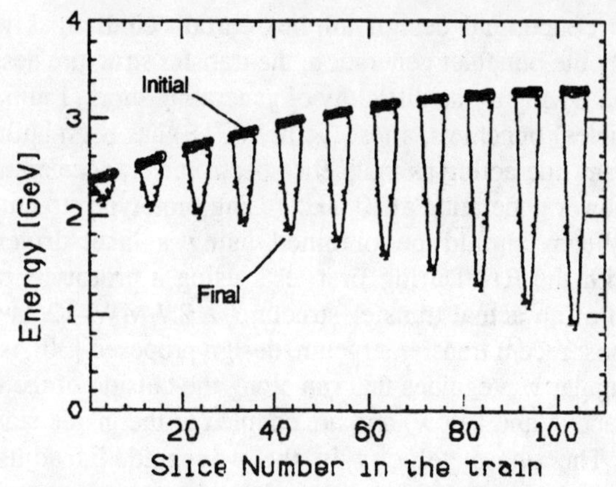

Fig. 14. Initial and final energy distributions in a drive-beam train

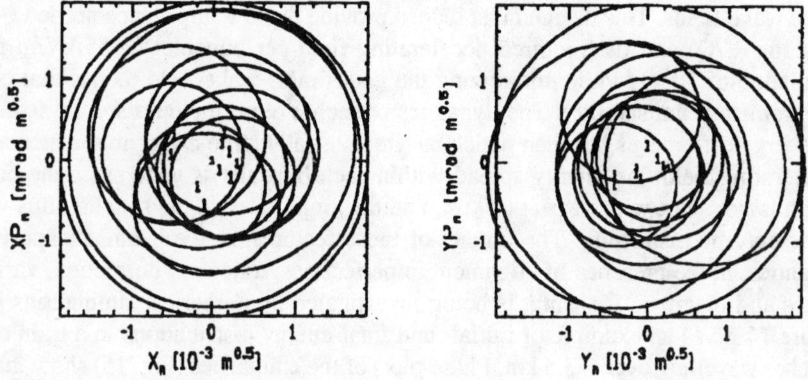

Fig. 15. Bunchlet emittances (H and V) at the drive-linac exit

5 Attempt at Defining Factory Characteristics

5.1 Linear Collider Schematic Layout

A possible layout often envisaged in the Linear Collider studies is sketched in Fig. 16. The injection systems, which we will not describe in this paper in any detail, could for instance be grouped together in a central position. They include two RF guns producing electrons, primary linacs accelerating the particles to an energy of the order of 1.5 to 3 GeV and damping rings running at these energies also. To produce the required positrons, a converter target is required that generates photons then transformed into e^+e^- pairs. A second linac then accelerates the collected positrons to the same energy as the main electrons. For reasons of simplicity, the converter target and second linac are not represented on the side of the e^+ injection that must begin with a primary acceleration of e^- up to the target. The very high rate production of the particles is often a challenge, mainly for the positrons (~ 10^{14} $e^\pm$/sec in CLIC, for instance). Damping rings are imposed, at least for the positrons, in order to reach the tiny transverse phase-space dimensions needed. Longitudinal dimensions required are also small and one or two bunch compressors have to be added at the exit of the damping rings to further reduce the bunch length. The first compressors acting at the ring energies are followed by pre-accelerations to something like 9 to 16 GeV in preliminary linacs (at 3 GHz), in order to limit the energy spread in the second compressors. It has been proposed to put the second compressors at the extremities of a kind of very long trombone (implying transfer lines behind the preliminary linacs) which could be extended by stages the day the final energy of the collider is raised. At the exit of these compressors and the injection into the main linac, the main beams must have the required characteristics that lay in the following interval approximately

$$\sigma_z \cong 0.1\text{–}0.2 \ \text{mm}$$
$$\gamma\varepsilon_x \cong 2\text{–}5 \times 10^{-6} \ \text{rad} \cdot \text{m}$$
$$\gamma\varepsilon_y \cong 5\text{–}20 \times 10^{-8} \ \text{rad} \cdot \text{m}$$

considering X-band colliders based on small spot sizes at the I.P. and small beam power. In this case, the main beams are then accelerated to the final energy through X-band linacs (10–30 GHz) with high gradient (50–100 MV/m). At the end of the linacs, the beams traverse the final-focus system and are focused at the I.P. to dimensions of the order of σ_x = 70–300 nm and σ_y = 3–16 nm. Multibunch operation (above 1 and up to 90 bunches say) might be necessary to reach the desired luminosity and the corresponding beam power would be around 2–4 MW/beam. The experimental areas and the beam dumps could possibly be located in the neighbourhood of the injection systems, as sketched in Fig. 16.

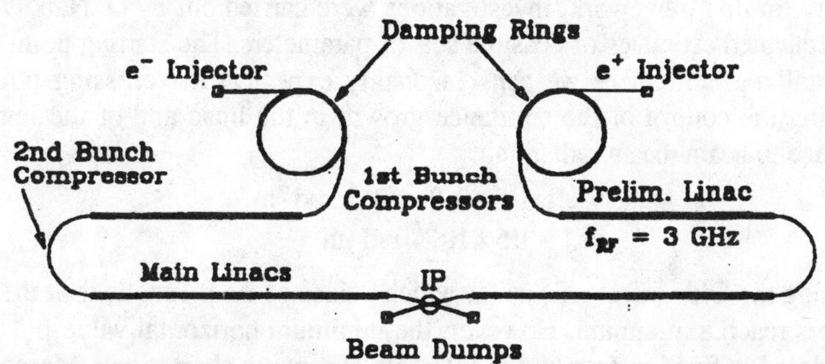

Fig. 16. Sketch of linear collider layout

5.2 Guessed Parameters and Performance for Linear Colliders as Factories

For the purpose of this paper, and to give indications to the reader about what looks possible today, guesses were specifically made on the basis of the CLIC feasibility studies, mainly for the final-focus system [2]. For the estimates of the relevant parameters and the projected performance, three levels of centre-of-mass energy have been retained, in agreement with the interests of the experimental physics:

1) $\sqrt{s} = 100$ GeV Z^0 factory
2) $\sqrt{s} = 250$ GeV   Top factory
3) $\sqrt{s} = 500$ GeV   Higgs factory

The main linacs would have increasing lengths, using the idea of the trombone, with the same injection and interaction point complex, for instance. Assuming an accelerating gradient of 80 MV/m, the active length of each linac would be between 625 m for case 1 and 3125 m for case 3. Adding approximately 20% for the unavoidable focusing quadrupoles and for the drift spaces would raise this length to 750 m and 3750 m, respectively. Ultimately, adding the space required by the final-focus system (400 to 450 m) would bring the length per "trombone" to something like 1200 m in case 1 and 4200 m in case 3.

Assuming that the properties of the beam at the exit of the linacs are those required, the linear collider performance depends on the adjustment capability of the final-focus system, for the energies considered. We have already mentioned the two main parts of this system, the telescope with a large demagnification ($\tilde{>}$ 50) and the chromatic correction sections with cells containing sextupoles and a phase shift of π. Aberrations and momentum bandwidths limit the maximum β-value acceptable in the quadrupoles, hence the minimum β-value (β^*) at the crossing point. Next, the synchrotron radiation in the last quadrupole before the interaction means that the vertical beam size $\sigma_y{}^*$ at the I.P. (smaller than $\sigma_x{}^*$ by choice) goes through a minimum even if β_y^* is further reduced (this phenomenon known as the Oide effect limits the benefits of strong focusing). Finally, limiting beamstrahlung energy-spread implies, as we have seen, flat beams and conditions on $\sigma_y{}^*$ to reach the desired luminosity. In this framework, investigations were carried out by O. Napoly [32] to provide educated estimates of possible sets of parameters. The starting point was the two normalized emittances we can reasonably expect at the crossing point after considering the control of the emittance growth in the linac and of the luminosity dilution due to beam–beam radiation,

$$\gamma\varepsilon_x \cong 1.5-2.0\times10^{-6} \text{ rad}\cdot\text{m}.$$
$$\gamma\varepsilon_y \cong 1.5\times10^{-7} \text{ rad}\cdot\text{m}$$

Analysing the Oide effect tells us for which values of the β-functions at the I.P. the beam sizes reach a minimum. However, the minimum horizontal value β_x^* that can be considered is fixed in fact by the tolerable chromatic aberrations. Moreover, the control of the average energy spread during collision (associated with beamstrahlung) implies the retention of β_x^*-values depending on the centre-of-mass energy. Therefore, the β_x^* are all different for the energy levels we are interested in and for the energy resolutions δ_E we are aiming at. Then, in the vertical plane, the condition that the β_y^*-value must not be smaller than the bunch length σ_z (0.17 mm in CLIC) imposes its value to ~ 170 μm. Hence, both β^* are above the Oide limit in all cases.

When all these final-focus parameters are selected, it is possible to calculate the nominal luminosity per bunch using the simple formula

$$\mathcal{L}_b^n \sim \frac{f_{rep}\, N^2}{4\pi\, \sigma_x^*\, \sigma_y^*} \tag{18}$$

On the one hand, correct tracking that includes aberrations, synchrotron radiation in quadrupoles, real particle distributions and losses tends to give luminosity values lower than $\mathcal{L}_b^n$ (18), on the other hand, pinch effect associated with the focusing due to fields from the other interacting bunch enhances the luminosity with respect to $\mathcal{L}_b^n$ by a factor of the order of 2 for flat beams. Altogether, the nominal value $\mathcal{L}_b^n$ gives a slightly pessimistic estimate, too low by only 20% say, a difference that can be neglected in our approximate projections.

Using the final-focus criteria quoted above and the equation (18) for a first estimate of the luminosity per bunch, it is possible to make lists of guessed parameters and performance for the three "factories" imagined above (Table IV). For the lower-energy factories (Z^0 and top) physics may require good energy resolution and small energy spread δ_E; therefore the price to pay in luminosity when reducing δ_E is also indicated in Table IV. For the Higgs factory however, there is no point in reducing δ_E below the limit of about 5% coming from the always-present 2γ events.

Table IV. List of guessed parameters at different energies

	Z^0 factory	Top factory	Higgs factory
$\sqrt{s}$ (GeV)	100	250	500
σ_y^* (nm)	16	10	8
σ_x^* (nm)	70	70	90
β_x^* (μm)	327	600	2200
δ_E (%)	4.2	7.3	5.9
$\mathcal{L}_b^n$ ($cm^{-2}s^{-1}$)	4×10^{32}	7×10^{32}	6.8×10^{32}
σ_x^* (nm)	280	300	–
β_x^* (mm)	5.2	11	–
δ_E (%)	0.28	0.55	–
$\mathcal{L}_b^n$ ($cm^{-2}s^{-1}$)	1.1×10^{32}	1.6×10^{32}	–

Speculating about the possibility to reduce by staggered tuning, in multibunch mode, the effects of the transverse wake fields on the follower-bunches of a train in a CLIC-type collider, one could imagine to have up to 11 bunches per RF pulse. These bunches would be separated by 10 RF periods perhaps, last for a maximum of 3.3 ns (compared with 11 ns filling time), and increase proportionally the total luminosity. In the most optimistic case, the total luminosities would range from 1.2 to 4.4×10^{33} $cm^{-2}s^{-1}$ for Z^0-factories, from 1.8 to 7.7×10^{33} for top factories and reach 7.5×10^{33} for so-called Higgs factories. These approximate projections made as an educational exercise show that the performance one can hope to expect is in promising agreement with the physics requirements.

6 Conclusions

The choice of the RF-frequency in a high-gradient normal-conducting linac will probably fall in the 10–30 GHz interval and result from a compromise between the opposing requirements of saving power and minimizing harmful effects. A lot of studies and simulations improved the knowledge of the mechanisms involved in wake field effects and beam stability in the presence of one bunch or several bunches per beam. They revealed a number of promising correction possibilities aiming at low energy-spread and small transverse emittances at the exit of the linac. The question of the tight alignment tolerances required remains challenging, but looks solvable if good alignment strategies are defined. The idea of using microwave quadrupoles for stabilizing the beam in the presence of strong wake fields has been reinforced by numerical simulations. Model work on high-frequency accelerating structures proved that they could be manufactured by industry and recent studies indicate that the risk of dark currents decreases at higher frequency. Microwave klystrons for 250-GeV and 50-MV/m linacs seem feasible in the near future, as do the companion RF-pulse compression systems. Prospects for the peak power requirements at 500 GeV and 100 MV/m are however more distant, since technical limitations have to be overcome. For the CLIC scheme based on a two-stage accelerator, significant progress has been made on the study of the drive-beam dynamics and the design of the transfer structures. In this case, challenging issues are the generation of the required short and dense bunchlets, and the development of efficient transfer structures that are least harmful to the drive beam. Workable experimental conditions seem achievable and designs of final-focus systems have been worked out. Projected performance of linear colliders as factories was estimated for centre-of-mass energies corresponding to the production of Z^0, top and Higgs particles. It gives the encouraging result that the physicists' expectations can possibly be satisfied.

Acknowledgements

The author is particularly grateful to K. Berkelman, C. Fischer, N. Holtkamp, T.L. Lavine, O. Napoly, K.A. Thompson, L. Thorndahl and I. Wilson who have kindly made their most recent results available to him.

References

[1] P. Zerwas, Proc. ECFA Workshop on Linear Colliders, Garmisch-Partenkirchen, 1992.

[2] O. Napoly, CLIC Note 144, CERN, 1991; CLIC Notes 155 and 166, CERN, 1992; O. Napoly et al., Proc. Part. Acc. Conf., San Fransico, 1991.

[3] K. Berkelman, CLIC Note 154, CERN, 1991; CLIC Notes 164 and 168, CERN, 1992.

[4] R.B. Palmer, SLAC-PUB-4295, 1987; W. Schnell, SLAC/AP-61, 1987.

[5] K.L.F. Bane, SLAC-AP-76, 1989; G. Guignard and C. Fischer, Proc. Part. Acc. Conf., San Francisco, 1991, Vol. 5, p. 3231.

[6] V.E. Balakin, Inst. of Nucl. Phys., Novosibirsk, Preprint 88-100, 1988.

[7] H. Henke and W. Schnell, CERN-LEP-RF/86-18, 1986.

[8] W. Schnell and I. Wilson, Proc. Part. Acc. Conf., San Francisco, 1991, Vol. 5, p. 3237.

[9] G. Guignard, CERN SL/91-19 (AP), 1991.

[10] K.A. Thompson, SLAC AAS Note 71, 1992.

[11] R. Ruth, SLAC-PUB-4541, 1988.

[12] R.B. Neal (editor), The Stanford Two-Mile Accelerator (W.A. Benjamin Inc., New York, 1968).

[13] R.B. Palmer, Proc. DPF Summer Study, Snowmass, SLAC-PUB-4542, 1988.

[14] H. Deruyter et al., Proc. Linear Acc. Conf., Albuquerque, 1990, p. 132.

[15] T. Higo et al., Proc. Linear Acc. Conf., Albuquerque, 1990, p. 147; T. Taniuchi et al., KEK Preprint 91–152, 1991.

[16] K.A. Thompson and J.W. Wang, Proc. Part. Acc. Conf., San Francisco, 1991, Vol. 1, p. 431.

[17] J.W. Wang et al., Proc. Part. Acc. Conf., San Francisco, 1991, Vol. 5, p. 3219.

[18] N. Holtkamp, Private Communication; T. Weiland (spokesman) et al., DESY 91-153, 1991.

[19] I. Wilson, W. Wuensh, C. Achard, Proc. EPAC 90, Nice, 1990, p. 943; I. Wilson and W. Wuensh, CERN-SL/90-103 (RFL), 1990.

[20] R. Parody, Private communication, CERN, 1991.

[21] T. Raubenheimer and R. Ruth, SLAC-PUB-5355, 1990; T. Raubenheimer, SLAC-387, UC-414, 1991.

[22] C. Fischer and G. Guignard, Proc. EPAC 92, Berlin, 1992.

[23] C. Fischer, Private Communication.

[24] L.N. Arapov, V.E. Balakin, Y. Kazakov, Proc. EPAC 92, Berlin, 1992.

[25] R.B. Palmer, W.B. Herrmannsfeldt, K. R. Eppley, Part. Acc. 30, 197–209, 1990.

[26] T.L. Lavine, Proc. EPAC 92, Berlin, 1992.

[27] W. Schnell, CERN-LEP-RF/88-59, 1988.

[28] L. Thorndahl, CLIC Note 152, CERN, 1991.

[29] Y. Baconnier et al., Proc. Linear Acc. Conf., Albuquerque, 1990, p. 733.

[30] G. Carron and L. Thorndahl, Proc. EPAC 92, Berlin, 1992.

[31] G. Guignard, CERN SL/92-22 (AP), 1992.

[32] O. Napoly, Private Communication.

Injection

J. Le Duff

Laboratoire de l'Accélérateur Linéaire, IN2P3 - CNRS
et Université de Paris-Sud, 91405 ORSAY, France

1 Beam current requirements in factories

1.1 Projects overview

A general feature of electron-positron factories is the high luminosity ($\sim 10^{33}$ cm^{-2} s^{-1}) obtained with small β^*'s (~ 1 cm) and conservative tune shifts (~ 0.04). Consequently the charge per bunch is similar in all cases. The total current, the number of bunches and the bunch spacing are however quite different since they depend on the circumference and the eventual choice of a crossing angle.

Table 1 summarizes the main design parameters of three different projects, with respect to the beam energy: the Phi-factory under construction in Frascati [1], the τ-charm factory under consideration for Spain [2], and the SLAC/LBL Beauty factory under detailed design [3].

1.2 Beam-beam lifetime

In a very high luminosity machine the beam lifetime is strongly affected by the beam-beam interaction itself [4]. This is due to the high rate of electron-positron inelastic collisions (bremsstrahlung). The corresponding beam lifetime is :

$$\tau_{bb} = \frac{N_b}{L_b \sigma_t n_c}$$

where
n_c = number of collisions points
L_b = luminosity per bunch
N_b = number of particles per bunch
σ_t = total cross section of e^+e^- bremsstrahlung

The total cross section σ_t can be expressed as :

$$\sigma_t = 2\alpha r_e^2 \left\{ \left(\frac{8}{3} Ln \frac{1}{\varepsilon_{RF}} - \frac{5}{3}\right) \left(Ln 4\gamma^2 - \frac{1}{2}\right) + \frac{3}{4}\left(Ln\frac{1}{\varepsilon_{RF}}\right)^2 - 1 - \frac{4\pi^2}{9} \right\}$$

Table 1. Design parameters of some e^+e^- factories

	Phi	τ-charm	e^+ Beauty	e^-
Energy [GeV]	0.5	2	3	9
Luminosity [$cm^{-2}s^{-1}$]	0.5 10^{33}	1 10^{33}	3 10^{33}	
Collision frequency [MHz]	368.2	25	238	
Beam current [A]	5.2	0.57	2.18	1.48
Particle/bunch	8.9 10^{10}	1.4 10^{11}	5.9 10^{10}	4.1 10^{10}
Number of bunches/beam	120	30	1658	
Circumference [m]	98	360	2199.32	
Revolution period [μs]	0.326	1.2	7.3	
Number of rings	2	2	2	
Crossing angle	YES	NO	NO	
Dispersion at I.P.	NO	YES	NO	

showing a weak variation with energy and RF acceptance. Applying these formulae to the different projects, for a 1% RF acceptance and a single crossing point, leads to the beam lifetimes listed in Table 2. The case of the τ-charm

Table 2. Beam-beam lifetime for different factories

	Phi	τ-charm	Beauty
Flat beams	1780 min	291 min	1718 min
Round beams		146 min	

factory was treated also with round beams [5], $\beta_x^* = \beta_y^* = 1$ cm and the same luminosity was obtained with half the stored current per beam compared to the flat beam case, leading to a lifetime down by a factor 2.

It should also be pointed out that increasing the number of crossings will reduce linearly the lifetime.

Finally the worst case here is the τ-charm factory since the high luminosity is obtained with a smaller number of bunches.

1.3 Touschek lifetime

The physical process here comes from large-angle Coulomb scattering within the bunch, in which energy is transferred from the transverse betatron oscillation into the longitudinal direction. Particles are lost when the energy transfer exceeds the energy acceptance of the ring [6, 7, 8]. The corresponding lifetime can be expressed as :

$$\frac{1}{\tau_T} = -\frac{1}{N}\frac{dN}{dt} = \frac{N r_e^2 c}{8\pi\sigma_x\sigma_y\sigma_z}\frac{\lambda^3}{\gamma^2}D(\xi)$$

$$\lambda^{-1} = \left(\frac{\Delta E}{E}\right)_{RF}$$

$$D(\xi) \approx 0.3$$

Applying these formulae to the previous projects leads to the results of Table 3.

Table 3. Touschek lifetime for different factories

	Phi	τ-charm	Beauty
Flat beam	180 min	120 min	> 6000 min

Here the τ-charm machine is taken with a monochromatization scheme which implies a small horizontal emittance[9, 10, 11].

1.4 Integrated luminosity

Although the peak luminosity, just after injection into the storage ring, is a usefull measure of the storage ring performances, the integrated luminosity remains the quantity of real interest for the users since it determines the amount of usefull events.

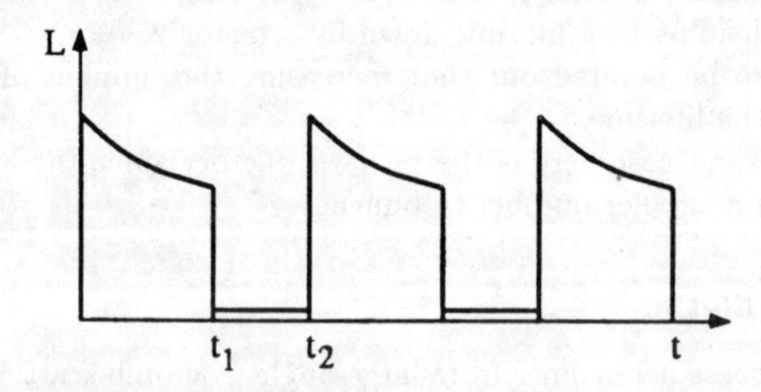

Fig. 1. Luminosity versus time

Figure 1 shows a typical running diagram where t_1 is the starting time for a new injection and t_2 is the time at which the two beams are put into collision. The difference t_2 - t_1 corresponds to the effective injection time.

As seen from previous sections the beam lifetime is current dependant. Considering for instance the beam-beam lifetime :

$$\frac{1}{I}\frac{dI}{dt} = -\frac{1}{\tau}$$

$$\tau = \frac{I}{e f_r \sigma_t L}$$

$$L = \frac{I^2}{e^2 f_r S}$$

where I is the beam current, f_r the storage ring revolution frequency and $S = 4\ \pi\sigma_x^*\sigma_y^*$ the effective beam cross section at the interaction point.

The integrated luminosity over one running period is :

$$L(t_2) = \int_0^{t_2} L dt = \frac{1}{e^2 f_r S}\int_0^{t_1} I^2(t) dt$$

$$L(t_2) = \frac{N_0}{\sigma_t}\left[1 - \frac{1}{1 + t_1/\tau_0}\right]$$

where N_0 and τ_0 are respectively the initial number of particles per beam and the initial lifetime at time $t = 0$. It is seen that the integrated luminosity depends essentially on the initial stored current ; the initial lifetime τ_0 only determines the speed at which the maximum integrated luminosity is reached.

It is interesting now to estimate the time t_1 where the next injection should start, knowing the time it will take to inject, $T_{inj} = t_2 - t_1$. From the average luminosity :

$$< L >= \frac{N_0}{\sigma_t}\frac{\left(1 - \frac{1}{1 - t_1/\tau_0}\right)}{(t_1 + T_{inj})}$$

and the cancellation of its derivative with respect to t_1, $\frac{\partial <L>}{\partial t_1} = 0$, one gets the optimum value :

$$t_1 = \sqrt{\tau_0 T_{inj}}$$

As a matter of fact, the shorter the initial lifetime is the smaller should be the time between successive injections.

1.5 Injection rates

As can be seen from previous sections the beam lifetime is rather short in most of these high luminosity e^+e^- factories, so it is necessary to aim at quite fast injection rates.

Since in general, for low and medium energies machines, the positron production is much less efficient compared to electron production, one can concentrate only on the positron rate and corresponding filling time. In the following a positron filling time $\leq$ 10 min. will be used as a design example for any injector type.

An ideal injection scheme would use a bunch to bunch transfer from injector to storage ring. Hence the injector must produce a bunch train which fits the useful buckets in the storage ring as shown on Figure 2 where T_r is the storage ring revolution period and k_b the total number of equally spaced bunches in each

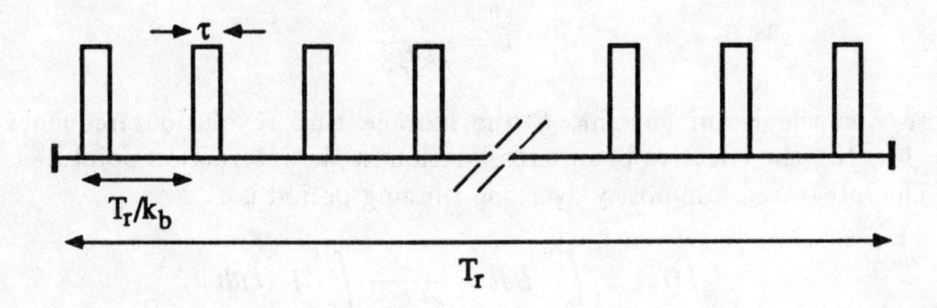

Fig. 2. Bunch train from injector

beam. The injector micropulse length, τ, needs to be matched to the storage ring RF bucket.

With this simple scheme one can concentrate on the filling time of each single bunch. For a 100 % overall transfer efficiency one gets the required performances of Table 4. Let's now carry on with this ideal scheme by introducing the following

Table 4. Ideal positron rates for different factories

	Φ	τ-charm	Beauty
N_b^+	8.9 10^{10}	1.4 10^{11}	5.9 10^{10}
e^+ rate [s^{-1}]	1.5 10^8	2.3 10^8	1.10^8

realistic constraints :

1. Overall transfer efficiency $\sim$ 20%
2. Transverse damping time in the ring. The repetition rate at which pulses are stored is determined by the rate at which the newly injected beam is damped. This suggests that the injection energy E_{inj} should be set at the nominal one.

Table 5 shows the corresponding parameters and the positron requirement at the target output for the different machines. It also gives the required electron gun performances under the assumption that the electron beam impinging the target has an energy of 200 MeV [5] and that the effective corresponding convertion efficiency is 5 10^{-3}. The numbers which have been obtained this way show that the overall scheme is realistic. It may not be optimum.

Table 5. Injection parameters

	Φ	τ-charm	Beauty
Transv. damping time [ms]	36	23	37
Inject. repetition rate [Hz]	28	43	27
Number of e^+/μ pulse at target	2.7 10^7	2.7 10^7	1.8 10^7
Numb. of e^-/μ pulse on target	5.4 10^9	5.4 10^9	3.6 10^9
e^- Gun: peak current [A]	0.86	0.86	0.58
e^- Gun: μ pulse length [ns]	1	1	1

2 Injection schemes

2.1 General

The previous “ideal injection scheme” favours a full energy linac with long pulses since :

- it provides high repetition rates
- it permits a bunch-to-bucket transfer over the whole ring circumference, taking into account the bunch frequency.

However long macropulses (0.3 to 7 μs) lead to rather small accelerating gradients ($\leq$ 15 MV/m) since RF compression pulse is not usable, and hence to a relatively long linac (Fig. 3). This solution is expensive unless it already exists on the site.

In the case of a τ-charm factory, with a small number of bunches, an alternative scheme with successive bunch filling can be considered with pulse compression and high gradient.

Another alternative consists of having a pre-injector linac followed by a full-energy synchrotron booster (Fig. 4). The efficiency for such a scheme can be enhanced by adding a damping accumulator ring.

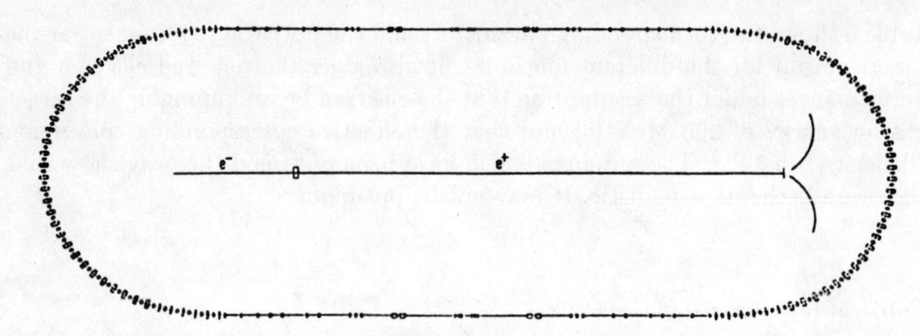

Fig. 3. Full-energy linac

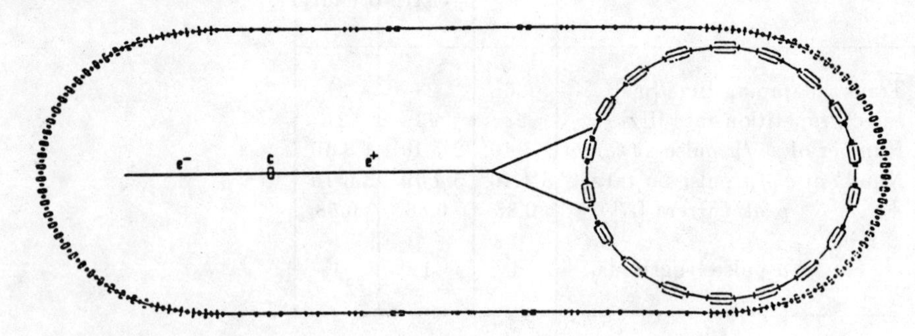

Fig. 4. Full-energy booster and pre-injector linac

2.2 High-gradient, full-energy linac with pulse compression

The RF pulse compression, of the SLED type [12], only permits the acceleration of short pulses. Then one can consider filling the ring, bunch after bunch, if the number of bunches is not too large. Table 6 gives the main parameters of such a dedicated injector for the particular case of a τ-charm factory operating with 30 bunches in each beam. Compared to an ideal case the loss factor on the e^+ rate is 6 if the linac is operated at 200 Hz. Hence the number of e^+ per micropulse should be increased by that factor. The electron gun performances given in Table 6 are realistic provided a sub-harmonic buncher is used to improve the bunching efficiency.

The positron linac would use, as already mentioned, a pulse compression system to increase the accelerating gradient and reduce its length [13]. The gradient which is here considered has been explored routinely on different types of accelerating structures and does not seem to create any reliability problem [14]. Table 7 lists the main parameters of such a positron linac.

2.3 Pre-injector linac and full-energy booster

It is believed that, as the injection energy increases, a booster would have a cost advantage compared to a linac. The expected cost behaviour is shown on Figure 5.

Table 6. Main parameters of a full-energy linac

Repetition rate [Hz]	200
Number of micropulses	1
Micropulse length [ns]	≤ 0.5
Electron linac :	
Energy [MeV]	200
Gun peak current (2 ns micropulse) [A]	2.3
Bunching efficiency [%]	100
Peak current on target [A]	9.3
Converter :	
Resolved efficiency	$4.8\ 10^{-3}$
Positron linac :	
Energy [MeV]	2500
Peak current [mA]	45
Number of e^+/micropulse	$1.4\ 10^{8}$
Output energy spread [%]	$\pm\ 0.2$
Output emittance [mm.mrd]	$1\ \pi$

Table 7. High-gradient linac characteristics

Energy [MeV]	2500
Klystron peak power [MW]	40
Klystron pulse length [μs]	5
Accelerating section length [m]	2.5
Accelerating gradient [MeV/m]	36.5
Number of structures/klystron	2
Total number of structures	28
Total number of klystrons	14
Overall linac length [m]	100

But a booster has also clear disadvantages. For instance using a metallic vacuum chamber will limit the repetition rate ($\leq$ 10 Hz). Moreover a simple booster design will almost look like a circle (see Fig. 4) with a corresponding circumference much smaller compared to a racetrack storage ring. Consequently, compared to an ideal scheme, there is a loss factor in the filling time, unless a damping/accumulator ring is used at the end of the pre-injector linac, in order to transform a high linac repetition rate into a low booster repetition rate. For comparison with the previous scheme of section 2.2, let's consider the case of a τ-charm factory with a 2.5 GeV injection energy. The characteristics of a simple booster with missing magnets to provide enough free space for diagnostics and vacuum equipments are listed in Table 8. It is seen with this example that a factor 3 is lost due to the difference between booster and ring circumferences. It can be recovered however using a 3-turn injection scheme into the booster, so that 30 micropulses from the pre-injector linac will fill 10 active buckets in the booster. Another loss factor of 4 remains, due to the lower booster repetition

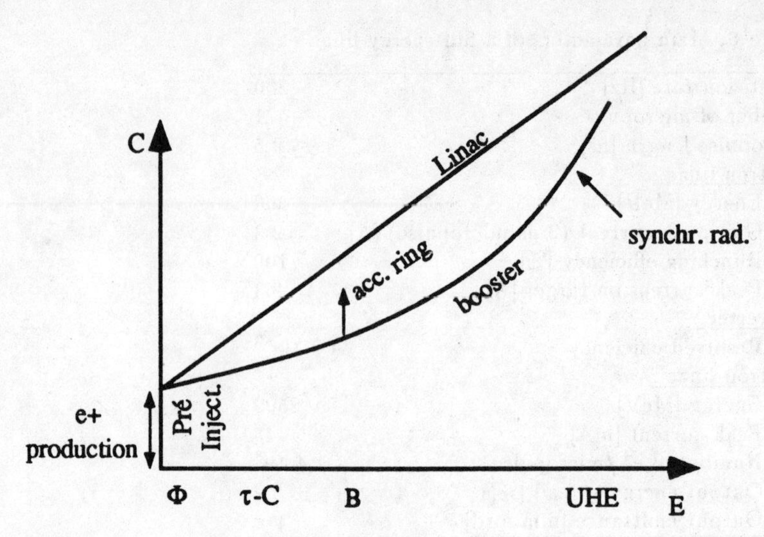

Fig. 5. Cost comparison between injector types

rate, which puts more constraints on the gun.

Finally, possible design parameters for the pre-injector linac are shown in Table 9, where it can be seen that the positron efficiency is reduced by a factor 2 compared to an ideal case to take account of the lower positron linac energy which automatically leads to higher emittance and energy spread not entirely captured by the booster. This lower efficiency is also called the resolved efficiency.

2.4 Damping/accumulator ring

One way to overcome the loss factor due to the low booster repetition rate consists of locating between the pre-injector linac and the booster itself a damping/accumulator ring as seen on Figure 6. This small ring is designed to have fast transverse damping to accept the possible high repetition rate from the pre-injector linac ; hence, the beam is accumulated while waiting for the complete booster cycle. Let's call τ_{acc} the transverse damping time of the accumulator.

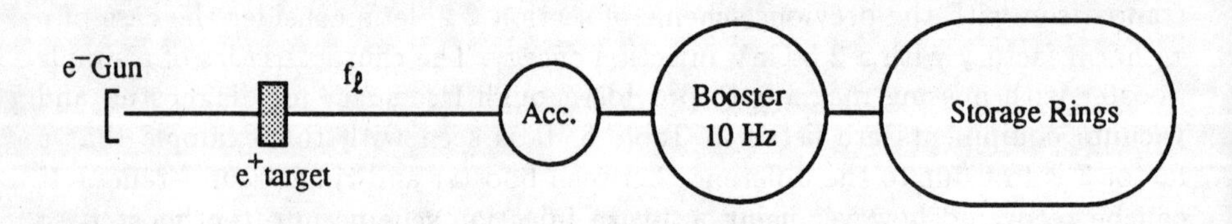

Fig. 6. Injector chain with damping ring

The linac repetition rate f_ℓ can be as high as :

$$f_\ell = \tau_{acc}^{-1}$$

The ring damping time is such that :

$$\tau_{acc} \propto \frac{T_r}{E^2 B}$$

where T_r is the revolution period, E the energy and B the bending field of the accumulator. Since the energy should be kept low for cost reasons, a good damping ring will require a high field and a small circumference (compact ring). In practice one can aim at a few ms damping time which allows the linac to operate at 200 Hz or more, depending on a reasonable compromise on linac power consumption.

For filling times much below 10 minutes in the factory itself, it is wise to include a damping/accumulator ring in the project.

3 Positron production

The most conventional way of generating a positron beam consists of bombarding a heavy material target with a powerful electron beam. Bremsstrahlung photons will lead to e^+e^- pair production and the positrons will be selected by proper adjustement of the phase of the accelerating structure located just behind the converter target.

The general principle is illustrated on Figure 7, where the target is immersed in a magnetic field to improve the capture efficiency. The important parameter

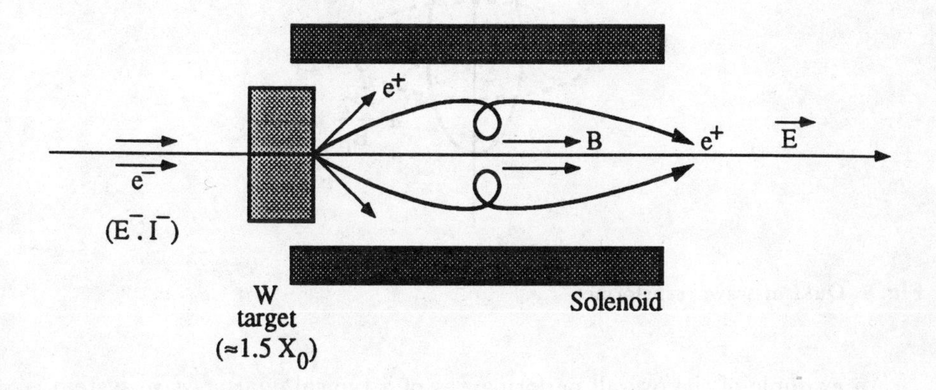

Fig. 7. General positron production scheme

is the positron production efficiency :

$$\eta = N^+/N^-$$

which turns out to be proportional to the incident electron beam energy. It is quite common to give the value of η for a 1 GeV electron beam.

The positrons are generated with large angles, hence producing a large emittance. They are also produced with a broad energy spectrum. To take care of the large angle a matching system, using solenoids, is generally used, which rotates the phase-space ellipse to be acceptable for the iris opening $2a$ of the first accelerating structure. In other words a small size beam with a large divergence is transformed into a bigger size beam with a smaller divergence.

Two different types of matching systems are used in practice [15]. The so-called "Quarter-wave Transformer" (Fig. 8) in which a high pulsed magnetic field is produced at the target and followed by a lower d.c. field over a longer distance. The second method uses a tapered solenoid which leads to an adiabatic field variation (Fig. 9).

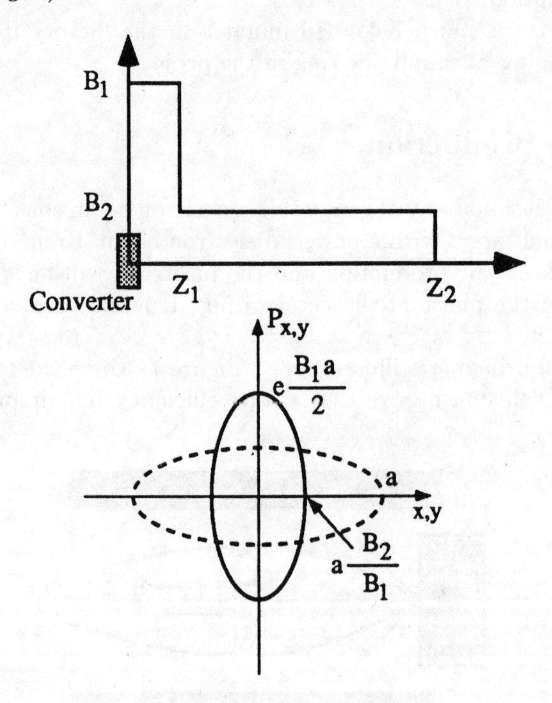

Fig. 8. Quarter-wave transformer

An example of the overall performances of a typical quarter-wave system is given in Table 10, corresponding to the CERN LIL pre-injector e^-/e^+ linac [16]. Notice that in this table LIPS means "pulse compression system".

The second method corresponding to the SLC project at SLAC, Stanford [17, 18], is illustrated in Table 11. This data was collected at the time of the design specifications. Today, practice shows that the positron yield at the target is about 4 at 30 GeV, and is reduced by a factor 3 at the exit of the linac, so that the resolved efficiency is roughly 0.04/GeV.

that the resolved efficiency is roughly 0.04/GeV.

Table 12, taken from reference [15], after correcting the SLC data, summarizes the situation for different machines in the world.

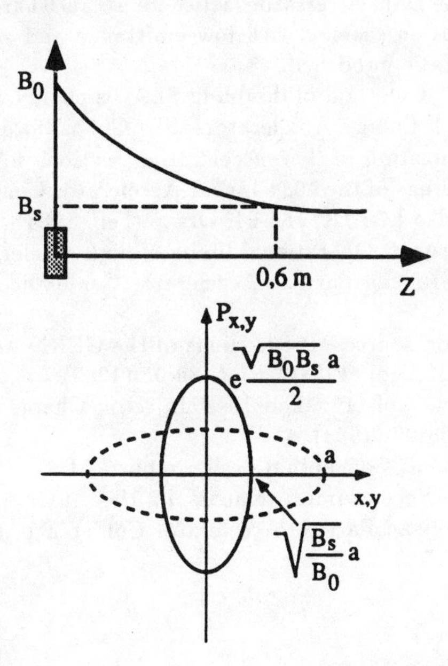

Fig. 9. Adiabatic system

References

1. M.E. Biagini, DAϕNE Status Report, EPAC 92, Berlin, 24-28 March, 1992, p. 60.
2. Y. Baconnier et al., A Tau-Charm factory laboratory in Spain. CERN/AC/90-07.
3. "An asymmetric B factory based on PEP", Conceptual Design Report, LBL PUB 5303, SLAC 372, February 1991.
4. J. Haïssinski, Bremsstrahlung simple dans les collisions negatons-positrons. RT/5-64, Anneaux de stockage, Laboratoire de l'Accélérateur Linéaire, Orsay (1964).
5. J. Gonichon et al., Preliminary study of a high luminosity e^+e^- storage ring at a c.m. energy of 5 GeV. LAL/RT/90-02, Laboratoire de l'Accélérateur Linéaire, Orsay, January 1990.
6. C. Bernardini et al., Phys. Rev. lett. **vol. 10** (1963) 407.
7. J. Haïssinski, Calcul des pertes de particules dues aux collisions à l'intérieur d'un même paquet. Cas d'ADA. RT/41-63. Anneaux de stockage, LAL, Orsay (22 April 1963).
8. J. Le Duff, Single and multiple Touschek effects. Proc. of CERN Accelerator School, Berlin, 14-25 September 1987. CERN 89-01,(1989), 114.
9. A. Zholents, Polarized J/ψ meson at a Tau-Charm Factory with a monochromator scheme. CERN SL/92-27 (1992).

10. P. Beloshitsky, A magnet lattice for a Tau-Charm Factory suitable for both standard scheme and monochromatization scheme, LAL/RT/92-09, Laboratoire de l'Accélérateur Linéaire, Orsay, July 3, 1992.
11. A. Faus Golfe, J. Le Duff, A versatile lattice for a Tau-Charm Factory that includes a monochromatization scheme with low emittance and a standard scheme with high emittance. To be published.
12. Z.D. Farkas et al., A method of doubling SLAC's energy, Proc. 9th International Conference on High Energy Accelerators, SLAC, Stanford, Ca, May 27, 1974.
13. J. Le Duff, Optimization of TW accelerating sections for SLED type modes of operation. Proceedings of the 1984 Linear Accelerator Conference, Seeheim, FRG, May 7-11, 1984. Also LAL/RT/84-01, Orsay (Feb. 1984).
14. G. Bienvenu, P. Brunet, Operational limits of high accelerator gradient. Proceedings of the 2nd European Particle Accelerator Conference, EPAC 90, Nice, June 12-16, 1990.
15. R. Chehab, Positron sources. Proceedings of the CERN Accelerator School, Salamanca, Spain, 19-30 Sept. 1988, CERN 89-05 (1989).
16. LEP design report. Vol. 1 : The LEP Injector Chain, CERN-LEP/TH/83-29 (1983), LAL/RT/83-09 (1983).
17. SLAC Linear Collider, Conceptual design report, SLAC-229 (June 1980).
18. P. Krejcik et al., Recent improvements in the SLC Positron System Performance. Third European Particle Accelerator Conference. EPAC 1992, Berlin, 24-28 March, 1992, 527.

Table 8. Booster performances and main parameters list

Nominal energy [GeV]	2.5
Circumference [m]	120
Revolution frequency [MHz]	2.5
Number of periods	18
Physical radius [m]	19.1
Tunes Q_x/Q_y	5.23/5.36
Momentum compaction	$4.1\ 10^{-2}$
Natural horizontal emittance [m.rd]	$1.9\ 10^{-7}$
r.m.s. energy spread	$7.2\ 10^{-4}$
Dipoles :	
Number	18
Bending radius [m]	8.33
Bending angle [degree]	20
Magnetic length [m]	2.91
Nominal field [T]	1.
Gap [mm]	35
Quadrupoles :	
Number	36
Magnetic length [m]	0.3
Nominal gradient [T/m]	14
Bore radius [mm]	40
Sextupoles :	
Number	18
Magnetic length [m]	0.15
Nominal G' [T/m^2]	105
Bore radius [mm]	40
Radio frequency :	
R.F. frequency [MHz]	500
Harmonic number	200
Number of circulating bunches	8
Loss per turn at 2.5 GeV [keV]	415
Peak voltage at 2.5 GeV [MV]	1.2
Natural r.m.s. bunch length at 2.5 GeV [cm]	2.7
Cavity shunt impedance [M Ω/m]	25
Active 5-cell cavity length [m]	1.5
Number of 5-cell cavities	1
Accelerating gradient [MV/m]	0.8
Output klystron power [kW]	40

Table 9. Pre-injector linac performance

Repetition rate [Hz]	10
Macropulse length [μs]	1.2
Number of micropulses	30
Micropulse length [ns]	≤ 2
Electron linac :	
Energy [MeV]	200
Gun peak current [A]	7.7
Transfer efficiency [%]	50
Peak current on target [A]	3.8
Converter :	
Resolved efficiency	$2.4\ 10^{-3}$
Positron linac :	
Energy [MeV]	250
Peak current [mA]	9.2
Number of e^+/micropulse	$1.15\ 10^{8}$
Energy spread [%]	± 1
Emittance [mm.mrd]	$10\ \pi$

Table 10. LIL pre-injector performance

General		
Frequency	2.99855	GHz
Klystron power	35	MW
No. of sections/klystron (LIPS)	4	
No. of sections/klystron (non-LIPS)	2	
Repetition rate	100	Hz
Beam pulse length	12	ns
High-current electron linac		
No. of sections (incl. buncher)	5	
Input current	6	A
Input energy	90	keV
Output current	2.5	A
Output energy	200	MeV
Converter		
Type	Tungsten + $\lambda/4$ solenoid	
Conversion efficiency, e^+e^-	0.04	GeV^{-1}
Low-current positron linac		
No. of sections	12	
Output energy	600	MeV
Positron operation		
Input energy (mean)	8	MeV
Resolved output current[a)]	12	mA
Emittance[b)]	≤ 4	π-mm-mrad
Energy spread[b)]	≤ 0.02	

a) Defined as current within 2 % energy spread

b) Defined as containing > 80 % output current

Table 11. SLC positron source specifications

A. Extraction system	
Beam : Electron energy	33 GeV
Pulse energy	194 Joules
Power	35 kW
Intensity	3.7 x 10^{10} electrons/pulse
Pulse rate	180 Hz
Size	0.4 to 1.0 mm (standard deviation)
B. Target system	
Material	74 % W - 26 % Re
Energy deposition	39 Joules/pulse
Pulse temperature rise	200-480°C
Maximum compressive stress	60.000-140.000 psi
Length	6 radiation lengths = 21 mm
Power deposition	7 kW
Maximum pulse temp.	800-1100°C
Steady-state temp.	600°C
C. Collection system	
Beam : Energy range at target	2 - 20 MeV
Emittance	2 mm x 2.5 MeV = 5 MeV-mm
Yield	$e^+/e^-_{in} = 4.8$
Initial Magnetic Focusing :	

	Length (m)	Field Range (kG)
Pulsed tapered solenoid	0.15	100 – 15
DC tapered solenoid	0.40	15 - 5
DC uniform solenoid	6	5

Table 12. Positron production systems in the world

Laboratories	I^-A or(N^-)	E^-GeV	Target	Matching system	Total yield e^+/e^- GeV^{-1}	Linac output e^+/e^- GeV^{-1}
LEP(LIL)	2.5	0.2	W	QWT 18 - 3 kGauss		2.5×10^{-2}
DESY	1.4	0.28	W	QWT 20 - 3 kGauss		4×10^{-2}
ALS (Saclay)	0.006	0.1	W - Re	QWT 12.5 - 3 kGauss		2×10^{-2}
LAL (Orsay)	0.8	1	W	AD 12.5 -1.8 kGauss	3 x 10^{-2}	2×10^{-2}
SLC	(2 x 10^{10})	33	W - Re	AD 50 - 5 kGauss	12 x 10^{-2}	4×10^{-2}

Vacuum Systems for e^+e^- Storage Rings

Nari B. Mistry

Newman Laboratory of Nuclear Studies, Cornell University, Ithaca NY 14853-5001, USA

1 Introduction

The design of vacuum systems for storage rings requires broad application of many different diverse fields. It is no longer possible to be an artist and put together a system of pipes and pumps on the spot and launch it on a ten-year career without much further attention! The design of an ultra-high vacuum (UHV) system for an electron/positron storage ring requires a knowledge of surface science; materials science; thermal and mechanical stress analysis; beam-induced RF fields; x-ray absorption, reflection and fluorescence; scattering of beam particles and production of secondaries; and above all, a flair for the art of magic to put all this together and make a UHV system out of narrow beam pipes with very low conductance, no space for pumps, and permission denied to make large holes for pumping ports! UHV systems for e^+e^- collider factories have to handle large circulating beam currents in the ampere range containing hundreds of short intense bunches. Dynamic pressures are required to be in the low 10^{-9} Torr range or even lower, to allow beam storage times from a few hours to as much as 24 hours.

Some of the problems presented for UHV design are:

1. Very high intensity of SR incident on chamber walls, absorbers, beam stops and x-ray windows.
2. Large rates of gas desorption induced by the SR.
3. Short intense beam bunches give rise to beam-induced RF fields and power losses that can cause damage to sensitive vacuum components such as bellows, ceramic windows, ceramic chambers, etc.
4. Severe problems arise in collider interaction regions (IR) due to synchrotron radiation and particle loss backgrounds which interfere with the HEP detectors.
5. Positive ion trapping in the potential well of the electron beam exacerbates all the problems of residual gas density in the beam path. Positron beams are happily immune to this, so that synchrotron light sources can avoid this problem altogether by using only positron beams.

6. In light sources, severe constraints are introduced by small, variable-aperture insertion devices and by the requirement for many x-ray beam lines with crotches, beam-stops, windows, etc.

We will briefly describe the solutions to some of these problems adopted for collider rings proposed as "B Factories" and use as illustrations some solutions used in new light sources currently under construction.

2 Typical Storage Ring Parameters

We show below some parameters relevant to vacuum system design. Table 1 lists typical parameters for B Factory proposals at Cornell[1], CERN[2] and SLAC[3]. Table 2 lists similar parameters of a light source (the European Sycnchrotron Radiation Facility, ESRF[4]).

Table 1. Typical parameters for B Factories

Parameter	CESR-B @($\mathcal{L} = 3 \times 10^{33} cm^{-2}s^{-1}$)		CERN/PSI @($\mathcal{L} = 1 \times 10^{33} cm^{-2}s^{-1}$)		SLAC/LBL @($\mathcal{L} = 3 \times 10^{33} cm^{-2}s^{-1}$)	
E_o, beam energy (GeV)	3.5	8.0	3.5	8.0	3.1	9.0
I_{beam}, nominal current (A)	2.0	1.0	1.28	0.56	2.14	1.50
ρ, bend radius in arcs (m)	20	87	65	65	30.5	165
P_{SR}, radiated power (kW/m)	10.6	7.6	0.64	7.65	2.1	5.1
ϵ_c, critical energy (keV)	4.75	13.1	1.5	17.5	2.2	9.8
Absorber material	copper		copper		copper	
$\dot{q}$, dynamic gas load (torr-ℓ/s/m)	6.7×10^{-6}	1.8×10^{-6}	0.83×10^{-6}	0.5×10^{-6}	1.7×10^{-6}	0.63×10^{-6}
$\dot{q}_T$, thermal gas load (torr-ℓ/s/m)	0.2×10^{-6}	0.2×10^{-6}	-	-	0.2×10^{-6}	0.13×10^{-6}
P_o, ave. pressure in arcs (ntorr)	2.7	3.6	1.7	1.0	7.5	5
S, ave. pumping speed (ℓ/s/m)	2555	583	500	500	225	125

Table 2. Typical Light Source Parameters

E_0 (positron beam)	: 6 GeV nominal
I (beam current)	: 0.2 A maximum
n_B	: 992 bunches maximum
Circumference	: 850 m
Total SR power	: 0.900 MW
ϵ_c (critical SR energy)	: 19.2 keV typical
SR power density on discrete absorbers	: ~ 400 W/mm^2
Beam-induced gas load	: $\sim 8 \times 10^{-5}$ torr-ℓ/s
Thermal gas load	: $\sim 6 \times 10^{-5}$ torr-ℓ/s
Total gas load	: $\sim 1.4 \times 10^{-4}$ torr-ℓ/s
Total pumping speed for 10^{-9} torr	: 1.4×10^5 ℓ/s

The intense power of SR striking chamber walls and absorbers is illustrated by the examples above. In the SLS, because of the short length of the magnets, discrete absorbers and localized massive pumping are particularly suited to the geometry. In the B Factories, solutions include both continuous and specially designed discrete absorber concepts in various parts of the rings to accomodate a range of linear SR power densities. For example in CESR-B, the B Factory proposed at Cornell, the SR power density ranges from about 0.3 kW/m in a soft-bend magnet to about 25.2 kW/m in a hard-bend magnet of the high energy ring (HER). Similarly, gas loads in the rings vary over a wide range of values in different sections. For example, in CESR-B the linear gas load evolved is about 3.3×10^{-7} Torr-ℓ/s/m in the soft-bends of the HER and about 6.8×10^{-6} Torr-ℓ/s/m in the hard-bend magnets of the low energy ring (LER).

These parameters call for new and ingenious solutions tailored to each situation.

3 SR-Induced Gas Desorption

The main gas load in the arcs is produced by desorption induced by the SR incident on the absorber walls and by scattered x rays on the rest of the chamber. Photon fluxes on absorber walls are typically in the range of about 10^{18} photons/s-m. Thermal outgassing is a small fraction of the SR-induced gas load except in the long straight sections such as the interaction region and RF straights. Hydrogen, CO and CO_2 are the main components of the gas desorbed by SR from clean UHV chamber walls. Although H_2 usually makes up almost 50% of the residual gas, the effects on beam lifetime and detector background are both dominated by CO and CO_2 because of the Z dependence of the scattering processes.

For estimating the SR-induced gas load, the parameter of interest is the desorption coefficient η_{SR} for CO and CO_2 expressed in molecules desorbed per incident photon. It is well known from CESR, PEP, PETRA and other existing electron rings that beam-scouring produces the most effective means of reducing the desorption coefficient. The desorption rates for aluminum alloy, stainless steel and copper have been carefully measured and compared [5, 6, 7, 8, 9, 10] and the art of predicting outgassing rates has been turned into a science[11, 12]. All available data indicate that at very high doses (e.g., above 10^{24} photons per meter) the desorption coefficients for aluminum, stainless steel and copper are very close to each other, although initial desorption is considerably higher for cleaned aluminum than for copper or stainless steel. However, only a few of the experiments listed above have measured η_{SR} at doses large enough to be directly usable for predictions of desorption rates after many amp-hours of operation. Recently, η_{SR} for copper has been measured at doses large enough to be directly usable for predictions of ultimate desorption rates after many amp-hours of operation.

Figures 1 and 2 show the desorption coefficients for aluminum, stainless steel, and copper measured at Orsay[5] and Fig. 3 shows data obtained at BNL by a collaboration of the Brookhaven Light Source (NSLS) and Livermore Lab (LLNL)[13].

Note that the total quantity of desorbed gases reaches a limit of around 10^{-1} Torr-liter for each species so that it is possible to plan on the finite total pumping capacity of installed getter pumps like NEG or TiSP.

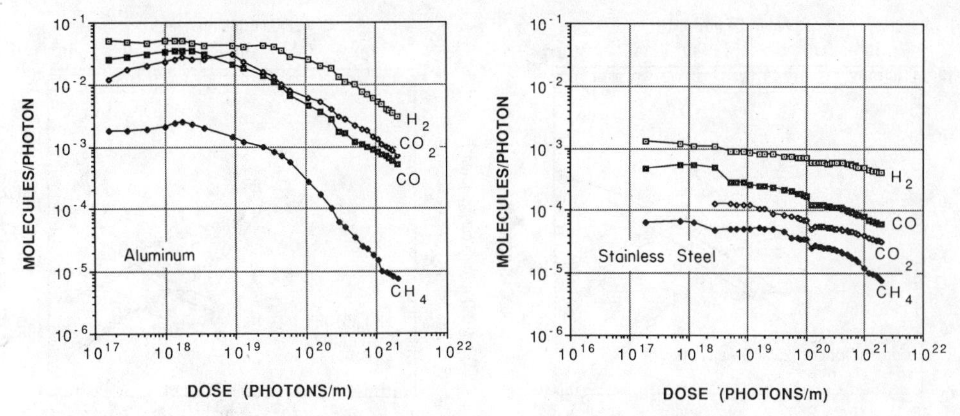

Fig. 1. The photon-induced neutral gas desorption coefficients for aluminum and stainless steel as a function of the photon dose. Data from reference [5].

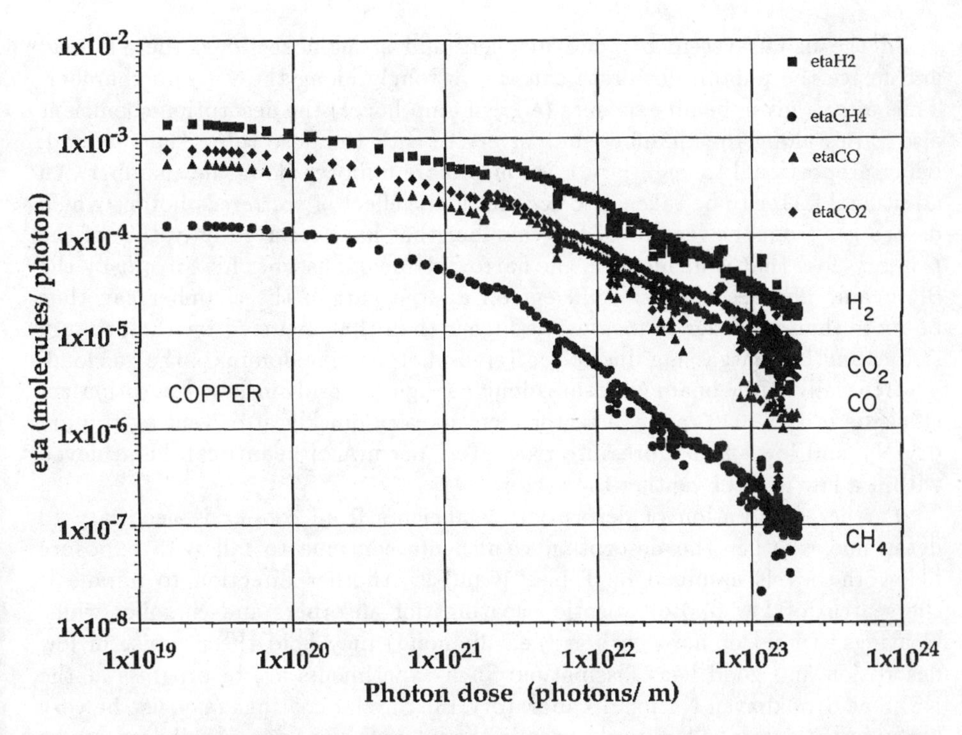

Fig. 2. The photon-induced neutral gas desorption coefficients for Cu for large photon dose. Data from reference [5].

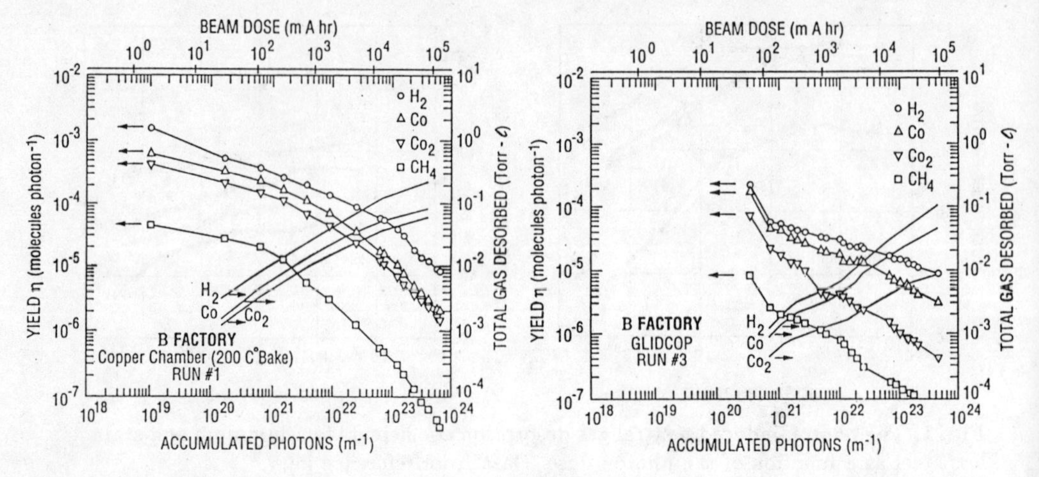

Fig. 3. The photon-induced neutral gas desorption coefficients for Cu for large photon dose. Data from reference [6].

In the usual system of bend magnets and straight sections, the SR load and hence the photon dose rate can vary strongly along the vacuum chamber. Thus after a given beam exposure (e.g., in amp-hours) the desorption coefficient also varies along the chamber, but inversely with the dose rate. The gas load, being proportional to $(\eta_{SR} \times n_\gamma)$, is much more uniform along the chamber. An additional factor to be taken into account is the effect of *scattered* photons which desorb gas from the regions of the chamber that have received only 10% of the primary dose that is incident on the narrow stripe of absorber hit *directly* by the SR beam. The desorption coefficient on a large part of the chamber can thus be more than an order of magnitude higher than that assumed for the primary stripe, and the outgassing due to scattered photons may dominate the gas load!

Generally, after beam exposures long enough to result in low desorption coefficients in the 10^{-6} range, a sector recovers very quickly after venting to pure dry N_2, and low specific pressure rises (Torr per mA of beam) can be achieved within a few days of venting the sector.

Further exploration of desorption coefficients from copper is necessary to determine whether the desorption coefficients continue to fall with exposure below the levels assumed in Tables 1 and 2. Another direction to pursue is the search for low photodesorption coatings for absorbers and chamber walls. Coatings of gold or hard carbon (i.e., diamond) may hold the promise of low desorption and good heat dissipation. Such experiments are in progress at the NSLS at Brookhaven National Laboratory[13]. Similar coatings may also be very useful in the design of SR masks used in the interaction region of colliders, where these SR masks have to absorb high power densities without scattering too much into the detector.

4 Vacuum Chamber Geometry and Pumping Techniques

The actual configuration of the vacuum chamber is dictated by the two constraints of safe absorption of the intense synchrotron radiation power emitted in the bend regions and sufficient pumping to meet the pressure requirements. A further constraint due to the short bunch length and high current is the need for a smooth profile in the beam chamber to minimize parasitic higher-order-mode (HOM) power losses and induced fields.

The conventional solution is to absorb the SR on a continuous water-cooled wall of the chamber and install distributed pumping in a parallel pump chamber on the opposite side of the beam. Pumping slots provide a fairly large conductance to the pump chamber. This solution has been used in SPEAR, CESR, PEP, PETRA and most recently in LEP. New light sources have been designed with discrete SR absorbers placed between bend magnets, with massive discrete pumping to take the main gas load. The Argonne APS uses an "antechamber" to allow the SR fan to pass through and incorporates distributed pumping in the antechamber. However, to make full use of the antechamber concept one has to incorporate distributed pumping on *both* sides of the beam channel to take advantage of the isolation provided by the slot to the antechamber where the SR is absorbed.

Figure 4 shows vacuum chamber profiles from CESR and from the APS synchrotron light source.

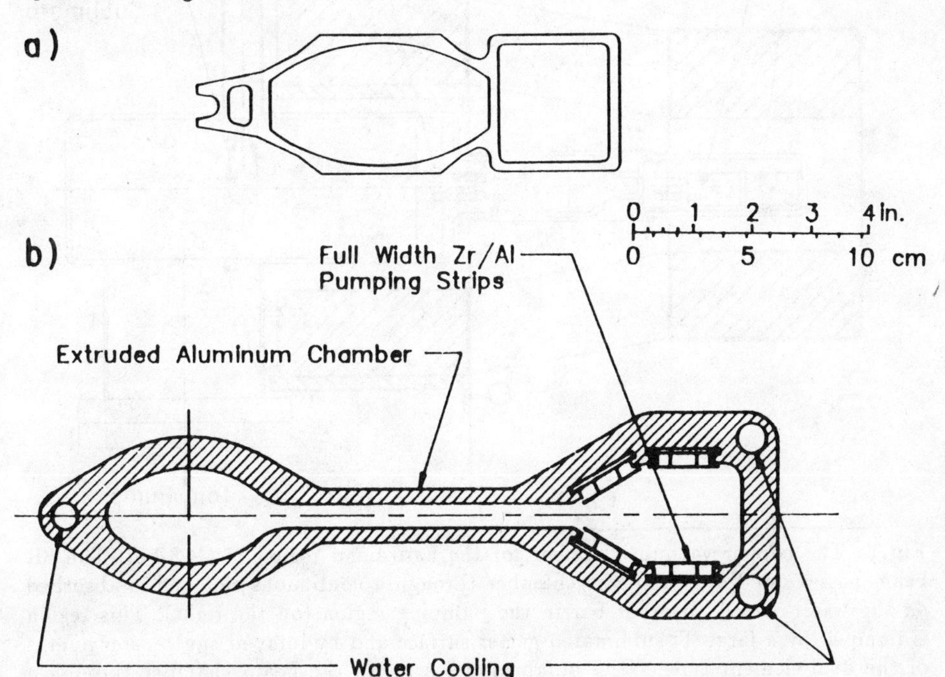

Fig. 4. Typical vacuum chamber profiles. (a) Extruded aluminum chamber from CESR. (b) Extruded aluminum chamber for the Argonne APS light source ring.

A unique chamber profile has been adopted for the hard-bend ($\rho = 45$ m) regions of the arcs flanking the IR in the B-Factory CESR-B proposed at Cornell, where the pressure must average 1×10^{-9} Torr because of beam-gas background constraints in the experiment. In these regions of intense SR and consequent high gas load per meter, it is necessary to allow the SR to pass through a continuous slot and impinge upon an absorber bar within a large pumping chamber. (See Fig. 5.) This configuration provides the necessary high conductance for gas molecules between the absorber and the sublimated getter material on the walls of the pump chamber. Differential pumping is provided by NEG (non-evaporable getter) modules on the opposite side of the beam chamber.

A similar solution is adopted for the LER, but the absorber bar and enlarged pump chamber lie in the drift space between arcs of a half-cell as shown in Fig. 6. This configuration is intermediate between the continuous wall and discrete absorber concepts.

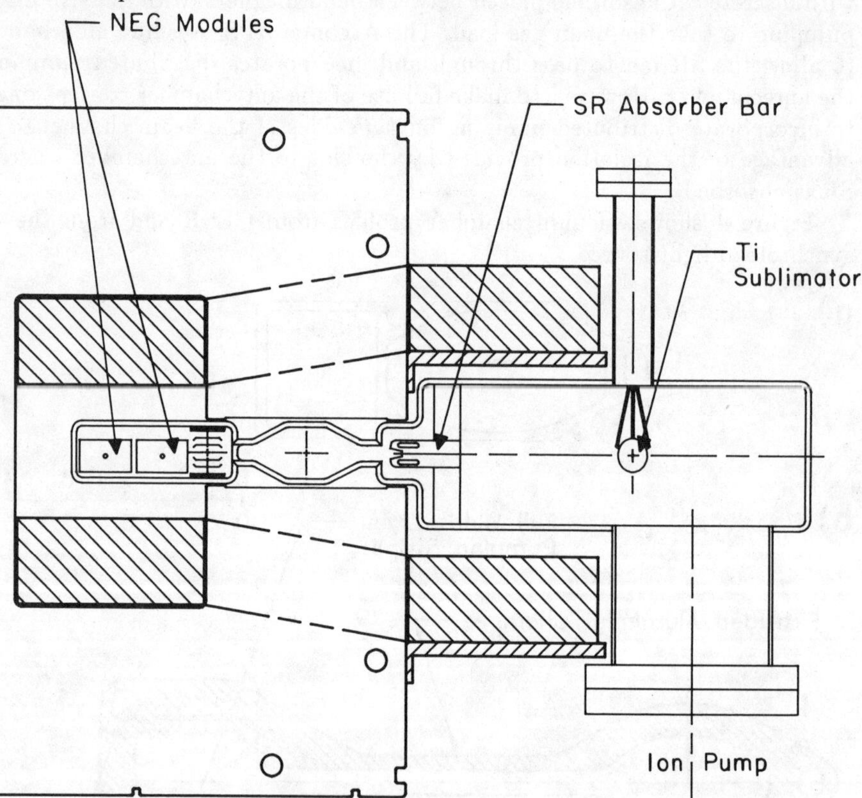

Fig. 5. The copper vacuum chamber for the hard-bend region of CESR-B. The SR beam passes out from the beam chamber through a continuous slot and is absorbed on the water-cooled absorber bar in the pumping region (on the right). This region is pumped by a large Ti-sublimated getter surface and by lumped sputter-ion pumps of the dual-element type. NEG pump modules pump the beam chamber through a symmetric slot on the opposite side, providing a dynamic pressure around 1×10^{-9} Torr in spite of the intense gas load.

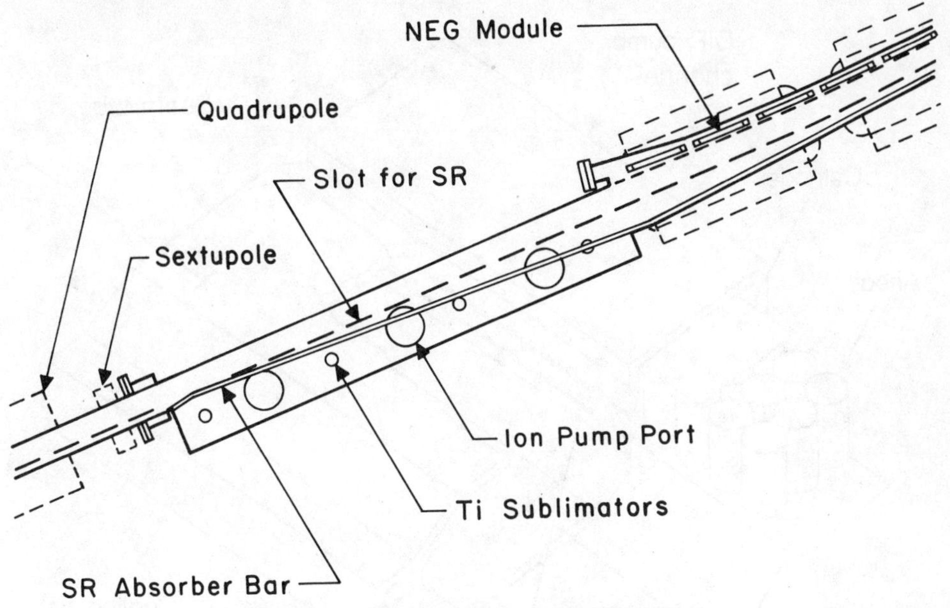

Fig. 6. Copper vacuum chamber for the LER arcs of CESR-B.

The SR beam is absorbed on the inclined water-cooled absorber bar in the enlarged pumping region between the bend magnets. This region is pumped by a large Ti-sublimated getter surface and by lumped sputter-ion pumps of the dual element type. NEG pump modules pump the beam chamber in the bend magnets. The average dynamic pressure is about 3×10^{-9} Torr in the arcs.

In the SLAC B Factory design, the conventional configuration is used for the HER, as shown in Fig. 7. However, the LER uses a different approach, with straight pumping chambers of double-walled design as shown in Fig. 8. The inner beam chamber is a copper water-cooled extrusion and the outer pumping manifold is a stainless-steel tube with lumped sputter-ion pumps. Most of the SR power is absorbed in these chambers, which lie downstream of the shorter bend-magnet chambers.

In LEP as also in the APS and ESRF light sources, and in the proposals for B Factories from KEK and CERN, the distributed pumping is provided by NEG strips. In CESR-B, distributed pumping will be provided by Ti-sublimation and NEG surfaces as described above. (Distributed sputter-ion pumps are limited to about 125 ℓ/s/m at these pressures and cannot provide enough pumping speed in these rings.) However, the getters pump only chemically active gases, and thus lumped sputter-ion pumps must be installed throughout the rings to provide sufficient pumping speed for noble gases and for non-getterable gases such as methane. Indeed, one must be careful that sufficient conductance is provided to these pumps so that the average partial pressure of methane does not become the limiting factor in beam lifetimes.

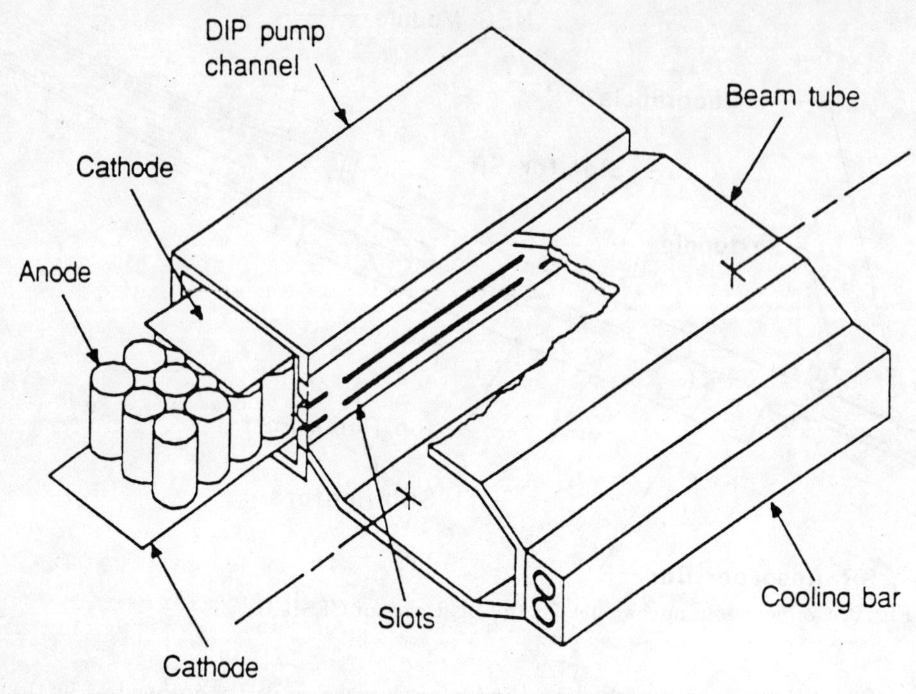

Fig. 7. The copper vacuum chamber for the HER arcs of the SLAC B-Factory.

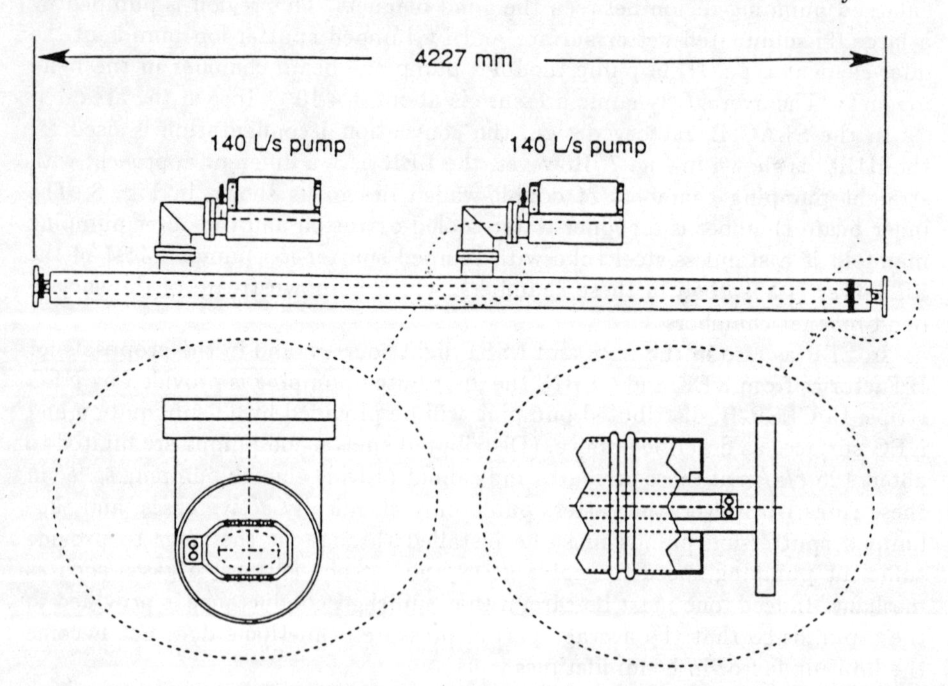

Fig. 8. The pumping chamber which is downstream of the bend magnets in the LER arc cells of the SLAC B-Factory.

In the SLAC B Factory design, because of the large ring radius and consequent lower linear gas load, distributed sputter-ion pumps operating at 125 ℓ/s/m are sufficient to maintain the required pressure profile in the HER arcs.

As described above, the ultimate desorption coefficients for aluminum, stainless steel and copper appear to be very similar. This is reflected in the diverse choice of materials in various designs. The APS chamber is made of extruded aluminum, and the ESRF chamber is fabricated from stainless steel, as is the chamber proposed for the Φ Factory, DAΦNE. Most of the B Factory proposals favor copper in order to limit the radiation damage to magnet and tunnel components due to scattered x rays. Copper provides sufficient shielding to absorb most of the scattered photons at these energies.

5 Interaction Region Vacuum System

In colliders, the residual gas pressure and composition in the region immediately flanking the interaction point (IP) are of particular concern because this region is a prime source for lost particles which create spurious background in the detector[14]. This region includes the straight sections of each ring around the crossing point and the "soft-bend" magnets immediately adjacent. The pressure in this region should be no higher than 1×10^{-9} Torr of CO and CO_2 combined. The end of the last soft-bend magnet is a source of bremsstrahlung-induced background and could profitably be kept at an even lower pressure.

Another concern is the masking of SR x rays arising from the IR quads and the nearest bend magnets. A thin small-diameter central beryllium pipe is usually prescribed for the IP at the detector, to get the first measurement of tracks at the smallest radius possible and to minimize multiple scattering. The inner wall is coated with a thin layer (25 mm) of copper or gold to absorb scattered x rays which could otherwise easily pass through the thin beryllium beam pipe in the detector and become a serious source of spurious hits and current in a vertex detector. The design must also incorporate the shadowing of upstream chamber wall surfaces that can "shine" with reflected (Rayleigh scattered) high-energy synchrotron x rays into the thin beryllium beam pipe at the IP. This phenomenon has been observed at CESR[15]. For this reason, the SR-absorbing surfaces of the last few magnets leading into the IR straight must be angled inward so that they are not visible from the thin beryllium pipe or from the tips of the SR masks within the detector region.

A typical configuration of the masking system in the detector for CESR-B is shown in Fig. 9. The masks must be smoothly tapered to avoid inducing large HOM fields, and the central beryllium pipe must be adequately cooled to withstand the power losses arising from both HOM and image currents. A double-walled beryllium tube appears to be a satisfactory solution. The SLAC design uses gaseous helium as the coolant, and at CESR-B the pipe is cooled by water. The inner and outer walls are tied together by beryllium ribs for strength, and the inner wall is coated with 25 μm of gold to absorb scattered x rays.

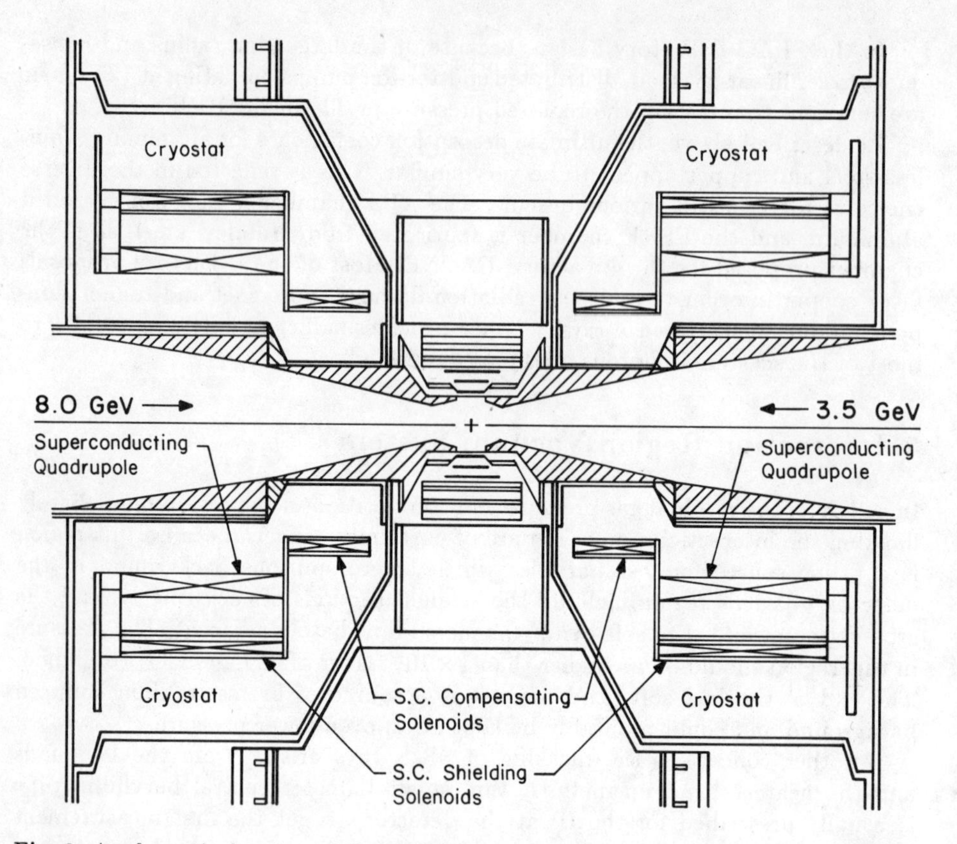

Fig. 9. A schematic drawing of the CESR-B interaction point in the detector, showing the masking scheme. The horizontal scale is highly foreshortened.

The vacuum chamber in the IR region of CESR-B is shown schematically in Fig. 10. The average pressure is maintained at or below 1×10^9 Torr by sputter-ion and Ti-sublimation (TSP) pumps as shown. Sublimation pump chambers surround the beam pipe near the regions where significant SR power from the off-center quads in the outgoing beams is expected to be absorbed.

6 SR Absorbers, Beam Stops, Crotches, Windows

The high intensity of the SR fans around the rings calls for special designs for the safe absorption of unwanted radiation on the walls, at crotches and at beam-stops and windows in x-ray lines. For the new light sources, special discrete absorbers and crotches have been designed. Figure 11 shows the inclined vee-shaped crotch used in the APS. Materials such as dispersion-strengthened copper are proposed for the absorbers, as the thermal stresses may exceed the yield strength of copper.

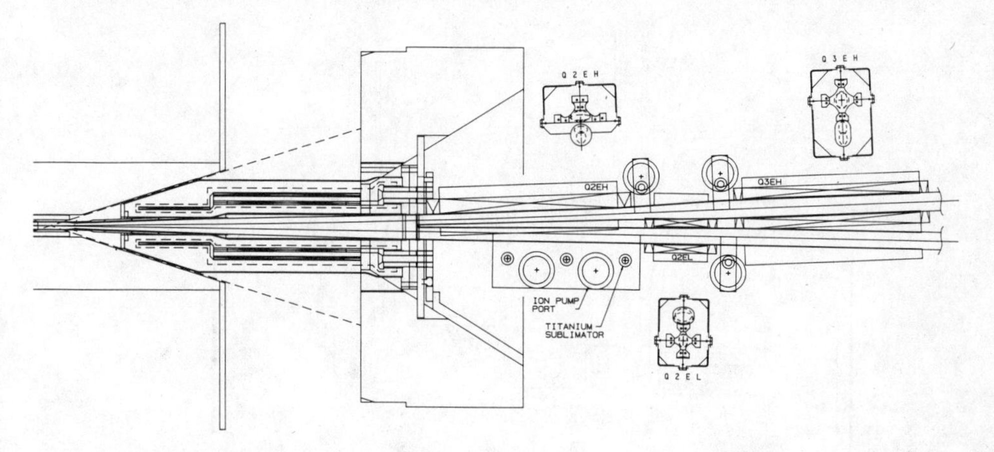

Fig. 10. Schematic view of the CESR-B IR vacuum system. The horizontal scale is highly foreshortened.

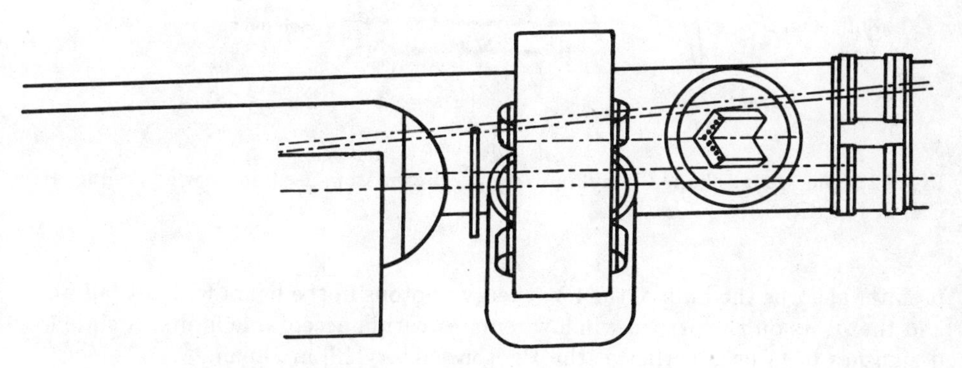

Fig. 11. The vee-shaped inclined crotch for the APS light source.

Beam-stops for undulator and wiggler lines may have to intercept power densities of up to about one kW/mm^2. This is twice the power density absorbed on a wedge-shaped beryllium/copper beam-stop[16] at the CHESS facility at CESR. (See Fig. 12.) This absorber is designed to stop radiation from a 25-pole wiggler delivering 17 kW of total power at a power density of 500 W/mm^2. The beryllium serves to diffuse the intense stripe of SR and to scatter an appreciable fraction (about 25%) of the power out of the beam-stop into the surrounding water-cooled walls.

Beryllium exit windows in x-ray beam lines have to withstand very severe thermally induced stress without failure. Windows in use at CHESS[17] have been designed for the wiggler beam described above where a linear power density of 3.8 kW/cm may be expected at 6-GeV, 200-mA operation. A 250-μm-thick vacuum-tight brazed Be window is preceded by a 500-μm-thick Be prefilter. This

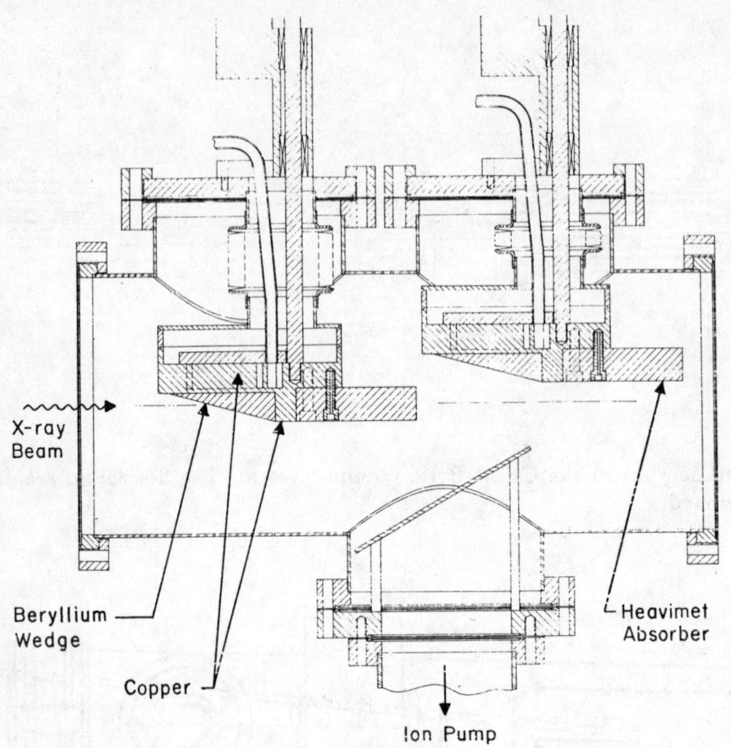

Fig. 12. The wedge-shaped beryllium/copper beam-stop used in a wiggler line at CHESS/CESR.

prefilter absorbs the bulk of the low-energy photons in the beam without failure, and the stress on the brazed window remains within acceptable limits. A similar design has been used earlier at the Photon Factory[18] in Japan.

To absorb higher power levels, it may be necessary to cool the Be windows to 77 K, to take advantage of higher thermal conductivity, lower thermal expansion coefficient and higher mechanical strength at low temperatures. An alternative is to use a thin diamond window or diamond-coated beryllium windows.

7 Summary

We have explored some of the challenges that face vacuum system design for high luminosity e^+e^- colliders and for high intensity synchrotron light sources. Many of these challenges are common to the two types of storage rings. The new light sources have been designed to solve some of the problems, but the proposed B Factory colliders will have to use ingenious methods to achieve the required performance.

References

1. CESR-B, Conceptual Design for a B Factory Based on CESR, Cornell Report CLNS 91-1050 (1991).
2. O. Gröbner et al., Design Considerations for the Vacuum System of a Beauty Factory in the ISR Tunnel, CERN Vac. Note AT-VA/OG (90-56) (1990), unpublished.
3. An Asymmetric B Factory Based on PEP, SLAC-372 (1991).
4. B. A. Trickett, The ESRF Vacuum System, in *Vacuum Design of Advanced and Compact Synchrotron Light Sources*, AIP Conf. Proc. 153 (1988), *Ed.* H. J. Halama, J. C. Scuchmann and P. M. Stefan.
5. A. Mathewson et al., Comparison of SR Induced Neutral Gas Desorption from Aluminum, Stainless-steel and Copper Chambers, in *Vacuum Design of Synchrotron Light Sources*, AIP Conf. Proc. 236 (1990), *Ed.* Y. Amer, S. Bader, A. Krauss and R. Niemann.
6. C. L. Foerster et al., Desorption Measurements of Copper and Copper Alloys for PEP-II, Preprint BNL-48367 (1992); to be published in *Vacuum.*
7. M. Andritschky et al., Differences in Synchrotron Radiation-induced Gas Desorption from Stainless Steel and Aluminum Alloy, *Proc. 1989 Part. Accel. Conf.*, Chicago, p. 563, IEEE (1990).
8. M. Andritschky et al., *Vacuum* **38**, 933 (1988).
9. H. J. Halama and C. L. Foerster, *Vacuum* **42**, 185 (1990).
10. S. Ueda et al., in *Proc. 11th. Int. Vac. Congress*, Köln, 1989; *Vacuum*, **41**, 1928 (1990).
11. O. Gröbner et al., *J. Vac. Sci. Technol.* **A7**, 223 (1989).
12. A. Mathewson, A Survey of $e^{\pm}$ Storage Rings and a Comparison of Their Vacuum Characteristics, LEP Vacuum Note, April 1989.
13. C. L. Foerster and G. Korn, The Search for Low Photodesorption Coatings, in *Vacuum Design of Synchrotron Light Sources*, AIP Conf. Proc. 236 (1990), *Ed.* Y. Amer, S. Bader, A. Krauss and R. Niemann.
14. H. DeStaebler, Interaction Region Considerations for a B-factory, SLAC-PUB-5299 (1991).
15. T. Letson et al., Study of Beam Related Backgrounds in CLEO-II, CLEO Internal Report CBX 90-67, Cornell (1990).
16. M. J. Bedzyk et al., *Rev. Sci. Instrum.* **60**, 1460 (1990).
17. Q. Shen et al., *Rev. Sci. Instrum.* **60**, 1464 (1990).
18. S. Sato et al., *Nucl. Instrum. Methods* **A246**, 177 (1986).

Ion Trapping and Clearing

A. Poncet

CERN, 1211 Geneva 23, Switzerland

Abstract

Ion trapping in the electron beams of future factories is one of the major issues to be addressed. After a brief review of the adverse effects of neutralization in particle storage rings, the basic topics of ion production by ionization of the residual gas are recalled: ion production rate, natural clearing rates, ion kinematics, and conditions of trapping for bunched and unbunched particle beams with positive or negative space charge. Different methods of clearing are described and their performance discussed, namely, d.c. clearing electrodes, empty buckets (in electron storage rings) and beam shaking. Examples of neutralization effects and diagnostics are taken from CERN machines.

1 Introduction

In accelerators and storage rings, ions created by the circulating particles from neutral molecules of the residual gas may be trapped in the beam space-charge potential, and may generate all sorts of ill effects: reduced beam lifetime (increased pressure), emittance growth and losses through excitation of resonances, and coherent beam instabilities. Whilst they can occur in proton beams (e.g. CERN ISR trapping electrons), these neutralization phenomena mainly affect machines with negative beam space charge, such as electron storage rings, and antiproton accumulators.

Low-energy machines are more subject to ion trapping because of their small size, which leaves little space between bunches for ions to escape the beam potential, and suffer most because of their inherent high sensitivity to space charge effects. To illustrate this point, the incoherent space charge tune shift can be written as [1]

$$\Delta Q = \frac{N}{k}\left(\eta - \frac{1}{\gamma^2}\right), \tag{1}$$

(where k relates to beam transverse dimensions and the bunching factor, γ is the relativistic factor, η the neutralization coefficient defined as the ratio of trapped charges to beam charges, and N is the number of particles in the beam).

ΔQ can be unacceptably large if γ is small (low energy), and/or η is high. For instance: ΔQ is ~ $0{\cdot}1\ \eta$ in the CERN 600 MeV EPA (Electron Positron

Accumulator [2]) for its nominal electron beam, where in the absence of clearing η can reach values close to one. This has to be compared with the value $\Delta Q \cong 0.1/\gamma^2$, i.e. ~ 10^{-7} for perfect clearing. A large tune shift is accompanied by a large tune spread, owing to the non-linearity of the ion focusing forces on the beam particles. This results in the excitation of a large number of resonances, as can usually be seen in the tune diagram. Figure 1 below illustrates this effect as seen from the beam intensity and stationary emittances in the EPA.

Ion trapping is one critical issue for the high-intensity electron beam of the particle factories proposals and projects [3]. The very high luminosity aimed at in e^+e^- collisions in these future machines is limited by a maximum allowable beam–beam tune shift of currently 0.06. The effect of ions on the e^- beam is similar to the effect of one beam upon the other during collisions (beam–beam interaction). One must thus place an upper limit on the ion density in the e^- beam by demanding that the ion-induced tune shift be substantially less than the maximum beam–beam tune shift, for instance 0.01. This number puts a very severe upper limit of 10^{-4} or less on acceptable neutralization levels in these machines, difficult to achieve with present clearing means [4].

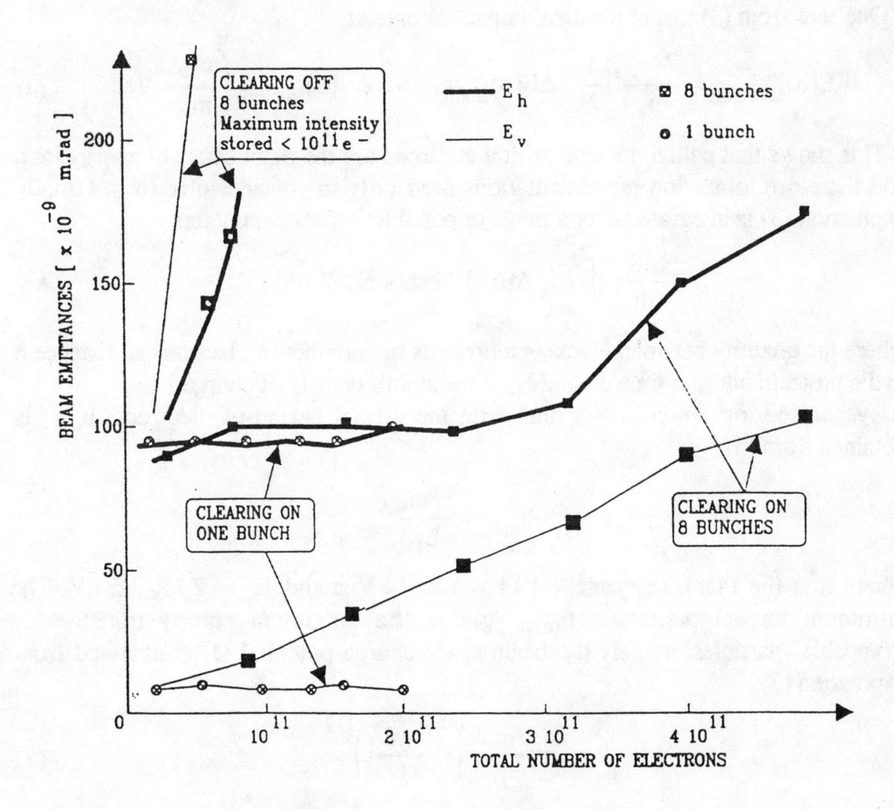

Fig. 1. EPA emittances with and without clearing

2 Ion Production

Beam particles with relativistic velocities interact with nuclei and the electrons of the gas atoms via the Coulomb forces. The energy transfer $\Delta E(b)$ in a collision, which depends on the impact parameter b (distance between the target and the projectile), may be sufficiently large to break the liaison between the nucleus and the electrons, leading to ionization:

$$\Delta E(b) > Z' \times 13.5\,\mathrm{eV}\,(\text{ionization potential } E_1)\,. \tag{2}$$

In S.I. units [5]

$$\Delta E(b) = \frac{e^4}{8\pi^2 \varepsilon_0{}^2 \beta^2} \, \frac{Z^z Z'^2}{m_0} \, \frac{1}{b^2}\,(J) \tag{3}$$

where Z and Z´ are the atomic numbers of the projectile and the target, respectively, m_0 the mass of the target, e the elementary charge, c the speed of light, and ε_0 the vacuum permittivity. If the target is an electron: $m_0 = m_e$, $Z' = 1$; for a nucleus $m_0 \sim 2m_p Z'$.

One sees from (3) that at identical impact parameter:

$$\Delta E(b)_{electron} = \frac{2m_p}{m_e} \, \frac{1}{Z'} \, \Delta E(b)_{nucleus} >> \Delta E(b)_{nucleus} \left(\frac{m_p}{m_e} \sim 1836 \right). \tag{4}$$

This shows that collisions with orbital electrons are the main cause of energy loss, and therefore ionization rate calculations need only to consider electrons. Usually expression (3) is integrated over a range of possible impact parameters

$$\frac{dE}{dt} = \int_{b_{min}}^{b_{max}} \Delta E(b) \; [2\pi b \beta c N_a Z' \, db]\,, \tag{5}$$

where the quantity between brackets represents the number of electrons at distance b to the projectile during time dt, and N_a is the atom's density of charge Z´.

By considering the collision time with the orbital period of electrons, b_{max} is obtained from [5]

$$b_{max} = \frac{\gamma\beta c h}{E_1} \tag{6}$$

where h is the Plank constant = 4.14×10^{-15} eV·s, and $E_1 = Z' \times 3.5$ eV. The minimum impact parameter b_{min} (giving the maximum energy transfer for 'trappable' particles, namely the beam space charge potential U) is obtained from expression (3):

$$b_{min} = \frac{r_0 c Z}{\beta} \left(\frac{2m_0 Z'}{U} \right)^{1/2}, \tag{7}$$

since species created with $\Delta E(b) > U$ can escape to the vacuum chamber wall, and are therefore not trappable.

Table 1. Values of parameters r_0, m_0, and Z' for different beam trapping events. r_p and r_e are the classical proton and electron radii ($r_{p,e} = e^2/4\pi\varepsilon_0 m_{p,e}c^2$)

Trappable particle	Electron (positive beam)	Proton (e^- or antiproton beam)	Ion (e^- or antiproton beam)
r_0	r_e	r_p	r_p
m_0	m_e	m_p	$2m_p$
Z'	1	1	Z'

The production rate is therefore:

$$R_p = \frac{1}{E_0}\frac{dE}{dt} = \frac{2\pi m_e c^3 r_e^2}{\beta} \frac{N_a Z'}{E_0} \ell n \left(\frac{\gamma\beta^2 h}{E_1 r_0 Z} \frac{U}{2m_0 Z'} \right)^{1/2} , \quad (8)$$

E_0 being the average energy for the formation of an ion–electron pair (~35 eV).

A consequence of this expression is that — everything being equal (energy, beam potential) — a positive space-charge beam (e.g. protons) will trap less electrons than a negative one will trap positive ions, since the fraction of electrons produced in the ionization process with sufficient energy to escape the beam potential is larger.

This consideration, together with expression (4), illustrates the fact that positive ions are created with much less energy than electrons (in fact with quasi thermal energies of < 0.04 eV), and are therefore generally all trappable. Indeed a quick numerical application of expression (3) would show that the impact parameter has to be 4.7×10^{-14} m, i.e. a quasi head-on collision with the nucleus, and therefore highly improbable — for the H_2^+ ion to be produced with 10 eV energy (the electron would obtain 20 keV).

The neutralization coefficient of a beam is the ratio of the ion production rate R_p to the clearing rate R_c [s^{-1}]. Since the ions are virtually produced at rest, the production rate of ions can therefore be obtained from experimentally-determined ionization cross sections σ_i:

$$R_p = \sum_{i=1}^{n} \sigma_i \ N_{mi} \ \beta c , \quad (9)$$

N_{mi} being the molecular density of gas species i.

As a typical example, the values for the EPA (600 MeV, $\gamma = 1200$ for electrons) at $P = 10^{-9}$ mbar [6] are shown in Table 2:

Table 2. EPA ionization cross sections

Gas	σ_i (m^2)	R_p (s^{-1})	Ionization time (s)
H2	0.4×10^{-22}	0.4	2.5
CO	1.54×10^{-22}	1.5	0.7

Owing to a logarithmic dependence on the energy of the primary particle, these cross sections would only grow by 20% at 6 GeV/*c*.

2.1 Beam Heating

Distant collisions with a large impact parameter — much more probable than close ones leading to ionization — are important, since they feed energy differentially to ions. In some circumstances (neutralization pockets) this may be a clearing mechanism, i.e. when the trapped species get enough energy to escape the beam potential:

$$R_c = \frac{1}{eU}\frac{dE}{dt} = \frac{1}{eU}\int_{b_{min}}^{b_{max}} \Delta E(b)\left(2\pi b\beta c N_p db\right). \qquad (10)$$

This represents the 'natural' clearing rate for a singly-charged species. The expression between brackets is the number of projectiles passing at distance b to the ion target during time dt. N_p is the projectile density of charge $Z = 1$.

In a good approximation, b_{max} and b_{min} can be chosen to have the same values as the ion and nucleus radii respectively, leading to [5]:

$$R_c = \frac{2\pi\ m_0 c^3 r_0^2}{\beta}\ \frac{N_p Z'}{eU}\ \ell n\left(3\times 10^4 \times Z'^{-2/3}\right), \qquad (11)$$

with m_0 and r_0 being m_e, r_e, if the trapped species is an electron; m_p, r_p for a proton; and $2\ m_p$, r_p for an ion of charge Z'.

As an example, typical clearing times for the EPA machine with 6×10^{11} electrons (300 mA), and 1 mm beam radius, giving a beam potential of ~50 V, are shown in Table 3:

Table 3. EPA natural ion clearing rates

Gas	Clearing rate R_c (s^{-1})	Clearing time τ_c (s)
H^+	3×10^{-3}	350
H_2^+	6×10^{-3}	166
CO^+	0.04	25
CO_2^+	0.07	15

The process is thus slow compared with typical ionization rates, but may be important to explain why, in some circumstances (pockets, very low gas pressure: 5×10^{-11} Torr, [Ref. 7]), fully ionized light ions can chase heavy ones, and accumulate up to a dangerous level.

2.2 Gas Cooling

Seldom taken into account, gas cooling could perhaps be an important process for high pressures and long ion sojourn times. In addition, charge-exchange phenomena by which a positive ion captures an electron from a gas molecule may occur at ion energies of only a few eV. The new ion is created with the primary molecule's energy, while the newly created neutral species carries away the initial ion energy. Resonant capture cross sections between an ion and its own neutral molecule can be very high at low energy:

$$\sigma = 1.2 \times 10^{-15}\,\mathrm{cm}^2 \quad \text{for He}^+ \text{ in He}\,, \tag{12}$$

for an ion energy of 3 eV [8].

2.3 Limits on Ion Accumulation

In the vast majority of cases (electron storage rings with typical pressures of 10^{-9} mbar, and ionization times of a second or less), ionization is, however, the dominant effect in the absence of any clearing mechanism. The production rates are:

– for singly-ionized species (density N_i^+):

$$\frac{d\left(N_i^+\right)}{dt} = N_m N_p \sigma_i c - N_i^+ N_p \sigma_i c = N_p \sigma_i c \left(N_m - N_i^+\right), \tag{13}$$

– doubly ionized:

$$\frac{d\left(N_i^{++}\right)}{dt} = N_i^+ N_p \sigma_i c - N_i^{++} N_p \sigma_i c = N_p \sigma_i c \left(N_i^+ - N_i^{++}\right), \tag{14}$$

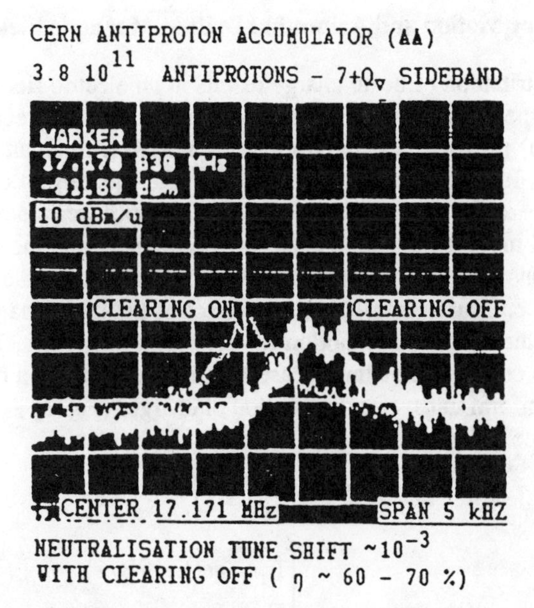

Fig. 2. AA transverse vertical Schottky scan showing the band $(7 + Q_v)\, f_{rev}$ with clearing electrodes turned OFF (full neutralization), and ON (little neutralization)

etc. until (Z' being the total number of electrons of the gas atom):

$$\frac{d\left(N_i^{Z'+}\right)}{dt} = N_i^{(Z'-1)+} N_p \sigma_i c \,. \tag{15}$$

In the steady state (assuming constant ionization cross sections)

$$N_i^{+} = N_i^{++} = \dots = N_i^{(Z'-1)+} \le N_m \,. \tag{16}$$

Therefore partially ionized ions can, at most, reach the molecular density N_m. Only the fully ionized state $N_i^{Z'+}$ could get close to the particle density (usually much larger than the gas density) divided by the final charge state: $N_i^{Z'+} \le N_p/Z'$ corresponding to full neutralization of the particle beam.

The degree of neutralization of a particular beam can be estimated from the incoherent tune shift. Almost full neutralization has been measured in the CERN AA when all the clearing electrodes are turned off (Fig. 2).

3 Ion Dynamics (d.c. Beam)

Produced with near thermal velocities, ions are generally not free to simply drift in the beam. Their motion is mostly governed by the beam space-charge potential when the beam is not fully neutralized, and by external forces such as magnetic fields in dipoles and quadrupoles. The Lorentz force due to the beam magnetic field is weak, and can be neglected in practical cases.

3.1 Ion Oscillatory Motion and Azimuthal Drift in Magnetic Field-free Regions

The transverse distribution of beam charge results in an electric field E to which the charged trapped species is sensitive. This centripetal force (directed towards the centre of the beam) provokes a radial oscillatory motion with an amplitude equal to the radius at birth. Its frequency is proportional to the square root of the local field derivative times the ion charge, divided by its mass (bounce frequency).

The longitudinal modulation due to changing beam sizes, and the varying vacuum chamber dimensions, give rise to longitudinal fields which drive the oscillatory ions around the machine. Usually negligible, the effect of the beam magnetic field may also contribute to the longitudinal motion for very high intensities. To illustrate this with numbers, we consider the simple case of a round beam in a circular vacuum chamber, with a uniform transverse distribution of charges (Fig. 3)

The Lorentz Force acting on the ion is:

$$\bar{F} = qe\left(\bar{E} + \bar{v} \times \bar{B}\right), \tag{17}$$

$$\text{with components} \quad \begin{cases} F_r = qe\left(E_r - B \cdot \dot{z}\right) \\ F_\theta = 0 \\ F_z = qe\left(B \cdot \dot{r} + E_z\right) . \end{cases} \tag{18}$$

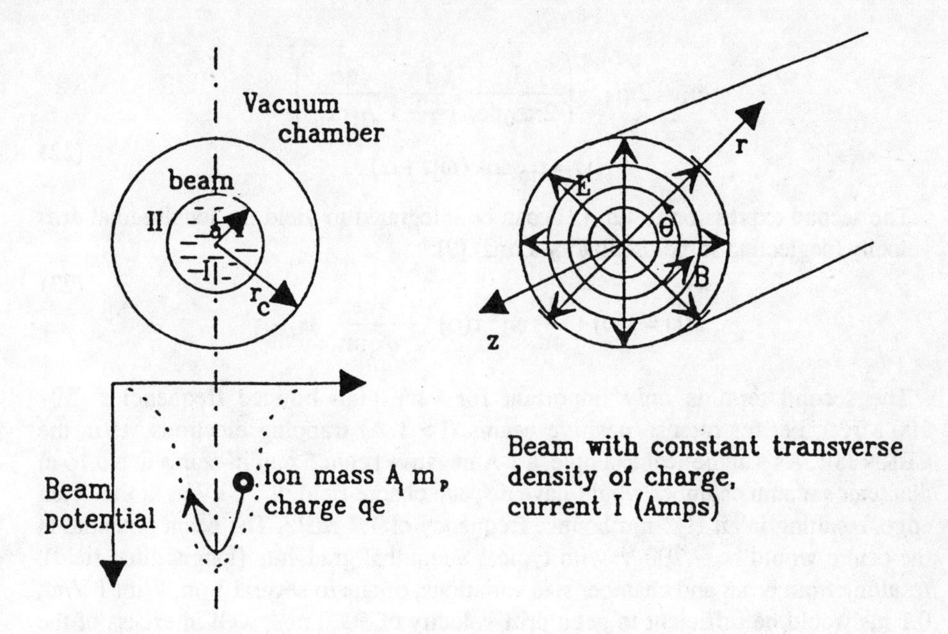

Fig. 3. Beam potential, and magnetic and electric fields of a uniform beam

From the Gauss theorem, the electric field can be written as

$$\text{in region I:} \quad E_r(r) = \frac{I}{2\pi\varepsilon_0\beta c}\,\frac{r}{a^2}$$

$$\text{in region II:} \quad E_r(r) = \frac{I}{2\pi\varepsilon_0\beta c}\,\frac{1}{r}\,. \tag{19}$$

The magnetic field is simply

$$B = \frac{\beta E_r}{c}\,.$$

The beam potential at the centre is [1]:

$$U = \frac{I}{2\pi\varepsilon_0\beta c}\left(\ell n\,\frac{r_c}{a} + \frac{1}{2}\right). \tag{20}$$

The forces on the ion reduce to:

$$F_r = A_i m_p \ddot{r} = \frac{qe\,I}{2\pi\varepsilon_0\beta c a^2}\left(1 - \beta\frac{\dot{z}}{c}\right) r$$

$$F_z = A_i m_p \ddot{z} = \frac{qe\,I}{2\pi\varepsilon_0\beta c^2}\,\dot{r}\,r + qeE_z\,. \tag{21}$$

As the ions can only have a maximum potential energy equal to the beam potential times their charge, i.e. typically up to a few hundred eV, they are non-relativistic. Equations (21) can be uncoupled by neglecting $\beta\,\dot{z}/c << 1$. The ion motion is thus transversally oscillatory with a frequency ('bounce' frequency) of:

$$\omega_i = 2\pi f_i = \left(\frac{I}{2\pi\varepsilon_0\beta c} \; \frac{1}{a^2} \; \frac{qe}{A_i m_p} \right)^{1/2}$$

$$r(t) = r_m \cos\left(\omega_i t + \alpha\right) . \tag{22}$$

The second expression in Eq. (21) can be integrated to yield the longitudinal drift velocity (neglecting some oscillatory terms) [9]:

$$\dot{z}(t) = \dot{z}(o) + \frac{1}{4c}\, \omega_i{}^2\, r(o)^2 + \frac{qe}{A_i m_p}\, E_z t . \tag{23}$$

The second term is only important for very high bounce frequencies (50–100 MHz), i.e. for intense positive beams (I > 1 A) trapping electrons, as in the CERN ISR. As a numerical example, a 1 A negative beam 5 mm in radius in a 0.16 m diameter vacuum chamber would have a space charge field of 12 kV/m at the beam edge, resulting in an H_2^+ ion bounce frequency of 1.7 MHz. The beam potential at the centre would be ~ 200 V with typical azimuthal gradients (longitudinal field), resulting from beam and chamber size variations, of one to several V/m. With 1 V/m, 0.1 ms would be sufficient to get a drift velocity of 5000 m/s, well in excess of the thermal velocity $\dot{z}_0$. This illustrates the fact that the ion motion in field-free regions is governed by the beam space charge. In particular, so-called neutralization pockets may exist in places where the beam potential is deepest (bellows, chamber enlargments etc.): ions created there do not have enough energy to overcome the potential barriers to eventually reach the nearest clearing electrodes.

Neutralization pockets — or potential barriers — may also be created by highly insulating ceramic vacuum chambers becoming electrostatically charged [7]. Metallization of their inner surface is therefore important, as is the necessity to keep the vacuum chamber cross section as uniform as possible (shielded bellows, screens, etc.), thus, in this respect, joining the conditions imposed by impedance considerations. A precise knowledge of the potential variation around a machine is therefore important, as this also determines the strategic locations of the clearing electrodes. Better expressions exist for more precise calculations of the beam potential of a beam at any location in a rectangular vacuum chamber. For elliptic beams with Gaussian transverse distributions, closed solutions have been given for the electric field [10] and the beam potential [11].

Using the beam envelope (Twiss) parameters, and the vacuum chamber dimensions as input to a computer program, these formulae can be used to produce a plot of the beam potential, maximum electric field, ion bounce frequencies, etc. around a machine [12], useful for locating clearing electrodes.

3.1.1 Transverse Distribution of Ions

In contrast to a widely used assumption, it has recently been shown that the transverse distribution of ions produced at rest is not a replica of the beam's distribution, but that instead it is composed of a central core narrower than the beam, with tails, if ion–ion and ion-neutral gas molecule forces can be neglected [13]. By explicity solving the Liouville equation in a one-dimentional model, assuming a transverse beam Gaussian distribution, it can be shown that the transverse phase space distribution of ions produced at rest and oscillating in the beam space charge is greatly diluted at the beam edges when projected on its horizontal transverse axis (Fig. 4).

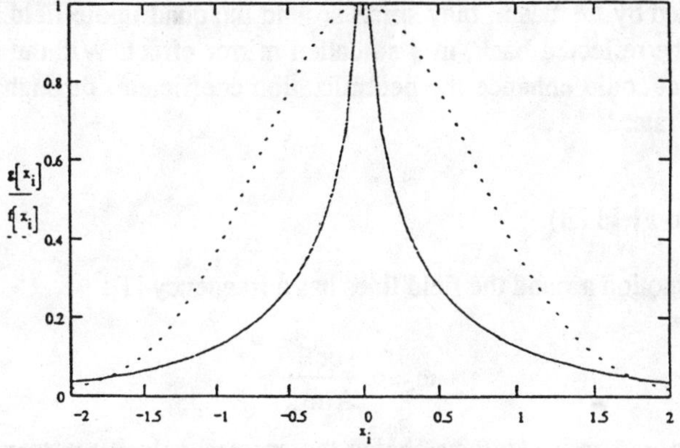

Fig. 4. Ion-cloud profile for ions created at rest (no thermal energy spread). The dashed line is the electron beam; the solid line is the ion cloud.

3.2 Drift Motion in Dipoles and Quadrupoles, Magnetic Mirror (Containment Effect)

The motion of a non-relativistic trapped species of charge qe and mass $A_1 m_p$ is well covered in Ref. [1] and classical electrodynamics textbooks, but will now be summarized for the more usual magnetic fields.

3.2.1 Quadrupole Fields (Gradient dB/dr)

The main effect is a longitudinal drift (perpendicular to the field lines) with velocity:

$$V_D = W_{kin} \frac{1}{qeB^2}\left(\frac{dB}{dr}\right), \tag{24}$$

neglecting the small additional term caused by the field lines curvature. W_{kin} is the ion's kinetic energy perpendicular to the beam axis. For ions produced at the centre of the beam, the drift can be very slow indeed.

Example: minimum ion energy (thermal): $W_{kin} = 4 \times 10^{-2}$ eV (ion produced at the centre of the beam)

$$B = 1\text{ T}, \quad \frac{dB}{dr} = 10\text{ Tm}^{-1}, \tag{25}$$

yields a drift velocity of $V_D = 0.4$ m/s (for $W_{kin} = 10$ eV, $V_D = 400$ m/s).

Because of this possibility of relatively high neutralization, clearing electrodes are sometimes installed in quadrupoles, possibly combined with beam position electrostatic pick-ups. This is the case of the CERN AA, in which another effect has been experimentally seen when measuring the ion current drawn by these electrodes: namely an enhanced ionization rate possibly caused by a containment effect of electrons. As hinted in Ref. [1], electrons from the primary ions, of the same charge and thus repelled by the beam, may spiral around the quadrupole field lines towards the poles and be reflected back, in a so-called mirror effect. Without clearing, this additional effect could enhance the neutralization coefficient, through an increased ion production rate.

3.2.2 Uniform Field (B)

The cyclotron motion around the field lines has a frequency [1]:

$$\omega_{ci} = \frac{qeB}{A_i m_p}, \tag{26}$$

with a radius $r_i = A_i m_p v_\perp / qeB$, $v_\perp$ being the species' velocity perpendicular to the field lines.

The motion of the radius of gyration along the lines of force ($V_{//}$) is not affected; the spinning particle vertically, in dipoles, behaves as in field-free regions.

Owing to the combined action of the beam's space charge field $E_{(r)}$ and B, the longitudinal cross-field drift velocity is independent of the charge and mass of the ion, to a first approximation:

$$V_0 = \frac{E(r)}{B}. \tag{27}$$

If one takes into account the non-linearity of the electric field $E_{(r)}$, there is a slight dependence on the ion mass: heavy ions drift out more slowly than light ones [14].

Trapped species drift in opposite directions on each side of the beam, and the velocity at the centre is zero. As an example, for a 0.1 A beam with a radius of 5×10^{-3} m in a field of 1 T, the field at the beam edge is 1.2×10^3 V/m, and the drift velocity 1200 m/s. But the fact that the drift velocity falls to zero at the centre of the beam may result in a relatively high neutralization in dipoles. Experiments of beam shaking (see Section 6) in the CERN AA seem to support this fact, as hinted by the behaviour of the clearing currents drawn at the extremities of long bending magnets [15], the only ones around the ring to react to the shaking perturbation. In addition, the observation that neutralization effects can be reduced mainly with a vertical shaking of the beam near ion bounce frequencies (the CERN AA and the EPA) concords with the hypothesis that it affects ions in dipoles, where the vertical motion is the only degree of freedom that a coherent beam force can excite.

3.2.3 Dipole Fringe Fields

It has been shown [16] that slow ions with low kinetic energy drifting from no-field regions, can be reflected by the longitudinal gradient $\partial B/\partial z$ of the dipole field (fringe field). An ion oscillating horizontally in the beam space charge potential, and drifting towards a dipole, starts a cyclotronic motion around the vertical field lines, with a decreasing radius as it proceeds towards this magnetic field. Motion reversal may occur for some initial conditions. Dipole fringe fields can therefore represent potential barriers for trapped species, and straight sections between dipoles must incorporate clearing electrodes. This effect, added to the field reversal, could be very detrimental in terms of ion trapping for undulators in electron storage rings.

4 Ion Dynamics (Bunched Beams)

4.1 Field-free Regions

Up to now we have only considered unbunched beams. Obviously bunching does not change the ion production rate but with our neutralization coefficient definition inside the bunch:

$$\eta_{max}\left(N_i = N_p\right) = \frac{1}{B} ,$$

with the bunching factor defined as

$$B = \frac{2\pi R}{n l_b} >> 1 ,$$

R being the machine radius, n the number of bunches, and l_b the bunch length.

A trapped ion will therefore be submitted to the repetitive bunch space charge centripetal (focusing) force, independent of neutralization. If the ion motion is slow, and/or the ion heavy, or if the bunch spacing is uniform and small with respect to the machine circumference, the ion will mainly respond to the d.c. component of the Fourier expansion of the passing charge. In this case, on the average, the ion dynamics will be that of an unbunched beam, plus some stable oscillations. If these conditions are not met, the ions may perform resonant oscillations, thus becoming unstable and lost to the chamber wall. The ion motion in fact obeys a Hill's equation, similarly to the beam particles in a synchrotron. Analogously, the beam bunches represent thin focusing lenses for the ions. With uniform bunch repartitions, and in the linear approximation of a uniform beam, the analysis leads to the concept of a critical mass (or rather mass-to-charge ratio) above which ions perform stable oscillations and thus can be trapped [1]:

$$A_c \geq \frac{N_p r_p}{\beta n^2} \quad \frac{\pi R}{b^2\left(1 + \frac{a}{b}\right)} . \tag{28}$$

Equation (28) ensures vertical stability of ion motion in a uniform elliptic beam of horizontal and vertical sizes b and a, respectively. For a similar horizontal stability criterium, b and a have to be interchanged in (28). Since usually $b < a$, the vertical stability criterium is the most stringent one. This expression shows that the critical mass increases (i.e. neutralization is likely to be less severe) with the number of beam particles, with reduced beam emittances, and with a smaller number of bunches. Figure 5 relevant to the CERN EPA machine, confirms this behaviour of the critical mass with varying numbers of bunches. For a given beam intensity, emittances (and neutralization) are lower for a small number of equidistant bunches, which corresponds to an increased critical mass.

On the other hand, for a given number of bunches, the calculated critical mass (circled numbers) remains more or less constant as the number of beam particles increases. This is because the increase in N_p in Eq. (28) is compensated by an increase in beam sizes a and b: neutralization therefore remains more or less constant as intensity increases, meaning more ions in absolute numbers, i.e. an increasing strength of non-linear forces on individual particles. This leads to increasing emittances as particle accumulation proceeds.

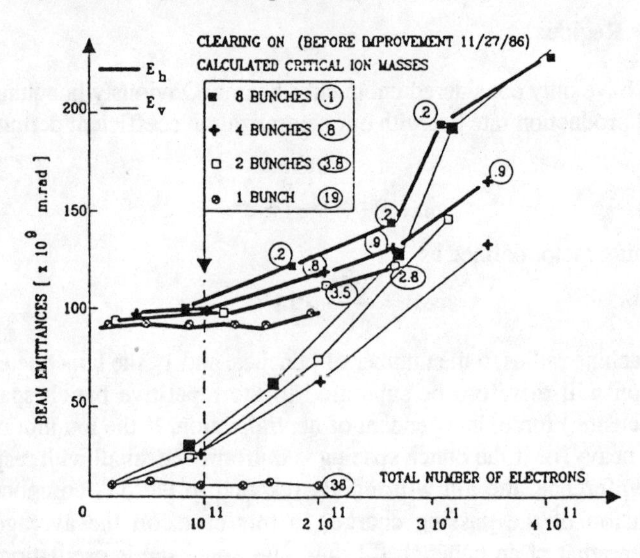

Fig. 5. Beam emittances and ion critical masses in the EPA for equidistant electron bunches

For completeness, it must be stressed that so far the criterium of ion stability has assumed linear and regular forces. Computer simulations can easily include the non-linearity of the force, important at large amplitude, and show that a large fraction of ions which ought to be stable according to the linear theory are in fact unstable if created with large initial position and velocity (r_0, $\dot{r}_0$). In addition, bunches are rarely equally populated, and this represents random gradient errors for the ion motion, in a similar way to synchrotrons, leading to 'enlarged' ion stopbands [17]. Finally, ions represent additional thick defocusing lenses for themselves, and a detailed analysis shows that this influences the limit of accumulation (although usually only weakly).

4.2 In Magnets (Bunched Beams)

We have seen that in uniform fields (dipoles), the vertical motion of the centre of gyration is unaffected by the presence of the magnetic field B. Therefore the vertical stability criterium (Eq. (28)) holds.

In addition, in some rare cases and rather fortuitously, there may be a resonant condition between the horizontal cyclotron motion of the ions and the frequency of the passing bunches, leading to horizontal instability. In quadrupoles, the same argument holds to first order along the field lines with, in addition, an increased longitudinal drift velocity as resonant ions gain energy (see Eq. (24)).

5 Clearing Means

5.1 Clearing Electrodes

Clearing electrodes consisting of negatively polarized plates fitted into the vacuum chamber provide a transverse electric field which diverts beam-channelled ions onto them, where they are neutralized and return into the gas phase. Figure 6 qualitatively represents the potential variation across a vacuum chamber of radius r_c, with an electrode on one side with a potential of U_{ce}.

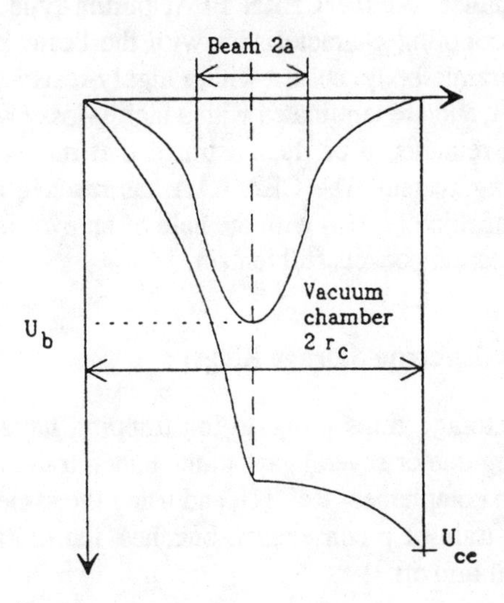

Fig. 6. Transverse potential distribution across a vacuum chamber, due to the beam space charge and a clearing electrode

A minimum condition for capture of the passing ion is that the transverse field provided by the electrode equals the maximum beam space charge field:

$$\frac{U_{ce}}{2r_c} > E(a) \sim \frac{I}{2\pi\beta c\varepsilon_0 a} . \qquad (29)$$

In fact, because ions may have transverse and longitudinal velocities corresponding to energies of up to a few eV, and because a clearing electrode is necessarily limited in size, the field provided by the electrode must usually be larger than the calculated beam field. For instance, in the EPA ring, where electrodes are of the button type with a diameter of 20 mm and are installed flush to the beam, transverse clearing in field-free regions is complete with the following parameters [18]:

$$U_{ce} = -6 \text{ kV} \tag{30}$$

electrode field: 30 kV/m on beam axis
beam max. field: 12 kV/m for I = 0.3 A
and ~10^{-8} mrad horizontal emittance with 10% coupling.

The number and the optimum location of electrodes are in principle dictated by the tolerable degree of residual neutralization. In practice, even with a large number of electrodes, uncleared pockets always remain, and contribute to typical residual neutralizations of a fraction to a few per cent.

No small electron storage ring exists which has reached a fully satisfactory ion-free situation, even with clearing electrodes. Perhaps one reason for this is that up to now clearing systems have not been very complete, partly owing to the fact that clearing electrodes complicate mechanical design of the vacuum chamber and may contribute to the machine impedance. At the CERN EPA, button-type clearing electrodes presenting negligible coupling characteristics with the beam have been designed. They are made of a ceramic body, coated with a highly-resistive glass layer (thick-film hybrid technology), and are terminated with a highly-lossy wide-band filter [19].

Nevertheless, it still remains to be demonstrated that full beam clearing can be achieved with a clearing system. The CERN AA has reached a low neutralization level for a d.c. beam machine (< 1%) with the help of an ever increasing number of electrodes, but neutralization pockets still remain.

5.2 Missing Bunches (Electron Storage Rings)

Many small electron storage rings prone to ion trapping have partly solved their problems by introducing one or several gaps in the bunch train, by not filling certain buckets at injection. To complement Ref. [1], and using the same notations, over one revolution period of a train of p consecutive bunches, the motion of a trapped ion (vertical here) is the solution of:

$$\begin{bmatrix} y \\ \dot{y} \end{bmatrix}_1 = M_{T_y} \begin{bmatrix} y \\ \dot{y} \end{bmatrix}_0 , \tag{31}$$

with the transfer matrix

$$M_{T_y} = \left(\begin{bmatrix} 1 & 0 \\ -a_y & 1 \end{bmatrix} \begin{bmatrix} 1 & \frac{1}{f} \\ 0 & 1 \end{bmatrix} \right)^p \begin{bmatrix} 1 & (h-p)\frac{1}{f} \\ 0 & 1 \end{bmatrix} . \tag{32}$$

The terms in parenthesis represent the linear kick received at the p bunch passages where [1]:

$$a_y = \frac{N}{n} \frac{r_p c}{b\left(1+\frac{a}{b}\right)} \frac{1}{A_i}, \tag{33}$$

interleaved with drifts of duration 1/f, f being the radio frequency. The period of p successive kicks plus drifts is terminated by the drift in the time interval (h – p) 1/f, where h is the cavity harmonic, i.e. the maximum number of bunches that the machine can handle. The Floquet's condition of stability for the ion of mass-to-charge A_i:

$$-2 < \mathrm{Tr}\left(M_{T_y}\right) < 2\,, \tag{34}$$

does not lead to a simple criterion defining which A_i are stable. Rather, the trace of the transfer matrix (Eq. (32)) is of the order p in N, the total number of circulating particles. This means that there are p stable bands of ion mass-to-charge ratios for a given N, or that a given ion will be stable or unstable, depending on the number of beam particles, or on its location around the ring.

Figure 7 illustrates the conditions of linear stability for various ions in the EPA ring, as a function of the number of beam particles and consecutive bunches [14]:

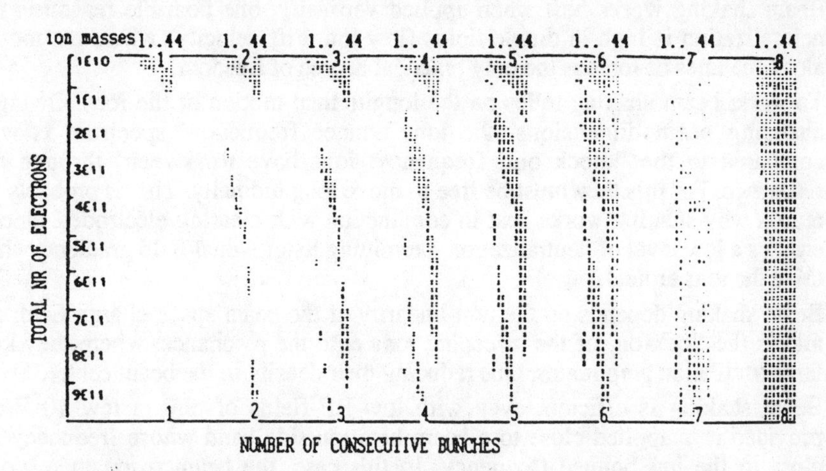

LINEAR STABILITY OF IONS OF MASS=1.2.12.16.18.20.22.28.40.44
AS FUNCTION OF THE NR OF CONSECUTIVE BUNCHES & TOTAL NR OF ELECTRONS
RMS HOR.&VERT. BEAM SIZES (mm)=.75 .25
(EPA average nominal beam dimensions with 10% coupling
and 9 10^{-8} m.rad nominal emittance (0 coupling)
(dotted vertical lines indicate stability for the given ions...)

Fig. 7. Linear stability of ions in the EPA as a function of total beam intensity and the number of consecutive bunches for nominal beam emittances

Although not always absent, ion trapping is indeed less severe in the EPA with missing bunches. This stems from the fact that since a given trapped ion can be further ionized, it then has a good chance of falling in an unstable band, and thus of being cleared.

6 Resonant Transverse Shaking of the Beam

Quite recently, neutralization effects have been considerably reduced when exciting vertical coherent oscillations with a transverse kicker at a given frequency [20], both in electron storage rings (bunched) and Antiproton Accumulators (unbunched). This technique of 'RF knock out' has been determinant in solving ion problems in the CERN AA, where it has been studied both theoretically [21] and experimentally [22], and where it is permanently implemented, with the following parameters:

shaking: vertical
shaking frequency: 490 kHz
sideband frequency (fractional tune q_v): 480 kHz
length of kicker electrodes: 0.6 m
kicker field: ~ 20 V/cm.

Although still at an early stage of both understanding and development, some beam shaking experimental observations can be summarized as follows:

1) Beam shaking works best when applied vertically: one possible reason is that neutralization is high in dipole fields (low ion drift velocity) where the motion along the lines of force is the only practical degree of freedom.
2) To work, beam shaking relies on the longitudinal motion of the ions. Owing to changing beam dimensions, the ion 'bounce frequency' spectrum is wide compared to the 'knock out' frequency: ions have to 'sweep' through this resonance. For this they must be free to move longitudinally. This is probably the reason why shaking works best in conjunction with clearing electrodes, since it ensures a low level of neutralization, permitting longitudinal field gradients which drive the ions around.
3) Beam shaking depends on the non-linearity of the beam space-charge field: this allows the 'lock-on' of the sweeping ions onto the resonance, where they keep large oscillation amplitudes, thus reducing their density in the beam centre.
4) Beam shaking is efficient even with low RF fields of only a few 10 V/cm, provided it is applied close to a beam betatron side band whose frequency lies close to the ion bounce frequency. In this case, the beam resonant response ensures sufficiently large non-linear forces on the ion. Experimentally it is found that for a weakly exciting RF field, shaking works best above a band (n + Q) or below a band (n – Q). This observation of assymetry of weak resonant shaking is important in that it validates the non-linear character of the ion motion and the 'lock-on' conditions.

To illustrate this in a simple way, we use a quasi-linear description of the two-body resonant conditions (from D. Möhl, see also Ref. [21]) for an unbunched beam. We consider only one ion species i, of mass-to-charge ratio A_i, with the following definitions:

Ω: circular revolution frequency of circulating beam ($\Omega = 2\pi f_i$)
$Q_i = 2\pi f_i / \Omega$ the ion bounce number in the beam potential well

$$\Omega_i Q_p^2 = \frac{2N_p r_p c^2}{\pi b(a+b)\gamma R}. \qquad (35)$$

If Q_v is the beam particle unperturbed incoherent tune, and Q_p the beam particle bounce number in the ion-potential well where

$$\Omega^2 Q_p^2 = \frac{2N_i r_p c^2}{\pi b(a+b)\gamma R}, \qquad (36)$$

and

$$Q = (Q_v^2 + Q_p^2)^{1/2}$$

is the perturbed beam tune, then a beam particle and an ion obey the coupled set of linear differential equations

$$\text{particle: } \left(\frac{\partial^2}{\partial t^2} + \Omega \frac{\partial^2}{\partial \Theta^2}\right) y_p + Q^2\Omega^2 y_p - Q_p{}^2\Omega^2 \bar{y}_i = Fe^{i\omega t} \qquad (37)$$

$$\text{ion: } \left(\frac{d^2}{dt^2}\right) y_i + Q_i{}^2\Omega^2 \left(y_i - \bar{y}_p\right) = 0,$$

where the bar on y denotes the average vertical position of each beam, and the $Fe^{i\omega t}$ term is the harmonic of the external driving force close to beam and ion resonance:

$$\omega \sim (n \pm Q)\,\Omega \sim Q_i\,\Omega.$$

Assuming solutions of the form:

$$y_p = \xi_p e^{i(n\Theta + \omega t)} \qquad (38)$$

$$y_i = \xi_i e^{i\omega t}$$

the ion amplitude becomes:

$$y_i = \frac{Q_i{}^2 \dfrac{F}{\Omega^2} e^{i\omega t}}{\left(x^2 - Q_i{}^2\right)\left((n+x)^2 - Q^2\right) - Q^2{}_p Q_i{}^2} \qquad (39)$$

with

$$x = \frac{\omega}{\Omega} \rightarrow (n \pm Q) \sim Q_i\,. \qquad (40)$$

Therefore, shaking works when y_i becomes large, i.e. when the denominator $\rightarrow 0$. But, as shown in Fig. 8, as the ions gain large amplitude, we have non-linear detuning such that $x^2 > Q_i{}^2$ (lock-on). Therefore for y_i to become large, requires that:

$$(n+x)^2 - Q^2 > 0,$$

i.e.:

$x < n - Q$ for excitation near a 'slow wave' beam frequency: $\omega \cong (n - Q)\Omega$: $n > Q$
$x > n + Q$ for a 'fast wave' frequency: $\omega \cong (n + Q)\Omega$: $n > Q$

This asymetry has been verified both in the CERN and Fermilab antiproton accumulators, and in the EPA ring [20].

To conclude on beam shaking as a means to suppress ion effects, it must be stressed that this technique is still at an early stage of development although already applied permanently to Antiproton Accumulators [22]. In the CERN EPA, shaking has made it possible to overcome a neutralization threshold which no other method (clearing electrodes, transverse kicks, etc.) could achieve. From tune-shift measurement, clearing electrodes reduce the neutralization coefficient down to typically 2–3%. Shaking the beam reduces it further to well below 1%.

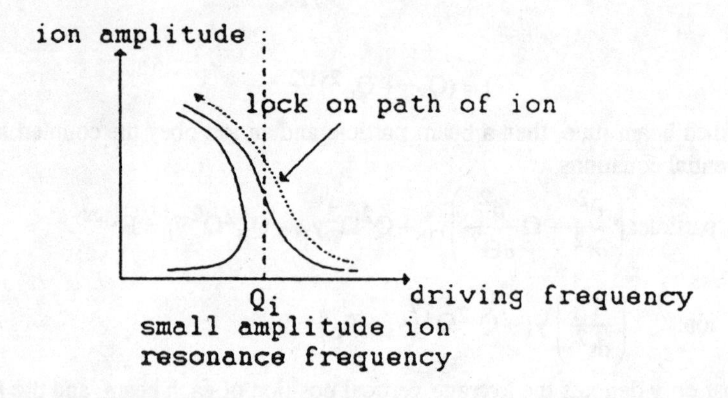

Fig. 8. Qualitative amplitude response curve of an ion versus the driving frequency, near resonance

7 Conclusion

Often very detrimental in their effects, ions may represent a real challenge to machine designers. We have seen that they can be partly eliminated with clearing electrodes, but this requires a careful design if one wants to avoid increasing the machine impedance. Empty buckets on bunched machines and, more recently, beam shaking, are other means which are used to further reduce their numbers. The latter is still at the experimental stage but represents a real hope for some future machines where the very small beam sizes and high intensities may render very difficult the design of clearing electrodes.

References

[1] Y. Baconnier, Neutralization of Accelerator Beams by Ionization of the Residual Gas, CERN/PS/PSR/84–24, CERN Accelerator School, Gif-sur-Yvette, France, 1984, CERN 85–19, pp. 267-288.

[2] S. Battisti, M. Bell, J.-P. Delahaye. A. Krusche, H. Kugler, J.H.B. Madsen and A. Poncet, 12th Int. Conf. on High Energy Accelerators, Fermilab, p. 2050, 1983.

[3] S. Kamada, Survey of Particle Factories Proposals and Projects, KEK Preprint 92–119, XVth Int. Conf. on High Energy Accelerators, Hamburg, Germany, 1992.

[4] D. Sagan, Ion Trapping in the CESR B-Factory, unpublished, CBN 91–2, 1991.

[5] J.D. Jackson, *in* Classical Electrodynamics (John Wiley & Sons, Inc., New York, 1962).

[6] F.F. Rieke and W. Prepejchal, Ionization Cross Sections of Gaseous Atoms and Molecules for High-Energy Electrons and Positrons, Phys. Rev. **A6** (1972).

[7] F. Pedersen and A. Poncet, Proton-Antiproton Instability in the AA, CERN PS/AA/ME Note 81, 1981.

[8] S.C. Brown, *in* Basic Data of Plasma Physics (Technology Press, J. Wiley & Sons, Inc., New York, 1959).

[9] B. Innert, Unpublished Note, ISR/VA/tn, 1977.

[10] M. Gygi-Hanney and B. Zotter, Field Strengh in a Bi-Gaussian Beam, CERN/LEP/Theory Note 44, 1987.

[11] R. Alves-Pires, Conformal Mapping for Two-dimensional Electrostatic Beam Potential Calculations, CERN PS/87–66 (AA), 1987.

[12] R. Alves Pires, Beam Dimensions and Beam Potential in the CERN Antiproton Accumulator Complex , CERN PS/87–70 (AA), 1987.

[13] P. Tavarès, Transverse Distribution of Ions Trapped in an Electron Beam, CERN-PS/92–55 (LP), 1992.

[14] A. Poncet, Trapping of Ions in the EPA Electron Beam: Stability Conditions and Diagnosis, CERN/PS 88–14 (ML), 1988.

[15] F. Pedersen, A. Poncet and L. Søby, The CERN Antiproton Accumulator Clearing System, CERN PS/89–17 (ML), 1989, 1989 Particle Accelerator Conf., Chicago, 1989.

[16] Y. Miyamara, K. Takayama and G. Horikoshi, Dynamical Analysis on the Longitudinal Motion of Trapped ions in Electron Storage Rings, Nucl. Instrum Methods in Phys. Res., **A270** (1988) 217.

[17] G. Brianti, Y. Baconnier, O. Gröbner, E. Jones, D. Potaux and H. Schönauer, Proceedings of the Workshop on $p\overline{p}$ in the SPS, SPS $p\overline{p}$. 1, p. 121, 1980.

[18] J.-C. Godot, K. Hübner and A. Poncet, Comparison of the Electric Field in the Beam to the Available Clearing Fields in EPA, Note PS/LPI/87–25, 1987.

[19] F. Caspers, J.P. Delahaye, J.C. Godot, K. Hübner and A. Poncet, EPA Beam-Vacuum Interaction and Ion Clearing System, CERN PS/88–37 (ML), Proc. EPAC, Rome, 1988.

[20] Y. Orlov, The Suppression of Transverse Instabilities in the CERN AA by Shaking the $\overline{p}$ Beam, CERN PS/89–01 (AR), 1989, Proc. Particle Accelerator Conf., Chicago, 1989.

[21] R. Alves-Pires et al., On the Theory of Coherent Instabilities due to Coupling between a Dense Cooled Beam and Charged Particles from the Residual Gas, CERN/PS/89–14 (AR), Proc. 1989 Particle Accelerator Conf. Chicago, 1989.

[22] J. Marriner, D. Möhl, Y. Orlov, S. Van der Meer and A. Poncet, Experiments and Practice in Beam Shaking, CERN/PS/89–48 (AR), 1989, Proc. 14th Int. Conf. on High Energy Accelerators, Tsukuba, Japan, 1989.

Backgrounds at e^+e^- B Factories

*Richard D. Ehrlich**

Laboratory of Nuclear Studies, Cornell University, Ithaca, NY 14853, USA

1 Introduction

B-factory designers and prospective users have learned that the viability of such facilities is crucially dependent upon control of machine-related backgrounds in the interaction region (IR). Photons from synchrotron radiation (SR), and the debris that follows loss of stored beam, can compromise the performance and thwart the mission of the elaborate and expensive experimental detectors. This danger is relevant to tau-charm and phi factories as well. My goal is to make clear to non-experts why this is so, and to explicate the underlying physical processes and important defensive measures. We shall stay close to reality by comparing simulation to actual experience at the CLEOII detector— CESR storage ring complex at Cornell.

Why should one expect background difficulties at B factories? First, backgrounds are not negligible now, at luminosities of $2 \cdot 10^{32}$ cm^2sec^{-1}; future facilities will require currents and luminosity 10 to 50 times greater. Second, backgrounds rise faster than linearly with beam current; since gas evolution is driven by SR, the beam-gas backgrounds will grow quadratically with current. High beam energies (for asymmetric colliders) and the very strong quadrupole fields required for small β^* and rapid beam separation exacerbate the problems posed by SR from magnets near the IR. The experiments will employ thin beryllium beampipes at (typically) 2.5 cm radius, with little inherent resistance to penetration by x rays. Finally, the HEP physics is "high-precision physics," especially vulnerable to corruption by backgrounds.

2 An Overview of the Disease

The two dominant sources of machine-related background are SR x rays and debris from interaction of stored beam with residual gas molecules. Both Coulomb scattering and bremsstrahlung can divert electrons and gamma rays into material near the IR; these may then shower in that material, sending secondary electromagnetic or hadronic particles into the experiment, or the primary may penetrate the beampipe directly. In all these cases the radiation can influence the experiment in three ways: through radiation damage to the detector elements, through spurious occupancy , i.e. hits in the detector which pre-empt or obscure the desired data, or through extra trigger rate should the spurious hits satisfy the trigger requirements. In the typical B factory detector, the elements most impacted by radiation are 1) the inevitable silicon

* Work supported by the U.S. National Science Foundation

vertex detector just outside the IR beampipe, used to pin down the event interaction point, and 2) the subsequent gas-proportional tracking devices which provide momentum reconstruction in the superposed magnetic field. The two background sources have different signatures in the detector and differ in their dependence on machine conditions. Table 1 shows the qualitative differences in capsule form.

Table 1. Comparison of synchrotron radiation and beam gas characteristics

	SR x rays	Beam-gas debris
Energy (typical)	1keV to >100keV	1MeV to GeV
Deposition in gas	full energy	~2keV/cm
Deposition in silicon	full energy	~100keV/ layer
Spatial distribution	pointlike	can make tracks
Beam energy dependence	rapid rise vs. beam energy	weak dependence
Lattice dependence	yes	yes
Vacuum dependent	yes	no
Radiation damage?	yes	yes

Limits for both sources were first formulated by T. Browder and M. Witherell at the 1990 Snowmass Workshop [1], based on tolerable radiation damage and occupancy.

3 Synchrotron radiation in detail

3.1 Nature of produced radiation

The physics of SR production is well covered in many texts. I have leaned most upon Jackson [2] and Sands [3]. SR is produced whenever a charged particle is bent in a magnetic field; its spectrum and intensity depend on the particle's instantaneous radius of curvature, ρ, and energy-to-mass ratio, γ. The radiation is very dominantly produced in the plane containing the velocity and acceleration. The angle out of this plane, ϕ, is limited to a few times γ^{-1}. Thus, the radiation pattern resembles that of a train's headlight, tangent to the train's trajectory.

The energy spectrum of the radiation is exactly describable in terms of Bessel functions of third-integral order; in principle the spectrum depends on ϕ . Though it can be calculated easily enough on modern computers, it is useful here to work with more memorable approximate forms. DeStaebler [4] provides a compendium of simplified forms for integral and differential number and power spectra. The most important parameter characterizing these distributions is the critical energy, u_{crit} , above which lies half the SR power. Above an energy of several times u_{crit}, the photon spectrum falls (approximately) exponentially with $(u_{crit})^{-1}$ as decay rate.

The critical energy can be expressed as

$$u_{crit}\,(\mathrm{keV})=2.\,2[(E_{beam}\,(\mathrm{GeV})]^3\,[\rho\,(\mathrm{meter})]^{-1}. \qquad (1)$$

The photon number-spectrum (per unit bend angle, ω) is a function of the scaled photon energy, $\zeta=u/u_{crit}$, the number of electrons per bunch, N_b, and the fine-structure constant, α :

$$\frac{d^2N}{d\zeta d\omega}=\frac{4}{9}\,\alpha\gamma\frac{S(\zeta)}{\zeta}. \qquad (2)$$

About $20/\rho$ (meters) photons are emitted, per electron and per meter of path. Their average energy is roughly $0.3u_{crit}$. The cubic dependence of u_{crit} on beam energy together with the exponential falloff of the spectrum, $S(\zeta)/\zeta$, explains the extreme sensitivity of background rates to beam energy and bend strength. Consider two beams of different energies, 3 Gev and 9 GeV, traversing the same bending magnet. If the low energy beam (LEB) has u_{crit} = 1 keV, the high energy beam(HEB) will have u_{crit} = 9 keV. The number of photons (per unit bending angle) above, say, 27 keV will be greater for the latter by a factor of about $e^{-3}/e^{-27}=e^{24}$! The high energy beam is vastly more threatening to an experiment, since protection against x rays below some energy like 10 keV is relatively easy. A little thought also shows that phi factories should not have to worry much about SR penetration: the beam energy is just too low.

3.2 How SR enters the experiment. Absorption and scattering.

Figure 1 suggests the typical manner by which the SR, emitted parallel to the particle orbits in both quads and bending magnets, leaves the vacuum chamber without being absorbed by the indicated protective masks. It does so by *scattering*. The ray labeled a scatters nearly backward off the upper-left mask. It may scatter again in the thin beampipe, thus reducing its projected path length through the high-Z coating needed to give some absorptive power to the central beryllium beampipe. Ray b also scatters, but only by penetrating the innermost "tip" of the upstream mask. The favored forward scattering process here is Rayleigh scattering, which has a large cross section at small momentum transfer, with all atomic electrons scattering coherently. Ray c reminds us that SR may scatter into the experiment off distant surfaces, unless these are screened from direct line-of-sight by the masks or other obstacles. The most potent magnets in B factory designs, like separation bends and off-axis IR quads, dump tens of kilowatts of SR on vacuum chamber walls and "crotches" within 15 meters of the IP.

The various scattering and absorption processes have distinctive material and x-ray energy dependent features. Appreciation of these will help to develop intuition about IR designs.

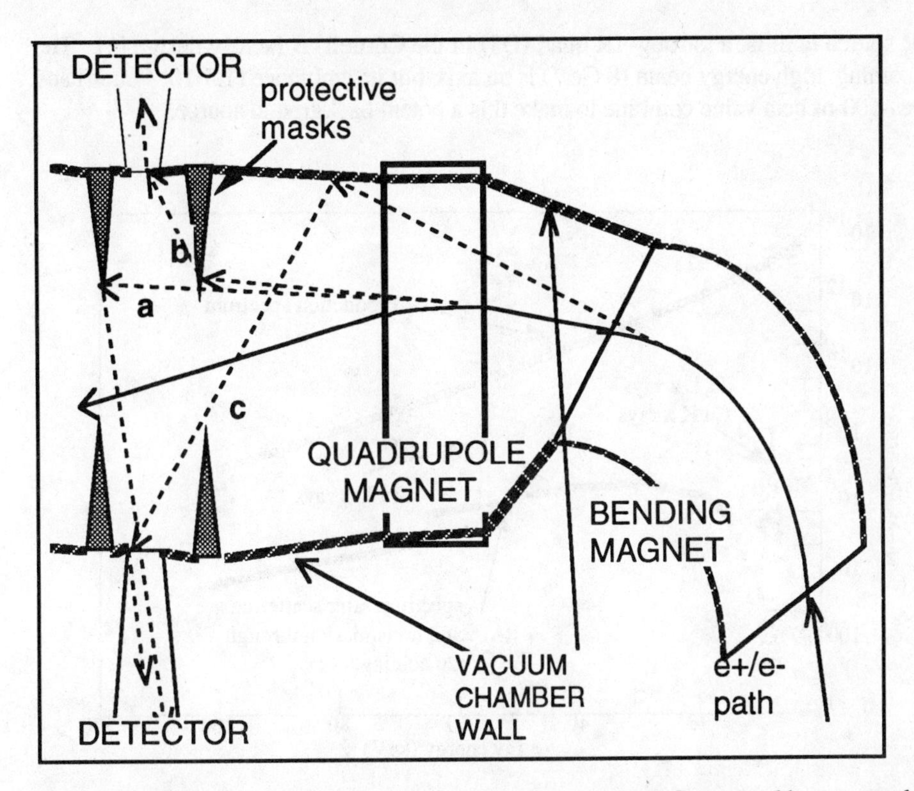

Figure 1 Schematic means by which SR x rays enter experiment. Rays a and b are examples of "backscattering" and "tipscattering" respectively. Ray c scatters off the vacuum chamber walls first. Both quads and bending magnets can be SR sources.

Photoelectric absorption is the only important absorptive process for x-ray energies, 1 keV<u<100 keV, relevant to B, phi or tau-charm factories. The cross section is a strong function of Z and u, varying like $Z^5/u^{3.5}$ for u above the K-edge of the material with atomic number Z. Thus, high Z is necessary to stop high u, but high energy x rays are not easily stopped. (Remember we must keep the thickness of the central beampipe down so as to minimize multiple scattering of B decay products.)

Rayleigh scattering in the forward direction (scattering angles less than 30°) is very important for high Z and low u. It is damped at large angles by form factors but is still comparable to Compton scattering, the incoherent scattering from bound electrons. Compton scattering at these low energies differs significantly from the high energy process. It vanishes in the forward direction but is relatively isotropic elsewhere, and is a weak function of u.

Absorbed x rays yield K,L...fluorescence daughter x rays with probabilty, f_K, $f_{L...}$ — the fluorescence yields. These are emitted isotropically. A fraction 1–f of the time we get instead *Auger electrons,* via internal conversion. These have negligible range in the absorbing material. Figure 2 shows the characteristic sculpting of a smooth input spectrum by the specific material properties, as well as distinctive peaks from fluorescence of both the tantalum protective mask and copper beampipe coating. The

SR source here is a strong IR quad (Q3) in the Cornell B factory design [5]. The incoming high energy beam (8 GeV) is on axis, but its emittance ($1.5 \cdot 10^{-7}$ mrad) and the ~100-m beta value combine to make this a potent background source.

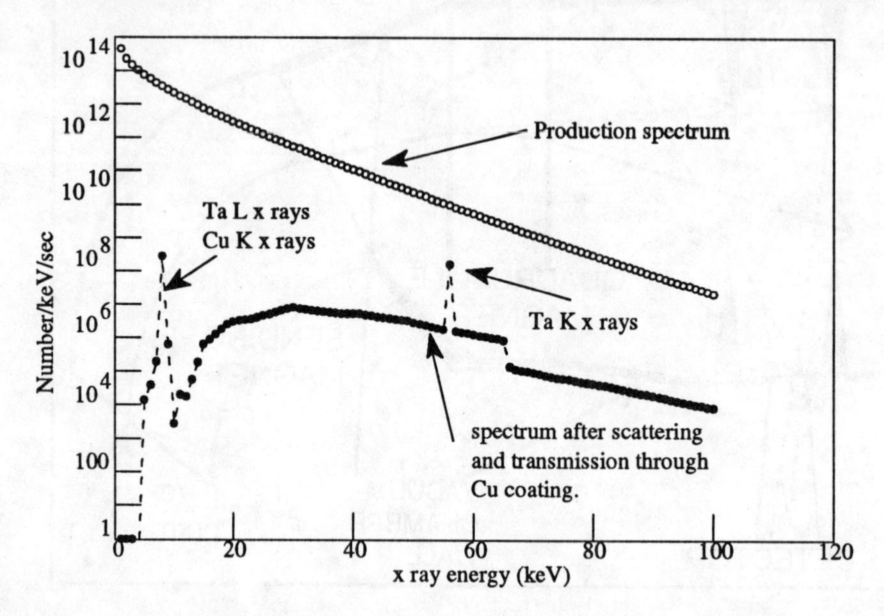

Figure 2 Energy spectrum of SR emitted by the incoming HEB in Q3, incident on down-stream Ta mask (top) and surviving photons (bottom), including fluorescence after scattering and penetrating 25-μ Cu beampipe coating. From Cornell CESR B design study.

3.3 Simulation of SR backgrounds at B factories

3.3.1 Generation from the lattice

Given a lattice, emittance and apertures, one tracks the central orbit of each beam through each relevant magnet, evolving the Twiss parameters and the local radius of curvature simultaneously, using their equations of motion. At each longitudinal position, we integrate over transverse beam size, sending all "colors" of SR along tangents to the orbit. We know the probability distribution for x,x', y,y' in terms of Twiss parameters.and equilibrium positions x_0 and y_0. For x, for example:

$$\frac{d^2P}{dxdx'} = \frac{1}{2\pi\varepsilon}\exp\left(-\left(\frac{(x-x')^2}{2\sigma^2} + \frac{(x' - Cx)^2}{2\delta^2}\right)\right). \qquad (3)$$

With neglect of dispersion, $\sigma=(\varepsilon\beta)^{1/2}$ and $\delta=(\varepsilon/\beta)^{1/2}$. C correlates x and x' and is $=\alpha/\beta$. Exact formulae can be derived even with dispersion present. Photon spectra are then generated according to the local ρ and u_{crit} and are accumulated on "receiving surfaces," from which they may then scatter. The weighted spectra are binned in one-keV intervals. It's crucial to sample the beam tails well, since even particles at 5 sigma may be significant. It should be emphasized that even without an elaborate generation program, a physicist with a straightedge can check simulation results and avoid design disasters, by employing (1), using the exponential approximation for the spectrum, and drawing tangents to the orbit.

3.3.2 Photon interaction code

The SR spectrum generated above provides a set of weights used at the next stage. For each areal element of the receiving surface and for each SR energy bin,photons are scattered or absorbed according to the tabulated properties of the material encountered [6]. The propagation code is a full-blown Monte Carlo calculation, with energy deposition in the various detector elements as its terminus. The beginner or checker can use a multiple scattering approximation to get started. For instance, let μ_{abs} be the absorption length and μ_{scat} be the scattering length in the mask material. Then the number of photons scattered back into solid angle $d\Omega$ is approximately equal to $[\mu_{scat}/(\mu_{abs}+\mu_{scat})]\bullet(d\Omega/4\pi)\bullet$Number incident.

Similar formulae can be deduced for fluorescence, tip scattering, and penetration through the beam pipe, and are valuable for checking or fixing fancier programs. To get things right at the "factor of two" level or better, one must use realistic interaction properties.

3.4 Representative results for a B factory at Cornell

CESR B, with one 8-GeV beam and one 3.5-GeV beam is, from SR considerations, threatened only by the HEB. It uses a ±12-mr crossing angle to achieve beam separation, hence has no strong "separation bend" to worry about, in contrast to the SLAC and KEK designs discussed at this school. It employs the conventional "soft bend" to protect the IR from more damaging upstream radiation and it has the IR quads on axis in both incoming beams. Nevertheless, the design is not immune to SR.

The horizontally focussing IR quad, Q3, is roughly 4 meters away from the IP. With a strength, K, of 0.47 m^{-2} and a horizontal beam sigma of ~5 mm, this magnet implies a ρ of 100 m for 4-sigma particles, corresponding to a u_{crit}>10 keV. This is, as it turns out, enough to produce about 0.3 % occupancy and several krad/ year in the Si vertex detector. If the emittance can be reduced to 10^{-7} m-rad (2/3 the assumed value), the Q3 contribution will drop by a factor of sixteen, simply because far fewer particle see the high-quad fields which can give penetrating radiation.

A scheme that achieves this result is shown in Figure 3. It differs somewhat from that proposed in [5]. The radiation from the HEB soft bend (ρ=500 m) is controlled by avoiding any backscattering component. The sense of bending is crucial here. It is chosen so that the SR "fan" from this magnet first encounters the mask at the upper left and then heads away from the upper right mask at an angle just less than the 12-mrad crossing angle. Tip scattering is minimized in both cases by locating the lower-left tips further(20 cm) from the IP than the lower right ones, which shield the less potent 3.5-GeV beam.

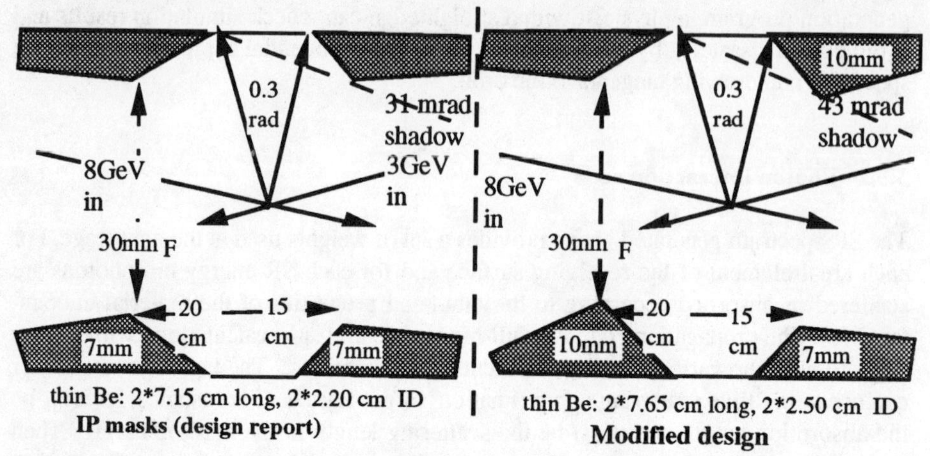

Figure 3 Two mask designs in plan view. The HEB is incident from top left in both cases. The design on the right is 1) almost immune to backscattering of the HEB, and 2) better shadowed from remote sources by the 10-mm-deep tip at upper right.

The depth of the tip at 20 cm from the IP is greater in the modified design. It sweeps the downstream 7-mm-deep mask nearly free of Q3 radiation. An additional improvement comes from the deeper mask at the upper right. This still escapes almost all soft-bend radiation, thanks to the crossing-angle effect, but shadows much more of the downstream vacuum chamber than the first version. That region is subjected to very high fluxes of SR from the *outgoing* 8-GeV beam, which is off axis in the IR quads. In the presence of misaligned masks, the backscattered "crotch" radiation could be troublesome. The price paid for these improvements is increasing the beampipe radius from 2.2 cm to 2.5 cm. Table 2 summarizes the relative merits of these two schemes in terms of occupancy in the silicon vertex detector. Note that the relatively small geometry changes have large rate consequences.

Table 2. Comparison of two masking shemes, Si layer 1 occupancy rates

Source	Design report backscatter	Design report tipscatter	Modified design backscatter	Modified design tipscatter
Q3 (HEB)	1.8%	0.15%	10^{-6}%	0.28%
Soft bend	0.004%	0.018%	0.002%	0.025%
8-GeV crotch	0.3%	------	0.004%	---------
Misalign masks by1 mm	2.8% (Q3) 0.42% (crotch)	------	0.001% (Q3) 0.004% (crotch)	---------

Other machine designs discussed at this school have their unique means of coping with SR. Both the SLAC [7] and KEK [8] designs employ eccentric incoming quads and strong bending magnets just before the IP to effect separation. They require masks to approach very close to the beams to prevent direct SR radiation of the beam-pipe. Tip scattering is dominant in these cases; the separation bending sense ("S-bend") protects the downstream masks from irradiation in a manner analogous to the way that the crossing angle helps the Cornell design. The larger rings differ most dramatically from CESR B in their reduced emittances. The KEK design makes use of a tiny horizontal emittance ($1.9 \cdot 10^{-8}$ m-rad) and an asymmetrically shaped beampipe to assure that SR rays emitted from within six sigma can strike neither the central beam pipe nor the IP masks. Their only SR backgrounds should come from twice-scattered radiation. The bottom line on SR seems to be: all proposed machines can co-exist with both silicon and wire-chamber devices. Attendant radiation doses are all <10 krad/year and Si occupancies are <1% for 1-μsec memory time.

4 Beam-gas backgrounds

4.1 Physical processes

These backgrounds and the underlying physical processes are more familiar to HEP people than the SR lore. The bremsstrahlung process (radiative Coulomb scattering) produces both degraded electrons and hard γ's from collisions with the residual gas. The photons produced in the last 1 to 2 mrad of bend can impinge on the IR masks just like the SR from the same magnet. However, since the energy spectrum of brems photons is almost flat, the typical one has an energy of several GeV. The electrons from the same process will be steered into some part of the vacuum chamber unless they have lost sufficiently little energy to be captured by the RF system. One can speak of the energy loss as engendering a cosine-like betatron oscillation which interferes with the obvious off-energy orbit. The electrons will tend to leave the clear aperture at points of high β and, perhaps, high dispersion.

Elastic Coulomb scattering, in either plane, engenders sine-like betatron motion. Again losses will tend to be at high-β points among which the IR quads are counted.

Poorly understood processes like dust or ion trapping can also produce high rates of beam loss. With bad luck, the ions or dust may be located at a place from which the scattered particles are transported efficiently to the IR. This phenomenon has been seen at CESR and is flagged by occurring only with e^- beams, since these provide an attractive potential well for positive ions. Nonlinear resonances can cause bad life-time and beam loss, but can be avoided by astute accelerator operation.

4.2 The scale of the problem

Unlike SR x rays, few-GeV particles are hard to stop in simple protective masks. Experience and simulation show that it is unwise to make them angry; i.e., the IP must be protected by massive shielding, several radiation lengths thick tranverse to the beam. It is hard to do this. In current machines lifetimes of 4 to 5 hours can be attained, if residual gas pressures are kept in the 10^{-9} torr range. At currents of 1 amp and ring circumferences of about 1 km, > 10^9 particles are lost each second. The IR can perhaps be protected from all but 10^{-4} of these to keep the detector content, but only if the aforementioned pressure is attained. Since pressure at high beam currents tends to be proportional to current via SR outgassing and desorption, high-performance vacuum technology is vital.

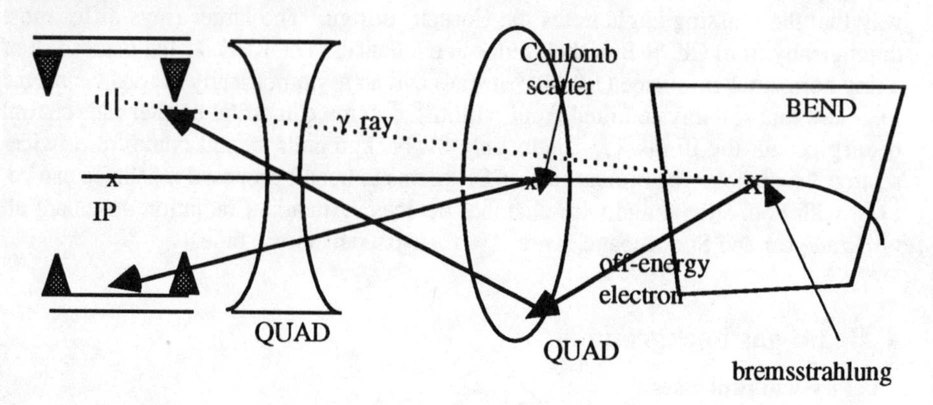

Figure 4. Cartoon illustrating several ways that GeV particles reach the IR. Shown are bremsstrahlung in a bending magnet, andCoulomb scattering in a quad. Two charged particles and one photon impinge on the IR.

4.3 A picture is worth many words

Figure 4 suggests schematically what can happen when beam-gas interactions cause particle loss, but pictures from an actual experiment make a more memorable impression. For reasons of space, we show only two examples from CLEOII. In Figure 5, we see the typical character of electromagnetic shower-remnants in the inner tracking chambers. The energy loss and ionization peak at small radius. The innermost circle is at 35 mm radius and bounds CLEOII's straw-tube chamber. The outermost circle divides the intermediate tracking chamber from the central drift chamber at abut 25 cm radius. This most common sort of event may superpose itself on another legitimate event, but will not by itself trigger the readout of data.

Figure 6 is an example of a background event which will make a trigger. All six tracks seen in the central drift chamber are likely protons, which result from a photonuclear interaction in or near the beam pipe. To simulate such events requires a reliable generator of both hadronic and electromagnetic processes.

CLEOII people believe that it is necessary to sample these sorts of events through random trigger sampling. Embedding Monte Carlo events in a random trigger should provide useful estimates of the effects of beam-gas on resolution, efficiency, etc.

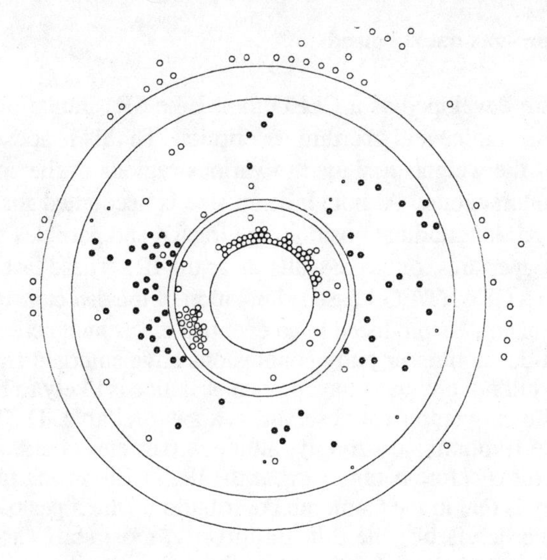

Figure 5. Typical example of shower debris in inner tracking chambers of CLEOII. The innermost chamber starts at 35 mm radius. The picture extends to about 25 cm radius.

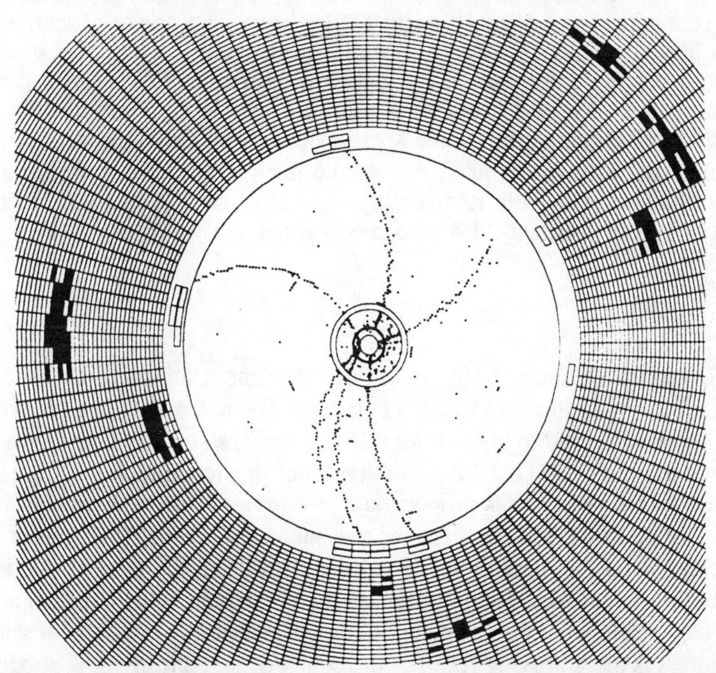

Figure 6. An example of a trigger-generating, hadronic interaction of shower debris near the IP. All or most of the particles in the central drift chamber are protons of less than 1 GeV/c.

4.4 Simulation of beam-gas backgrounds

CESR simulations were developed as a CLEO tool. Like SR simulations they start with a complete machine lattice and aperture description. To this is added a pressure profile which governs the weights assigned to various regions of the accelerator as sources of beam-gas interactions. Although beam size is accounted for, it does not greatly affect results. All longitudinal positions are treated and particles tracked until they are lost to distant apertures or strike walls near the IR. These last "spent electrons" are sent along to a GEANT/EGS based simulation of the detector. It follows the secondary and tertiary daughters produced in an eventual electromagnetic shower.

Very few intuition-developing or pithy conclusions have emerged from this process; in particular it is still not obvious that a particular lattice is likely to be especially good or bad. The following empirical observation appear reliable. 1) The first fifty meters upstream of the IP dominate strongly, although particles could, in principle, travel for one or more turns before being lost near the IR. 2) The strong radial dependence of the occupancy is due to the Coulomb contribution, which peaks at small angles. 3) Although some debris hits the thin Be directly, most stuff showers in upstream material. Choice and depth of this shielding (W, Pb, Ta...) is important. Make it as thick and uniform as possible just outside the beam aperture. 4) The longest-range shower daughters, photons of a few MeV, reach the time-of-flight counters with fair probability, thus influencing the trigger rate; 50-MeV deposits in the endcap CsI crystals are also not uncommon. 5) Fine segmentation of the silicon vertex detector is invaluable. 98% of 10-track B-B-bar events are free of obscuration by background hits. 6) CESR B differs somewhat from PEPII, where the strong separation-bends concentrate the rate on the inside of the rings. 7) The HEB contributes dominantly bremsstrahlung background; the LEB mostly Coulomb scattering. One suspects that phi factories may have some trouble with Coulomb scattering backgrounds, since the rate scales as E^{-2}, and the designs have long race-track straights. 8) The final news is good. Remarkably, dose rates are similar to those due to SR. We expect <1 krad/ year and < 1% silicon occupancy.

4.5 Checks from CESR

In 1994 CLEOII will install a new 2-cm-radius beampipe and a silicon vertex detector. We are also preparing to increase CESR's current to 600 mA and to operate with small crossing angles. We need to know if all this will work, so we have been pursuing a simulation of present (1992) conditions to verify our prediction tools.

We use the fact that the SR rate is proportional to current, I, while the beam-gas rates scale as $aI + bI^2$, to check the ratio of the two sources. We relate the current in the straw tube chamber to its measured gas gain, and read out the current in each layer in 11 azimuthal segments. Inputting beam current, vacuum, materials, etc., and unleashing our computing tools we calculate *ab initio* the chamber currents for various beams. Figure 7 is the gratifying result of the comparison. Agreement is good to within a factor of 1.5. Both SR and beam-gas contributions are needed to get the right azimuthal distribution.

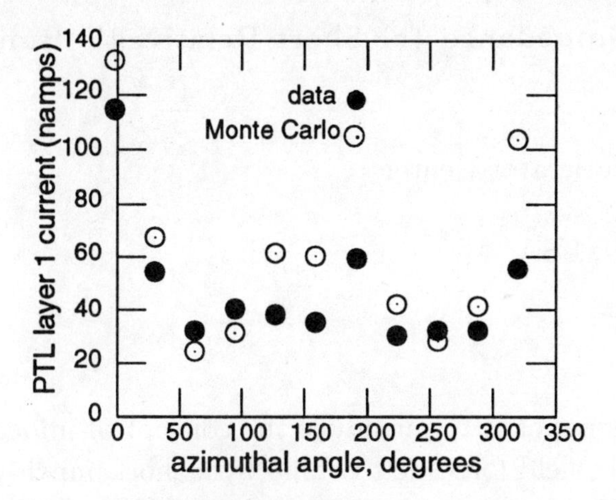

Figure 7. Current in the 11 azimuthal segments of the first layer of CLEOII's straw.tube chamber in nanoamps plotted against azimuth for both data and Monte Carlo. A single positron beam was in CESR.

Last, the simulations tell us that by adding a 2-cm-thick W shield inside the last micro-beta quads of CESR, we can reduce the beampipe to the desired 2 cm radius, increase beam current to 0.6 A, yet see a fivefold improvement in background rate.

5 Conclusions and acknowledgments

The outlook for control of synchrotron radiation and beam-gas backgrounds at B factories looks promising if vacuum techniques deliver the requisite pumping power. Close cooperation between detector users and machine physicists must continue to ensure success as designs evolve. My colleagues at CESR and CLEOII have my sincere thanks and admiration for their diligence and talents in pursuing that goal. Tom Browder, Dave Cinabro, Dan Coffman, and Nari Mistry have all contributed greatly to the simulations and CESR/CLEO checks.

References

[1] Browder, T. and Witherell, M. Limits on backgrounds for the B-factory detectors, Cornell CLNS **90-1019** (1990)
[2] Jackson, J.D. Classical Electrodynamics, Wiley, New York(1962) pp. 477-488
[3] Sands, M. The physics of electron storage rings, SLAC-121 (1970)
[4]DeStaebler, H. Interaction region considerations for a B-factory, SLAC-PUB-**5299** (1990)
[5] CESR-B conceptual design, Cornell CLNS **91-1050** (1991)
[6] Storm, E. and Israel, H. Photon cross sections from .001 to 100 MeV for elements 1 through 100,Los Alamos **LA-3753** (1967)
[7] An asymmetric B factory based on PEP, SLAC **372** (1990)
[8] Accelerator design of the KEK B-factory, KEK report **90-24** (1991)

Introduction to Impedance for Short Relativistic Bunches*

Phil L. Morton

Stanford Linear Accelerator Center
Stanford University
Stanford, CA 94309, USA

1. Introduction

The purpose of this paper is to introduce the concept of impedance to calculate the wake-field forces left behind by a short bunch which travels at relativistic speed through a structure with discontinuities.[1] We will try to be as intuitive as possible and leave the more rigorous derivations to the second paper on this subject by J. Wang.

2. Representation of Cavity by Equivalent Circuit

We will consider the cavity shown in Fig. 2.1 which has rotational symmetry about the z axis and is excited by the beam current, I_B, passing through the gap. For the time being, we will consider only one mode of excitation for the cavity; namely, the mode where the magnetic field $\vec{B}$ is azimuthal around the beam direction as shown. The current I_L flows in the outer cavity wall in the direction shown to oppose the magnetic field in the cavity excited by the beam current. This current I_L causes a build-up of positive and negative charges on the exit and entrance plane of the gap as shown. This charge build-up produces an increasing electric field in the direction opposite to the direction of the beam current.

We will use Gaussian units and write Maxwell's equation as

$$\nabla \times \vec{B} = \frac{4\pi}{c}\vec{J} + \frac{1}{c}\frac{\partial \vec{E}}{dt} \tag{2.1}$$

where $\vec{J}$ is the current density, $\vec{B}$ the magnetic field, $\vec{E}$ the electric field, c the speed of light, and t the time. The coefficient $(\frac{4\pi}{c})$ is

* **Work supported by Department of Energy contract DE-AC03-76SF00515.**

equal to Z_0, the impedance of free space, which in practical units is 377 Ω. We integrate both sides of Eq. (2.1) over the surface area of a plane perpendicular to the direction of the beam current as shown in Fig. 2.1. The area of integration includes the walls of the cavity. The term $\int\int(\nabla \times \vec{B}) \cdot d\vec{A} = \oint \vec{B} \cdot d\vec{l} = 0$, since the boundary of the surface is in the walls of the cavity where $B = 0$. The integral of $\vec{J}$ over the area gives the current $I_B + I_L$. The integral of the time variation of the electric field is defined as a displacement current

$$I_C = \frac{1}{4\pi} \int \frac{\partial \vec{E}}{\partial t} \cdot d\vec{A} \ . \tag{2.2}$$

The surface area integral of Eq. (2.1) yields Kirchhoff's law

$$I_B + I_L + I_C = 0 \ . \tag{2.3}$$

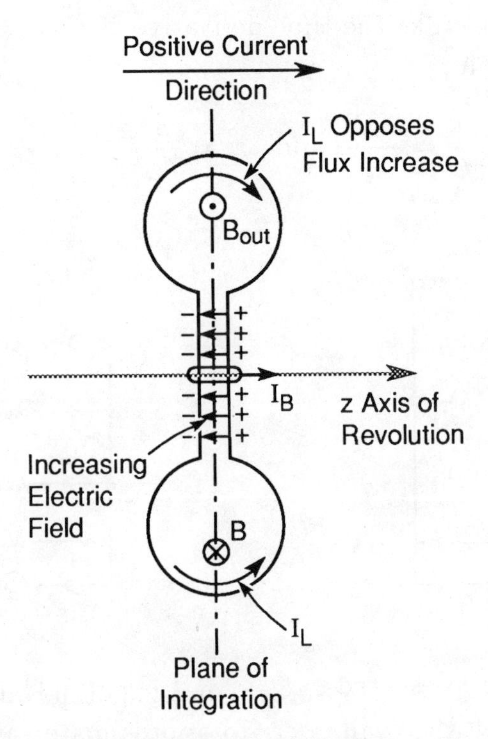

Figure 2.1. Cavity excited by Beam Current I_B.

We can represent this cavity by an equivalent circuit with I_B a source current as shown in Fig. 2.2. Note that the wall resistance of the cavity has been included as a resistance in series with the inductance. This comes about because the finite conductivity of the wall produces a non-zero electric field parallel to the cavity wall which is proportional to the current I_L. The circuit shown in Fig. 2.2 can be excited to large voltages when the time variation of the exciting current I_B is near the resonant frequency $\omega^2 = (1/LC)$. If, at this frequency, the series resistance is small compared to the inductive reactance, i.e. $r << \sqrt{L/C}$, then the circuit in Fig. 2.2 can be well represented by the circuit shown in Fig. 2.3 with a shunt resistance $R = (L/rC)$. For this circuit Kirchhoff's law becomes

$$C\frac{dV}{dt} + \frac{1}{R}V + \frac{1}{L}\int V dt = -I_B \, . \tag{2.4}$$

Capacitive +	Resistive +	Inductive =	Driving
term	term	term	term

It is quite common to take the time derivative of Eq. (2.4) and use the following notation

$$\frac{1}{C} = \frac{\omega_r R}{Q} \quad \text{and} \quad \omega_r^2 = \frac{1}{LC} \tag{2.5}$$

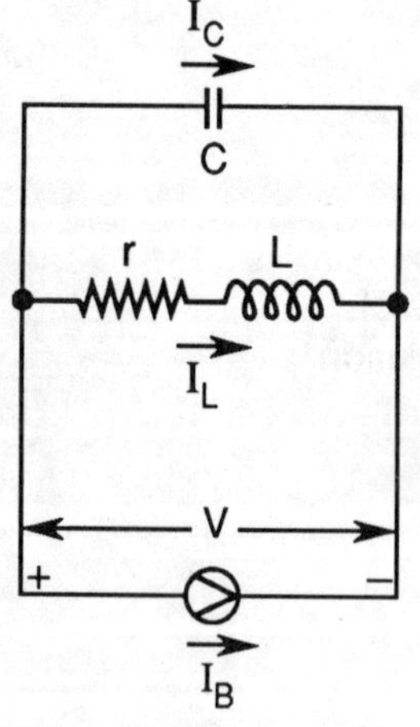

Figure 2.2. Cavity represented by equivalent circuit with wall losses included by series resistance.

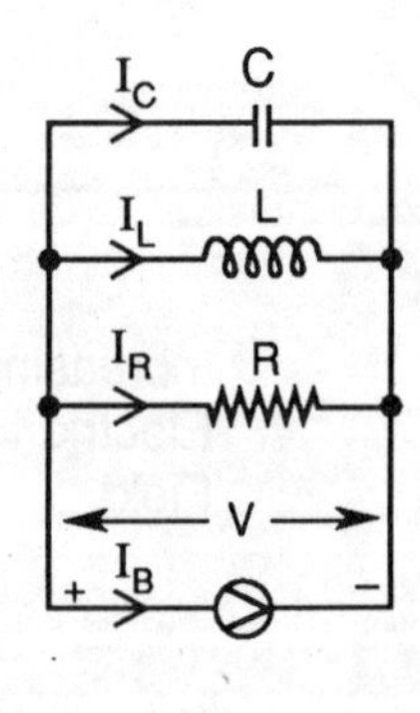

Figure 2.3. Equivalent circuit to approximate cavity with parallel resistance.

to arrive at the second-order differential equation

$$\ddot{V} + \frac{\omega_r}{Q}\dot{V} + \omega_r^2 V = -\frac{\omega_r R}{Q}\dot{I}_B \,. \tag{2.6}$$

So far, we have considered an equivalent cavity with only one possible mode. Most cavities have many modes of excitation. We can represent these cavities by a generalization of the equivalent parallel circuit as shown in Fig. 2.4. We obtain a separate equation for the voltage V_n of each mode n excited by beam current I_B,

$$C_n \ddot{V}_n + \frac{\dot{V}_n}{R_n} + \frac{V_n}{L_n} = -\dot{I}_B \,. \tag{2.7}$$

The total voltage is given by the sum of the voltages over all N modes of the cavity

$$V_t = \sum_{n=1}^{N} V_n \,. \tag{2.8}$$

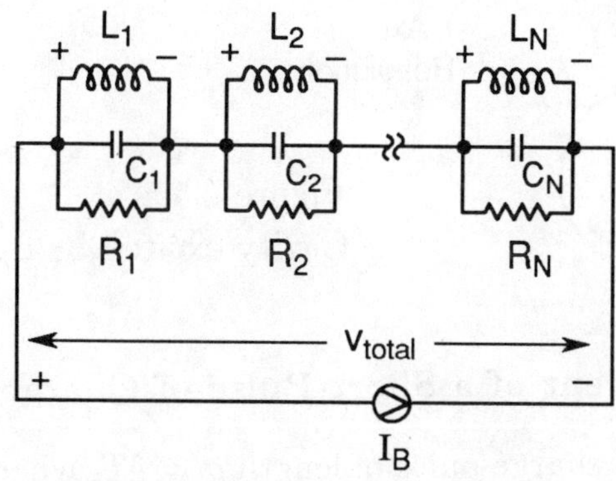

Figure 2.4. Generalization of equivalent circuit for multiple mode cavity.

While the main purpose of this paper is to discuss how the beam excites wake fields in vacuum structures, it is useful, for completeness, to illustrate how a cavity can be driven in its fundamental mode from an external source with a coupling loop as shown in Fig. 2.5. Kirchhoff's law, Eqs. (2.3-2.6), still holds. However, we need to include the generator current $-I_G$ in the loop for the driving term along with the beam current, so

$$C\frac{dV}{dt} + \frac{1}{R}V + \frac{1}{L}\int V\,dt = I_G - I_B \tag{2.9}$$

where we have chosen the direction of the current I_G to produce an accelerating voltage. This circuit equation is used extensively to describe the cavity voltage in a steady state, or in a slowly varying amplitude and phase approximation, and will be discussed in the papers by P. B. Wilson and F. Pedersen. Our purpose is to illustrate their connection to single bunches traveling through different structures in the ring vacuum chamber.

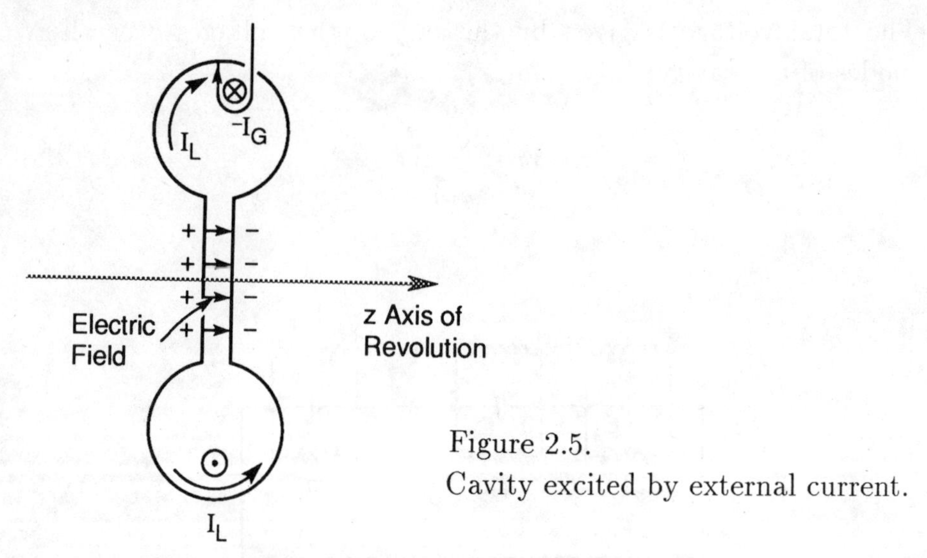

Figure 2.5.
Cavity excited by external current.

3. Driving Current of a Short Pulse of Charge

We will consider a charge pulse of length $\sigma = vT$, where the center of the pulse passes through the cavity center $z = 0$ at time $t = 0$. The time duration of the pulse is T, and the pulse velocity is v, which we will assume is close to the speed of light. The linear charge density of the pulse can be represented by $\lambda(s)$ with $s = (vt - z)$, the position of the charge in the bunch relative to the center of the bunch. Note that the front of the bunch passes through the cavity at $s < 0$ as shown in Fig. 3.1. This illustrates that bunch density profile is what one would see on an oscilloscope trace. The reader should be careful to note that other authors may display a snapshot of the bunch density profile at a fixed time so that the front of the bunch would be reversed from the convention of this paper. The beam current, which is to be used in the equivalent circuit of the previous section, is given by

$$I_B(t) = v\lambda(vt - z) = v\lambda(s)\,. \tag{3.1}$$

The pulse $\lambda(s)$ can also be represented by its Fourier transform $\tilde{\lambda}(\omega)$ with

$$\lambda(s) = \frac{1}{2\pi}\int \tilde{\lambda}(\omega)e^{-i\omega s/c}d\omega \tag{3.2}$$

and

$$\tilde{\lambda}(\omega) = \frac{1}{v}\int \lambda(s)e^{i\omega s/c}ds\,. \tag{3.3}$$

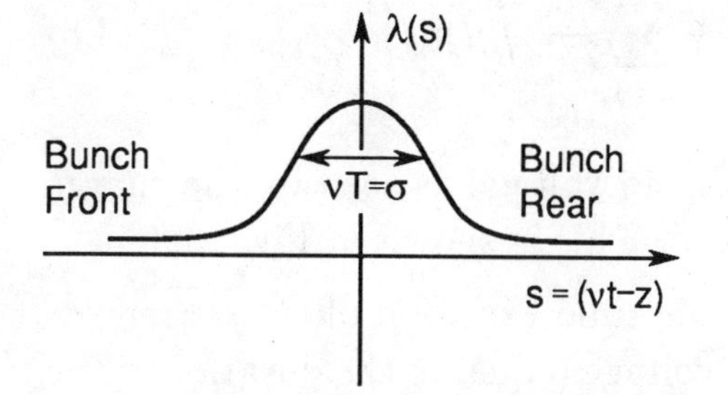

Figure 3.1.
Linear density profile of bunch.

One of the most common pulse distributions considered is the Gaussian distribution

$$\lambda(s) = \frac{Q}{\sqrt{2\pi}\sigma}e^{-s^2/2\sigma^2} \tag{3.4}$$

where Q is the total charge in the bunch. This Gaussian distribution has a Fourier transform

$$\tilde{\lambda}(\omega) = Qe^{-\omega^2\sigma^2/2c^2}\,. \tag{3.5}$$

The spectrum for $\tilde{\lambda}(\omega)$ falls off rapidly for frequencies $\omega > c/\sigma$ as shown in Fig. 3.2. Of course, when the density distribution is

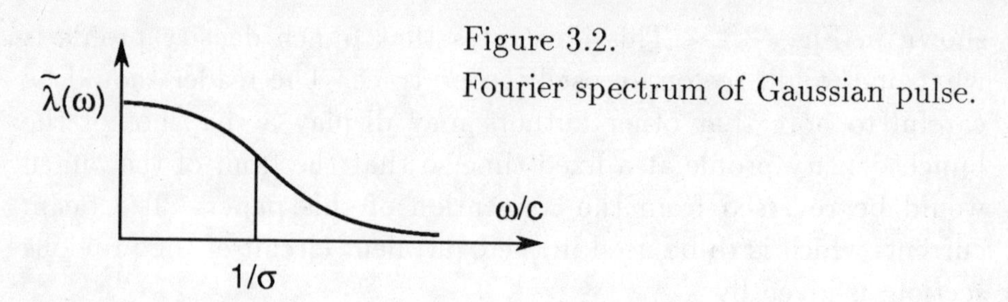

Figure 3.2.
Fourier spectrum of Gaussian pulse.

influenced by the wake fields, the assumption that the distribution is Gaussian is suspect.

It is instructive to consider the case when the time variation of the current or the time of interest for the cavity voltage is small compared to the resonant period of the cavity modes, i.e. when the bunch is short enough or the mode frequencies of the cavity low enough that $\omega_n T << 1$. For this case, we can ignore the second and third terms on the left-hand side of Eq. 2.7 and approximate the wake voltage in the cavity as

$$V(t) = \sum_{n=1}^{N} V_n = -\sum_{n=1}^{N} \frac{1}{C_n} \int I_B(t)dt \,. \tag{3.6}$$

We denote this form of the wake-field voltage as seen by the charge in the bunch as a *Capacitive Wake.* This is shown in Fig. 3.3.

Next we consider the case when the time variation of the current or the time of interest for the cavity voltage (equal to the duration of the bunch passage through the cavity) is large compared to the resonant period of the cavity modes, i.e. $\omega_n T >> 1$. We also assume that the Q of the cavity is sufficiently high that the fields do not decay appreciably during the passage of the bunch through the cavity. For this case, the third term is the dominant term on the left side of Eq. (2.7), and we can approximate the wake voltage in the cavity as

$$V(t) = \sum_{n=1}^{N} V_n = -\sum_{n=1}^{N} L_n \frac{dI_B}{dt} \,. \tag{3.7}$$

We call this form of the wake-field voltage as seen by the charge in the bunch an *Inductive Wake.* This type of wake is shown in Fig. 3.4 ,

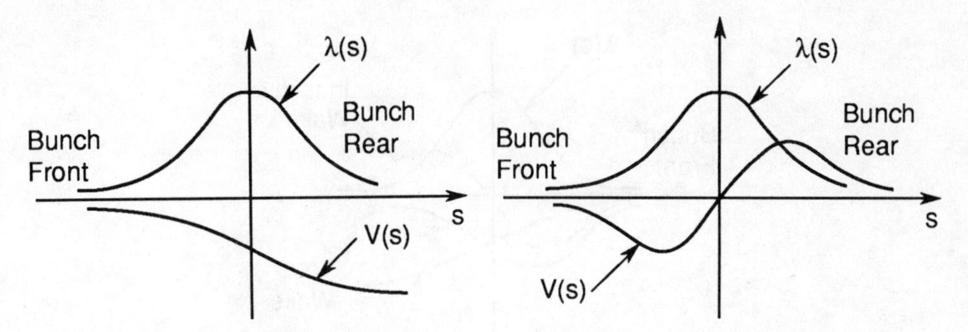

Figure 3.3. Wake voltage for short bunch (capacitive).

Figure 3.4. Wake voltage for long bunch (inductive).

where we see that the energy lost by the front of the bunch is gained by the rear of the bunch, so that, for a pure inductive wake, the net energy lost by the bunch is zero.

There is one other case which we should consider for completeness; namely, the case of a very long bunch or a very lossy cavity where the fields decay in a time much shorter than the time it takes for the bunch to pass through the cavity. In this case, $\omega_n T >> Q$, and we can approximate the cavity voltage as

$$V(t) = \sum_{n=1}^{N} V_n = -\sum_{n=1}^{N} \frac{1}{R_n} I_B \quad . \tag{3.8}$$

This is called a *Resistive Wake* and is shown in Fig. 3.5 , where we see that the cavity voltage and bunch density are exactly in phase.

For the case of high-Q cavities and intermediate bunch lengths, the total wake field is the sum of the capacitive and inductive parts of

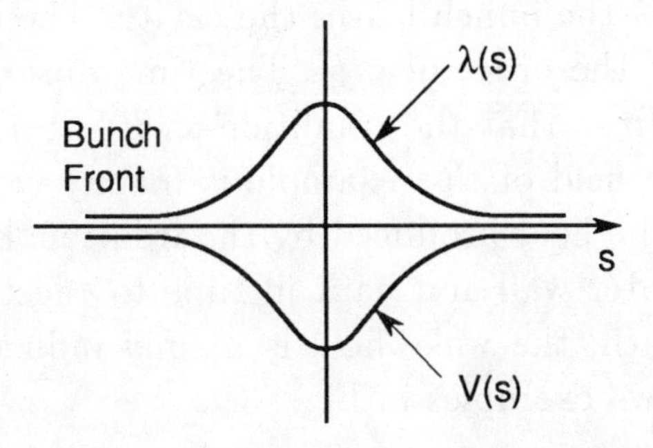

Figure 3.5. Wake voltage for very long bunch (resistive).

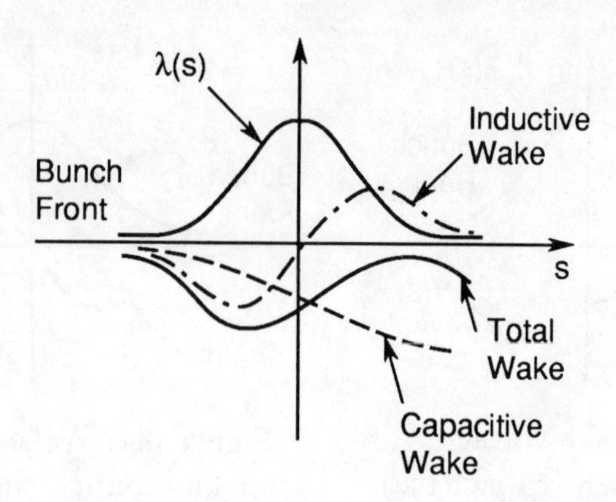

Figure 3.6. Total wake field for intermediate bunch lengths.

the wake as shown in Fig. 3.6. Note that even for an infinite shunt resistance (i.e. an infinite Q), there is a net loss of energy for the bunch, just as in the case when a resistive term is present.

In addition to the bunch length, both the transverse distance to the outer wall and the length of the gap determine whether the cavity is capacitive or inductive. Three examples of cavity shapes and bunch lengths are shown in Fig. 3.7 ; one illustrates a capacitive wake, and the others inductive wakes. For the first example, shown in Fig. 3.7 , we see that if

$$(g + l/2)l < 2(b - a)^2 \tag{3.9}$$

with g the gap length, l the bunch length, a the radius of the inner wall, and b the radius of the outer wall, the fields produced by the front of the bunch do not have time to propagate to the outer cavity wall and back before the bunch leaves the cavity. The cavity modal frequency, ω_n, is of the order of c/b. The time duration T of the pulse is given by l/c so that the condition $\omega_n T << 1$ satisfies Eq. (3.9), and the wake field of this example is mainly capacitive. On the other hand, if the field produced by the front of the bunch can propagate to the outer wall and back in time to affect most of the particles in the bunch, the wake field is mainly inductive. This is shown by the last two examples in Fig. 3.7.

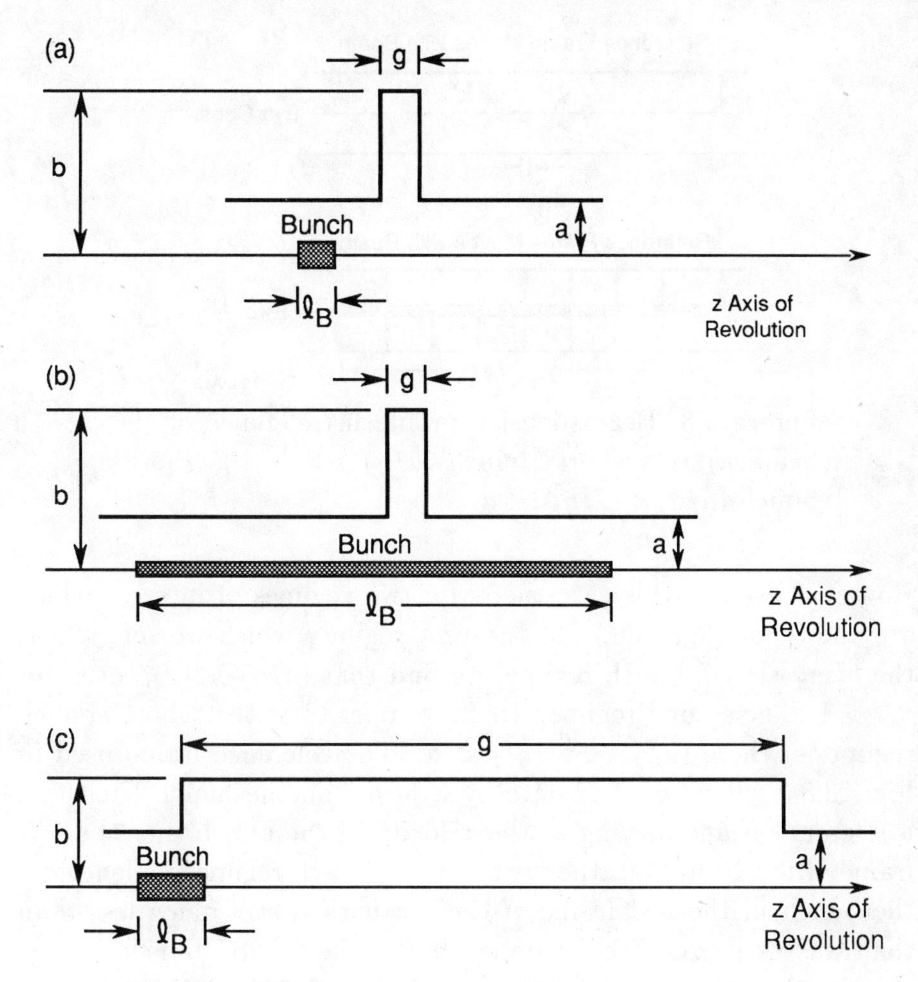

Figure 3.7. Three examples of beam cavity interactions: a) capacitive wake, b) & c) inductive wake.

In order to properly define a "short" bunch it is necessary to consider the relativistic energy parameter of the beam γ which is defined as the ratio of the particle energy to the rest mass energy. We assume that the particles in the bunch travel near the speed of light, and we can substitute c for v except when the difference is required, and then we can substitute $(c - v)$ by $c/(2\gamma^2)$. As discussed above, the bunch is short when the cavity gap and the bunch length are much smaller than the transverse size of the cavity.

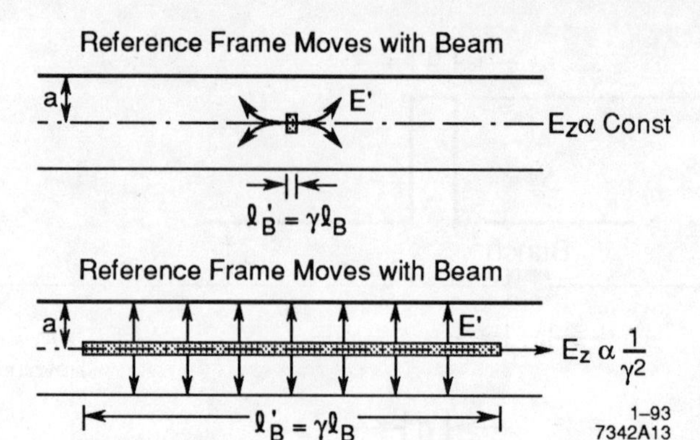

Figure 3.8. Beam density profile in reference of the beam: a)"very short" bunch $(a/\gamma) >> l_B$; b) "short" bunch $(a/\gamma) << l_B << a$.

However, even in this case there are two regimes of interest which are shown in Fig. 3.8. In the first regime, which we refer to as the "very short" bunch regime, we find that $(a/\gamma) >> l_B$, even for $\gamma >> 1$. The second regime, which we refer to as the "short" bunch regime is where $(a/\gamma) << l_B << a$. The reference frame used in Fig. 3.8 , which we designate by a prime on the bunch length, is a reference frame moving at the velocity of the bunch (i.e. the rest frame of the beam). In the "very short" bunch regime the length of the bunch in the rest frame of the particles is still much less than the transverse size of the chamber, but in the "short" bunch regime, the length of the bunch in the rest frame of the particles is much greater than the transverse size of the chamber. The first regime is where many Free Electron Lasers operate, while the second regime is where the high energy colliders considered in this school operate. In the following, we will consider only the "short" bunch regime, so that only wake fields which are due either to the finite conductivity or to the discontinuities of the vacuum chamber are considered.

4. Connection Between Wake Potential and Impedance

In the previous section, we found that the wake voltage V could be written as a function of the driving current (or its equivalent the linear charge density of the bunch). We write

$$V(s) = -\int_{-\infty}^{\infty} W_{\|0}\left(s - s'\right) \lambda(s') ds' \tag{4.1}$$

where we have chosen V positive for an accelerating voltage.[2] For the capacitive, resistive and inductive wakes discussed in the previous section, the potential kernels are given by the following expressions:

$$W_{\|0}\left(s - s'\right) = \sum_{n=1}^{N} \frac{1}{C_n} H\left(s - s'\right) \tag{4.2a}$$

$$W_{\|0}\left(s - s'\right) = c \sum_{n=1}^{N} \frac{1}{R_n} \delta\left(s - s'\right) \tag{4.2b}$$

and

$$W_{\|0}\left(s - s'\right) = c^2 \sum_{n=1}^{N} L_n \frac{d\delta\left(s - s'\right)}{ds} \tag{4.2c}$$

where $H(s)$ is the unit step function equal to one for $s > 0$ and equal to zero for $s < 0$, while $\delta(s)$ is the Dirac delta function. When Eqs. (4.2a-4.2c) are substituted into Eq. (4.1), one obtains Eqs. (3.6), (3.7) and (3.8). It is useful to define the longitudinal impedance by the Fourier transform of the wake function $W_{\|0}(s)$

$$Z_{\|}(\omega) = \frac{1}{c} \int W_{\|0}(s) e^{i\omega s/c} ds \tag{4.3a}$$

and

$$W_{\|0}(s) = \frac{1}{2\pi} \int Z_{\|}(\omega) e^{-i\omega s/c} d\omega \tag{4.3b}$$

where the impedance can be written as the sum of the real part (resistance) and imaginary part (reactance)

$$Z_{\|}(\omega) = R_{\|}(\omega) + iX_{\|}(\omega) \; . \tag{4.4}$$

From causality, $W(s)$ is zero for $s < 0$, so that $R_{\|}$ is an even function, and $X_{\|}$ an odd function of ω. The amount of energy lost by the bunch, which travels through a cavity, divided by the charge in the bunch is called the loss factor and is given by

$$k_{\|} = -\frac{1}{Q} \int \lambda(s) V(s) ds \tag{4.5a}$$

or in terms of the impedance

$$k_{\|} = \frac{1}{Q} \int \widetilde{\lambda}^2(\omega) R_{\|}(\omega) dw \tag{4.5b}$$

where we have used the fact that the term $X_{\|}(\omega)$ integrates to zero. A more general expression for the impedance may be given[3] by expanding in powers of $(1/\sqrt{\omega})$:

$$\begin{aligned} Z(\omega) = &-i\omega L + B\sqrt{\omega} + R \\ &+ \frac{Z_1}{\sqrt{\omega}} + \frac{1}{\omega C} + \cdots \end{aligned} \tag{4.6}$$

The first term is the inductive term caused by bellows, slots, ports, etc. in the vacuum chamber, and is only valid for the lower frequencies below cutoff. This term is discussed in the previous section. The second term, proportional to $\sqrt{\omega}$, is due to the finite conductivity of the chamber wall and is present even in a smooth chamber. This term is valid up to frequencies where the displacement current in the wall becomes comparable to the ohmic current. The third term is a resistive term with R the usual dc resistance, also discussed in the previous section. The fourth term, proportional to $1/\sqrt{\omega}$, is due to the diffraction of the electromagnetic field at sharp discontinuities in the chamber. This term is valid for the frequencies driven by the beam in the "short bunch" regime. The last term is the usual capacitive term which is higher order in $1/\sqrt{\omega}$ and can be ignored for high frequencies. Both the resistive wall term and the diffraction term are discussed below.

5. Resistive Wall Impedance

In order to understand physically the wake field, which is responsible for the resistive wall impedance, consider the following argument.[4] As the beam passes by a given point in the vacuum wall, a surface current is induced on the wall. Subsequently, if the conductivity of the wall is finite, this current diffuses into the metal giving rise to wake fields. This process can best be illustrated by a simple example. We will examine the currents in the wall for the case of a charged particle pulse traveling parallel to an infinite metallic plane as shown in Fig. 5.1c. Imagine that the pulse of particles is made up of two semi-infinite beams, one positive and one negative, as shown in Figs. 5.1a and 5.1b. The image charges and currents are also shown for the case of a perfectly conducting wall. Because the wall conductivity is infinite, no current can exist inside the metal, and the induced currents stay on the surface of the wall. The wall currents and charges due to the (+) and (−) beams have the same magnitudes but opposite signs; by superposition, they cancel each other in the region behind the pulse as shown in Fig. 5.1c. Hence, no current is left in the wall for the case of a perfectly conducting wall. However, if the wall conductivity is finite, the surface currents can diffuse toward the inside of the metal. The diffusion of the image currents of the (+) and (−) beams is shown in Figs. 5.2a and 5.2b. Because these image currents are turned on at different times, the wall current corresponding to the (+) beam has diffused farther into the metal than that of the (−) beam at the same point along the wall. This gives rise to currents in the wall in the region behind the pulse as shown

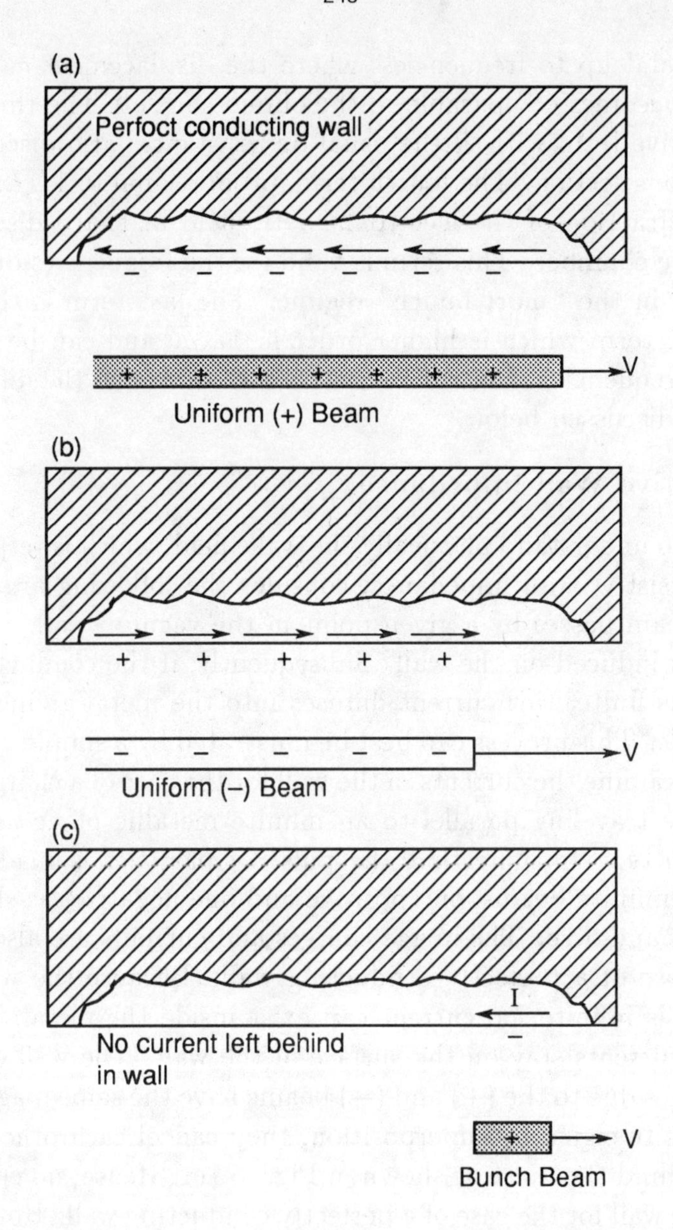

Figure 5.1. An illustration of resistive wall effects for a perfectly conducting wall: a) a semi-infinite (+) beam, b) a semi-infinite (–) beam, and c) a bunched (+) beam.

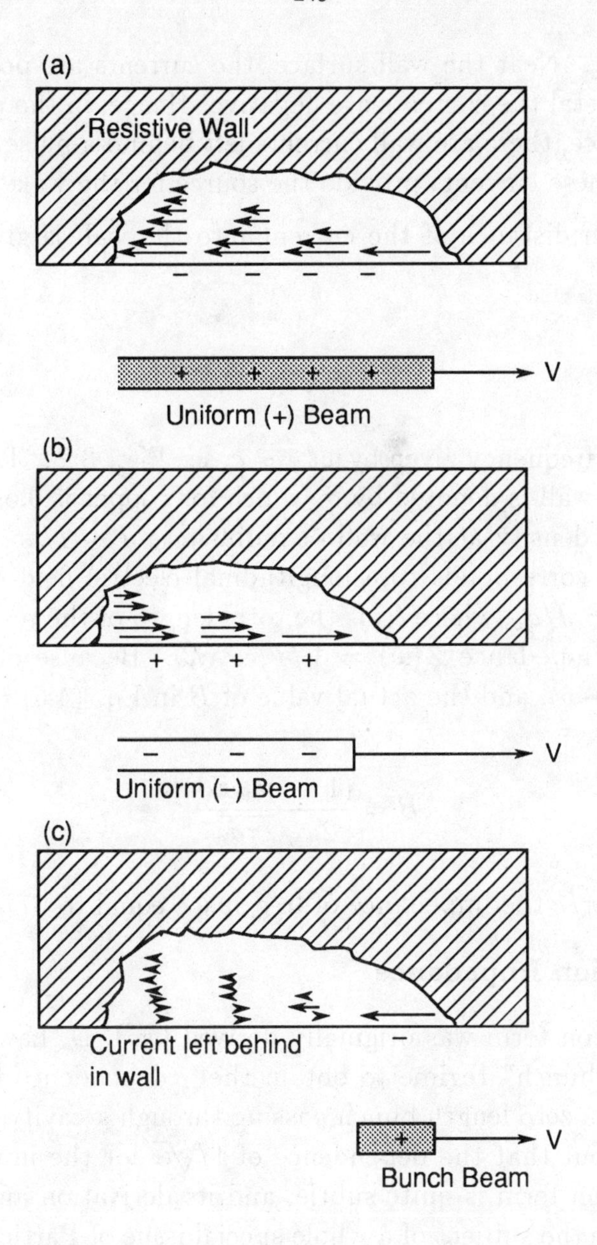

Figure 5.2. An illustration of resistive wall effects for a lossy wall: a) a semi-infinite (+) beam, b) a semi-infinite (–) beam, and c) a bunched (+) beam.

in Fig. 5.2c. Near the wall surface, the currents are positive, and inside the metal the currents are negative. Hence, in the presence of wall resistance, there are wall currents left behind a pulse of charged particles. These currents provide the source for the wake fields.

The diffusion distance of the current into the wall is given by the skin depth

$$\delta = c/\sqrt{2\pi\omega\sigma_c}$$

where ω is a frequency given by $\omega\ell_B \sim c$, see Fig. 3.2. The conductivity of the wall is denoted by σ_c which for copper is 0.5×10^{18}/sec. The current density at the wall of radius a is $J \sim I/a\delta$. This current density corresponds to a longitudinal electric field component $E_z = V/\ell_w = J/\sigma_c$, where ℓ_w is the total length of the resistive wall. Hence, the impedance $Z(\omega) = V/I \propto \sqrt{\omega}$. Because of causality, $Z(\omega) = Z^*(-\omega)$, and the actual value of B in Eq. (4.6) is given by

$$B = \frac{(1-i)\ell_w Z_0}{2a\sqrt{2\pi\sigma_c}}$$

with $Z_0 = 4\pi/c$ the impedance of free space equal to 377 Ω.

6. Diffraction Impedance

The diffraction term was originally derived by J. D. Lawson in the "very short bunch" regime to obtain the γ dependence for the energy lost by a zero length bunch passing through a cavity.[5] It should be pointed out that the dependence of $1/\sqrt{\omega}$ for the impedance of the diffraction term is quite subtle, and its derivation and range of validity were the subject of a whole special issue of Particle Accelerators devoted to a workshop attended by the "experts" in the field of impedances for short bunches. A beautiful physical argument to derive this result has been given by R. B. Palmer.[6] With his kind permission, we have repeated his treatment here.

We consider the cavity excited by a short bunch as shown in Fig. 6.1. When the bunch enters the cavity, the electromagnetic field which

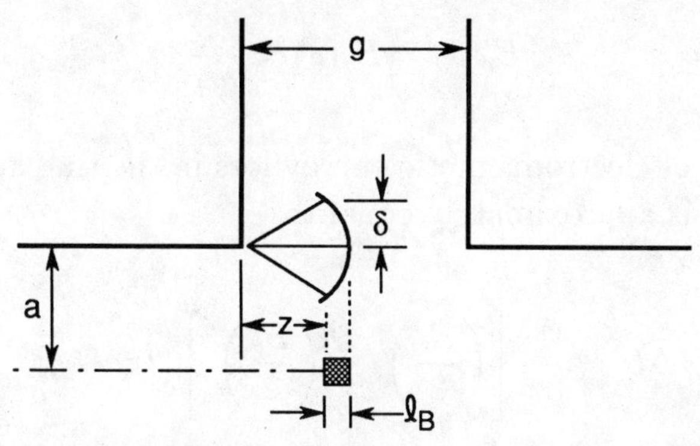

Figure 6.1. Diffraction of electromagnetic energy of beam passing through a short gap.

had been contained within the pipe of radius a starts to diffract away from the cavity edge as shown. When the rear of the bunch is at position z from the edge of the cavity, the electromagnetic field at radius a, which started at the front of the bunch, has diffracted transversely a distance δ such that the field at $r = a+\delta$ and $r = a-\delta$ is at a distance ℓ_B behind the field at $r = a$. We will consider the case where $\delta << a$ and $z >> \ell_B$. The transverse distance δ is related to the longitudinal distance z by

$$\sqrt{z^2 + \delta^2} = z + \ell_B$$

or

$$\delta \approx \sqrt{2\ell_B z} \ . \tag{6.1}$$

When the distance z equals the gap width g, the beam enters the pipe, and the portion of the field that has diffracted by the amount $\delta \approx \sqrt{2\ell_B g}$ is retarded such that it can not catch up with the beam. This portion of the energy is lost. In the "very short bunch" regime, we must substitute $\ell_B = a/\gamma$, while in the "short" bunch regime, ℓ_b is the length of the bunch. The electromagnetic field at radius a in the pipe is given by

$$E_r(a) = B_\theta(a) = \frac{2Q}{a\ell_b} \quad . \tag{6.2}$$

The amount of electromagnetic energy loss in the ring at $r = a$ with thickness 2δ is approximately equal to

$$\Delta U \approx \frac{1}{8\pi}\left[\left(\frac{E_r}{2}\right)^2 + \left(\frac{B_\theta}{2}\right)^2\right][2\pi a\delta\ell_B] \tag{6.3}$$

which for the "very short bunch" is

$$\Delta U = \frac{Q^2 g^{1/2}\gamma^{1/2}}{\sqrt{2}a^{3/2}} \tag{6.4}$$

while for the "short bunch" regime

$$\Delta U = \frac{Q^2 g^{1/2}}{\sqrt{2}a\ell_b^{1/2}} \quad . \tag{6.5}$$

The result for the "very short bunch" originally derived by Lawson shows the $\sqrt{\gamma}$ dependence on the energy loss. We, of course, are only interested in the "short bunch," so we will only consider Eq. (6.5). We use the definitions for the voltage and current $V = \Delta U/Q$ and $I = Qc/\ell_B$ to obtain the impedance at the frequency $\omega = c/\ell_B$ given by

$$Z \approx \sqrt{\frac{gc}{2\omega}\frac{Z_0}{4\pi a}} \quad . \tag{6.6}$$

Again we must use causality with $Z(\omega) = Z^*(-\omega)$. The value of the term in Eq. (4.6) for a Gaussian bunch is given by

$$Z_1 = \frac{(1+i)Z_0}{2a}\sqrt{\frac{cg}{\pi^3}} \quad .$$

7. Summary

The main purpose of this paper is to give an introduction to the jargon used in discussing wake fields excited by a short bunch of particles passing through a structure. Often, terms such as the inductive or resistive part of the wake are used to describe the characteristics of the field or voltage which acts on the particles in the bunch. We have tried to illustrate how some of these terms relate to the common parallel circuit of a cavity. Many people will refer to a resistive portion of the total wake voltage shown in Fig. 3.6 when a portion of the wake voltage is in phase with the beam density profile. This is because the wake voltage for a resistive impedance is in phase with the beam density as shown in Fig. 3.2. Of course, we know that this portion of the wake in Fig. 3.6 comes from the capacitive part of the impedance. To the particles in the beam, however, this portion of the impedance has a resistive effect, since it produces a net energy loss. If, after reading this paper, the student has a more physical feel for the wake field in different regimes and finds the more advanced papers easier to read, then this paper has fulfilled its purpose.

Acknowledgements

I am indebted to a large number of colleagues who have patiently attempted to explain the relationship between impedance and wake potential. The discussion in Sec. 2 was previously used by Richard Cooper to illustrate why the cavity can be replaced by an equivalent circuit. In particular, I have had many friendly discussions with Karl Bane, Sam Heifets, Bob Palmer, and Perry Wilson. However, these people should not be held responsible for any misunderstanding or mistakes made in the translation of their lectures.

References

1. Much of the original work on wake fields and their influence on the particle beam dynamics was done by L. J. Laslett, V. K. Neil, and A. M. Sessler, see for example Rev. Sci. Instr. Vol. 32, No. 3, March, 1961, p. 256-279. For example, on page 279, Eq. (4.8) gives the famous criteria for the energy spread in a

beam, as a function of the impedance, necessary to stabilize the microwave instability. The first use of the term wake field to describe fields left behind a pulse of charge was in a paper on resistive wall by P. L. Morton, V. K. Neil and A. M. Sessler, J. Appl. Phys. Vol. 37, No. 10, Sept. 1966. The use of the equivalent circuit to describe the beam-cavity interaction was introduced by K. W. Robinson in a CEA report CEAL-1010, Feb. 1964, to describe the beam-cavity stability. A more recent report, on a workshop held at LBL on Impedance Beyond Cutoff, Part. Acc. Vol. 25, No. 2-4, contains some of the most up-to-date results on this subject.

2. The notation used in this section has been chosen to agree with that of K. Bane and M. Sands in the workshop on Impedance Beyond Cutoff, Ref. (1), p. 73.
3. S. Heifets, Broad Band Impedance of the B-factory ABC-60, Note SLAC, 1991.
4. Original work on this subject was by P. L. Morton, V. K. Neil, and A. M. Sessler, Ref. (1), and K. W. Robinson, SLAC Report No. 49, 1965, p. 32. More recent papers discussing this subject are in the 1982 SLAC summer school on Physics of High Energy Particle Accelerators by A. W. Chao, AIP Conf. No. 105, 1982, p. 361, and in the 1983 BNL/SUNY summer school on Physics of High Energy Particle Accelerators by K. Bane and P. Wilson, AIP Conf. No. 127, p. 903.
5. J. D. Lawson, workshop on Impedance Beyond Cutoff, Ref. (1), p. 107.
6. R. B. Palmer, workshop on Impedance Beyond Cutoff, Ref. (1), p. 97.

COHERENT-INSTABILITY-INDUCED RADIATION

Jiunn-Ming Wang

Brookhaven National Laboratory, Upton, NY 11973, USA

1 Introduction

We explore here the possibility of using the coherent instability in a storage ring as the source of a coherent light. A coherent instability is always associated with a loss of beam revolution energy. If the cause of the instability is an evanescent impedance, for example, the RF cavity modes below the beam pipe cutoff, then the energy lost by the beam is deposited into the impedance source. On the other hand, if the instability is caused by radiation impedance [1], then the energy lost by the beam is turned into radiation that propagates around the storage ring in synchrony with the coherent beam oscillation. The radiation mode discovered by the authors of ref.[1] has a nonvanishing longitudinal component of the electric field. Hence, the radiation inevitably transfers energy back to the particle beam. If this feedback mechanism causes the amplitude of the beam coherent motion to grow, we have a coherent instability. The amplitude of the induced field will grow with the same growth rate.

We shall consider only the radiation impedance. It is the purpose of this paper to calculate the coherent-instability-induced radiation power assuming that the instability is started by the shot noise of the beam.

We shall restrict our discussion to the following "microwave" region:

$$\lambda \ll l_W \ll \sigma \ , \tag{1}$$

where λ is the perturbation wavelength (carrier wavelength), l_W is the wake-length, and σ is the bunch length. All lengths are in units of radians. The wake-length l_W is a measure of the distance within which two particles can interact with each other through the wakefield. In terms of the carrier wave number $n_0 = 2\pi/\lambda$ and the impedance bandwidth $b \equiv (4\pi - l_W)/2l_W \cong 2\pi/l_W$, the above condition is equivalent to

$$n_0 \gg b \gg 2\pi/\sigma \ . \tag{2}$$

We shall refer to $2\pi/\sigma$ as the bunch bandwidth.

It has been conjectured [2] and proved [3] that the "microwave"-instability condition can be obtained simply by (i) writing down the coasting beam instability condition corresponding to n_0, and (ii) replacing the average current I_{av} that appears in the coasting beam condition by the local beam current of the bunched beam.

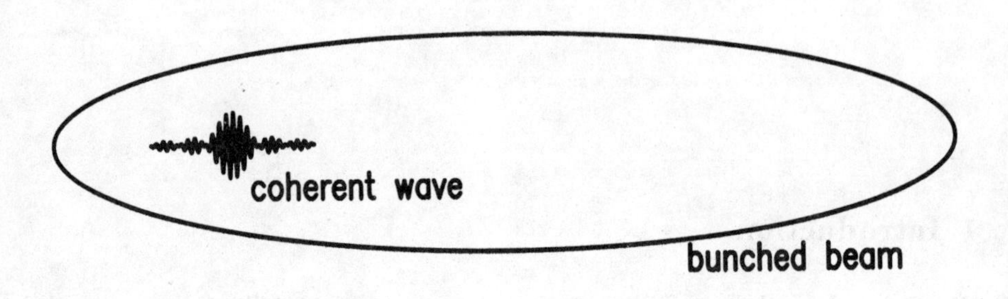

Fig. 1. A coherent wave localized in the bunch

This can be understood as follows: a coherent wave corresponding to microwave instability is localized in a small region within the bunch as depicted in Fig. 1, where the width of the wave envelope is of the order of the wakelength. Therefore, only the current density at the location of the coherent wave, not the average current, contributes to the instability condition. Also, if the distance between two beam particles is greater than the wakelength, then the effect of the wakefield induced by one of the particles on the other is negligible. This means that the two particles can oscillate coherently with each other only if they are less than a wakelength apart. As a consequence, the width of the coherent wave in Fig. 1 should be the same as the wakelength. In other words,

$$\text{coherence length} = \text{wakelength} \quad .$$

This paper is organized as follows: In Section 2, we start from the appropriate Vlasov equation in the microwave region (1) and (2). We show that the Vlasov equation together with an initial condition is equivalent to a non-linear integral equation. We then linearize the integral equation to the first order in the "centered" shot noise distribution function. In Section 3, we show that the linear integral equation derived in Section 2 is equivalent to a homogeneous linear integro-differential equation for the current density. We refer to this equation as the basic equation. The differential operator of the basic equation is second order in time.

In Section 4, we introduce an orthonormal set of functions to a finite dimensional linear space in preparation for the discussion of a solvable model to be introduced in the next section. In Section 5, we introduce a model impedance and the corresponding basic equation for the current. We show that the basic equation can be diagonalized using the orthonormal set of functions discussed in the previous section. We also solve the initial value problem of the equation with the current and the induced radiation field expressed in terms of the initial perturbation current.

In Section 6, we investigate how the radiation power grows in time because of the coherent instability. We assume that the initial perturbation of the instability is due to the grainy characteristic of the shot noise in the beam. The graininess of the shot noise is averaged over in the expression for the radiation power.

2 Vlasov Equation

2.1 Equation of Motion and Vlasov Equation

We use the dynamical variables ϕ and ϵ to describe the beam particle motion:

$$\phi = \theta - \omega_0 t \ ; \qquad \epsilon = E - E_0 \ ,$$

where θ describes the position around the ring, and E_0 and ω_0 are, respectively, the nominal energy and the (angular) revolution frequency of the beam. It has been demonstrated [3,4] that the effect of the synchrotron frequency on the microwave instability is negligible. Hence we can write the equations of motion as

$$\dot{\phi} = -\hat{\alpha}\,\epsilon \ ; \qquad \dot{\epsilon} = ce\,\mathcal{E}(\phi, t) \ ,$$

where $\mathcal{E}$ is the longitudinal electric field induced by the coherent signal of the beam, and $\hat{\alpha}$ is related to the momentum compaction α by

$$\hat{\alpha} \equiv \alpha\omega_0/E_0 \ .$$

The corresponding Vlasov equation is

$$\frac{\partial}{\partial t}\Psi(\phi, \epsilon, t) - \hat{\alpha}\,\epsilon\frac{\partial}{\partial\phi}\Psi + ce\,\mathcal{E}(\phi, t)\frac{\partial}{\partial\epsilon}\Psi = 0 \ , \tag{3}$$

where $\Psi(\phi, \epsilon, t)$ is the distribution function in (ϕ, ϵ) space. We are interested in the transient solution of this equation. So, we next discuss the initial condition we impose on the Vlasov equation.

2.2 Initial Condition and an Integral Representation

We assume that the coherent instability starts at $t = 0$, and that the beam has no energy spread initially:

$$\Psi(\phi, \epsilon, t = 0) = F(\phi)\,\delta(\epsilon) \ .$$

Subject to the above initial condition, the Vlasov equation (3) is equivalent to

$$\Psi(\phi, \epsilon, t) = F(\phi)\,\delta(\epsilon) - ce\int_0^t dt'\,\mathcal{E}[\phi + \hat{\alpha}\epsilon(t - t'), t']\,\frac{\partial}{\partial\epsilon}\Psi[\phi + \hat{\alpha}\epsilon(t - t'), \epsilon, t'] \ . \tag{4}$$

This equation obviously satisfies the above stated initial condition; it is straightforward to verify that (4) implies (3).

Note that the equation (4) is not linear in Ψ since the induced field $\mathcal{E}$ depends on Ψ. In what follows we shall deal with this integral equation instead of the Vlasov equation itself.

2.3 Shot Noise and Linearization

The beam is composed of N particles. Denote the initial position of the j-th particle by ϕ_j and represent the initial distribution by $F(\phi;\phi_1,\phi_2,...\phi_N)$. The granularity of the distribution can be thought of as shot noise. We assume that the partial coherence of the shot noise is entirely responsible for starting the coherent instability.

The shot noise can be represented by

$$F(\phi;\phi_1,\phi_2,...\phi_N) = \frac{1}{N}\sum_{j=1}^{N}\delta(\phi-\phi_j) \ . \tag{5}$$

Assuming that the probability density of each ϕ_j is $\rho(\phi_j)$ and averaging F over ϕ_j, we have

$$< F(\phi) > \equiv \prod_{j=1}^{N}\int d\phi_j\,\rho(\phi_j)\,F(\phi;\phi_1,\phi_2,...\phi_N) = \rho(\phi) \ . \tag{6}$$

$\rho(\phi)$ is normally referred to as the line density; here it is normalized to unity. We assume $\rho(\phi)$ to be a smooth function of ϕ, e.g., a Gaussian function with variance σ^2.

It is convenient to define the "centered" distribution function

$$f(\phi;\phi_1,\phi_2,...\phi_N) = F(\phi;\phi_1,\phi_2,...\phi_N) - \rho(\phi) \ , \tag{7}$$

so that

$$\int d\phi\, f(\phi;\phi_1,\phi_2,...\phi_N) = 0 \ ,$$

and

$$< f(\phi) > = \prod_{j=1}^{N}\int d\phi_j\,\rho(\phi_j)\,f(\phi;\phi_1,\phi_2,...\phi_N) = 0 \ .$$

To linearize the integral equation, let us regard ρ as the zeroth-order term, and f the first order. We treat (4) up to $O(f)$. In order to linearize the integral equation, we iterate the equation once by substituting the first term on the right of (4) into the last term. Since $\mathcal{E}$ is a beam-induced field, it is linear in Ψ. Hence it is linear in both ρ and f. As mentioned in the introduction, in our region of interest, $\mathcal{E}$ satisfies $\lambda \ll \sigma$, where σ is the bunch length of the smooth function ρ. However, the amount of such a small-wavelength component of $\mathcal{E}$ induced by a smooth long bunch represented by ρ is negligible. Hence we conclude

$$\mathcal{E} = O(f) \ .$$

As a consequence, the linearized integral equation can now be writen as

$$\Psi(\phi,\epsilon,t) = F(\phi)\,\delta(\epsilon) - ce\int_0^t dt'\,\mathcal{E}[\phi+\hat{\alpha}\epsilon(t-t'),t']\,\frac{\partial}{\partial\epsilon}\{\rho[\phi+\hat{\alpha}\epsilon(t-t')]\delta(\epsilon)\}\ . \tag{8}$$

3 The Basic Equation

We show in this section that the integral equation (8) is equivalent to a homogeneous linear integro-differential equation for the current. The beam current is related to Ψ by

$$I(\phi,t) = eN\omega_0\int d\epsilon\,\Psi(\phi,\epsilon,t)\ .$$

If we combine this equation with (8), integrate the result over ϵ, and then perform an integration by parts, we obtain

$$I(\phi,t) = 2\pi I_{av}\,f(\phi;\phi_1,\phi_2,...\phi_N) + 2\pi I_{av}\,\hat{\alpha}\rho(\phi)ce\int_0^t dt'\,(t-t')\,\frac{\partial}{\partial\phi}\mathcal{E}(\phi,t)\ , \tag{9}$$

where $I_{av} = eN\omega_0/2\pi$. A term $O(\rho)$ has been discarded in (9), since as mentioned above, such a term does not contribute to radiation satisfying (1) or (2). Differentiating (9) with respect to t, we have

$$\frac{\partial}{\partial t}I(\phi,t) = 2\pi ceI_{av}\,\hat{\alpha}\rho(\phi)\int_0^t dt'\,\frac{\partial}{\partial\phi}\mathcal{E}(\phi,t)\ , \tag{10}$$

$$\frac{\partial^2}{\partial t^2}I(\phi,t) = 2\pi ceI_{av}\,\hat{\alpha}\rho(\phi)\,\frac{\partial}{\partial\phi}\mathcal{E}(\phi,t)\ . \tag{11}$$

From (9) and (10), we have the following initial conditions:

$$I^{(0)}(\phi) \equiv I(\phi,t=0) = 2\pi I_{av}\,f(\phi;\phi_1,\phi_2,...\phi_N)\ , \tag{12}$$

$$\dot{I}^{(0)}(\phi) \equiv \frac{\partial}{\partial t}I(\phi,t=0) = 0\ . \tag{13}$$

We note that (11) relates the current I to the beam-induced longitudinal electric field $\mathcal{E}$. In the remainder of this section, we transform this equation into an equation which relates I to the impedance or, equivalently, to the wakefield.

Let us Fourier transform the current I,

$$I(\phi,t) = \sum_n\int d\Omega I_n(\Omega)\exp(in\phi-i\Omega t) = \sum_n I_n(t)\exp(in\phi)\ ,$$

$$I_n(t) = \int d\Omega I_n(\Omega)\exp(-i\Omega t)\ ,$$

where the sum over n is from $-\infty$ to ∞. The induced electric field can now be written as

$$\mathcal{E}(\phi,t) = -\frac{1}{2\pi R}\sum_n \int d\Omega I_n(\Omega) Z_n(n\omega_0+\Omega)\exp(in\phi - i\Omega t) \ , \tag{14}$$

where $Z_n(\omega)$ is the longitudinal impedance and R is the average ring radius.

We are dealing with an impedance with a large bandwidth [see condition (2).] Let us further assume that the bandwidth of Z is much larger than the bandwidth in Ω of $I_n(\Omega)$. Then, in the region of Ω where $I_n(\Omega)$ is appreciable, $Z_n(n\omega_0+\Omega) \cong Z_n(n\omega_0)$. In terms of the shorthand notation $Z_n \equiv Z_n(n\omega_0)$, we can write (14) as

$$\mathcal{E}(\phi,t) \cong -\frac{1}{2\pi R}\sum_n I_n(t) Z_n \exp(in\phi) \ . \tag{15}$$

We now define the wake functions W and U by

$$W(\phi) = \frac{1}{2\pi}\sum_n Z_n \exp(-in\phi) \ , \tag{16}$$

$$U(\phi) = W'(\phi) = \frac{1}{2\pi}\sum_n U_n \exp(-in\phi) \ , \tag{17}$$

with $U_n \equiv -inZ_n$. The beam-induced electric field can now be written as

$$\mathcal{E}(\phi,t) = -\frac{1}{2\pi R}\int_{-\pi}^{\pi} d\phi' \, W(\phi'-\phi)\, I(\phi',t) \ , \tag{18}$$

$$\frac{\partial}{\partial\phi}\mathcal{E}(\phi,t) = \frac{1}{2\pi R}\int_{-\pi}^{\pi} d\phi' \, U(\phi'-\phi)\, I(\phi',t) \ , \tag{19}$$

and the equation (11) becomes

$$\frac{\partial^2}{\partial t^2} I(\phi,t) = \kappa \int_{-\pi}^{\pi} d\phi' \, K(\phi,\phi') I(\phi',t) \ , \tag{20}$$

with

$$\kappa = e\omega_0\hat{\alpha} I_{av} = e\alpha\omega_0^2 I_{av}/E_0 \ ,$$

and

$$K(\phi,\phi') = \rho(\phi)U(\phi'-\phi) \ . \tag{21}$$

We shall refer to (20) with (21) as the basic equation.

We have to solve the basic equation (20) subject to the initial conditions (12) and (13). For a coasting beam, $\rho(\phi) = 1/2\pi =$ constant, the right of (20) is a convolution integral. Hence the equation can be diagonalized by a simple Fourier transform, and the solution to the initial value problem is immediate. The complication for a bunched beam arises from the ϕ dependence of ρ which causes different n's within the bunch bandwidth $1/\sigma$ to be coupled [3].

In order to treat the microwave instability satisfying the condition (2), we introduce in the next section a set of orthonormal functions which will later be used to describe the modulation of the coherent wave, and to diagonalize the operator (21).

4 Modulation Space and Modulation Basis

Let $\mathcal{S}$ be the ∞-dimensional space which consists of the functions of the variable ϕ, $-\pi < \phi < \pi$. This space is spanned by the basis $\{\exp(in\phi), n = 0, \pm 1, \pm 2, ... \pm \infty\}$. Define a $(2b+1)$-dimensinal subspace $\mathcal{M} \subset \mathcal{S}$ which is spanned by by the basis $\{\exp(i\nu\phi), \nu = 0, \pm 1, \pm 2, ... \pm b\}$.

We now introduce another orthonormal basis $\{\Gamma_\alpha\}$ of $\mathcal{M}$ which will be used in the next section to describe the modulation envelope of the coherent wave on the bunch. The integer b here will be identified with the b in the condition (2). In accordance with the condition, we assume b to be large.

Divide the interval $-\pi < \phi \leq \pi$ into $2b+1$ equal parts with the lattice points

$$\phi = \phi_\alpha \equiv l_W\, \alpha/2 \ , \qquad (\alpha = 0, \pm 1, \pm 2, ... \pm b) \ , \tag{22}$$

where

$$l_W = 4\pi/(2b + 1) \ .$$

Definition. For $\alpha = 0, \pm 1, \pm 2, ... \pm b$,

$$\Gamma_\alpha(\phi) = [\, 2\pi(2b+1)\,]^{-1/2} \sum_{\nu=-b}^{b} \exp[i\nu(\phi - \phi_\alpha)] \ , \tag{23}$$

$$= [\, 2\pi(2b+1)\,]^{-1/2} \frac{\sin[(b + 1/2)(\phi - \phi_\alpha)]}{\sin[\,(\phi - \phi_\alpha)/2\,]} \ . \tag{24}$$

$\Gamma_\alpha(\phi)$ is a function peaked at $\phi = \phi_\alpha$, the first zeros of the function are at $\phi = \phi_\alpha \pm l_W/2$. The functions Γ_α corresponding to different α's are of identical shape, but they are shifted from each other in ϕ by integer multiples of $l_W/2$. Neighboring Γ's have non-negligible overlaps; the peak of a Γ and the first zero of the next Γ coincide. A few Γ's are depicted in Fig. 2. Note that if we take one of the $\Gamma_\alpha(\phi)$'s and multiply it with the carrier wave with wavelength λ satisfying (1), we obtain the coherent wave depicted in Fig 1.

We refer to the space $\mathcal{M}$ as the modulation space, to the set $\{\Gamma_\alpha\}$ as the modulation basis, and to the function $\Gamma_\alpha(\phi)$ as the modulation function. We now state a few theorems relating to the modulation basis.

Thoeorem 1 (orthonormality).

$$\int_{-\pi}^{\pi} d\phi \Gamma_\alpha(\phi)\Gamma_\beta(\phi) = \delta_{\alpha\beta} \ . \tag{25}$$

Theorem 2 (completeness in $\mathcal{M}$).

$$\sum_{\alpha=-b}^{b} \Gamma_\alpha(\phi)\Gamma_\alpha(\phi') = \frac{1}{2\pi} \sum_{\nu=-b}^{b} \exp[i\nu(\phi - \phi')] = \sqrt{\frac{2b+1}{2\pi}}\, \Gamma_0(\phi - \phi') \ . \tag{26}$$

This theorem implies that $\{\,\Gamma_\alpha\,\}$ spans the space $\mathcal{M}$, and that

$$\int_{-\pi}^{\pi} d\phi' \left[\sum_{\alpha=-b}^{b} \Gamma_\alpha(\phi)\Gamma_\alpha(\phi') \right] f(\phi') = f(\phi), \qquad \forall\, f \in \mathcal{M} \ .$$

In other words, the operator (26) is the unity operator on $\mathcal{M}$. If f is orthogonal to $\mathcal{M}$, then the above integral vanishes.

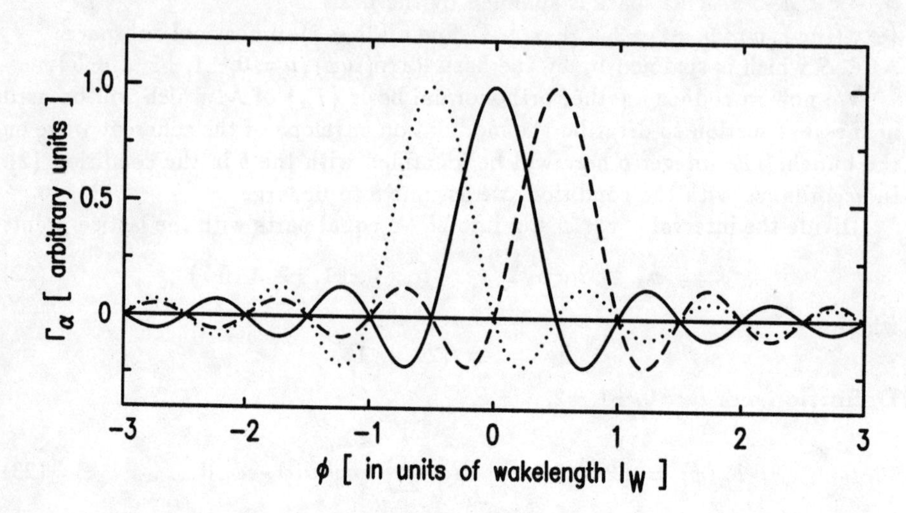

Fig. 2. Three neighboring modulation functions

5 A Solvable Model

In this section we introduce an impedance which makes the operator (21) diagonalizable in terms of the modulation basis $\{\Gamma_\alpha\}$ introduced in the last section. This makes the basic equation (20) solvable.

In what follows the symbols n_0, b, λ, σ and l_W carry the same meaning as in the introduction. They are assumed to satisfy the conditions (1) and (2).

5.1 Model Impedance and Kernel

For $n > 0$, the model impedance is defined by

$$U_n = \begin{cases} \bar{U} & \text{if } n_0 - b \leq n \leq n_0 + b, \\ 0 & \text{otherwise.} \end{cases} \tag{27}$$

If we ignore terms of order $O(b/n_0)$, (27) is equivalent to

$$Z_n = \begin{cases} \bar{Z} & \text{if } n_0 - b \leq n \leq n_0 + b, \\ 0 & \text{otherwise,} \end{cases} \tag{28}$$

with $\bar{U} = -in_0\bar{Z}$.

The impedances for $n < 0$ can be obtained by using

$$U_{-n} = U_n^*, \qquad Z_{-n} = Z_n^* ,$$

where * indicates the complex conjugate.

We now calculate the corresponding model kernel $K(\phi, \phi')$. Substituting (27) into (17), we obtain

$$U(\phi) = [\bar{U} \exp(-in_0\phi) + \text{c.c.}] \sqrt{\frac{2b+1}{2\pi}} \Gamma_0(\phi) ,$$

where (23) with $\alpha = 0$ and the relation $U_{-n} = U_n^*$ have been used, and c.c. stands for complex conjugate. Combining the above equation with (26) and (21), we have

$$K(\phi, \phi') = \{ \bar{U} \exp[in_0(\phi - \phi')] + \text{c.c.} \} \rho(\phi) \sum_{\alpha=-b}^{b} \Gamma_\alpha(\phi) \Gamma_\alpha(\phi') .$$

By assumption (2) and equation (24), $\rho(\phi) \cong \rho(\phi_\alpha)$ within the range of ϕ where $\Gamma_\alpha(\phi)$ is appreciable; namely, within the range $|\phi - \phi_\alpha| < \lambda_W/2$. Hence, introducing the notation

$$\rho_\alpha \equiv \rho(\phi_\alpha) ,$$

we can approximate

$$\rho(\phi)\Gamma_\alpha(\phi) \cong \rho_\alpha \Gamma_\alpha(\phi) , \tag{29}$$

and within the same approximation,

$$K(\phi, \phi') \cong \{ \bar{U} \exp[in_0(\phi - \phi')] + \text{c.c.} \} \sum_{\alpha=-b}^{b} \rho_\alpha \Gamma_\alpha(\phi) \Gamma_\alpha(\phi') .$$

Note that the kernel is now diagonalized in the modulation space. This is the form of the kernel we use below.

5.2 Diagonalization of the Basic Equation

We are now ready to solve the basic equation (20). First, let us collect together here some of the relevant formulae:

$$\frac{\partial^2}{\partial t^2} I(\phi, t) = \kappa \int_{-\pi}^{\pi} d\phi' \, K(\phi, \phi') \, I(\phi', t) , \tag{30}$$

$$K(\phi, \phi') = \{ \bar{U} \exp[in_0(\phi - \phi')] + \text{c.c.} \} \sum_{\alpha=-b}^{b} \rho_\alpha \Gamma_\alpha(\phi) \Gamma_\alpha(\phi') , \tag{31}$$

$$I^{(0)}(\phi) \equiv I(\phi, t = 0) = 2\pi I_{av} f(\phi; \phi_1, \phi_2, \ldots \phi_N) , \tag{32}$$

$$\dot{I}^{(0)}(\phi) \equiv \frac{\partial}{\partial t} I(\phi, t = 0) = 0 , \tag{33}$$

$$I(\phi, t) = \sum_{n=-\infty}^{\infty} I_n(t) \exp(in\phi) . \tag{34}$$

Decompose $I(\phi, t)$ into three mutually orthogonal components,

$$I(\phi, t) = \exp(in_0\phi) \, J(\phi, t) + \exp(-in_0\phi) \, J^*(\phi, t) + \check{I}(\phi, t) ,$$

where $J \in \mathcal{M}$ ($\mathcal{M}$ was defined in Sect 4.) Note that the first component is the contribution of $I_n(t)$ with $n \in [n_0 - b,\ n_0 + b]$, the second component is the contribution from $n \in [-n_0 - b,\ -n_0 + b]$, and $\check{I}$ is defined to be the contribution from n outside both these bands. These three components are orthogonal to each other since they belong to non-overlapping bands of n. For example,

$$\int_{-\pi}^{\pi} d\phi \, \exp(in_0\phi)\, J(\phi,t)\, \check{I}^*(\phi,t) = 0 \ .$$

We observe right away from (30) and (31) and the definition of $\check{I}$ above that

$$\frac{\partial^2}{\partial t^2}\check{I}(\phi,t) = 0 \ .$$

This, together with (33), implies that $\check{I}(\phi,t)$ is independent of t. We shall therefore ignore $\check{I}$ in the remainder of this paper, and write

$$I(\phi,t) = \exp(in_0\phi)\, J(\phi,t) + \exp(-in_0\phi)\, J^*(\phi,t) \ . \tag{35}$$

Note that in this equation the carrier wave $\exp(in_0\phi)$ is modulated by $J(\phi,t)$.

Now combining (30), (31) and (35), we obtain an equation in the modulation space $\mathcal{M}$,

$$\frac{\partial^2}{\partial t^2}J(\phi,t) = \kappa \bar{U} \sum_{\alpha} \rho_\alpha \Gamma_\alpha(\phi) \int_{-\pi}^{\pi} d\phi' \Gamma_\alpha(\phi') J(\phi',t) \ . \tag{36}$$

Since $\{\Gamma_\alpha\}$ is an orthonormal basis of $\mathcal{M}$, this equation is already diagonalized, the eigenfunction being $\Gamma_\alpha(\phi)$. Let

$$J(\phi,t) = \sum_{\alpha=-b}^{b} J_\alpha(t)\, \Gamma_\alpha(\phi) \ , \tag{37}$$

then

$$\frac{d^2}{dt^2}J_\alpha(t) = \kappa\rho_\alpha\, J_\alpha(t) \ . \tag{38}$$

The coherent frequency Ω_α of the mode α can readily be obtained from (38). Let

$$J_\alpha(t) \sim \exp(-i\Omega_\alpha t)$$

then

$$\Omega_\alpha^2 = -\kappa\rho_\alpha \bar{U} \ . \tag{39}$$

Note that for each eigenvector $\Gamma_\alpha(\phi)$ of (36) there are actually two solutions for $J(\phi,t)$,

$$\Gamma_\alpha(\phi)\exp(-i\Omega_\alpha t) \quad \text{and} \quad \Gamma_\alpha(\phi)\exp(i\Omega_\alpha t) \ . \tag{40}$$

This is a reflection of the fact that the basic equation (30) is second order in time.

If we take the well-known expression for the coherent frequency of the coasting beam instability corresponding to the mode $n = n_0$ and then replace I_{av} in the expression with the local current density, $2\pi\rho_\alpha I_{av}$, of the bunched beam at the position $\phi = \phi_\alpha$, we would also obtain (39). This amounts to proof of the Boussard conjecture. It is worth repeating here, for emphasis, what was stated in the introduction: The realization of the Boussard conjecture is a consequence of the coherent wave of the microwave instability being localized in the bunch, even though the coasting beam coherent wave $\exp(in_0\phi)$ covers the whole ring.

5.3 Matching the Initial Condition

We have just found that corresponding to each mode number α there are two solutions (40) for $J(\phi,t)$. In order for these two solutions to satisfy the initial condition (33), they must combine to give

$$J(\phi,t) \sim \Gamma_\alpha(\phi)\cos\Omega_\alpha t \ .$$

Now adding up the contributions from all α, and including the contribution from the J^* term in (35), we obtain

$$I(\phi,t) = \sum_{\alpha=-b}^{b} \Gamma_\alpha(\phi)\left[I_\alpha^{(0)}\exp(in_0\phi)\cos\Omega_\alpha t + \text{c.c.}\right] , \qquad (41)$$

where

$$I_\alpha^{(0)} = \int_{-\pi}^{\pi} d\phi\Gamma_\alpha(\phi)I^{(0)}(\phi)\exp(-in_0\phi) , \qquad (42)$$

so that the initial condition (32) is satisfied.

Having found the solution (41) to the initial value problem of our model, we now calculate the longitudinal electric field $\mathcal{E}$ induced by $I(\phi,t)$.

Combining the equations (15), (28) and (41), and then using the relation $Z_{-n} = Z_n^*$, we obtain

$$\mathcal{E}(\phi,t) = -[\,\bar{Z}\sum_{\alpha=-b}^{b} \Gamma_\alpha(\phi)\,I_\alpha^{(0)}\exp(in_0\phi)\cos(\Omega_\alpha t) + \text{c.c.}\,]/(2\pi R) . \qquad (43)$$

Let us close this section by the following comments: We have succeeded in solving the transient problem of our model by expressing $I(\phi,t)$ and $\mathcal{E}(\phi,t)$ in terms of the initial current $I^{(0)}(\phi)$. However, the initial current $I^{(0)}(\phi)$ as given by (32) is a messy collection of grainy shot noise f (cf. Sect 2.3.) We calculate in the next section the radiation power, using the results of this section. The effects of the shot noise will be statistically averaged in the expression for the radiation power.

6 Radiation Power

The power lost by the beam (per radian of the beam distribution) is

$$P(\phi, t) = -R\,\mathcal{E}(\phi, t)\, I(\phi, t) \ . \tag{44}$$

From conservation of energy, this is also the radiation power. The total radiation power is then

$$P_{tot}(t) = \int_{-\pi}^{\pi} d\phi\, P(\phi, t) \ . \tag{45}$$

Substituting (41) and (43) into (44) and ignoring the fast oscillating terms involving $\exp(\pm i 2 n_0 \phi)$, we obtain

$$P(\phi, t) = \frac{1}{2\pi} \sum_{\alpha,\beta} [\, \bar{Z}\, I_\alpha^{(0)}\, I_\beta^{*(0)} \cos \Omega_\alpha t \cos \Omega_\beta^* t \, + \, \text{c.c.}\,] \Gamma_\alpha(\phi) \Gamma_\beta(\phi) \ , \tag{46}$$

where both α and β are summed from $-b$ to b.

6.1 Averaging over Shot Noise

We use the notation $< ... >$, as we did in Sect. 2.3, to indicate averaging over shot noise:

$$< P(\phi, t) > = \frac{1}{2\pi} \sum_{\alpha,\beta} [\, \bar{Z} < I_\alpha^{(0)}\, I_\beta^{*(0)} > \cos \Omega_\alpha t \, \cos \Omega_\beta^* t \, + \, \text{c.c.}\,] \Gamma_\alpha(\phi) \Gamma_\beta(\phi) \ . \tag{47}$$

From (42),

$$< I_\alpha^{(0)}\, I_\beta^{*(0)} > = \int_{-\pi}^{\pi} d\phi \int_{-\pi}^{\pi} d\phi' \exp[i n_0 (\phi' - \phi)]\, \Gamma_\alpha(\phi) \Gamma_\beta(\phi') < I^{(0)}(\phi) I^{(0)}(\phi') > . \tag{48}$$

We must now evaluate $< I^{(0)}(\phi) I^{(0)}(\phi') >$ or equivalently, $< f(\phi) f(\phi') >$. The shot noise in a bunched beam is correlated because the bunch distribution function $\rho(\phi)$ is not uniform. In anticipation of the final results, we make the following remarks: Since $\Gamma_\alpha(\phi)$ is appreciable only within a width $l_W \cong 2\pi/b$ around $\phi = \phi_\alpha$, we see from (48) that we have to take the average only in this region. From the assumption $l_W \ll \sigma$, the bunch distribution $\rho(\phi)$ is nearly constant within the width l_W of $\Gamma_\alpha(\phi)$; we can therefore safely take the uniform shot noise average, assuming the noise density to be $N \rho_\alpha$.

We start from the definition

$$< f(\phi) f(\phi') > = \int_{-\pi}^{\pi} \prod_{j=1}^{N} [\rho(\phi_j) d\phi_j\,] f(\phi; \phi_1, \phi_2, ... \phi_N)\, f(\phi'; \phi_1, \phi_2, ... \phi_N) \ .$$

Using (5) and (7) on the above equation, we have

$$< f(\phi) f(\phi') > = \, = \frac{1}{N} [\, \rho(\phi) \delta(\phi' - \phi) \, - \, \rho(\phi) \rho(\phi')\,] \ . \tag{49}$$

The last term of this equation reflects the effects of the correlation induced by non-uniformity of the bunch. Let us ignore this term for now and show later that the contribution of this term is indeed negligible in our region of interest given by (1) and (2).

If we ignore the last term of (49), then (48) yields

$$< I_\alpha^{(0)} I_\beta^{*(0)} > = [(2\pi I_{av})^2/N] \int_{-\pi}^{\pi} d\phi \rho(\phi) \Gamma_\alpha(\phi) \Gamma_\beta(\phi) \ .$$

With use of the approximation (29), the above equation becomes

$$< I_\alpha^{(0)} I_\beta^{*(0)} > = [(eN\omega_0\rho_\alpha)^2/(N\rho_\alpha)]\delta_{\alpha,\beta} \ , \tag{50}$$

where Theorem 1 has been used. Substituting (50) into (47), we obtain

$$< P(\phi,t) > = \sum_{\alpha=-b}^{b} \Gamma_\alpha(\phi) < P_\alpha(\phi,t) > \ , \tag{51}$$

with

$$< P_\alpha(\phi,t) > = 2\bar{\mathcal{R}} \left[(eN\omega_0\rho_\alpha)^2/(2\pi N\rho_\alpha) \right] |\cos \Omega_\alpha t|^2 \Gamma_\alpha^2(\phi) \ , \tag{52}$$

where $\bar{\mathcal{R}}$ is the resistive part of $\bar{Z}$. Had we assumed the initial beam to consist of uniform shot noise with density ρ_α we would have obtained the same result.

The calculation of averaged total radiation power from $< P(\phi,t) >$ above is straightforward. The result is

$$< P_{tot}(t) > = \sum_{\alpha=-b}^{b} < P_{tot,\alpha}(t) > \tag{53}$$

with

$$\begin{aligned} < P_{tot,\alpha}(t) > &= \int_{-\pi}^{\pi} d\phi < P_\alpha(\phi,t) > \\ &= 2\bar{\mathcal{R}} \left[(eN\omega_0\rho_\alpha)^2/(2\pi N\rho_\alpha) \right] |\cos \Omega_\alpha t|^2 \ . \end{aligned} \tag{54}$$

If we write

$$\Omega_\alpha = \Omega_{\alpha,R} + ig_\alpha \ , \tag{55}$$

where $\Omega_{\alpha,R}$ is the real coherent frequency shift and g_α is the growth rate of the mode α, then, for large t,

$$|\cos \Omega_\alpha t|^2 \cong \frac{1}{4} \exp(2g_\alpha t) \ . \tag{56}$$

It is interesting to approximate the summation in (53) by an integral. If we set

$$\alpha \longrightarrow \frac{2b+1}{2\pi}\phi \cong \frac{b}{\pi}\phi \ , \qquad \sum_{\alpha=-b}^{b} \longrightarrow \frac{b}{\pi}\int_{-\pi}^{\pi} d\phi \ ,$$

$$\Omega_\alpha \longrightarrow \Omega(\phi) = \Omega_R(\phi) + ig(\phi) \ , \qquad \rho_\alpha \longrightarrow \rho(\phi) \ ,$$

then (53) and (54) give

$$< P_{tot}(t) > = \frac{1}{2\pi} \bar{\mathcal{R}} \int_{-\pi}^{\pi} d\phi \frac{[eN\omega_0\rho(\phi)]^2}{N\rho(\phi)l_W} \exp[2g(\phi)t] \ , \tag{57}$$

where (56) has been used. In the integrand of the last equation, $eN\omega_0\rho(\phi)$ in the numerator is the local current density, and the denominator $N\rho(\phi)l_W$ is the number of particles in a coherence length. The denominator reflects the fact that, as the number of particles in a coherence length increases, the partially coherent signal of the initial shot noise responsible for the startup of the coherent instability decreases.

We have so far ignored the effect of the last term of (49), the effect of the non-uniformity of the shot noise in a bunched beam. Let us verify now that it is indeed ignorable. The last term of (49) adds to (50) a term proportional to

$$\frac{1}{N} \int_{-\pi}^{\pi} d\phi\rho(\phi)\Gamma_\alpha(\phi) \exp(-in_0\phi) \int_{-\pi}^{\pi} d\phi'\rho(\phi')\Gamma_\beta(\phi') \exp(in_0\phi') \ .$$

If we apply the approximation (29) to the integrands above, then both integrals vanish, since $\exp(\pm in_0\phi)$ is orthogonal to $\Gamma_\alpha(\phi)$ and $\Gamma_\beta(\phi)$.

7 Conclusions

We have proposed that one can use the microwave instability in an electron or proton storage ring as a source of partially coherent radiation. Unlike the spontaneous radiation, the beam particle mass is not a decisive parameter of the coherent-instability-induced radiation power.

We constructed a model where the radiation power can be calculated precisely. In this model, the beam is assumed to have no initial energy spread. It would be worth while to investigate the effects of the initial beam energy spread on the radiation power.

ACKNOWLEDGEMENTS:
I have enjoyed and benefited from discussions with Eric Blum and Jim Murphy. This work was performed under the auspices of the US DOE.

References

[1] A.Falten and L.J.Laslett, Report BNL-20550 (1975).
R.L.Warnock and P.Norton, Particle Accelerators **25** (1990) 113.
K.Y.Ng, Particle Accelerators **25** (1990) 153.

[2] D.Boussard, CERN LABII/RF/INT/75-2, 1975.

[3] J.M.Wang and C.Pellegrini, Report BNL-51236 (1979).
J.M.Wang and C.Pellegrini, Proc. 11th Int Conf. on High Energy Accelerators, Geneva (1980), p. 554.
S.Krinsky and J.M.Wang, Particle Accelerators **17** (1985) 109.
J.M.Wang, 1985 SLAC Accelerator School, AIP Proc. **153** p. 697.

[4] R.Ruth, Ph.D. Thesis, Report BNL-51425 (1981).

Multibunch Instabilities

Flemming Pedersen

CERN, Geneva, Switzerland

1 Introduction and Summary

The theory for transverse and longitudinal multibunch instabilities is reviewed. The coherent beam modes are classified, and the various mode numbers defining the coherent modes are explained. Sacherer's longitudinal and transverse growth rate formulae are discussed and compared with the commonly used short-bunch approximation and Robinson's characteristic equation. Coupling impedances with long-range wakes are particular troublesome for the large high-current colliders planned for the next decade. These are the higher-order modes of the RF cavities, the fundamental mode of the RF cavities, and the transverse resistive wall impedance.

Table 1. Classification of coherent beam modes

	Coasting Beams	Bunched Beams
Longitudinal	n = azimuthal mode number = 1,2,3, ... ∞	n = coupled bunch mode number = 0, 1, 2, ... (M-1) m = phase plane periodicity = 1 (dipole), 2 (quadrupole), 3 (sextupole), ... (q = radial mode number)
		Mode coupling ⇒ Single-bunch "microwave" instability (turbulence).
Transverse	n = azimuthal mode number = -∞, ..., -1, 0, 1, 2, ... +∞ k = phase plane periodicity = 1 (dipole), 2 (quadrupole), 3 (sextupole), ...	n = coupled bunch mode number = 0, 1, 2, ... (M-1) m = head-tail mode number = ..., -2, -1, 0 1, 2, ... k = phase plane periodicity = 1 (dipole), 2 (quadrupole), 3 (sextupole), ..
		Mode coupling ⇒ Single-bunch, fast, head-tail instability (turbulence).

2 Classification of Coherent Beam Modes

Coherent beam modes are classified according to whether the coherent beam motion is longitudinal or transverse, and according to whether the beam is bunched or debunched (coasting beam), Table 1. In this way four main classes of coherent beam modes are defined. Beams are of course always bunched in e+/e- rings due to synchrotron radiation, but it is useful to classify the beam modes in this general way.

The complexity of the beam motion increases from longitudinal to transverse, and from coasting to bunched, such that the mode description requires an increasing number of mode numbers.

3 Longitudinal Bunched-Beam Modes

The general theory for coherent bunched beam modes and their interaction with the environment is due to Sacherer [1][2][3]. Basically two mode numbers describe the motion (see Fig. 1). The *coupled bunch mode number n* is defined as the *number of waves of coherent motion per revolution*, and resembles therefore the azimuthal mode number for coasting beams.

For bunched beams with M equidistant bunches, the bunch-to-bunch phase shift $\Delta\phi$ is related to the coupled bunch mode number n by $\Delta\phi = 2\pi n/M$. There are M coupled bunch modes numbered from 0 to (M-1). This is in contrast to the azimuthal mode number for coasting beams, where there is an infinite number of modes.

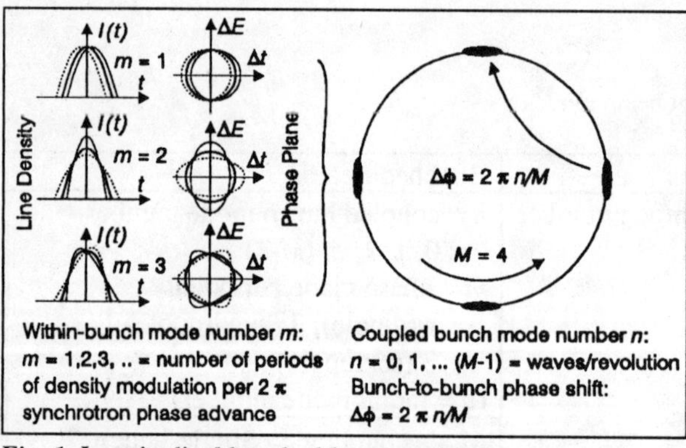

Fig. 1. Longitudinal bunched-beam modes

The within-bunch mode number m is the *number of periods of phase space density modulation per synchrotron period* in the longitudinal phase plane. The lowest mode number is $m = 1$, which corresponds to the dipole mode, $m = 2$ is the quadrupole mode, $m = 3$ is the sextupole mode and so on. The line density is the projection of the phase space distribution on the time axis. The observed pattern for a given bunch oscillates with m times the synchrotron frequency, and the pattern has m nodes along the bunch (see Fig. 1).

The theory for longitudinal bunched beam mode interactions [4][5] contains in addition a *radial mode number* $q = m$, m+2, ..., which is describing an infinity of orthogonal radial modes with different density variations versus synchrotron amplitude (= radius). The first higher-order radial mode $q = 3$ for the dipole mode

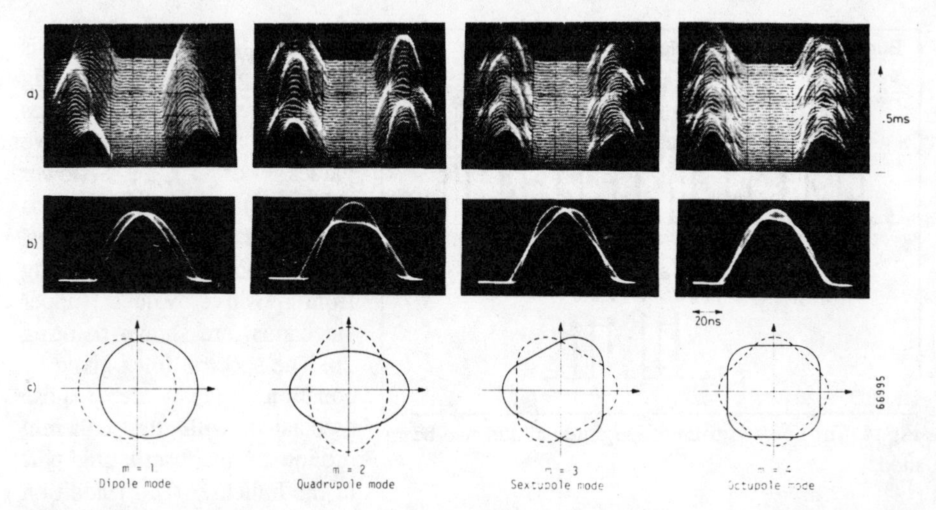

Fig. 2. Within bunch modes $m = 1$ to 4, coupled-bunch pattern $n = 4$. a) Mountain-range display of one synchrotron period; b) Superimposed; c) Phase space.

($m = 1$) has thus a line density pattern which looks like a sextupole mode ($m = 3$), but it oscillates with the synchrotron frequency and not three times this frequency. Normally only the lowest radial mode is observed. This is probably due to the fact that the higher radial modes have higher Landau damping thresholds [5].

An example of various different coupled bunch modes observed in the CERN PS Booster ($M = 5$, $n = 4$, $m = 1$ to 4) is shown on Fig. 2. Note the bunch-to-bunch phase shift of the coherent motion of $\Delta\phi = 360°*n/M = 288° = -72°$.

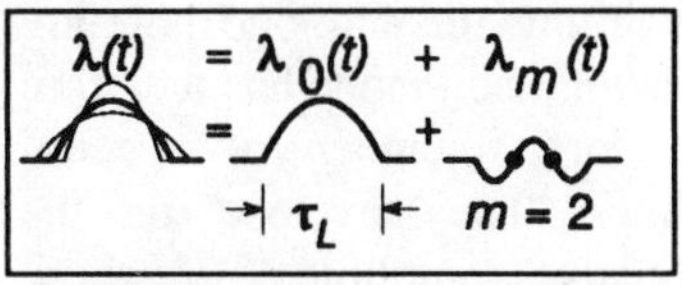

$\lambda(t) = \lambda_0(t) + \lambda_m(t)$

τ_L $\quad m = 2$

Fig. 3. Stationary and oscillating part of line density

It is apparent from figures 1 and 2 that the line density $\lambda(t)$ or current $I(t) = e\,\lambda(t)$ of a bunch can be decomposed into two parts, the stationary distribution $\lambda_0(t)$ plus an additional charge density $\lambda_m(t)$ oscillating at m times the synchrotron frequency. The oscillating part $\lambda_m(t)$ is an approximately sinusoidal standing-wave pattern with m fixed nodes along the bunch (Fig. 3).

By taking the Fourier transform of the bunch current $I(t)$, we get the frequency spectrum of the modes. It is a line spectrum, and the line frequencies for mode (n,m) are:

$$f_{nm,p} = (n + pM)f_0 + mf_s \qquad -\infty < p < +\infty \tag{1}$$

where f_0 is the revolution frequency, f_s is the synchrotron frequency, and p is an integer assuming both negative and positive values. It is convenient from a mathematical point of view to let the frequencies $f_{nm,p}$ assume both positive and negative values, although physically a frequency is always a positive quantity.

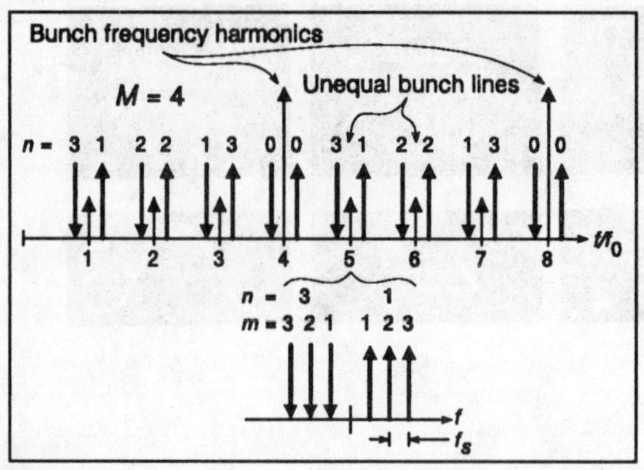

Fig. 4. The line spectrum of longitudinal bunched beam modes

This line spectrum is depicted graphically (see Fig 4), where the mode lines originating from negative values of $f_{nm,p}$ (lower sidebands) are shown as downward pointing arrows, while mode lines originating from positive values (upper sidebands) are shown pointing up. The spectral lines given by equation (1) are those associated with the coherent motion or the oscillating part of the bunch $\lambda_m(t)$. In addition there are lines associated with the stationary bunch spectrum or $\lambda_0(t)$. There are strong lines at the bunch frequency Mf_0 and harmonics thereof. In practice the M bunches will have slightly different intensities or there might be a gap in the bunch train, which causes spectral lines of lower amplitude to appear at the intermediate revolution harmonics.

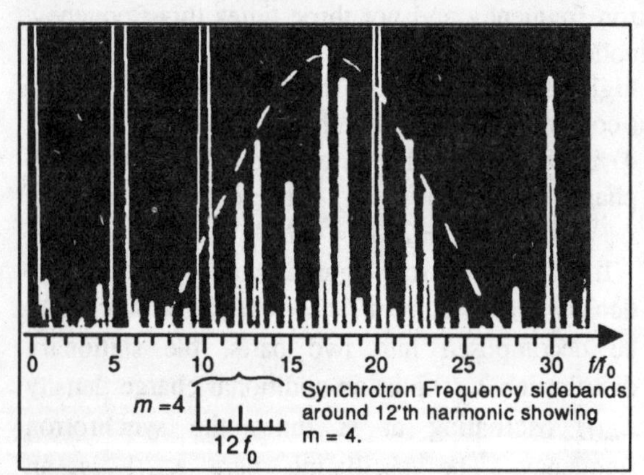

Fig. 5. Observed spectrum for mode $n = 3$, $m = 4$ in the CERN PSB. Bandwidth 300 kHz, range 0 - 50 MHz, linear scale, bunch length $\tau_L = 66$ ns.

Each bunched beam mode (m,n) has thus two lines appearing within each band Mf_0 wide between two bunch frequency harmonics, one pointing up and one pointing down. Each mode has therefore a large number of spectral lines. The envelope or the relative amplitude of those lines depends upon the within-bunch mode number m and the bunch length τ_L.

An example of a measured longitudinal mode spectrum from the CERN PS Booster [3] is shown on Fig. 5 ($M = 5$, $n = 3$, $m = 4$, octupole mode). Both coherent mode lines, bunch frequency harmonics, and unequal bunch lines are clearly seen.

3.1 Sacherer's Formula for Longitudinal Bunched Beam Modes

Due to electromagnetic interactions with the beam environment quantitatively described by the longitudinal coupling impedance $Z_L(\omega)$, the coherent mode frequencies are shifted a complex quantity $\Delta\omega_{mn}$ away from their low intensity values. This shift is obtained by a weighted sum (factor $F_m(f_p\tau_L)$) of the coupling impedance $Z_L(\omega_p)$ sampled at all those frequencies $\omega_p = 2\pi f_p$ that corresponds to a mode line (see Fig. 4, equation (1)) of mode (m,n), Sacherer's formula [1][2][3]:

$$\Delta\omega_{mn} = j\omega_s \frac{m}{m+1} \frac{I}{3B_0^2 hV \cos\phi_s} \sum_p F_m(f_p\,\tau_L) \frac{Z_L(f_p)}{p+n} \tag{2}$$

where ω_s is the synchrotron frequency, I is the total beam current in M bunches, B_0 the bunching factor ($= \tau_L/T_0$, where $T_0 = 1/f_0$ is the revolution period), h the harmonic number, V the total RF voltage, ϕ_s is the synchronous phase angle with the convention that $V\sin\phi_s$ is the energy gain per turn, $\phi_s < 90°$ below transition, $\phi_s > 90°$ above transition, and τ_L the full bunch length. The form factor F_m is the normalised power spectrum of the perturbed part $\lambda_m(t)$ of the line density $\lambda(t)$:

$$F_m(f_p\tau_L) = \frac{1}{MB_0} \frac{\left|\tilde{\lambda}_m(p)\right|^2}{\sum_p \left|\tilde{\lambda}_m(p)\right|^2} \tag{3}$$

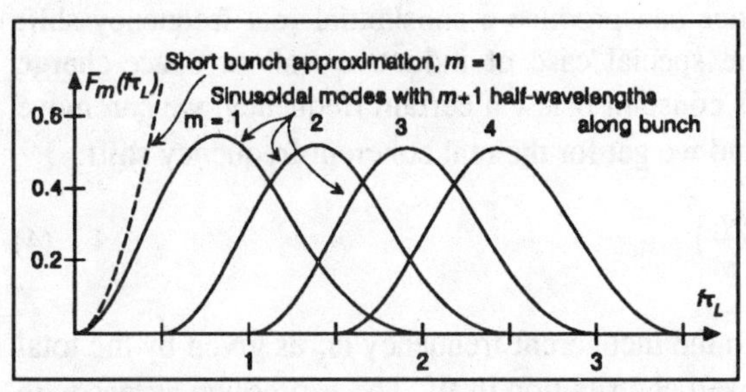

Fig. 6. Sinusoidal mode form factors (from [3]) and short bunch approximation

where $\tilde{\lambda}_m(p)$ is the Fourier transform of the perturbed part $\lambda_m(t)$ of the line density and M the number of equidistant bunches. For the sinusoidal type modes shown on Fig. 3, the form factors $F_m(f\tau_L)$ are plotted on Fig. 6.

The *growth rate* of mode (m,n) is $-\mathrm{Im}\{\Delta\omega_{mn}\}$ which is related to the *real* part of the coupling impedance $Z_L(\omega)$. The *real coherent frequency shift* of mode (m,n) is $\mathrm{Re}\{\Delta\omega_{mn}\}$, which is related to the *imaginary* part of the coupling impedance.

3.2 Discussion of Sacherer's Longitudinal Bunched Beam Formula

The real part of the coupling impedance $Z_L(\omega)$ is symmetric, ($\mathrm{Re}\{Z_L(\omega)\} = \mathrm{Re}\{Z_L(-\omega)\}$), so contributions to the sum in (2) associated with negative and positive values of f_p (mode lines pointing down and up on Fig. 4) enter with opposite signs due to the factor $1/(p+n)$. Broad-band and resonant impedances with bandwidths larger than the bunch frequency Mf_0 (decay time less than bunch separation) do therefore not cause any growth or damping rate. Each mode has two mode lines in each band between two bunch frequency harmonics, and their contributions to the sum of (2) tend to cancel each other.

For resonant impedances with bandwidths smaller than the bunch frequency Mf_0 (decay time longer than bunch separation), a single mode line for each mode dominates in the sum and one or several coupled-bunch modes grow. The growth rate is proportional to the total current I in M bunches, and independent of the number of bunches unless the bunch frequency is low enough to get partial cancellation from another mode line of the same mode falling within the resonator bandwidth.

The signs in (2) are such that *upper* sidebands are *unstable above* transition, and lower sidebands arc stable (assuming *passive* coupling impedances: $\mathrm{Re}\{Z_L\} > 0$). The opposite is true below transition. For $M \geq 3$ upper and lower sidebands belong to different coupled-bunch modes (except for $n = 0$ and $M/2$) so a resonator will drive one mode n_1 and damp the other complementary mode n_1-M. For a single bunch, or two bunches, upper and lower sidebands belong to the same coupled-bunch mode and therefore tend to cancel unless the impedance is very narrow-band (such as the fundamental RF resonance) and tuned asymmetrically with respect to the frequencies with mirror symmetry such as pMf_0 or $(p+½)Mf_0$.

The imaginary part of the coupling impedance $Z_L(\omega)$ is anti symmetric, $(\mathrm{Im}\{Z_L(\omega)\} = -\mathrm{Im}\{Z_L(-\omega)\}$, so contributions to the sum in (2) associated with negative and positive values of f_p (mode lines pointing down and up on Fig. 4) add up, so even a broad-band impedance can produce a substantial real frequency shift due to its imaginary part. For the special case of inductive-wall or space-charge impedance, where $\mathrm{Im}\{Z_L(\omega_p)/p\}$ is constant below a certain frequency, we can move $Z_L(\omega_p)/p$ outside the summation, and we get for the real coherent frequency shift:

$$\Delta\omega_{1n} = -\omega_s \frac{I}{6B_0^3 MhV\cos\phi_s} \mathrm{Im}\{\frac{Z_L}{p}\} \tag{4}$$

This is the coherent shift relative to the incoherent frequency ω_s as given by the total voltage V *including* the inductive-wall contribution itself. The expression agrees with the expression given in [6] except for a factor $(\pi/3)^2 = 1.097$. It is a single-bunch, short-range wake effect where all coupled-bunch modes n have the same shift, which is the same as saying that all individual bunches have their coherent frequencies shifted the same amount. As expected, the shift is proportional to the current per bunch I/M.

3.3 Turbulent Bunch Lengthening and Longitudinal Mode Coupling

For sufficiently high currents (or sufficiently high impedances), single bunches can become unstable, as was first observed by Boussard [7]. The instability is usually called the microwave instability in proton rings due to the high frequencies involved, or turbulent bunch lengthening in electron rings, probably due to difficulties involved in observing the even higher frequencies involved in electron rings. The instability threshold is given by the Keil-Schnell criterion [8] for coasting beams, but with local values of bunch current and momentum spread as suggested by Boussard.

It was shown by Sacherer [2], that this threshold is consistent with the mode coupling threshold for two higher order longitudinal modes m and $m+1$, as the real frequency shifts associated with the imaginary part of the coupling impedance cause their frequencies to cross. The mode coupling theory gives a more precise information about the threshold dependence upon resonator bandwidth than the Boussard criterion.

3.4 Short Bunch Approximation for Dipole Modes

Bunches in e+/e- rings are often so short that the frequencies of interest below the vacuum pipe cut-off frequency are below the peak of F_1. In this case the complex coherent frequency shift for the dipole mode $m = 1$ can be simplified to:

$$\Delta\omega_{1,n} = -j\frac{I\eta}{2Q_s\beta^2(E/e)}\sum_p f_p Z_L(f_p) \tag{5}$$

where $\eta = 1/\gamma_t^2 - 1/\gamma^2$, I is the total beam current, $Q_s = \omega_s/\omega_0$ is the longitudinal tune, E is the total energy, e the fundamental charge, β and γ the usual relativistic parameters, and γ_t is the transition energy. Note that the form factor F_1 does not appear in this equation. By using the expression for the synchrotron frequency ω_s:

$$\omega_s = \omega_0\sqrt{\frac{-\eta h V\cos\phi_s}{2\pi(E/e)\beta^2}} \tag{6}$$

and equating the coherent shifts given by equation (5) and (2) we get an expression for the rigid dipole mode form factor F_1 implied in equation (5):

$$F_1(f\tau_L) = 3(p+n)^2 B_0^2 = 3(f\tau_L)^2 \tag{7}$$

which is plotted on Fig. 6 together with the form factor for the sinusoidal modes. It corresponds to the initial parabolic behaviour of $F_1(f\tau_L)$, and is a valid approximation for $f\tau_L << 0.5$.

It is seen from Fig. 6, that the form factor corresponding to the short bunch formula (5) is slightly higher than the form factor computed from sinusoidal modes. This is because the sinusoidal modes assumed are slightly different from the rigid bunch motion assumed in (5).

4 Longitudinal Coupling Impedances

Longitudinal coupling impedances can be subdivided into *short-range-wake or broad-band* impedances (bandwidth > Mf_0), which have identical effects on all bunches and on all coupled-bunch modes n, and *long range wake or narrow band* impedances (bandwidth < Mf_0) which have very different effects on the different coupled bunch modes n due to the line structure of the spectrum. Note that the dividing criterion depends on the bunch frequency Mf_0 or bunch separation $1/Mf_0$, since the crucial point is whether the wake decays in the bunch interval.

Broad-band impedances are responsible for single-bunch effects, which have thresholds related to current per bunch, and where each bunch behaves independently of the others. Narrow-band impedances are responsible for multibunch effects, where the thresholds are related to total current, and the coupled-bunch mode number is important.

Next generation, high-current colliders (B, tau/charm and phi factories) typically have very high bunch frequencies, and even fairly well damped broad band resonators have significant long-range multibunch effects.

4.1 Parasitic Higher Order Modes

RF cavities are often the major source of long-range, narrow-band coupling impedances due to undesired higher-order modes (HOM's) between the fundamental RF resonance and the vacuum pipe cut-off frequency. For low total current and low bunch frequencies, it is usually possible to lower the Q and the shunt impedance of these cavities to a level where the coupled-bunch growth rates are below the synchrotron radiation damping rate.

For high-current colliders, sophisticated HOM damping schemes [9] are required to damp the higher-order modes by more than two orders of magnitude without significantly affecting the fundamental mode. Usually one or several wave guides with a cut-off frequency between the fundamental RF cavity resonance and the lowest frequency higher-order mode form a high-pass filter between the RF cavity and the HOM loads, both for normal conducting cavities [10][11] and superconducting cavities [12], where an enlarged portion of the beam pipe is used as a HOM wave guide.

Even with a sophisticated HOM damping scheme, the residual shunt impedance may still be large enough to cause longitudinal dipole-mode growth rates in excess of the synchrotron radiation damping rate, and a multibunch feedback system is required to stabilise all the modes. Operating with a high RF voltage, as required to obtain short bunches, appears to make the growth rate smaller (ω_s^{-1} scaling, equation (5)), but does in fact make it worse, as the total resonant HOM impedance Z_L is proportional to the number of cavities.

If the bandwidth of the HOM's is large compared with the synchrotron frequency, about half the coupled-bunch modes are unstable as discussed above. For one or two bunches growth rates are small due to cancellation, since upper and lower sidebands at each harmonic belong to the same mode n. For small rings with high synchrotron tune and little or no HOM damping, the HOM bandwidths may become small compared to the synchrotron frequency, and cancellation is no longer effective. This is the case for the SLC damping rings [13], where HOM's of the accelerating cavities drive the $n = 1$ mode ($M = 2$), also called the π mode, unstable. This fact can also be used with advantage to passively damp the instability. The instability has been cured by *adding* a passive cavity, tuned to interact mainly with the a lower sideband of the $n = 1$ mode.

4.2 Fundamental Resonance of RF Cavities

For small rings, the fundamental resonance interacts only with the $n = 0$ mode, which is easily stabilised (at least for electron rings with no complex RF feedback loops) by an appropriate tuning of the RF cavity. For large rings with high beam currents, other coupled-bunch dipole modes ($n = M - 1$, $M - 2$, etc.) may be driven unstable.

4.2.1 The Robinson Criterion and the $n = 0$, $m = 1$ Mode

A complete analysis of the stability of the interaction between the $n = 0$ dipole mode of the beam and the fundamental resonance of the RF cavity (without feedback) was first described by Robinson [14], and results in a characteristic equation of fourth order. The stability criterion can be solved analytically, and above transition we get:

$$\frac{2 I_0 \cos\phi_s}{I_B} < \sin 2\phi_z < 0 \tag{8}$$

where I_0 is the peak value of RF current required to drive the cavity to the operating voltage when in tune (the loss current), $I_B \cong 2\,I$ is the peak value of the fundamental RF component of the beam current, which for short bunches equals twice the beam DC current I, and ϕ_Z the impedance angle of the cavity impedance at the operating RF frequency. The notation used here is slightly different from the one used in Robinson's original paper. The two regions of instability corresponding to the two inequalities in (8) are depicted graphically on Fig. 7.

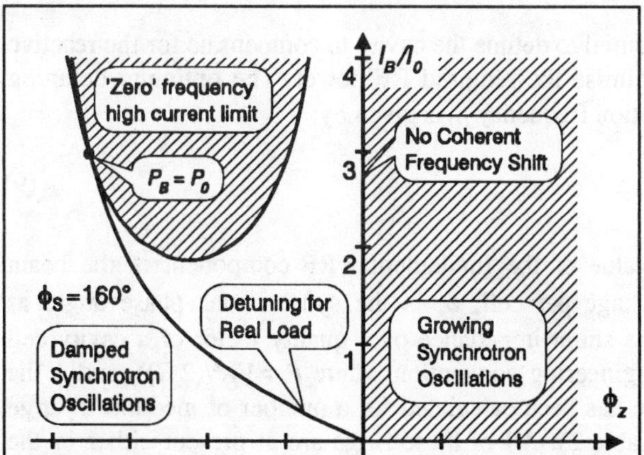

Fig. 7. Zones of instability versus normalised beam loading I_B/I_0, and cavity impedance angle $\phi_Z = \mathrm{Arg}(Z_L)$

For low currents the growth rate that is obtained from Robinson's characteristic equation by a root perturbation technique is in perfect agreement with the short bunch formula (5). If the cavity is tuned exactly to the operating RF frequency, the symmetry in the real part of the cavity impedance causes a perfect cancellation of the contributions from upper and lower sidebands, so no growth rate is induced of the $n = 0$, $m = 1$ mode. Since the imaginary part of the impedance is anti symmetric around the RF frequency, there is no real frequency shift either. By *detuning the cavity below the operating RF frequency* (above transition) this mode is damped. This corresponds to satisfying the right hand inequality in (8).

Since the fundamental of the beam current is out of phase with the voltage, the beam load represents a partly reactive load to the RF source, which results in an unnecessary increased power demand from the RF source. By detuning the RF cavity such that a *real total load* (beam plus cavity) is presented to the RF source, less power is needed. This optimum detuning is usually done automatically by a tuning loop. It is fortunate that the signs are such that this detuning is always to the same side as that which produces damping of the $n = 0$ mode. At high current a *zero-frequency instability occurs* even if the cavity is detuned to the 'stable' side. This instability threshold corresponds to the left hand side of the Robinson criterion (8). This instability occurs (for optimum detuning) when the *beam power* equals or exceeds the *dissipated power* in the cavity (including the equivalent loss in the RF source impedance). If the power source is *back matched* (e. g. circulator in transmission line), the RF power source is *matched to the total load* (beam plus cavity), and the *cavity optimally detuned* (real total load) the system is in principle always stable against small transients, although not by a large margin if the beam power is significant.

The stability margin can be increased by increasing the detuning beyond optimum detuning, and by adjusting the power source to cavity RF coupling above optimum coupling (overcoupling).

4.2.2 Coupled Bunch Modes $n \neq 0$ Driven by the Detuned Fundamental Resonance

As mentioned above, it is essential to detune the cavity to compensate for the reactive part of the beam load to minimise the required RF power. The optimum detuning Δf_{do} normalised to the revolution frequency f_0 is given by:

$$\frac{\Delta f_{do}}{f_0} = \frac{I_B \cos \phi_s}{2 V_C} \left(\frac{R}{Q}\right) h \tag{9}$$

where $I_B \cong 2I$ is the peak value of the fundamental RF component of the beam current, V_C is the cavity voltage per cell, ϕ_s is the synchronous phase angle as previously defined, R/Q is the shunt impedance over quality factor of a cavity cell (using the usual electrical engineering convention where $P = V_C^2/(2R)$), and h the harmonic number. This value has been calculated for a number of medium to large high-current synchrotrons (Table 2). All of these rings are at present either in the design stage or under construction.

The fundamental RF resonance is likely to create a serious instability problem for the $n = M - 1, M - 2,$ .. modes if the cavity detuning is in the same order of magnitude as the revolution frequency or larger. It is seen from equation (9) that this is likely to happen for large h (large ring, high RF frequency), high R/Q value (low stored energy), high beam current, and at low voltage per cell. Using superconducting cavities, which have a high voltage per cell and a low R/Q, tend to

Table 2. Normalised optimum detuning for reactive beam load compensation

Machine	$I_B[\mathrm{A_p}]$	V_C [kV$_p$]	R/Q [Ω]	h	$\Delta f_{do}/f_0$
PEP II	4.28	925	117	3'492	0.93
CESR B	3.96	3000	45	1'275	0.038
KEK B-fact. n/c	5.2	390	98	5'120	3.34
KEK B-fact. s/c	5.2	2750	43	5'120	0.21
SSC collider	0.14	500	125	104'544	1.77
SSC HEB	0.90	325	150	2'178	0.45
LHC	1.70	1500	43	36'540	0.89
KAON C	5.6	100	100	225	0.63
KAON D	5.6	50	100	225	1.26

reduce the problem as is apparent from the three rings in the table above with superconducting cavities (CESR B, KEK s/c, and LHC). This also helps to solve problems associated with transients induced by beam gaps [15].

The growth rate can be very large. For the PEP II LER (low energy ring) for example it is the same order of magnitude as the synchrotron frequency, and almost three orders of magnitude larger than the radiation damping rate.

An obvious solution to the coupled-bunch problem would be not to detune the cavities, such that the cavity impedance is symmetric around the RF frequency. In most cases with heavy beam loading, the required extra RF power due to the large amounts of reflected power makes this solution unattractive. In addition the stable zone near the $\phi_z = 0$ axis is very narrow (Fig. 7), and the required tuning tolerances difficult to achieve, as a rather precise RF vector sum representing the total RF current injected into the cavity must be made.

Reducing the shunt impedance by loading, as used for the parasitic higher-order modes, is also totally unacceptable due to large amounts of wasted power. The most attractive solution is to apply RF cavity feedback [16][17][18], by which the apparent beam impedance of the RF system can be reduced several orders of magnitude.

Another solution being considered for the KEK B factory is to substantially lower the R/Q of a normal conducting cavity by coupling it to a storage cavity operating in a very high Q mode such that the lower R/Q is achieved without much loss in shunt impedance [19].

5 Transverse Bunched-Beam Modes

The theory for the transverse coherent bunched-beam modes (Table 1, Fig. 8) and their interactions with the environment were again first described in its most general form by Sacherer [20][21]. As in the longitudinal case there are M coupled bunch modes characterised by the *integer number of waves n of the coherent bunch motion around the ring*. The coupled-bunch mode number therefore resembles the azimuthal mode number n for coasting beams.

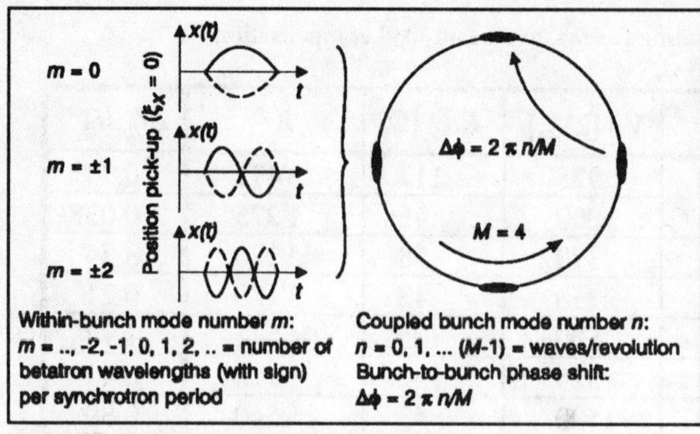

Fig. 8. Transverse bunched beam modes.

For bunched beams with M equidistant bunches, the bunch-to-bunch phase shift $\Delta\phi$ is related to the coupled-bunch mode number n by $\Delta\phi = 2\pi n/M$. There are M coupled-bunch modes numbered from 0 to (M-1). This is in contrast to the azimuthal mode number for coasting beams, where there is an infinite number of modes.

The within-bunch mode number m (also called the *head-tail mode number*) is the net *number of betatron wavelengths (with sign) per synchrotron period* at a given instant. At any given instant, a closed pattern of $|m|$ betatron periods corresponds to one synchrotron period with the betatron phase either advancing ($m > 0$) or retarding ($m < 0$) in the direction of advancing synchrotron phase. Unlike the longitudinal case the mode number m can thus assume both positive and negative integer values as well as zero. For $m = 0$ all particles in the bunch have the same betatron phase for zero chromaticity. The relation between longitudinal phase space co-ordinates (ΔE, Δt) and transverse motion is depicted for zero chromaticity on the 3-D surface plots on Figs. 9 and 10, as well as the directly observable average displacement along the bunch.

There are $|m|$ nodes along the bunch. The transverse displacements for various energies at those positions along the bunch continue to average out to zero, so no

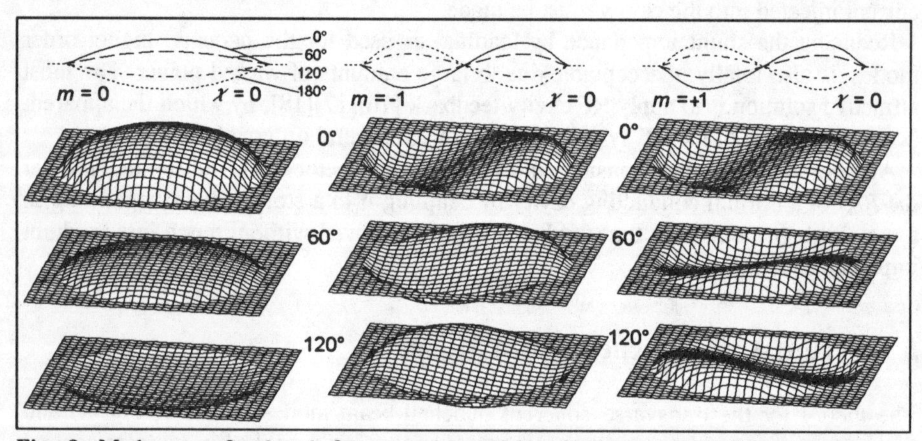

Fig. 9. Modes m = 0, -1, +1 for zero chromaticity. Surface plots indicating *transverse displacement times particle density* in longitudinal phase plane as function of longitudinal phase space co-ordinates Δt (left/right), and ΔE (front/back) for three successive values of betatron phase. Top trace is average transverse displacement times line density as observed by the Δ-signal of a transverse pick-up.

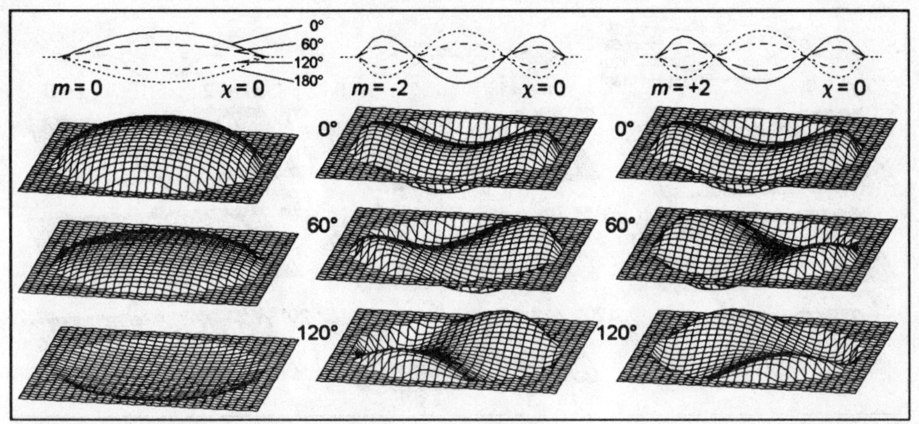

Fig. 10. Modes $m = 0, -2, +2$ for zero chromaticity

average displacement is observed on a pick-up, although there is much coherent transverse motion at different energies. The betatron phase pattern in the longitudinal phase plane appears to rotate either clockwise or counter-clockwise depending upon the sign of m. These two different senses of rotation are not immediately apparent from the observable, projected distribution, as they appear identical for mode $m = +1$, and mode $m = -1$. The difference is however apparent in the frequency domain, since the corresponding mode lines have slightly different frequencies, as the synchrotron frequency in one case is added to the betatron frequency, and in the other case subtracted.

For non-zero chromaticity ξ there is a tune modulation ΔQ associated with the momentum modulation Δp as a particle moves around the synchrotron orbit given by:

$$\frac{\Delta Q}{Q} = \xi \frac{\Delta p}{p} \tag{10}$$

This tune modulation results in a betatron phase advance χ during one half of the synchrotron period as the particle moves from head to tail of the bunch. This is compensated by a betatron phase retard $-\chi$ during the other half of the synchrotron period as the particle moves back from tail to head. The head-tail phase advance is proportional to the synchrotron amplitude and has its largest value for a particle with a synchrotron amplitude corresponding to the bunch length τ_L, and is given by

$$\chi = \frac{\xi}{\eta} Q \omega_0 \tau_L \tag{11}$$

where $\eta = 1/\gamma_t^2 - 1/\gamma^2$, ξ is the chromaticity as defined above, and ω_0 is the revolution frequency

This chromaticity dependent phase modulation is then superimposed upon the betatron phase pattern given by the mode number m, see Fig. 11. It is seen that the average transverse displacement signal contains the same $|m|$ nodes as with zero

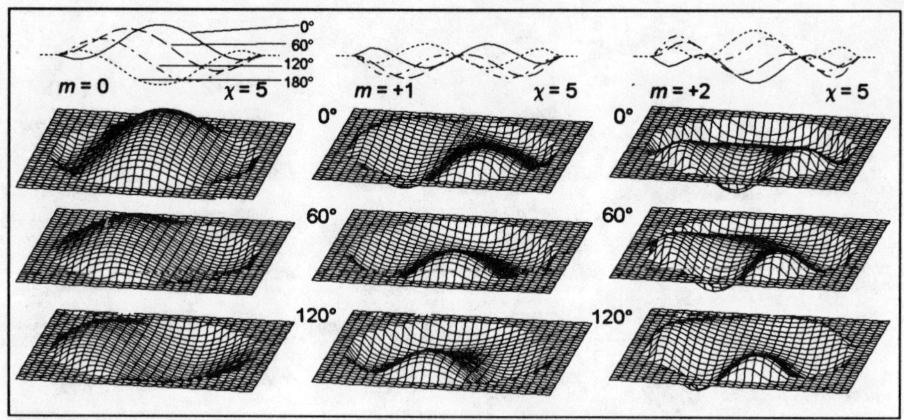

Fig. 11. Modes $m = 0$, +1, and +2 for a head-tail phase advance of $\chi = 5$ radians. Surface plots indicating *transverse displacement times particle density* in longitudinal phase plane as function of longitudinal phase space co-ordinates Δt (left/right), and ΔE (front/back) for three successive values of betatron phase. Top trace is average transverse displacement times line density as observed by the Δ-signal of a transverse pick-up.

chromaticity, and that a travelling-wave component is added to the standing-wave patterns observed for zero chromaticity.

Average transverse displacements superimposed and turn by turn mountain range displays are shown on Fig. 13 for three different chromaticities and for modes $m = 0$, +1, and +2. An example of several different vertical bunched-beam modes observed in the CERN PSB is shown on Fig. 12.

In addition (see Table 1) there is a third mode number $k = 1$ (dipole), 2 (quadrupole), 3 (sextupole), etc. which is the *number of periods of density modulation per betatron period*. In general, and as assumed in the preceding discussion, the dipole modes ($k = 1$) with one period of density modulation are observed. This corresponds to transverse displacements of the beam. The quadrupole modes ($k = 2$) corresponding to coherent beam width oscillations only interact very weakly with the vacuum chamber environment. The situation is different for very localised interactions such

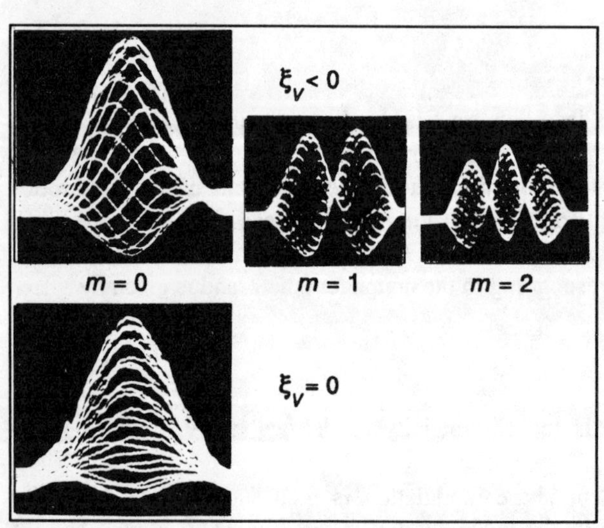

Fig. 12. Vertical head-tail modes observed in the CERN PS Booster.

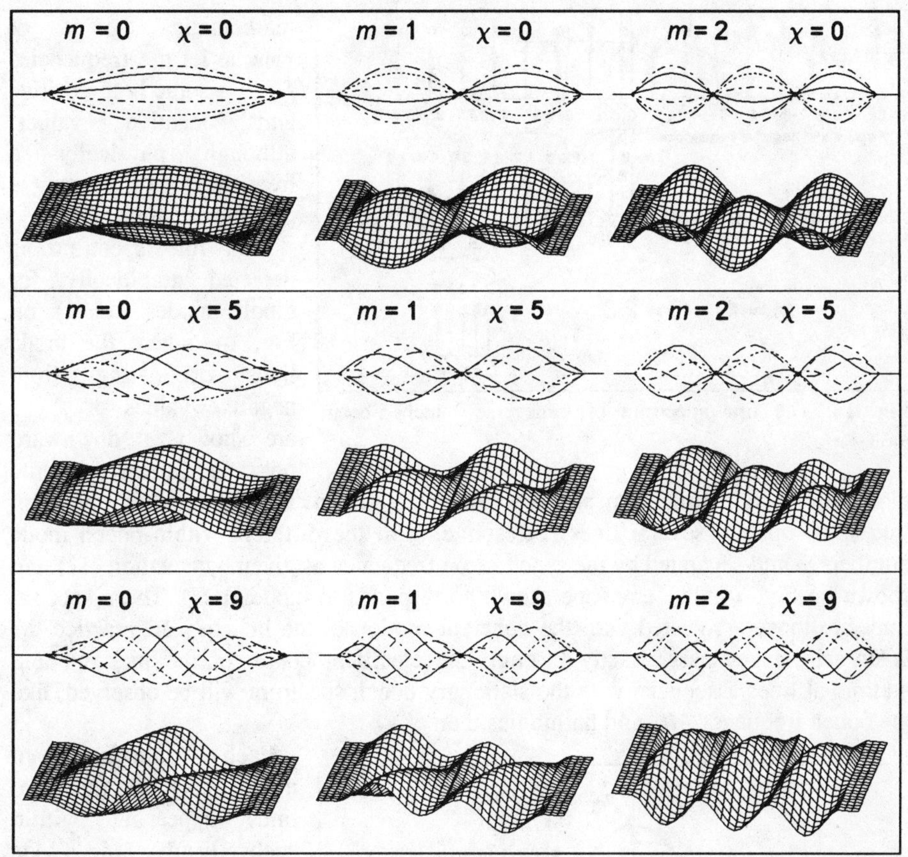

Fig. 13. Δ-signal of a single bunch versus time on separate revolutions for three different chromaticities, modes $m = 0, +1, +2$. Upper traces: every fourth turn, superimposed. Lower traces: every turn, mountain range display. Fractional tune: q = 0.0417, 15° phase advance per turn.

as coherent beam-ion [22] and coherent beam-beam interactions [23][24]. Transverse coherent beam-ion quadrupolar instabilities have been observed with coasting antiproton beams in the CERN AA [25], and can be expected to occur for bunched beams as well. A quadrupolar pick-up is required to observe such modes.

By taking the Fourier transform of the average transverse displacement times the bunch current (the signal seen by the Δ-signal of a transverse pick-up for dipole modes $k = 1$), we get the frequency spectrum of the modes. It is a line spectrum, and the line frequencies for mode (n, m, k) are:

$$f_{nmk,p} = (n + pM + kQ) f_0 + m f_s \qquad -\infty < p < \infty \tag{12}$$

where f_0 is the revolution frequency, f_s is the synchrotron frequency, and p is an integer assuming both negative and positive values. It is convenient from a

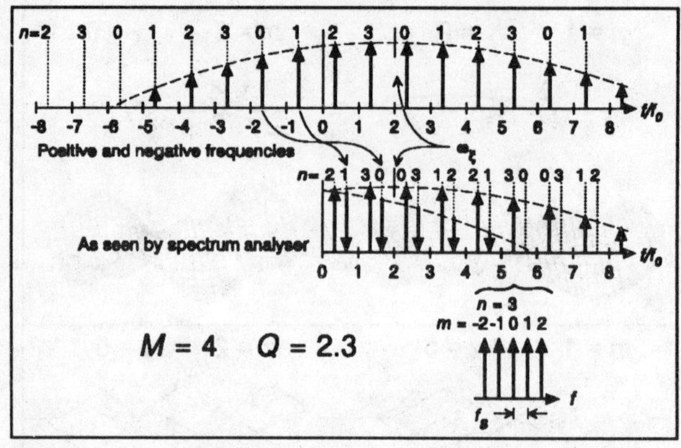

Fig. 14. The line spectrum of transverse bunched-beam modes

mathematical point of view to let the frequencies $f_{nmk,p}$ assume both positive and negative values, although physically a frequency is always a positive quantity.

This line spectrum is depicted graphically for dipole modes ($k = 1$) on Fig. 14, where the mode lines originating from negative values of $f_{nmk,p}$ are shown as downward pointing arrows, while mode lines originating from positive values are shown pointing up. Each betatron line splits up into several lines corresponding to the different within-bunch mode numbers m and separated by the synchrotron frequency as given by equation (12) and shown on Fig. 14. The envelope shown corresponds to mode $m = 0$. These spectral lines are those associated with the coherent motion of the bunches. In practice the beam will never pass exactly through the electrical centre of the pick-up, and additional lines associated with the stationary bunch spectrum will be observed, like the bunch frequency Mf_0 and harmonics thereof.

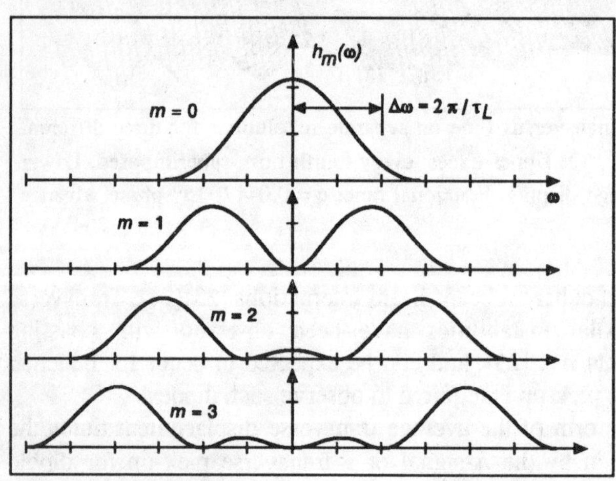

Fig. 15. Power-spectrum envelopes for modes m = 0 to 3 for zero chromaticity.

Each bunched-beam mode (m,n) has thus two lines appearing within each band Mf_0 wide between two bunch frequency harmonics, one pointing up and one pointing down. Each mode has therefore a large number of spectral lines.

The envelope or the relative amplitude of those lines depends upon the within-bunch mode number m, the bunch length τ_L, and the chromaticity ξ. The power spectrum envelopes $h_m(\omega)$ for zero chromaticity are shown on Fig. 15. The width of the envelope is inversely proportional to the bunch length τ_L. The chromaticity ξ will shift the centre frequency of those envelopes to the chromatic frequency ω_ξ given by:

$$\omega_\xi = \frac{\xi}{\eta} Q\omega_0 = \chi / \tau_L \tag{13}$$

where all symbols have been defined previously.

5.1 Sacherer's Formula for Transverse Bunched-Beam Modes

Due to electromagnetic interactions with the beam environment quantitatively described by the transverse coupling impedance $Z_T(\omega)$, the coherent mode frequencies are shifted a complex quantity $\Delta\omega_{mn}$ away from their low-intensity values. This shift is obtained by a weighted sum ($h_m(\omega)$) of the coupling impedance $Z_T(\omega_p)$ sampled at all those frequencies that corresponds to a mode line (see Fig. 14, equation (12)) of mode (m,n), Sacherer's formula [20][21]:

$$\Delta\omega_{m,n} = \frac{1}{m+1} \frac{je\beta I_0}{2Q\omega_0 \gamma m_0 L} \frac{\sum_p Z_T(\omega_p) h_m(\omega_p - \omega_\xi)}{\sum_p h_m(\omega_p - \omega_\xi)} \tag{14}$$

where β and γ are the usual relativistic parameters, e the electron charge, m_0 the particle rest mass, I_0 the current per bunch and Z_T the transverse coupling impedance in ohms/m.

The *growth rate* of mode (m,n) is $-\mathrm{Im}\{\Delta\omega_{mn}\}$ which is related to the *real* part of the coupling impedance $Z_T(\omega)$. The *real coherent frequency shift* of mode (m,n) is $\mathrm{Re}\{\Delta\omega_{mn}\}$, which is related to the *imaginary* part of the coupling impedance.

5.2 Discussion of Sacherer's Transverse Bunched-Beam Formula

The real part of the transverse coupling impedance has odd symmetry: $\mathrm{Re}\{Z_T(\omega)\} = -\mathrm{Re}\{Z_T(-\omega)\}$ and $\mathrm{Re}\{Z_T(\omega)\} > 0$ for $\omega > 0$ for passive impedances. The negative mode lines contribute therefore to growth in (14) while the positive mode lines contribute to damping. For zero chromaticity there is therefore cancellation of positive and negative frequency contributions for broad-band impedances and resonant impedances with bandwidths larger than the bunch frequency Mf_0 (decay time less than the bunch separation) similarly to the longitudinal case.

For non-zero chromaticity the cancellation for broad-band impedances is no longer effective, as the power spectrum envelopes $h_m(\omega - \omega_\xi)$ are shifted in frequency by ω_ξ. For the naturally negative chromaticity above transition ω_ξ is negative, so all coupled-bunch modes n are driven unstable by the real part of the broad-band (short range) coupling impedance. This is the so-called "head-tail" effect and is a single-bunch effect proportional to the current per bunch. This effect can also be used with advantage to damp the instabilities by correcting the chromaticity to a positive value thus achieving a passive damping of all bunches and thus all coupled-bunch modes.

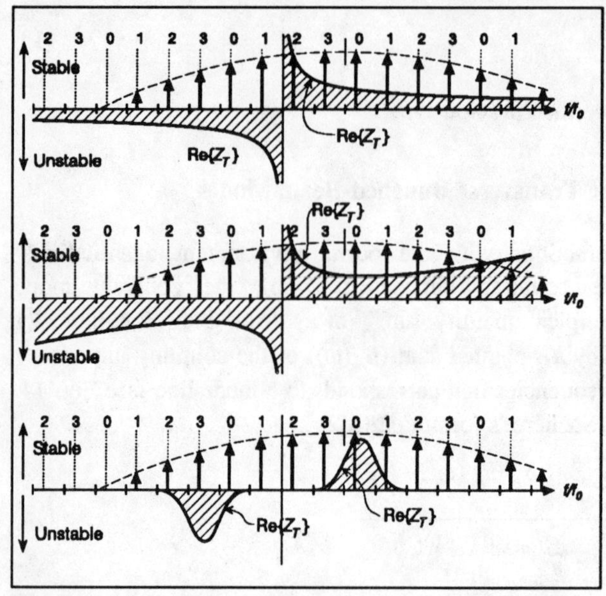

Fig. 16. Mode spectra for mode $m = 0$ and positive chromaticity (above transition) for typical examples of transverse coupling impedances with long-range wake. Top: Thick-wall, resistive-wall impedance; middle: same plus broad-band, high-frequency, coupling impedance; bottom: resonant narrow-band impedance.

For resonant impedances with bandwidths smaller than the bunch frequency Mf_0 (decay time longer than bunch separation), a single mode line for each mode dominates in the sum and one or several coupled-bunch modes are growing. The growth rate is proportional to the total current I in M bunches, and independent of the number of bunches unless the bunch frequency is low enough to get partial cancellation from another mode line of the same mode falling within the resonator bandwidth.

Some typical examples to illustrate this folding of the mode line spectrum with the frequency response of the coupling impedance are shown on Fig. 16 for a slightly positive chromaticity. On the top figure the long-range wake from the resistive-wall impedance drives the $n = 1$, $m = 0$ mode unstable, on the middle figure this mode is stabilised by the broad-band (single-bunch) interaction with the high-frequency impedance thanks to a positive ω_ξ. On the bottom figure a narrow-band resonator drives the $n = 3$, $m = 0$ mode unstable.

For a low number of bunches M the damping effect due to the head-tail effect and a positive chromaticity (above transition) may be effective in damping multibunch instabilities. This is much less so for large M as the damping effect is proportional to the single-bunch current while the growth rate due to a long-range wake (resistive wall or resonator) is proportional to the total circulating current.

5.3 Transverse Mode Coupling

In the transverse case the imaginary part of Z_T has even symmetry, $\text{Im}\{Z_T(\omega)\} = \text{Im}\{Z_T(-\omega)\}$, so contributions to the sum in (14) from positive and negative mode lines add up and give a real coherent frequency shift. Since $h_0(\omega)$ samples low frequencies with a predominantly inductive impedance while $h_1(\omega)$ samples much higher frequencies where the reactance may become capacitive, the real coherent shift of the $m = 0$ modes is different from the real coherent shift of the $m = \pm 1$ modes, in particular if the bunch length is short relative to the vacuum chamber. When the difference in frequency shift of the $m = 0$ mode and the $m = -1$ mode is equal to their separation by the synchrotron frequency, the two mode frequencies merge together. If this threshold is exceeded they become imaginary, an instability known as the fast head-tail or transverse mode coupling instability occurs [26][27]. It is a single-bunch instability since it is due to the broad-band or short-range wake characteristics of the transverse coupling impedance. It is often an intensity limiting factor for large rings with few short bunches.

6 Transverse Coupling Impedances

As in the longitudinal case coupling impedances can be subdivided into short-range wake or broad-band impedances and long-range wake or narrow-band impedances, depending on whether the bandwidth is larger or smaller than the bunch frequency Mf_0. The transverse coupling impedance Z_T (for dipolar modes, $k = 1$) is defined as the integral over one turn of the combined electric and magnetic deflecting field divided by the current I times displacement Δ, giving:

$$Z_T = j\frac{\oint\left[\bar{E} + \bar{v}\times\bar{B}\right]_\perp ds}{\beta I\Delta} \quad \text{[ohms / meter]} \tag{15}$$

where $\beta = v/c$, c is the speed of light, and $\perp$ indicates the component perpendicular to the beam direction.

6.1 Resistive-Wall Impedance

For frequencies where the skin depth δ:

$$\delta = \sqrt{\frac{2\rho}{\mu_0\omega}} \tag{16}$$

is smaller then the vacuum chamber thickness, the surface resistivity R_{surf} in ohms per square is given by:

$$R_{surf} = (1+j)\rho/\delta = (1+j)\sqrt{\frac{\mu_0\rho\omega}{2}} \tag{17}$$

where ρ is the specific resistivity of the chamber material and $\mu_0 = 4\pi\ 10^{-7}$ is the permeability of free space. The longitudinal coupling impedance due to the surface resistivity is simply the aspect ratio of the vacuum chamber surface times the surface resistivity:

$$Z_L = \frac{2\pi R}{2\pi b}R_{surf} = \frac{R(1+j)}{b}\sqrt{\frac{\mu_0\rho\omega}{2}} \tag{18}$$

where b is the vacuum chamber radius. It can be shown that for a circular chamber the following relation holds between the longitudinal and transverse coupling impedances:

$$Z_T(\omega) = \frac{2Z_L(\omega)c}{b^2\omega} \tag{19}$$

where c is the speed of light. From this equation it is seen that since $\mathrm{Re}\{Z_L\}$ is an even and $\mathrm{Im}\{Z_L\}$ an odd function of ω, the opposite will be true for the transverse coupling impedance: $\mathrm{Re}\{Z_T\}$ has odd symmetry (as depicted on Fig. 16) and $\mathrm{Im}\{Z_T\}$ has even symmetry. It also follows from equations (16), (17) and (18), that the real part of the transverse coupling impedance scales as $\omega^{-1/2}$, such that the coupled-bunch mode with a negative mode line frequency closest to the origin is driven unstable.

While the transverse resistive wall impedance can be extremely harmful due to the singularity near the origin, the longitudinal resistive wall impedance (17) is generally harmless since the mode $m = 0$ does not exist in the longitudinal case and the longitudinal form factor (7) for the lowest mode $m = 1$ scales as ω^2 at low frequencies. In the transverse case, even modes with $m \neq 0$ can be driven unstable by the resistive-wall impedance as non-zero chromaticity shifts the mode spectrum envelopes such that they overlap with the origin.

The resistive-wall impedance is particularly harmful for large rings with high current like PEP II, SSC and LHC. The frequency of the lowest frequency transverse mode line is very low, the aspect ratio of the vacuum chamber surface is high, and the beam current is high. For LHC it is a serious problem even with a very low resistivity, copper-coated, vacuum chamber at cryogenic temperatures.

6.2 Parasitic Higher-Order Modes

As in the longitudinal case the RF cavities are often the major source of long-range, narrow-band coupling impedances due to undesired higher-order deflecting modes. As in the longitudinal case two effects contribute to make these particularly harmful in high-current colliders, namely the ***high current and the high bunch frequency***.

For low bunch frequencies not much HOM damping is required before the HOM bandwidth becomes comparable to the bunch frequency, in which case *any coupled-bunch mode n* will have at least one positive and at least one negative frequency contribution to the sum in equation (14), of which the stabilising terms will dominate provided the chromaticity has the proper sign. For high bunch frequencies positive and negative contributions in (14) from the same mode n may be widely separated (up to $Mf_0/2$), and unstable coupled-bunch modes are driven unstable in spite of applying a positive chromaticity above transition.

7 Damping of Multibunch Instabilities

For moderate total currents (say 100 mA to 1 A), aggressive HOM damping of transverse and longitudinal modes in the RF cavities may keep the multibunch growth rates below the synchrotron radiation threshold. The resistive-wall instability is easier to damp if a tune just above an integer is used. Additional damping by the head-tail effect can be obtained by a positive chromaticity above transition, especially if the number of bunches is not too large. The longitudinal $n = M$-1 mode driven by the detuned fundamental RF resonance is normally not a problem except for very large rings (example SSC) or very low RF voltages.

For very high currents (for example LHC, PEP II, KEK B-factory) in the order of one ampere or more, all four sources (transverse: resistive-wall and HOM's, longitudinal: HOM's and detuned fundamental RF resonance) of multibunch instabilities cause growth rates in excess of synchrotron and/or Landau damping rates. In this case *longitudinal and transverse multibunch feedback systems* [3][28-33] become essential to maintain stability.

Full bandwidth (all coupled-bunch modes, bunch-by-bunch feedback for example) transverse dampers can in addition be used to raise the transverse mode coupling threshold. This is achieved by phasing it in such a way that it represents a *reactive impedance* which gives a real frequency shift of the $m = 0$ modes, which compensates the shift of the $m = 0$ modes from the broad band transverse machine impedance [34].

The detuned fundamental RF resonance produces such a fast growth rate that it becomes essential to first reduce the apparent cavity impedance by *local RF feedback* [16][17][18] to bring the impedance down to a level where the residual growth rate can be dealt with by means of multibunch feedback.

Conclusion

Colliders in the 70's and 80's typically had unseparated orbits, low number of bunches and therefore low bunch frequencies and were approaching single-bunch current limits. Short-range wake fields dominated the limiting phenomena: space-charge tune shift, beam-beam tune shift, transverse mode coupling and turbulent bunch lengthening. Examples are CESR, SPEAR, PEP SppS collider and the Tevatron collider.

The high-luminosity, two-ring colliders of the next decade (like PEP II, CESR B, KEK B-factory, τ/c factories, DAFNE, SSC, LHC) have very high bunch frequencies and high total beam currents. In addition to pushing single-bunch limits, multibunch instabilities and long-range wake fields will become important limiting phenomena. RF cavity feedback, aggressive HOM damping, and longitudinal and transverse multibunch feedback systems will become essential to achieve the specified design goals.

References

[1] F.J. Sacherer, A Longitudinal Stability Criterion for Bunched Beams, IEEE Trans. Nucl. Sci., NS-20, 825 (1973).

[2] F.J. Sacherer, Bunch Lengthening and Microwave Instability, IEEE Trans. Nucl. Sci., NS-24, 1393 (1977).

[3] F. Pedersen and F.J. Sacherer, Theory and Performance of the Longitudinal Active Damping System for the CERN PS Booster, IEEE Trans. Nucl. Sci., NS-24, 1396 (1977).

[4] F.J. Sacherer, Methods for Computing Bunched-beam Instabilities, CERN/SI-BR/72-5, (1972).

[5] G. Besnier, Contribution à la Théorie de la Stabilité des Oscillations Longitudinales d'un Faisceau Accéléré en Régime de Charge d'espace, Thesis, Université de Rennes, B-282, (1978).

[6] S. Hansen, H.G. Hereword, A. Hofmann, K. Hübner and S. Meyers, Effects of Space Charge and Reactive Wall Impedance on Bunched Beams, IEEE Trans. Nucl. Sci., NS-22, 1381 (1975).

[7] D. Boussard, Observations of Microwave Longitudinal Instabilities in the CERN PS, CERN Lab. II/RF/Int. 75-2 (1975).

[8] E. Keil and W. Schnell, Concerning Longitudinal Stability in the ISR, CERN-ISR-TH-RF/69-48 (1969).

[9] E. Haebel, RF Design (Higher-order Modes), these proceedings.

[10] P. Arcioni, and G. Conciauro, Feasibility of HOM-free Accelerating Resonators: Basic Ideas and Impedance Calculations, Particle Accelerators, Vol. 36, 177 (1991).

[11] R. Rimmer, M. Allen, J. Hodgeson, K. Ko, N. Kroll, G. Lambertson, R. Pendleton, H. Schwarz, and F. Voelker, An RF cavity for the B-factory, Proc. 1991 IEEE Part. Acc. Conf., San Francisco, 819 (1991).

[12] H. Padamsee, P. Barnes, C. Chen, W. Hartung, H. Hiller, J. Kirchgessner, D. Moffat, R. Ringrose, D. Rubin, Y. Samed, D. Saraniti, J. Sears, Q. S. Shu, and M. Tigner, Accelerating Cavity Development for the Cornell B-factory, CESR-B. Proc. 1991 IEEE Part. Acc. Conf., San Francisco, 786 (1991).

[13] Y. Chao, P.L. Corredoura, A. Hill, P. Krejcik, T. Limberg, M. Minty, M. Nordby, F. Pedersen, H. Schwarz, W.L. Spence, P.B. Wilson, Damping The pi Mode Instability in the SLC Damping Rings With a Passive Cavity, SLAC-PUB-5868, (1992) and Proc. 15th Int. Conf. on High-Energy Accelerators, Hamburg, Germany, Jul 20-24, (1992).

[14] K.W. Robinson, Stability of Beam in Radiofrequency System, CEA Report CEAL-1010, (1964).

[15] D. Boussard, RF Power Requirements for a High Intensity Proton Collider, Proc. 1991 IEEE Part. Acc. Conf., San Francisco, 2447 (1991).

[16] D. Boussard, Control of Cavities with High Beam Loading, IEEE Trans. Nucl. Sci. NS-32, 1852 (1985).

[17] F. Pedersen, A Novel RF Cavity Tuning Feedback Scheme for Heavy Beam Loading, IEEE Trans. Nucl. Sci. NS-32, 2138 (1985).

[18] F. Pedersen, RF Cavity Feedback, CERN/PS 92-59 (RF), and Proceedings of B factories, The State of the Art in Accelerators, Detectors and Physics, SLAC-400, 192 (1992).

[19] T. Shintake, Proposal of Accelerating RF-cavity Coupled with an Energy Storage Cavity for Heavy Beam Loading Accelerators, KEK Preprint 92-191.

[20] F.J. Sacherer, Transverse Bunched Beam Instabilities - Theory, Proc. 9th International Conference on High Energy Accelerators, Stanford, 347 (1974).

[21] F.J. Sacherer and B. Zotter, Transverse Instabilities of Relativistic Particle Beams in Accelerators and Storage Rings, CERN 77-13, 175 (1977).

[22] R. Alves Pires, D. Möhl, Y. Orlov, F. Pedersen, A. Poncet, S. van der Meer, On the Theory of Coherent Instabilities due to Coupling Between a Dense Cooled Beam and Charged Particles from the Residual Gas, Proceedings of the 1989 IEEE Particle Accelerator Conference, Chicago, 800 (1990).

[23] A.W. Chao and R.D. Ruth, Coherent Beam-beam Instability in Colliding Beam Storage Rings, SLAC PUB 3400 (1984).

[24] K. Hirata and E. Keil, Barycentre Motion of Beams due to Beam-beam Interaction in Asymmetric Ring Colliders, Proceedings of 2nd EPAC, Nice 1990.

[25] G. Carron, D. Möhl, Y. Orlov, F. Pedersen, A. Poncet, S. van der Meer, D.J. Williams, Observation of Transverse Quadrupole Mode Instabilities in Intense Cooled Antiproton Beams in the AA, Proceedings of the 1989 IEEE Particle Accelerator Conference, Chicago, 803 (1990).

[26] R.D. Kohaupt, Transverse Instabilities in PETRA, Proc. 11th Int. Conf. on High Energy Accelerators, Geneva 1980, 562 (1980).

[27] B. Zotter, Transverse Instabilities ue to Wall Impedances in Storage Rings, IEEE Trans. Nuclear Sci., NS-32, 2191 (1985).

[28] T. Kasuga, M. Hasumoto, T. Kinoshita and H. Yonehara, Longitudinal Active Damping System for UVSOR Storage Ring, Japanese Journal of Applied Physics, Vol. 27, No. 1, January, 100 (1988).

[29] D. Briggs, J.D. Fox, W. Hosseini, L. Klaisner, P. Morton, J.L. Pellegrin and K.A. Thompson, Computer Modelling of Bunch-by-bunch Feedback for the SLAC B-factory Design, Proc. 1991 IEEE Part. Acc. Conf., San Francisco, 1407 (1991).

[30] D. Briggs, P. Corredoura, J.D. Fox, A. Gioumousis, W. Hosseini, L. Klaisner, J.L. Pellegrin, K.A. Thompson, and G. Lambertson, Prompt Bunch by Bunch Synchrotron Oscillation Detection via a Fast Phase Measurement, Proc. 1991 IEEE Part. Acc. Conf., San Francisco, 1404 (1991).

[31] L. Vos, Transverse Feedback System in the CERN SPS, CERN SL/91-40 (1991)

[32] M. Ebert, D. Heins, J. Klute, R.D. Kohaupt, K.H. Matthiesen, J. Meinen, H. Musfeldt, S. Pätzold, K.H. Richter, J. Rümmler, H.P. Scholz, M. Schweiger, M. Sommerfeld, J. Theiss, Transverse and Longitudinal Multi-Bunch Feedback Systems for PETRA, DESY 91-036 (1991).

[33] F. Pedersen, Feedback Systems, CERN PS/90-49 (AR) (1990).

[34] S. Myers, Stabilization of the Fast Head-tail Instability by Feedback, Proceedings of the 1987 IEEE Particle Accelerator Conference, Washington, 503 (1987).

Fundamental–Mode rf Design in e^+e^- Storage Ring Factories*

P. B. Wilson

Stanford Linear Accelerator Center
Stanford University, Stanford, California 94309

1 Introduction

The difficulties arising in the design of the rf system for a factory-type storage ring lie mainly in two areas. First, a gap in the circulating beam current (on the order of 5% of the ring circumference) is required for ion clearing. Because of the high beam loading current, this gap produces a strong transient variation in the rf cavity voltage, which can in turn lead to a significant shift in the synchronous phases between bunches on either side of the gap. This phase shift would produce an unacceptable shift in the collision point, unless compensated by a corresponding shift in the bunch phases in the other ring. In order to work out the details of this compensation, the transient beam loading effects produced by the gap must be calculated quite carefully. A major goal of this chapter is to provide the insight and the basic analytic tools necessary for this analysis.

The second major problem for the fundamental mode rf design is also a consequence of the high average current (and the consequent large number of bunches) needed for a storage ring particle factory: longitudinal multibunch beam instabilities at sideband frequencies within the passband of the accelerating mode. These instabilities can be damped by an appropriate feedback system, as discussed elsewhere in these proceedings.[1] However, as background for this problem, we need to understand the phase and amplitude variations produced in the cavity voltage when the bunches undergo small-amplitude synchrotron oscillations. In the final section, the cavity voltage variation induced by such oscillations is calculated and applied to compute the Robinson damping time.

The emphasis throughout this chapter will be to provide a thorough understanding of beam loading effects. To this end, we begin in the next section with a calculation of the voltage induced in a cavity by a single point charge passing through it. The result will be a Green's function for beam loading problems. Once the solution for a point charge is known, the beam-induced voltage for a bunch with arbitrary longitudinal charge density profile, or for a train of such bunches, can then be constructed by an appropriate superposition.

*Work supported by Department of Energy contract DE–AC03–76SF00515.

2 Beam Loading by a Single Bunch

The derivation in this section relies on three basic assumptions. First, conservation of energy applies to the interaction between a moving charged particle and the fields in a cavity or accelerating structure. Second, we assume that superposition applies; that is, the net cavity field can be constructed as a vector (phasor) sum of component fields. Usually, this phasor is viewed in a reference frame rotating at either the cavity resonant frequency, or if there is an external generator driving the cavity, at the rf drive frequency. The third basic assumption is that the cavity fields are those for a single nondegenerate cavity mode which is orthogonal to all other modes. Thus a charge passing through a cavity independently deposits energy in each mode with which it can interact. We assume the conductivity of the cavity walls is sufficiently high so there is no significant coupling (overlap in impedance) with any other mode. We consider only the case of highly relativistic charged particles moving close to the speed of light. This has two consequences. First, the particle cannot change its velocity in response to beam-induced or generator-produced cavity fields. This allows a train of such particles to be modelled as a current generator in an equivalent circuit analysis of a beam-loaded cavity. Second, the cavity fields, summed over all cavity modes, must obey causality; that is, there is no net induced field ahead of a relativistic particle. This point will deserve further comment.

2.1 The Voltage and Energy Induced in a Cavity by a Point Charge

Assume that a charged particle moves through a cavity along the z-axis. In a given mode, the field at any point $E_z(z)$ is related to the energy U stored in the mode by

$$E_z(z) = \alpha(z)\, U^{1/2} . \tag{2.1.1}$$

A change in mode energy dU will produce a field change

$$dE_z(z) = \frac{\alpha^2(z)}{2E_z}\, dU .$$

On the other hand, a charge q moving through distance dz will lose energy

$$dU_q = -qE_z d_z .$$

This energy must go into energy stored in the cavity fields. The fields in this particular mode must then increase everywhere in the cavity during time $dt = dz/c$, even ahead of a particle moving at $v \approx c$. Causality does not apply to the cavity fields for a single mode, but it must of course be satisfied by a superposition of all modes. This is insured by the structure of Maxwell's equations, together with the cavity boundary conditions. By conservation of energy, dU (mode) $= dU_q$ (lost by charge), giving

$$dE_z(z) = -\frac{1}{2}\, q\alpha^2(z)\, dz . \tag{2.1.2}$$

This is the differential element of field induced by a charge in moving distance dz in the cavity. The minus sign indicates that the induced field opposes the motion of the charge. To calculate the net induced field, we must integrate the motion of the charge across the cavity, taking account of the fact that earlier induced differential field elements are rotating in phasor space according to $e^{j\omega_0 t}$, where ω_0 is the resonant frequency of the cavity. Calculating the net induced field at any time as the charge crosses the cavity is then a matter of adding up all of the field elements induced at earlier times. For convenience we choose a reference position $z = 0$ at the entrance to the cavity, where $E_z(0) \equiv E_0$ and $\alpha(0) \equiv \alpha_0$. Then, assuming the position of the charge is given by $z = ct$, the change in field at the reference position during time dt is

$$dE_0(t) \ = \ [\alpha_0/\alpha(z)]\, dE_z(z) \ = \ -\frac{1}{2}\, qc\,[\alpha_0\alpha(ct)]\ dt\ .$$

Using complex (phasor) notation, where a phasor quantity is denoted by a tilde, a field element induced at t' will ring as a function of time according to

$$\widetilde{dE_0}(t) \ = \ dE_0(t')\ e^{j\omega_0(t-t')}\ . \tag{2.1.3}$$

The net field at $z = 0$ at the time the charge exits from the cavity at $z = L$ and $t = L/c$ is then obtained as the superposition of all the differential field elements induced at earlier times, taking into account their proper phases:

$$\widetilde{E}_{0b}\,(t \ = \ L/c) \ = \ -\frac{1}{2} q\alpha_0 \int_0^L \alpha(z')\ e^{jk_0(L-z')}\ dz'\ , \tag{2.1.4}$$

where $k_0 = \omega_0/c$ and the subscript b indicates the beam-induced value. Note that $\widetilde{E}_{0b}$ is proportional to the charge times a factor that depends only on the geometry of the cavity mode, and not on the field amplitude. It will be useful to define a quantity k_ℓ, called the loss parameter for reasons which will become apparent, which depends on the mode configuration:

$$k_\ell \ \equiv \ \frac{\widetilde{V}\cdot\widetilde{V}^*}{4U} \ = \ \frac{V^2}{4U}\ . \tag{2.1.5}$$

Here $\widetilde{V}$ is the cavity voltage (and $\widetilde{V}^*$ the complex conjugate) seen by a test charge moving across the cavity according to $z = c(t - t_0)$ with $E_z(z,t) = E_z(z)\, e^{j\omega_0 t}$. The voltage seen in a frame of reference traveling with the particle is then

$$\widetilde{V} \ \equiv \ e^{j\omega_0 t_0} \int_0^L E_z(z')\ e^{jk_0 z'} dz' \ = \ V e^{j\omega t_0 + \theta}\ ;$$

$$V \ \equiv \ \left|\widetilde{V}\right| = \left(C^2 + S^2\right)^{1/2}\ ; \qquad \tan\theta = S/C\ ; \tag{2.1.6}$$

$$C \ = \int_0^L E_z(z)\ \cos k_0 z\ dz\ ; \qquad S \ = \int_0^L E_z(z) \sin k_0 z\ dz\ .$$

It is often convenient to define a reference plane at $z_r \equiv \theta/k_0$, such that the voltage gain of a test charge (electron) is given by $\widetilde{V} \ = \ V\, e^{j\omega_0 t_r}$, where t_r is the time at which the charge crosses the reference plane. Using the above definitions, together with Eq. (2.1.1) in Eq. (2.1.4) to eliminate $\alpha(z)$ and α_0, the beam-induced voltage becomes

$$\widetilde{V}_b \ = \ -2k_\ell q \left[e^{jk_0 L}\widetilde{V}^*/V\right]\ , \qquad V_b \ = \ \left|\widetilde{V}_b\right| \ = \ 2k_\ell q\ . \tag{2.1.7}$$

The quantity in brackets gives the phase of the beam-induced voltage with respect to the voltage defined by Eq. (2.1.6).

The voltage induced by a charge, as given by Eq. (2.1.7), is independent of any prior voltage present in the cavity. This is true because in equating dU (mode) to dU_q (lost by the charge) to obtain the differential beam-induced field element given by Eq. (2.1.2), both sides were proportional to the pre-existing field E_z, which therefore drops out of the final expression. The stored energy remaining in the cavity after the exit of the inducing charge must in general be calculated by first taking the vector (phasor) sum of the beam-induced voltage and any pre-existing voltage, and then

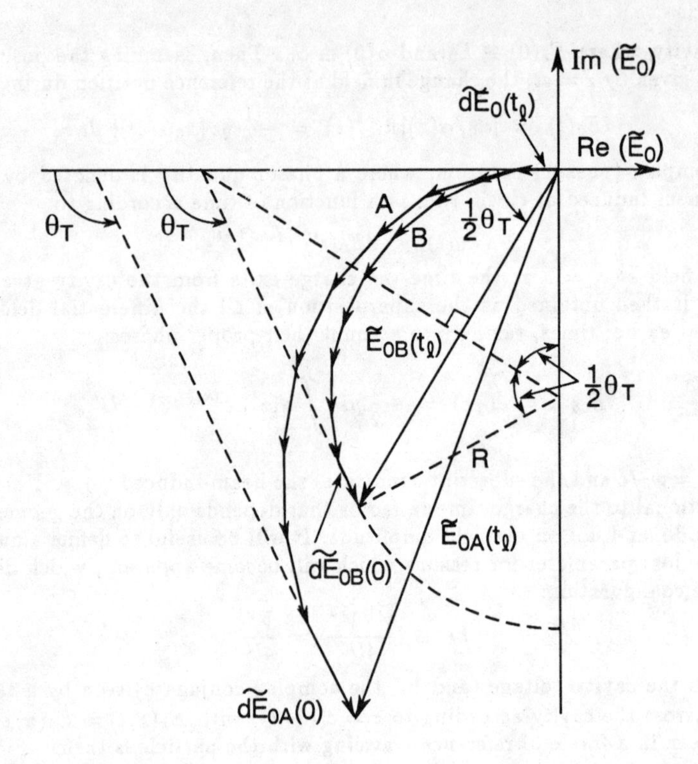

Figure 2.2.1. Differential superposition for two cases: (A) a mode in which E_z decreases along the particle trajectory; (B) a mode with constant E_z along the particle trajectory. In both cases, the field (for given stored energy) is assumed to be the same at $z = L$.

calculating the energy using Eq. (2.1.5). If there is no pre-existing cavity excitation, then the beam-induced energy is given by combining Eqs. (2.1.5) and (2.1.7) to obtain

$$U_b = \frac{V_b^2}{4k_\ell} = k_\ell q^2 \; ; \tag{2.1.8}$$

hence the name *loss parameter* for the quantity k_ℓ. The effective voltage V_q "seen" by the charge is the voltage necessary to extract energy U_b, or

$$V_q = U_b/q = k_\ell q \; . \tag{2.1.9}$$

This is just one-half of the induced voltage left behind in the cavity. This result is sometimes called the Fundamental Theorem of Beam Loading. Crudely stated, the charge does not experience any retarding voltage as it starts to cross the cavity, but it sees the full induced voltage as it leaves the cavity. On the average, it might then be expected to see one-half of the final induced voltage.

2.2 Differential Superposition

Figure 2.2.1 shows the geometry of the superposition of the beam-induced differential field elements for two cases. Case A shows an example in which the field function $\alpha(z)$ in Eq. (2.1.1) decreases with z, while case B is for a mode with a uniform electric field along the particle trajectory. The induced field elements are shown at time $t_\ell = L/C$, just as the particle leaves the cavity. Therefore the last induced field element lies along the negative real axis of the phasor diagram. Following Eq. (2.1.3), an earlier field element induced at time $t = t'$ will have rotated in phasor space by an angle $\omega_0(t_\ell - t')$. The first field element induced at $t' = 0$ will have rotated by an angle (the transit angle) $\theta_T = \omega_0 L/c$. For the case of constant E_z, it is seen that the reduction in the induced field due to the fact that the particle takes a finite time to cross the cavity, as compared to the induced field for a charge of infinite velocity, is just the ratio of the chord length to the arc length shown in the diagram:

$$T = \frac{E_{oB}}{R\theta_T} = \frac{2R \sin(\theta_T/2)}{R\theta_T} = \frac{\sin(\theta_T/2)}{\theta_T/2} ,$$

where T is just the usual transit angle factor. The phase of the net beam-induced field is seen to be rotated by an angle $\theta_T/2$ with respect to the final induced element. This is also the phase of the field induced at the center (symmetry plane) of the cavity.

2.3 Bunch Form Factor

The voltage induced by an arbitrary charge distribution can be related to the charge induced by a point charge using a bunch form factor. The voltage at time t induced at time t' by a charge element $dq = I(t')dt'$ is

$$\widetilde{dV}(t) = -2k_\ell \, I(t') \, e^{j\omega(t-t')} \, dt' .$$

The total voltage induced by the charge distribution can be set equal to that induced by a point charge, reduced by a factor F_b and located at time t_0 (or at phase $\omega t_0 = \phi$),

$$\widetilde{V}(t) = -2k_\ell \int_{-\infty}^{\infty} I(t') \, e^{j\omega(t-t')} \, dt' = -2q \, k_\ell \, F_b \, e^{j(\omega t - \phi)} .$$

Solving for F_b and ϕ,

$$F_b = \left(C_S^2 + C_A^2\right)^{1/2} ; \qquad \tan\phi = \frac{C_A}{C_S} , \tag{2.3.1}$$

where C_S and C_A are the symmetric and antisymmetric integrals

$$C_S = \frac{1}{q} \int_{-\infty}^{\infty} I(t') \cos\omega t' \, dt' ; \qquad C_A = \frac{1}{q} \int_{-\infty}^{\infty} I(t') \sin\omega t' \, dt' .$$

If a charge distribution having a time-width which is not negligible compared to the rf period is accelerated across a cavity, the average energy gain per electron in the bunch is reduced by the same form factor. If $V_0 \, e^{j\omega t}$ is the energy gain by a point particle crossing the cavity, then the charge-weighted average energy gain is

$$\widetilde{V}_{\text{ave}} = \frac{V_0}{q} \int_{-\infty}^{\infty} I(t') \, e^{j\omega(t-t')} \, dt' = V_0 F_b \, e^{j(\omega t - \phi)} , \tag{2.3.2}$$

where F_b and ϕ are again given by Eq. (2.3.1). It is important to note that, for any charge distribution, both the average accelerating voltage and the net beam-induced voltage are reduced by exactly the same factor with respect to a point charge. The position (phase) of an effective point charge which replaces the distribution is also the same.

Some useful bunch form factors are

$$F_b(\text{Gaussian}) = e^{-\omega_0^2 \sigma_t^2/2} ,$$

$$F_b(\text{rectangular}) = \frac{\sin(\omega_0 T_b/2)}{\omega_0 T_b/2} ,$$

where σ_t is the rms bunch length (Gaussian) and T_b is the full bunch width (rectangular).

2.4 Summary Comments on Single Bunch Beam Loading

In this section we have tried to give a reasonably thorough understanding of the physics underlying the voltage induced in a cavity by a single bunch. If we add the fact that this voltage will decay as a function of time according to e^{-t/T_F}, where $T_F = 2Q_L/\omega_0$ is the loaded cavity filling time, then we have a Green's function for calculating any beam loading problem. The voltage induced by a train of bunches with arbitrary charges and spacing is then calculated by superposition. In the general case, of course, the voltage produced by an external generator must be included by a further superposition. In a storage ring, a strong constraint is added by the fact that, after initial damping, the bunches adjust their phases with respect to the net cavity voltage in a way such that each of the bunches gains the same energy (to make up for synchrotron radiation and impedance losses). This can add considerable complexity to beam loading calculations when bunch charges or bucket spacings are not equal—for example, when there is a gap in the circulating beam.

3 Beam Loading by a Train of Equally Spaced Bunches

3.1. Beam-Induced Voltage for Small Bunch Spacing

Using the definition of cavity voltage in Eq. (2.1.6), we can now define a cavity shunt impedance R in terms of the power P_c dissipated in the cavity walls, $R \equiv V^2/2P_c$. We assume the usual definitions for the Q's of the unloaded cavity, $Q \equiv \omega_0 U/P_c$, and loaded cavity, $Q_L \equiv Q/(1+\beta)$. Here $\beta \equiv P_e/P_c$ is the usual coupling coefficient for a coupling aperture or loop, such that P_e is the power emitted from the aperture into a matched load when there is no incoming rf wave from an external source. Taking the bunch spacing as ΔT_b, we have the following relations and definitions:

$$k_\ell \equiv \frac{V^2}{4U} = \frac{\omega_0}{2}\left(\frac{R}{Q}\right); \qquad \frac{R}{Q} = \frac{V^2}{2\omega_0 U}; \qquad \tau \equiv \frac{\Delta T_b}{T_f};$$

$$T_f = \frac{2Q_L}{\omega_0} = \frac{2Q}{\omega_0(1+\beta)}; \qquad V_{b0} = 2k_\ell q = \omega_0\left(\frac{R}{Q}\right) I_0 \Delta T_b = \frac{2I_0 R}{1+\beta}\cdot\tau , \tag{3.1.1}$$

where $I_0 = q/\Delta T_b$ is the dc current assuming equal bunch spacing. For a bunch current distribution of non-negligible time width, both V and V_{b0} must be reduced by the bunch form factor, as discussed in Sec. 2.3. A time reference is chosen such that the voltage V_{b0} induced by each of the bunches (assumed to be equally spaced) passing through the cavity lies along the negative real axis, following the convention in Sec. 2.3. We now assume that the bunch spacing is related to an rf frequency ω, which may be different from the cavity resonant frequency ω_0, such that $\omega\Delta T_b = 2\pi b$, where b is an integer (the number of rf wavelengths between bunches). Between successive bunches, the induced cavity voltage slips in phase (relative to a phasor coordinate frame rotating as $e^{j\omega t}$) by an amount

$$\delta = (\omega_0 - \omega)\,\Delta T_b$$

and decays in length by a factor $e^{-\tau}$. The process of the build-up of the net beam-induced voltage is illustrated in Fig. 3.1.1, shown after a large number of bunches

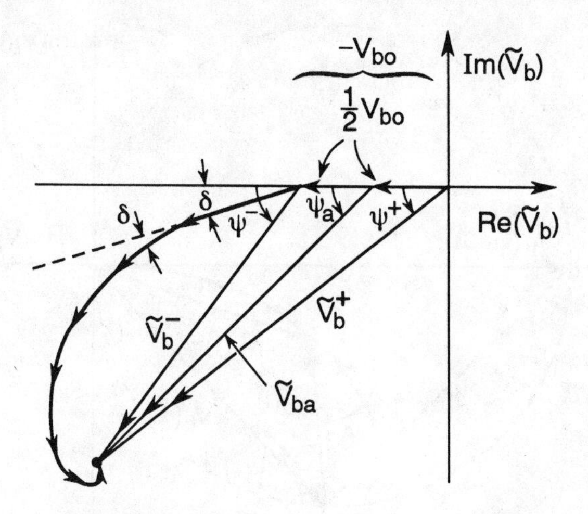

Figure 3.1.1. Diagram showing the buildup of the beam-induced voltage in a cavity by a train of equally spaced bunches.

have passed through the cavity. The net induced voltages just before and just after the bunch arrival time are denoted by $\widetilde{V}_b^-$ and $\widetilde{V}_b^+$ where

$$\widetilde{V}_b^+ = -V_{bo}\left(1 + e^{-\tau}\, e^{j\delta} e + e^{-2\tau}\, e^{2j\delta} + \ldots\right) = \frac{-V_{bo}}{1 - e^{-\tau}\, e^{j\delta}} \,. \tag{3.1.2}$$

Taking the limit $\Delta T_b/T_F \to 0$, such that $\tau, \delta \ll 1$,

$$\psi^+, \psi_a, \psi^- \longrightarrow \psi$$

$$\widetilde{V}_b^+, \widetilde{V}_{ba}, \widetilde{V}_b^- \equiv \widetilde{V}_b \longrightarrow -V_{bo} \cdot \frac{\tau + j\delta}{\tau^2 + \delta^2} = -\frac{2I_0 R}{1+\beta}\cos\psi e^{j\psi}$$

$$\tan\psi = \frac{\delta}{\tau} = (\omega_0 - \omega)\, T_f = \frac{2Q_L}{\omega_0}(\omega_0 - \omega) \,.$$

We see that ψ is just the usual *tuning angle*, which gives the variation in the phase of the beam-induced cavity voltage as the cavity is tuned off resonance. The magnitude of the induced voltage varies as $\cos\psi$, and therefore the tip of the phasor representing $\widetilde{V}_b$ follows a circle with diameter $2I_0R/(1+\beta)$ in phasor space as ψ is varied, as shown in Fig. 3.1.2. As is customary in complex notation, positive ψ is defined in the counter-clockwise direction.

If the series in Eq. (3.1.2) is summed to the nth term,

$$\widetilde{V}_b^+(n) = -\frac{V_{bo}\left(1 - e^{-n\tau}\, e^{jn\delta}\right)}{1 - e^{-\tau}\, e^{j\delta}} \,.$$

Again let $\delta, \tau \to 0$ and approximate n by $t/\Delta T_b = t/\tau T_f$. The above expression then becomes

$$\widetilde{V}_b(t) = -\frac{2I_0R}{1+\beta}\cos\psi\, e^{j\psi}\left[1 - e^{-t/T_F(1 - j\tan\psi)}\right] \,. \tag{3.1.3}$$

It is easy to show that transient variation of $\widetilde{V}_b(t)$, represented by the quantity in brackets, follows a logarithmic or equi-angular spiral (for example, see Ref. 2, Sec. 7.1).

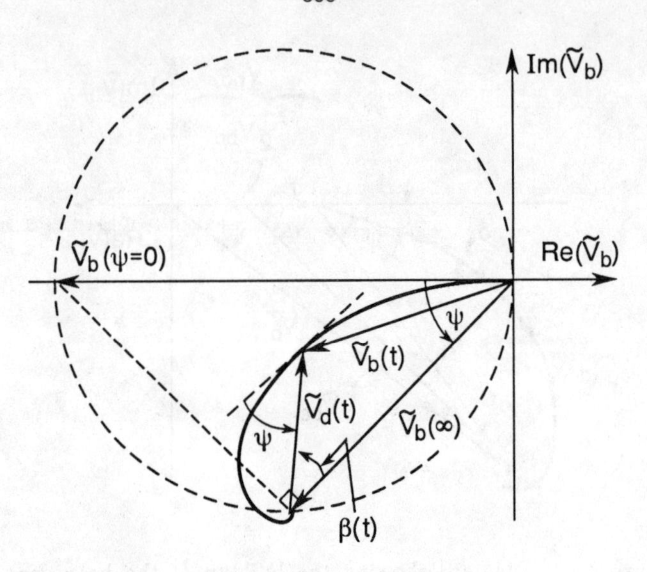

Figure 3.1.2. Diagram showing that the transient buildup of the beam-induced voltage $\widetilde{V}_b(t)$ follows an equi-angular spiral (solid curve), where angle $\beta(t) = (\omega_0 - \omega)t = (t/T_F)\tan\psi$ and $V_d(t) = V_b(\infty)\ e^{-t/T_F}$. The steady-state beam-induced voltage, $\widetilde{V}_b(\infty)$, follows a circle as cavity tuning is varied (dashed curve).

This is illustrated in Fig. 3.1.2, where a difference vector $\widetilde{V}_d(t)$ has been defined which connects $\widetilde{V}_b(t)$ to $\widetilde{V}_b(\infty)$. The tangent to the transient path followed by $\widetilde{V}_b$ always makes angle ψ with respect to $-\widetilde{V}_d$, which shrinks in time as $V_d(0)e^{-t/T_F}$ and rotates at a constant angular rate given by $e^{j(t/T_F)\ \tan\psi} = e^{j(\omega_0-\omega)t}$.

3.2 Relation to a Parallel–Resonant Equivalent Circuit

The result in Eq. (3.1.3) has been derived from basic principles, such as conservation of energy and superposition, with no reference to an equivalent circuit. However, this result is exactly what would be expected for the voltage induced by a current generator with rf current $\widetilde{I}_b = -2I_0$ across a parallel resonant circuit with shunt resistance R, shunt capacitance $1/C = \omega_0(R/Q) = 4k_\ell$, shunt inductance $\omega_0^{-1}(R/Q)$, and shunt resistance R/β to represent loading by the coupling network and external transmission line (assumed to be matched looking toward the generator; see Ref. 2, Sec. 3.5). Summarizing the results in the previous section, we have in the steady-state limit for the beam-induced voltage,

$$\widetilde{V}_b = -V_{br}\cos\psi\ e^{j\psi}\ , \qquad V_{br} = \tau V_{b0} = \frac{I_b R}{1+\beta} = \frac{2I_0 R}{1+\beta}\ , \tag{3.2.1}$$

where V_{br} is the magnitude of $\widetilde{V}_b$ at cavity resonance.

A current generator I_g can now be added to represent an external rf source driving the cavity. The rf power of the source is then identified as the available power from the generator, $P_g = I_g^2 R/8\beta$ (see, for example, the discussion in Ref. 2, Sec. 3.5). The voltage produced across the circuit is then

$$\widetilde{V}_g = V_{gr} \cos\psi \, e^{j\psi} , \qquad V_{gr} = \frac{I_g R}{1+\beta} = \frac{2\beta^{1/2}}{1+\beta} \cdot (2RP_g)^{1/2} . \tag{3.2.2}$$

Again note that if the bunch length is not small compared to the rf wavelength, both I_0 and I_g must be multiplied by the bunch form factor. From the form of Eq. (3.2.2), the tip of the phasor $\widetilde{V}_g$ also traces out a circle as the tuning angle ψ is varied, as shown for $\widetilde{V}_b$ in Fig. 3.1.2. If a step change is made in the driving generator voltage, $\widetilde{V}_g(t)$ also approaches a new steady-state value, $\widetilde{V}_g(\infty)$, along an equi-angular spiral. That is, the difference vector $\widetilde{V}_d(t) \equiv \widetilde{V}_g(t) - \widetilde{V}_g(\infty)$ shrinks in magnitude as e^{-t/T_F} and rotates in phase as $e^{j(\omega_0-\omega)t}$, in the same manner as $\widetilde{V}_d(t)$ in Fig. 3.1.2.

3.3 Bunch Spacing Comparable to the Cavity Filling Time

For a factory-type storage ring with a large number of bunches, the bunch spacing in time will be very small compared to the cavity filling time. There may, however, be occasion to calculate beam loading effects with only a few bunches in the ring (as is the case for most rings for high energy particle physics). The approximation $\tau \to 0$ and $\widetilde{V}_b^+ \approx \widetilde{V}_b^-$ cannot now be made. According to the Fundamental Theorem of Beam Loading, each bunch will experience the net voltage induced in the cavity by all bunches that have previously passed through it, $\widetilde{V}_b^-$, plus one-half of its own single-bunch induced voltage, $-\frac{1}{2}V_{b0}$. This is the voltage $\widetilde{V}_{ba}$ shown in Fig. 3.1.1. From Eq. (3.1.2),

$$\begin{aligned} \widetilde{V}_{ba} &= \widetilde{V}_b^- - \frac{1}{2}V_{b0} = \widetilde{V}_b^+ + \frac{1}{2}V_{b0} \\ &= -V_{b0}\left[\frac{1}{1-e^{-\tau}e^{j\delta}} - \frac{1}{2}\right] = -V_{b0}\left(F_R + jF_I\right) , \end{aligned} \tag{3.2.3}$$

$$F_R = \frac{1-e^{-2\tau}}{2(1-2e^{-\tau}\cos\delta + e^{-2\tau})} \xrightarrow{\tau\to 0} \frac{\tau}{\tau^2+\delta^2}$$

$$F_I = \frac{e^{-\tau}\sin\delta}{(1-2e^{-\tau}\cos\delta + e^{-2\tau})} \xrightarrow{\tau\to 0} \frac{\delta}{\tau^2+\delta^2} .$$

The quantities F_R and F_I give the steady-state values of the real and imaginary parts of the beam-induced voltage after an infinite succession of charges have passed through the cavity, as compared with the voltage induced by a single passage of the charge. The quantity

$$2F_R = \Re e[\widetilde{V}_{ba}] \Big/ -\frac{1}{2}V_{b0}$$

is sometimes called the resonance function, since it gives the net retarding voltage seen by a charge passing through a cavity with a resonant build-up of the beam-induced voltage, as compared to the voltage seen on a single passage. The resonance function is plotted and discussed in Ref. 2, Sec. 6.5.

In the limit of small τ, using $V_{b0} = \tau V_{br}$ and $\delta = \tau\tan\psi$, Eq. (3.2.3) approaches

$$\begin{aligned} \widetilde{V}_{ba} &\to -V_{br}\left[\tau F_R + j\tau F_I\right] = -V_{br}\cos\psi\, e^{j\psi} \\ \tau F_R &\to \cos^2\psi ; \qquad \tau F_I \to \cos\psi \sin\psi . \end{aligned} \tag{3.2.4}$$

4 Steady–State Beam Loading in a Storage Ring RF System

4.1 Basic Phasor Diagram

In the previous sections the beam-induced voltage in a resonant cavity was derived from first principles, without an external rf generator. In this case, it is reasonable to choose a reference phase such that the beam-induced voltage at resonance lies along the negative real axis. We will follow this same convention in drawing phasor diagrams for the general case in which an rf generator voltage component is present. This is at variance with the notation often used, which places the net cavity voltage along the positive real axis. There is not space here for a full discussion of the relative advantages and disadvantages of these alternative choices of a phasor reference frame. As a minimum, the reader will gain perspective by learning to view storage ring beam loading problems from a different vantage point.

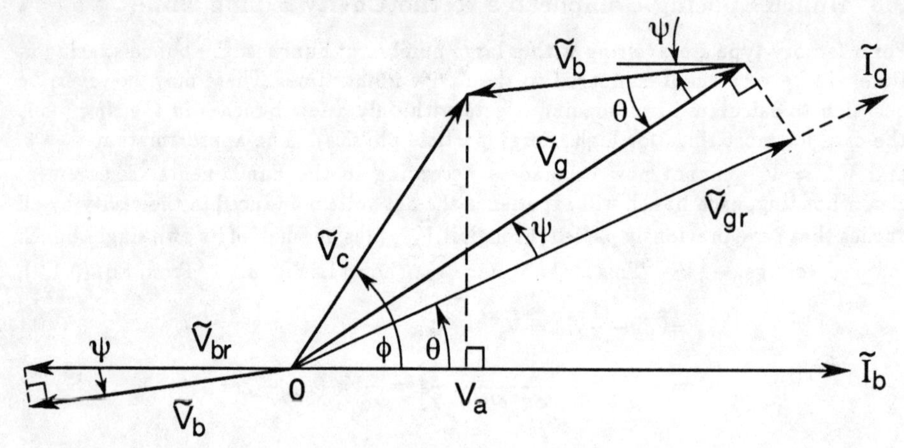

Figure 4.1.1. Diagram showing vector addition of generator and beam loading voltages in an rf cavity.

Figure 4.1.1 shows the basic phasor diagram in which the net cavity voltage, $\tilde{V}_c$, is obtained from the superposition of $\tilde{V}_g$ and $\tilde{V}_b$, as viewed in a reference frame rotating as $e^{j\omega t}$, where ω is the rf frequency and ω_0 the cavity resonant frequency. The beam current phasor lies along the positive real axis. The projection on this axis of the cavity voltage, lying at the synchronous phase angle ϕ, gives the accelerating component of the voltage. The generator current, which is colinear with the generator-induced voltage at resonance, $\tilde{V}_{gr}$, lies at an angle θ with respect to the real axis (and with respect to the beam current, I_b). As the cavity is tuned from resonance by a positive value of $\omega_0 - \omega$, both $\tilde{V}_g$ and $\tilde{V}_b$ rotate in the counter-clockwise direction through angle ψ, where $\tan\psi = 2Q_L(\omega_0 - \omega)/\omega$. It is also assumed that Q_L is relatively large .

From the diagram in Fig. 4.1.1, the real (accelerating) and imaginary components of the net cavity voltage $\tilde{V}_c$ are

$$\begin{aligned} V_A &= V_c \cos\phi = V_{gr} \cos\psi \cos(\theta+\psi) - V_{br} \cos^2\psi , \\ V_I &= V_c \sin\phi = V_{gr} \cos\psi \sin(\theta+\psi) - V_{br} \cos\psi \sin\psi . \end{aligned} \tag{4.1.1}$$

By eliminating $(\theta + \psi)$ from these two expressions (transfer the V_{br} components to the other sides of the equations, square, and add), then substituting for V_{gr} and V_{br}

using Eqs. (3.2.1) and (3.2.2), we obtain the required generator power in terms of V_c and ϕ for a given cavity tuning ψ and coupling β,

$$P_g = \frac{V_c^2}{2R} \cdot \frac{(1+\beta^2)}{4\beta} \cdot \frac{1}{\cos^2\psi} \left\{ \left[\cos\phi + \frac{2I_0 R}{V_c(1+\beta)} \cos^2\psi\right]^2 + \left[\sin\phi + \frac{2I_0 R}{V_c(1+\beta)} \cos\psi \sin\psi\right]^2 \right\} . \tag{4.1.2}$$

Angle θ is now fixed, and can be obtained, if desired, from either of Eqs. (4.1.1). In the general case when the bunch spacing is not small compared to the cavity filling time, the generator power can be obtained by substituting τF_R and τF_I for the factors $\cos^2\psi$ and $\cos\psi \sin\psi$ inside the brackets in Eq. (4.1.2), where F_R and F_I are given by Eq. (3.2.3); see Ref. 2, Sec. 6.4, for details.

4.2 Tuning Adjusted for Real Beam-Loaded Cavity Impedance

The reflected voltage from a beam-loaded cavity will look real (that is, it will have the same phase as the voltage reflected from the cavity at resonance without a beam) if the net cavity voltage $\widetilde{V}_c$ is colinear with $\widetilde{V}_{gr}$ (and therefore with $\widetilde{I}_g$). From Fig. 4.1.1, this implies that $\theta = \phi$. Using this condition, and applying the law of sines to the phasor triangle in Fig. 4.1.1,

$$\frac{V_b}{V_c} = \frac{V_{br}\cos\psi}{V_c} = \frac{\sin(\phi-\theta-\psi)}{\sin\theta} = -\frac{\sin\psi_0}{\sin\phi} ; \qquad \tan\psi_0 = -\frac{V_{br}}{V_c}\sin\phi . \tag{4.2.1}$$

By differentiating Eq. (4.1.2) with respect to ψ, we find that $\psi = \psi_0$ is also the condition for minimum generator power (and hence minimum power reflected from the cavity). Substituting for ψ in Eq. (4.1.2) using the condition in Eq. (4.2.1), we have at optimum cavity tuning,

$$V_{gr0} = V_c + V_{br}\cos\phi , \qquad P_{g0} = \frac{(1+\beta)^2}{4\beta} \cdot \frac{V_{gr0}^2}{2R} . \tag{4.2.2}$$

By differentiating P_{g0} with respect to β, we find the value of cavity coupling which minimizes the generator power:

$$\beta_0 = 1 + \frac{2I_0 R\cos\phi}{V_c} = 1 + \frac{P_b}{P_c} , \tag{4.2.3}$$

where $P_b = I_0 V_c \cos\phi$. Using $P_g = P_b + P_c + P_r$, where P_r is the reflected power, it is easy to show that $P_r = 0$ if Eq. (4.2.3) is satisfied. If it is not, but if the cavity tuning is optimum according to Eq. (4.2.1), then the reflected power is

$$\frac{P_r}{P_c} = \frac{(\beta-\beta_0)^2}{4\beta} . \tag{4.2.4}$$

As a practical example, consider the PEP–II B–Factory rf system design with parameters[3] for the 9–GeV high energy ring (values are per cavity assuming 20 cavities):

$$V_c = 0.925 \text{ MV} , \qquad I_0 = 1.5 \text{ A} , \qquad R = 3.5 \text{ M}\Omega ,$$
$$V_A = V_s + I_0 Z_{\text{hom}} \approx 0.192 \text{ MV} , \qquad \phi = \cos^{-1} V_A/V_c = 78.0^\circ ,$$
$$P_c = V_c^2/2R = 122 \text{ kW} , \qquad P_b = I_0 V_A = 288 \text{ kW} .$$

Here $V_s = 0.18$ MV per cavity is the loss to synchrotron radiation, and $Z_{\text{hom}} \approx 9\ k\Omega$ allows for losses to higher modes in the rf cavity and to the real part of the per cavity

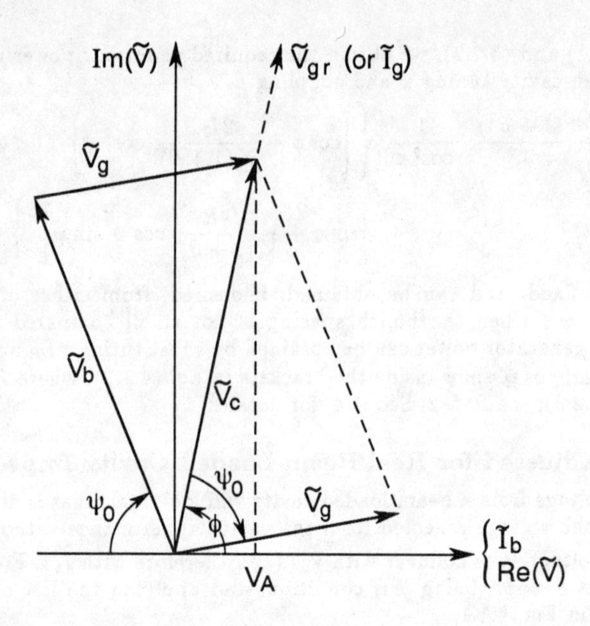

Figure 4.2.1. Phasor relationships for voltages in a PEP-II **rf** cavity at optimum tuning ($\tilde{I}_g$ colinear with $\tilde{V}_c$) and coupling (no reflected power).

share of the impedance of all other vacuum chamber components. The circulating current I_0 is set by the luminosity requirement, while the cavity voltage is set by bunch length and beam stability requirements, consistent with a reasonable klystron power. At optimum tuning and coupling, using (in order) Eqs. (4.2.3), (3.2.1), (4.2.2), and (4.2.1),

$$\beta_0 = 3.36\ , \qquad V_{br} = 2.41\ \text{MV}\ , \qquad V_{gr} = 1.425\ \text{MV}\ , \qquad P_g = 410\ \text{kW}\ ,$$

$$\psi_0 = -68.6^\circ\ , \quad V_b = V_{br}\cos\psi_0 = 0.88\ \text{MV}\ , \quad V_g = V_{gr}\cos\psi_0 = 0.52\ \text{MV}\ .$$

As a consistency check, note that the calculated generator power is just equal to the sum of the cavity wall losses and the power transferred to the beam, indicating that there is no reflected power at optimum coupling and tuning. In practice, the cavity coupling is often adjusted to be slightly greater than that given by Eq. (4.2.3), in order to make the rf system somewhat less sensitive to beam loading effects. For example, the cavity coupling might be set at $\beta = 4.0$ in the preceding example, rather than at 3.36. From Eq. (4.2.4), this increases the required generator power by 0.8% or 3 kW (the amount of the reflected power), but reduces V_{br} by 13% to 2.10 MV.

The phasor relationships in the example given above are plotted in Fig. 4.2.1. Note that if a klystron fails, the power dissipation in the cavity walls falls to $(0.88/0.925)^2 \times 122$ kW $= 110$ kW, while a reverse power $\beta P_c = 370$ kW is emitted from the cavity. If the beam should dump, but the klystron remain on, the power dissipated in the cavity is $(0.52/0.925)^2 \times 122$ kW $= 40$ kW, with a reflected power that is also equal to 370 kW. It is not a coincidence that this is exactly equal to the reverse power for the case of klystron failure. At optimum tuning and coupling, the wave emitted through the coupling aperture (or loop) by the beam-induced voltage component must exactly cancel the reflected wave due to the generator voltage component.

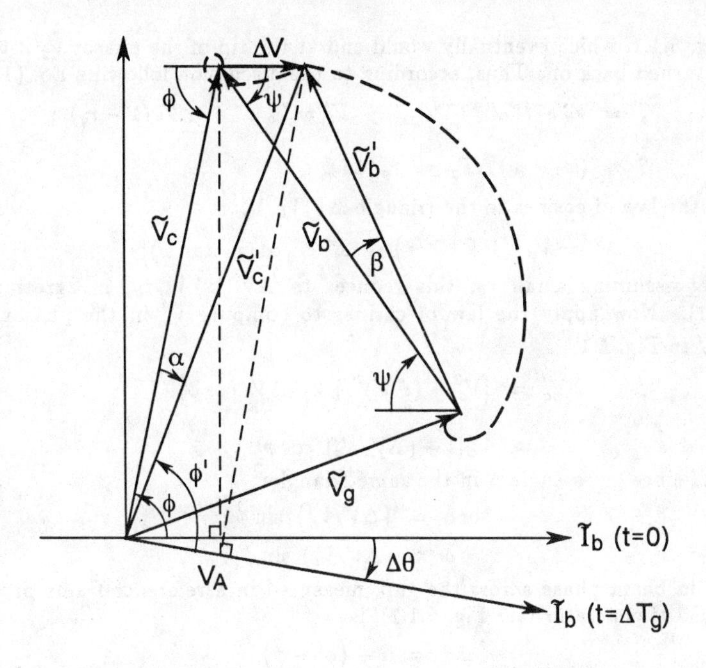

Figure 5.1.1. Phasor geometry for the case of a beam with a gap where $\tau_g \ll 1, \tau_b \gg 1$.

5 Beam Loading by a Circulating Beam with a Gap

5.1 Revolution Time T_0 Large Compared to the Cavity Filling Time

We first assume that the gap time, ΔT_g, is small compared to the cavity filling time, T_F, such that $\tau_g \equiv \Delta T_g/T_f \ll 1$. The basic assumption in this section is that the beam time, $T_b = T_0 - \Delta T_g$, is large compared to the cavity filling time such that $\tau_b \equiv T_b/T_F \gg 1$. This insures that the beam-induced voltage recovers very closely to its steady-state value after the passage of the gap. If this assumption is not met, the problem becomes considerably more complex and will be treated in Sec. 5.2. We will, however, be able to treat the more general case $\tau_g \approx 1$ later in this section, as long as the restriction $\tau_b \gg 1$ is kept.

The phasor geometry for the case $\tau_g \ll 1, \tau_b \gg 1$ is shown in Fig. 5.1.1. Here $\widetilde{V}_b$ and $\widetilde{V}_c$ are the steady-state values of the beam-loading and cavity voltages before the arrival of the gap, assumed to occur at time $t = 0$. A simple way to compute the voltage change ΔV is to assume that the actual ring current is continuous, but that a current of opposite sign, $-I_0$, is turned on at $t = 0$ for a time ΔT_g. The beam-induced voltage then must be along the positive real axis, as shown in Fig. 5.1.1, with a magnitude given by

$$\begin{aligned} \Delta V &= 2k_\ell q = \omega(R/Q)I_0'\Delta T_g = V_{br}\tau_g \\ I_0' &= I_0\left[T_0/(T_0-\Delta T_g)\right] \approx I_0\left(1+\Delta T_g/T_0\right) . \end{aligned} \tag{5.1.1}$$

As before, V_{br} and ΔV must be reduced by the bunch form factor, F_b in Eq. (2.3.1), in the case of long bunches. The magnitude of ΔV can be calculated in a second way, since we know that it is the beginning of the logarithmic spiral, shown by the dashed line in Fig. 5.1.1, which eventually would end at the tip of the phasor $\widetilde{V}_g$ if the beam were not turned back on. Thus, according to the discussion following Eq. (3.1.3),

$$\widetilde{V}_b' = \widetilde{V}_b \, e^{-\tau_g} e^{j(\omega_0-\omega)t} , \qquad V_b' = V_b \, e^{-\tau_g} \approx V_b (1-\tau_g) \; ; \tag{5.1.2a}$$

$$\beta = (\omega_0 - \omega) \, \Delta T_g = \tau_g \tan\psi \; . \tag{5.1.2b}$$

Applying the law of cosines to the triangle $\Delta V, \widetilde{V}_b, \widetilde{V}_b'$:

$$\Delta V = V_b \left[\left(1+e^{-2\tau_g}\right) - 2e^{-\tau_g} \cos\left(\tau_g \tan\psi\right) \right]^{1/2} . \tag{5.1.3}$$

Expanding assuming small τ_g, this reduces to $\Delta V \approx V_{br}\tau_g$, in agreement with Eq. (5.1.1). Now apply the law of cosines to compute V_c' in the phasor triangle $\Delta V, \widetilde{V}_c, \widetilde{V}_c'$ in Fig. 5.1.1:

$$V_c' = \left[V_c^2 + (\Delta V)^2 + 2V_c \Delta V \cos\phi \right]^{1/2} \tag{5.1.4a}$$

$$\approx V_c \left[1 + (\Delta V/V_c) \cos\phi\right] \; . \tag{5.1.4b}$$

The law of sines gives angle α in the same triangle:

$$\sin\alpha = \left(\Delta V/V_c'\right) \sin\phi \; , \tag{5.1.5a}$$

$$\alpha \approx (\Delta V/V_c) \sin\phi \; . \tag{5.1.5b}$$

The shift in beam phase across the gap, measured in a reference frame provided by the external rf generator (see Fig. 5.1.1), is

$$\Delta\theta = \alpha + \left(\phi' - \phi\right) \; , \tag{5.1.6}$$

where, using Eq. (5.1.4b),

$$\cos\phi' = V_A/V_c' \approx \cos\phi \left[1 - (\Delta V/V_c) \cos\phi\right] \; . \tag{5.1.7}$$

Using the trigonometric expression for $[\cos\phi' - \cos\phi]$ to expand Eq. (5.1.7), the shift in bunch phase given by Eq. (5.1.6) is

$$\Delta\theta \approx \frac{\Delta V}{V_c} \sin\phi + \left[\left(\tan^2\phi + 2\frac{\Delta V}{V_c} \cos\phi \right)^{1/2} - \tan\phi \right] , \tag{5.1.8}$$

where the term in brackets is the change in synchronous phase, $\phi' - \phi$. For ϕ near 90°, $\phi' \approx \phi$ and $\Delta\phi \approx \alpha \approx \Delta V/V_c$. For ϕ near zero, $\alpha \approx 0$ and $\Delta\phi \approx \phi' - \phi \approx (2\Delta V/V_c)^{1/2}$.

It is now easy to lift the restriction $\tau_g \ll 1$, although we will not be able to write an explicit expression for $\Delta\theta$. In Fig. 5.1.1, $\widetilde{\Delta V}$ is now a phasor which no longer lies in the positive real direction, but instead has its tip anywhere along the dashed spiral. Equation (5.1.3) can be used to find its magnitude. The angle between $\widetilde{\Delta V}$ and $\widetilde{V}_b$ will no longer be ψ, but something less, $\psi - \delta$ (see Fig. 5.1.2). This angle can be computed using the law of sines:

$$\sin\left(\psi - \delta\right) = \left(V_b'/\Delta V\right) \sin\beta = \left[V_b \, e^{-\tau_g}/\Delta V\right] \sin\left(\tau_g \tan\psi\right) \; .$$

Equation (5.1.4a) can now be used to compute V_c', replacing ϕ by $\phi + \delta$. Equation (5.15a) is then used to compute angle α, again replacing ϕ by $\phi + \delta$. Angle ϕ' is obtained as $\phi' = \cos^{-1}(V_A/V_c')$.

As a numerical example, consider the case of the PEP-II **B**-Factory with a 5% gap in the circulating beam. For PEP-II, some relevant rf parameters are[3] rf frequency, 476 MHz; $T_0 = 7.34\times10^{-6}$ sec; loss parameter $k_\ell = (\omega/2)(R/Q) \approx 1.74\times10^{11} V/C$; loaded cavity $Q, Q_L \approx 6,700$ for a cavity coupling coefficient of 3.5. Thus the filling time is $2Q_L/\omega \approx 4.5\times10^{-6}$ sec, and $\tau_g \approx 0.08$, $\tau_b \approx 1.6$. We see that the basic assumption of this section, $\tau_b >> 1$, is not very well met. If we proceed anyway to compute $\Delta\theta$, Eq. (5.1.3) gives $\Delta V \approx 0.19$ MV for a circulating current of 1.5 **A**. For a synchronous phase angle of 78° and a cavity voltage of 0.925 MV, Eq. (5.1.8) gives $\Delta\theta = 12°$.

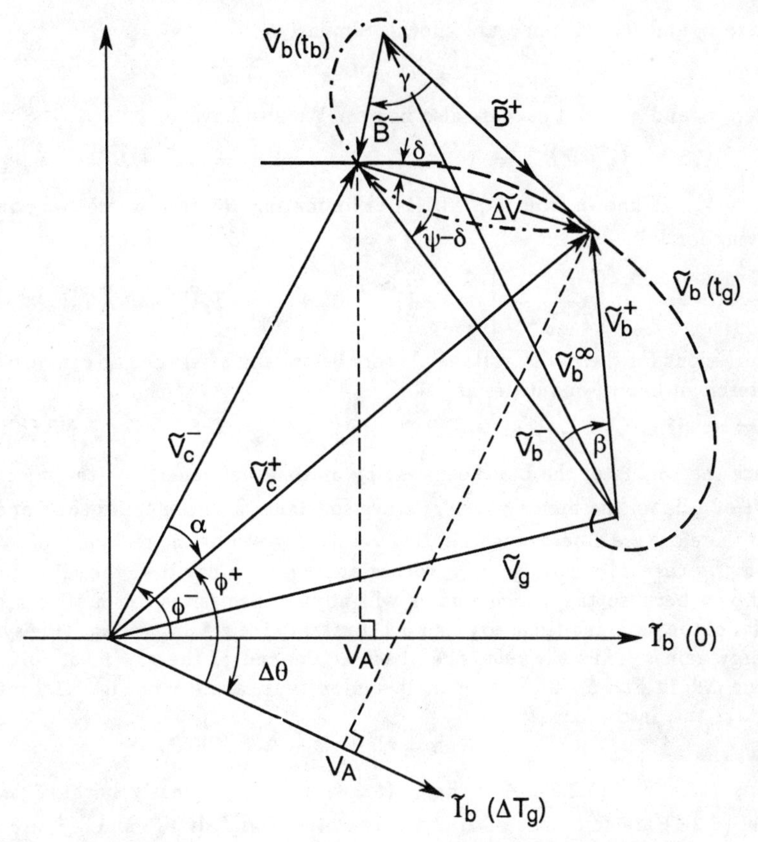

Figure 5.1.2. Phasor geometry for a beam with a gap with arbitrary τ_g and τ_b.

5.2 Gap with Arbitrary Values of τ_g and τ_b

Figure 5.1.2 shows the phasor geometry for the case of a beam gap where both the gap time and beam time are of arbitrary length with respect to the cavity filling time. The basic phasor reference frame is again chosen so that the phase of the last bunch before the gap begins, at time $t = 0$, lies along the positive real axis. During time $0 < t/T_f < \tau_g$ the beam-induced voltage component progresses along the logarithmic spiral $\widetilde{V}_b(t_g)$, shown by the dashed curve, starting at $\widetilde{V}_b^-$ at $t = 0$ and ending at $\widetilde{V}_b^+$ at $t/T_f = \tau_g$. During time $\tau_g < t/T_F < \tau_g + \tau_b$, the phasor $\widetilde{V}_b$ travels along the dot-dashed curve, driven by a current $I_0' = I_0\tau_0/\tau_b$, where $\tau_0 \equiv T_0/T_F = \tau_b + \tau_g$. At time $t = T_0$, the phasor is exactly back at $\widetilde{V}_b^-$. If the current were to continue, instead of the gap arriving again at $t = T_0$, the tip of the phasor would continue along the dot-dashed curve $\widetilde{V}_b(t_b)$ to the steady-state position, $\widetilde{V}_b^\infty$.

The transient beam loading problem would be relatively simple if we were dealing with a standing-wave linac cavity, driven by an incoming beam with a periodic gap. From Fig. 5.1.2 we could in this case write the following phasor relation:

$$\widetilde{V}_b^+ = \widetilde{V}_b^- \, e^{-\tau_g} \, e^{j\beta} \equiv \widetilde{T}_g \widetilde{V}_b^- \, , \tag{5.2.1}$$

where τ_g and $\beta = \tau_g \tan\psi$ are known. Similarly,

$$\widetilde{B}^- = \widetilde{B}^+ \, e^{-\tau_b} \, e^{j\gamma} \equiv \widetilde{T}_b \, \widetilde{B}^+ \, , \tag{5.2.2}$$

where τ_b and $\gamma = \tau_b \tan\psi$ are also known. We also have

$$\widetilde{V}_b^- = \widetilde{V}_b^\infty + \widetilde{B}^- = \widetilde{V}_b^\infty + \widetilde{T}_b \widetilde{B}^+ \, , \qquad \widetilde{V}_b^+ = \widetilde{V}_b^\infty + \widetilde{B}^+ = \widetilde{T}_g \widetilde{V}_b^- \, ,$$

where $\widetilde{V}_b^\infty$ is known from Eq. (3.2.1). Eliminating $\widetilde{B}^+$ from these two equations and solving for $\widetilde{V}_b^-$,

$$\widetilde{V}_b^- = \widetilde{V}_b^\infty \left[\frac{\widetilde{T}_b - 1}{\widetilde{T}_b \widetilde{T}_g - 1} \right] \equiv \widetilde{A} \widetilde{V}_b^\infty \, , \qquad \widetilde{V}_b^+ = \widetilde{T}_g \widetilde{V}_b^- = \widetilde{A} \widetilde{T}_g \widetilde{V}_b^\infty \, . \tag{5.2.3}$$

Expressions for the cavity voltage phasors before and after the gap can now be written in terms of known quantities as

$$\widetilde{V}_c^- = \widetilde{V}_g + \widetilde{V}_b^- = \widetilde{V}_g + \widetilde{A} \widetilde{V}_b^\infty \, , \qquad \widetilde{V}_c^+ = \widetilde{V}_g + \widetilde{V}_b^+ = \widetilde{V}_g + \widetilde{A} \widetilde{T}_g \widetilde{V}_b^\infty \, . \tag{5.2.4}$$

Since the timing of the bunches is set by an external generator (the injector) in the case of a linac, the bunch phase $\widetilde{I}_b$ stays constant. The bunches before and after the gap therefore see different accelerating voltages, given by the real parts of $\widetilde{V}_c^-$ and $\widetilde{V}_c^+$.

In the case of a storage ring with standing-wave cavities, the situation is more complex because the bunch phases will adjust themselves, on a time scale on the order of the damping time, to pick up a constant accelerating voltage (the synchronous energy gain). Thus the reference phase at the end of the gap must rotate through angle $\Delta\theta$, in Fig. 5.1.2. During the beam-on time, angle γ in Fig. 5.1.2 must change to take this into account:

$$\gamma = \tau_b \tan\psi - \Delta\theta \, . \tag{5.2.5}$$

Thus $\widetilde{T}_b$ in Eq. (5.2.2), $\widetilde{A}$ in Eqs. (5.2.3) and consequently both $\widetilde{V}_c^-$ and $\widetilde{V}_c^+$ in Eqs. (5.2.4) are functions of $\Delta\theta$. On the other hand, if $\widetilde{V}_c^-$ and $\widetilde{V}_c^+$ are given, the value of $\Delta\theta$ is readily calculated:

$$\begin{aligned} \Delta\theta &= \alpha + \phi^+ - \phi^- \, , \\ \phi^+ &= \cos^{-1}\left(V_A/V_c^+\right) \, , \qquad \phi^- = \cos^{-1}\left(V_A/V_c^-\right) \, , \\ \alpha &= \tan^{-1}\left[\Im m\left(\widetilde{V}_c^-\right) / \Re e\left(\widetilde{V}_c^-\right)\right] - \tan^{-1}\left[\Im m\left(\widetilde{V}_c^+\right) / \Re e\left(\widetilde{V}_c^+\right)\right] \, . \end{aligned} \tag{5.2.6}$$

Thus $\Delta\theta$ must be calculated by a self-consistent procedure: assume values for $\Delta\theta$ in Eq. (5.2.5), then carry through the preceding calculation for V_c^+ and V_c^- until the value for $\Delta\theta$ in Eq. (5.2.6) is in agreement with the initial assumed value.

We see, even in the simple case of a gap in a beam with bunches of equal charge, that just the calculation of the bunch phase shift across the gap has become quite complicated. If further information is desired, for example the phase positions of other bunches, or perhaps the effects due to unequal bunch charges, it would be difficult, or at least very awkward, to carry out the calculation analytically. It might then be best to resort to a simulation, in which the phase and energy of each bunch is tracked turn by turn. Such a program would show all the features of the longitudinal bunch dynamics, such as phase oscillations after injection, Robinson damping of these oscillations, and variations in the cavity voltage due to transient beam loading. A tracking program of this type has been written for the SLC damping ring to show transient effects at injection, and to determine the optimum injection phase and energy.[4]

5.3 Cures for the Gap-Induced Phase Shift

One obvious way to reduce the effect of a gap-induced phase shift on the longitudinal position of the collision point is to put a similar gap in the counter-rotating low-energy beam. However, the beam current, cavity voltage and synchronous phase angle are slightly different for the PEP-II low energy beam[3]: $I_0 = 2.15\text{A}$, $V_c = 0.95\text{MV}, \phi_s = 80.5°$, $P_b = 335$ kW. Using Eq. (4.2.3), the optimum cavity coupling coefficient is $\beta = 3.6$ for $P_c = 129$ kW, giving $\tau_g = .084$ for $Q_0 = 3.0 \times 10^4$. From Eq. (3.2.1), $V_{br} = 3.27$ MV and from Eq. (4.2.1) the optimum tuning angle is $-73.6°$. Then $V_b = V_{br} \cos\psi = 0.95$ MV, and Eq. (5.1.3) now gives $\Delta V = 0.26$ MV. Finally, Eq. (5.1.8) gives $\Delta\theta \approx 16°$. Since the high-energy beam has a phase shift of 12°, a residual phase shift of 4° remains. This is still large enough to produce a shift in the position of the collision point of 7 mm for $\lambda_{rf} = 63$ cm. This a substantial fraction of the bunch length, $\sigma_z = 10$ mm. This residual phase shift can be eliminated entirely if the current is reduced to 25% in the gap in the low-energy beam, instead of to zero.

Another possibility remains for reducing the residual phase shift across the beam gap. It is clear that if the rf generator voltage component $\widetilde{V}_g$ in Fig. 4.2.1 is jumped in phase and amplitude such that $\widetilde{\Delta V}_g = -\widetilde{V}_b$, then the transient effect of the gap is completely eliminated. However, this would require an increase in klystron power by a factor of $(V_c/V_g)^2 = (.925/.52)^2 = 3.2$, which is clearly not practical. However, we should calculate how much of a reduction in the gap phase shift can be obtained by a jump in klystron phase alone. Suppose the phase is shifted such that $\widetilde{V}_g' = \widetilde{V}_g e^{j\eta}$ at the beginning of the gap, where η is a counter-clockwise rotation of $\widetilde{V}_g$ in Fig. 5.1.1. We will not give all the details, but will only outline the calculation here. First a difference phasor $\widetilde{D}$ is defined such that $\widetilde{V}_g' + \widetilde{D} = \widetilde{V}_c$, where the angle between $\widetilde{V}_g'$ and $\widetilde{V}_c$ is $\eta - |\psi|$, where we assume $\widetilde{V}_c$ is colinear with $\widetilde{V}_{br}$. The law of cosines is used to calculate the magnitude of $\widetilde{D}$. During the gap period, the phasor $\widetilde{D}$ rotates through angle $\tau_g \; \tan\psi$ to position $\widetilde{D}'$, where $D' = De^{-\tau_g}$. The third side of this second phasor triangle is $\widetilde{\Delta V}$, which is calculated by the law of cosines. The angle opposite $\widetilde{D}'$, call it ψ', can now be computed by the law of sines. We will also need the angle opposite $\widetilde{V}_g'$ in the first phasor triangle, call it γ, where γ can also be computed using the sine law. Now establish a third phasor triangle, $\widetilde{V}_c, \widetilde{V}_c', \widetilde{\Delta V}$. The angle between $\widetilde{V}_c$ and $\widetilde{\Delta V}$, call it δ, is given by $\delta = \psi' - \gamma$. V_c' is now calculated by the cosine law, and angle α opposite $\widetilde{\Delta V}$ by the sine law. Angle ϕ' is now given by $\cos^{-1}(V_A/V_c')$, and the gap phase shift by $\Delta\theta = \alpha + \phi' - \phi$. Applying this procedure to the parameters of the PEP-II high energy ring, we calculate that the gap can be reduced to about 3° for $\eta \approx 100°$, with $\Delta V = 0.12$ MV and $V_c' = 0.83$ MV. This is a reduction by a factor of four from the 12° phase shift without the jump in generator phase. If a similar phase jump is carried out for the low-energy beam, the residual phase error would be reduced from 4° to about 1°. This produces a collision point shift of about $0.2\sigma_z$, which may be acceptable.

Finally, feedback can also be used to reduce the transient effects due to the gap (see Ref. 1).

6 Phase Stability and Phase Oscillations

6.1 Phase Stability

In an electron storage ring it is well known that, to be stable against phase perturbations, a particle must have a synchronous phase on the time-falling part of the rf wave. For example, a particle having too much energy compared to a synchronous particle will follow a longer path and will therefore receive less energy from the rf cavity on the next revolution. A particle that arrives at the rf cavity too early compared to a synchronous particle will get more than the synchronous energy gain, will consequently take a longer path and will arrive back at the cavity closer to the synchronous passage time. Using $V_A = V_c \cos[\omega(t - t_s) + \phi_s]$, the condition $dV_A/dt < 0$, evaluated at $t = t_s$, leads to $(-\omega V_c \sin\phi_s) < 0$, or $\sin\phi_s > 0$. Of course, ϕ_s must also be less than $\pi/2$ if V_A is to be positive. At high current, where the beam-induced voltage component is large, the situation is more complicated. As the arrival time varies due to phase oscillations, the beam-induced voltage component moves with the bunch and hence cannot contribute to phase stability; only the generator voltage component can provide a restoring force against phase perturbations. From Fig. 4.1.1 we see that the phase of the generator voltage component with respect to the beam is $\theta + \psi$, and hence the condition $dV_g/dt < 0$ at $t = t_s$ leads to $\sin(\theta + \psi) > 0$, or from Eq. (4.1.1),

$$2V_c \sin\phi + V_{br} \sin 2\psi > 0 \ . \tag{6.1.1}$$

This is the condition for the high-current limit on phase stability first derived by Robinson. Robinson's derivation involved setting up a set of linear equations in terms of slow (compared to the rf frequency) perturbations to the variables of the system. He then applied Routh's criterion to the determinant of the coefficients to test for exponentially growing solutions. However, the result is completely equivalent to the simple condition $dV_g/dt < 0$, which leads directly to Eq. (6.1.1) using the geometry of the basic phasor diagram in Fig. 4.1.1. If the cavity tuning is adjusted to make the beam-loaded cavity voltage look real, then Eq. (5.1.1), together with Eq. (4.2.1) gives

$$V_c > V_{br} \cos\phi \ . \tag{6.1.2}$$

If the cavity coupling is also optimized according to Eq. (4.2.3), then $V_{br} \cos\phi = V_c(\beta_0 - 1)/(\beta_0 + 1)$ and the condition in Eq. (6.1.2) is always satisfied.

6.2 Phase Oscillations

There is not space here for a complete derivation from first principles of the damping time for phase oscillations. A derivation emphasizing the time-domain behavior of phasor quantities subject to small perturbations is given in Ref. 2, Sec. 4.2. A more traditional derivation is given in, for example, Ref. 5. The result for the growth rate (inverse of the damping time) of the oscillation is

$$\frac{1}{t_d} = \frac{V_{br}\,\omega_s}{V_c \sin\phi} \cdot \frac{\xi\eta}{[1 + (\xi + \eta)^2][1 + (\xi - \eta)^2]} \ , \tag{6.2.1}$$

where $\xi = \tan\psi = (\omega_0 - \omega)T_f$ and $\eta = \omega_s T_F$. The synchrotron oscillation frequency is given by

$$\omega_s = \left[\frac{\alpha_m\,\omega\,V_c \sin\phi}{T_0\,E_0}\right]^{1/2} \ , \tag{6.2.2}$$

where E_0 is the beam energy in volts and α_m is the momentum compaction factor. From Eq. (6.2.1) we see that the oscillations are damped if $\tan\psi$ is negative; that is, if the rf frequency is greater than the cavity resonant frequency. This is the case if ψ is optimized according to Eq. (4.2.1) to produce a real beam-loaded cavity reflection coefficient. The damping rate given by Eq. (6.2.1) vanishes if either ξ or η approaches zero or infinity. For small η, the function on the right-hand side of Eq. (6.2.1) has a maximum value of 0.32 η for $\xi = 0.58$ ($\psi = 30°$). For large η, the function has a maximum value of 0.25 for $\xi \approx \eta$. Equation (6.2.1) can also be used to calculate the growth or damping rate of the coupled-bunch longitudinal instability due to a higher mode in the rf cavity or to a resonance in another vacuum chamber component. In this case, the growth or damping rate given by Eq. (6.2.1) must be multiplied by the ratio of the mode frequency to the frequency of the accelerating mode.

The physical origin of this damping (Robinson damping) can be traced to the inertia of the stored energy in the rf cavities. Because of the finite filling time, the beam-induced voltage cannot follow changes in the beam current instantaneously. A phase difference between the induced voltage and the driving current oscillation appears, which in turn leads to an energy interchange between the oscillation and the cavity fields. A similar effect is seen in other physical systems, for example, the excitation of mechanical oscillations is the walls of a superconducting cavity.[6]

References

1. F. Pedersen, these proceedings.
2. P. B. Wilson, "High Energy Electron Linacs: Applications to Storage Ring RF Systems and Linear Colliders," in *Physics of High Energy Particle Accelerators*, AIP Conference Proceedings No. 87 (American Institute of Physics, New York, 1982), pp. 450–563; also SLAC–PUB–2884 (1982).
3. "An Asymmetric B Factory: Conceptual Design Report," LBL PUB–5303; also SLAC–372 (1991).
4. P. Wilson and T. Knight, Internal SLAC Notes CN–38, 43, 74 and 86 (1981).
5. A. Hofmann in *Theoretical Aspects of the Behavior of Beams in Accelerators and Storage Rings* (Proceedings of the First Course of the International School of Particle Accelerators, Erice, November 1976); CERN 77–13 (1977), p. 139.
6. P. H. Ceperly, IEEE Trans. Nucl. Sci. **NS–19**, No. **2**, 217 (1972).

RF Design (Higher-order Modes)

Ernst Haebel

CERN, Geneva, Switzerland

1 Higher-order-mode (HOM) Impedance and Power

1.1 Sinusoidal beam

This lecture is about the possible methods to reduce the beam impedance of cavities. Cavities are not the only source of beam impedance in a machine. But if the beam tube is well designed i.e. without unnecessary irregularities, they are the dominant source. Let us just recapitulate what beam impedance means.

In a 'Gedankenexperiment' imagine a beam made up of positive and negative charges, which have the same speed v as the real beam particles but are arranged in a way to create a sinusoidally varying charge density. With such a beam we pass an ac current $I(t)$ of frequency ω through a reference plane in the cavity and excite a field oscillating with the same frequncy. Determine now the accelerating effect of this field on a very low intensity dc beam of test charges e (also with speed v) and define an accelerating voltage by $V(t) = \Delta E(t)/e$. Here $\Delta E(t)$ is the energy gain of test charges passing the reference plane at time t. Also $V(t)$ is sinusoidal but in general phase shifted against $I(t)$. Describing amplitude and phase of $V(t)$ by a phasor and dividing by the current amplitude I the longitudinal beam impedance $Z(\omega)$ is obtained.

$Z(\omega)$ depends on the particle speed and on the frequency of the beam current and is a maxima at the cavity resonances. There, the impedance becomes real, R_L. The beam loses power into this resistance:

$$P_L = \frac{1}{2} I V = \frac{1}{2} \frac{V}{R_L} V \tag{1}$$

The field oscillating in the cavity contains a certain stored energy U proportional to V.^{2}

$$\omega U = \frac{V^2}{2(R/Q)} \tag{2}$$

By this equation *we define the coupling impedance* (R/Q). For a given particle speed, (R/Q) depends only on the cavity's geometry. Adding the definition of a quality factor Q_L:

$$Q_L = \frac{\omega U}{P_L} \tag{3}$$

one finds

$$R_L = Q_L(R/Q) \tag{4}$$

There are two ways to reduce the beam impedance:

- To choose a cavity geometry which minimizes the (R/Q) of HOM in keeping the fundamental modes (R/Q) at an acceptable level. (Here superconducting (sc) cavities have a clear advantage. They make smaller fundamental mode (R/Q)s affordable).
- To diminish the HOM Q_L

Q_L is the loaded quality factor. One part P_0 of P_L goes into the lossy cavity walls, the other part into external dampers, P_{ex}.

$$\frac{1}{Q_L} = \frac{P_0 + P_{ex}}{\omega U} = \frac{1}{Q_0} + \frac{1}{Q_{ex}} \tag{5}$$

Q_0 and Q_{ex} are the unloaded and external quality factors respectively. For an unloaded cavity, i.e. $Q_L = Q_0$, R_L becomes identical to the shunt impedance R_{sh} of a mode.

The external damping is due to energy loss through openings within the cavity wall (of more or less complex form) communicating with elements able to carry away electromagnetic waves, as for example free space, transmission lines or waveguides. The beam tubes of a cavity form an example of this last category and, if constructed to absorb the extracted energy, may provide the simplest way of damping the cavity modes above their waveguide cut-off frequency.

The notions and formulae mentioned so far map perfectly on equivalents of linear circuit theory. (In fact we are in the realm of linear system theory and all its apparatus applies.)

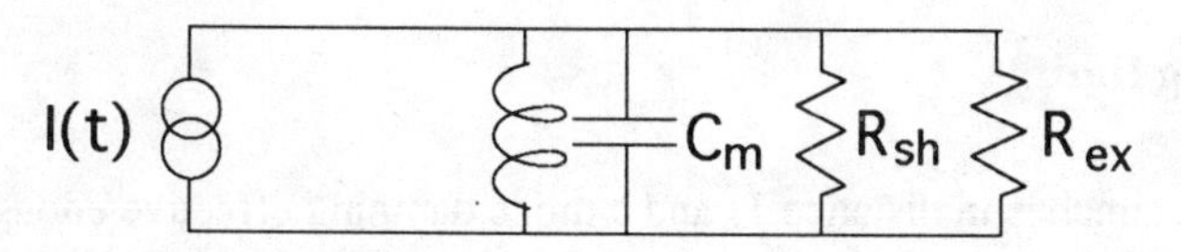

Fig. 1. Equivalent circuit of cavity mode and beam

If we depict the sinusoidally varying beam current by a pure current source and a cavity mode at ω_m by a LC resonator with parallel elements (Fig. 1), then

$$\sqrt{L_m C_m} = 1/\omega_m.$$

If, also, the voltage across C_m models the accelerating voltage and comparing Eq. (2) with the stored energy formula $U = V^2 C_m / 2$, one obtains:

$$(R/Q)_m = 1/(\omega_m C_m) = \sqrt{L_m / C_m} \tag{6}$$

The resistors R_{sh} and R_{ex} model the internal and external cavity losses. (The cavity's shunt impedance appears directly as resistor R_{sh} in the model.)

Guidance by this lumped element circuit analogy provides an easy access to most beam loading problems.

1.2 Bunched Beams

Real beams do not have a sinusoidally varying line density but are bunched. The current through the cavity is pulsed and so must be the pure current generator of the circuit model. For bunches which are short compared to the oscillation period, we may approximate the current $I(t)$ by a sequence of δ-pulses. For a single bunch of charge q:

$$I(t) = q\delta(t) \tag{7}$$

If this bunch passes an empty cavity (the condenser C_m is not charged) then, immediately after the pulse, the condenser is charged by q and its voltage and energy are

$$V_{bm} = q / C_m \tag{8}$$

$$U_{bm} = \frac{1}{2}\frac{q^2}{C_m} \tag{9}$$

$$U_{bm} = q(V_{bm}/2) \tag{10}$$

where U_{bm} is the energy lost by the bunch to a mode. When passing an empty cavity each mode decelerates the (short) bunch by an effective voltage which is one half of the accelerating voltage left by the bunch. (P. Wilson's fundamental theorem of beam loading.)

1.2.1 Strong damping limit

For a sequence of short bunches in distance T_b and a mode damping effective enough to dissipate U_{bm} before the next bunch arrives, ($\tau = 2Q_L/\omega_m < T_b$) the power lost by the beam becomes simply

$$P_m = \frac{1}{T_b}U_{bm} = \frac{1}{T_b}\frac{1}{2}\frac{1}{C_m}q^2 \tag{11}$$

and using Eq. (6)

$$P_m = \frac{1}{T_b}\frac{1}{2}\omega_m (R/Q)_m q^2 \tag{12}$$

Accelerator physicists use here the notion of a loss factor k_m

$$P_m = \frac{1}{T_b} k_m q^2 \tag{13}$$

and evidently

$$k_m = \frac{1}{2}\frac{1}{C_m} = \frac{1}{2}\,\omega_m (R/Q)_m \tag{14}$$

Finally, since mode fields are orthogonal we can sum powers so that the total dissipated power is, in this case of aperiodic damping,

$$P = \frac{1}{T_b} k\, q^2 \tag{15}$$

with

$$k = \Sigma k_m = \frac{1}{2}\sum \omega_m (R/Q)_m \tag{16}$$

Dampers have to cope with the larger part of this power. Depending on their geometry, single-cell cavities suitable for a tau-charm factory (τCF) machine have loss factors between 100 and 400 V/nC. Typical beam parameters are $T_b = 40$ ns and $q =$ 20 nC. P then lies in the range of 1 to 4 kW. For a 500 MHz cavity, HOM external Q values below 100 are required to realize the strong damping limit.

1.2.2 Weak-damping Limit

Calculation of P_m is also simple if a mode decays little within T_b ($\tau = 2Q_L/\omega_m > T_b$). The bandwidth of a mode resonance is then much smaller than the space $1/T_b$ between the spectral lines of a repeating current pulse. With a spectral line at ω and the mode near by at ω_m only this line has to be considered. With $\Omega = \omega/\omega_m - \omega_m/\omega$

$$P_m = \frac{1}{2} I_{RF}^2 \,\mathrm{Re}\,Z = \frac{1}{2} I_{RF}^2 \frac{Q_L (R/Q)_m}{1+(Q_L\Omega)^2} \tag{17}$$

$$I_{RF} = 2\frac{q}{T_b} = 2 I_{DC} \tag{18}$$

Taking again as illustration the τCF case and a 500 MHz sc cavity damped to a Q_L of 500 at the highest (R/Q) HOM ($(R/Q) = 10\Omega$) we get at resonance ($\Omega = 0$):

$$P_m \approx 2.5\mathrm{kW}$$

In contrast, in the strong damping limit and for the same mode (at about 800 MHz):

$$P_m \approx 250\mathrm{W}$$

Once we are obliged to assure beam stability by HOM damping but cannot exclude resonant excitation of modes, we should try to increase damping to the strong limit and get the benefit of reduced power-handling problems.

1.2.3 The General Case

A general expression for P_m, valid for all Q_L and tuning conditions, is obtained if one adds to $V_b/2$ (effective self-induced voltage of a bunch) the voltages left over from all preceding bunch passages [1]. The result is that Eq. (12) (strong damping limit) has to be complemented by a resonance factor. With $\tau_b = T_b/\tau$ and $\delta = (\omega_m - \omega_n)T_b$

$$P_m = \frac{1}{T_b} k_m q^2 \frac{\left(1-e^{-\tau_b}\right)\left(1+e^{-\tau_b}\right)}{1-2e^{-\tau_b}\cos\delta + e^{-2\tau_b}} \tag{19}$$

where δ describes the relative position of the mode frequency within the comb spectrum of the bunched beam, $\omega_n = n2\pi/T_b$. For gaussian bunches the comb spectrum's envelope function is $\exp(-(\omega\sigma)^2/2)$. The modal power formulas given above have to be multiplied by its square, the bunch form factor $\exp-(\omega_m\sigma)^2$.

1.3 Dipole Modes

In addition to damping of the longitudinal modes, which poses the problem of HOM power handling in the damping devices, the Q of dipole modes has to be reduced. For a cavity with cylinder symmetry this class of modes has no longitudinal electric field on the axis and therefore does not extract power from a centered beam. But once excited by a bunched eccentric beam, dipole modes also change the transversal momentum of centered particles which may lead to beam instability.

2 HOM Couplers

The first dipole mode frequencies of a typical accelerating cavity are near neighbours of the fundamental mode (fm). They have to be damped without coupling out excessive power from the fm field and this is one of the major problems encountered when designing HOM couplers. They must have filter properties. For cavities which make use of a dipole mode, the "crab cavity" for instance, a notch filter is required, but for accelerating cavities working in the TM_{010} mode a simple highpass characteristic is appropriate.

2.1 Waveguide HOM Couplers

Waveguides provide a high pass characteristic naturally but if dimensioned to have the required cut-off frequency they become very bulky; too bulky to be used on a sc 500 MHz cavity if they are not part of the construction anyway as is the case for the beam tubes. If one increases the beam tube diameter, more and more HOM will become propagating. As an additional advantage, the HOM (R/Q) diminish. But is there a beam tube diameter where all HOM will propagate but not the fm,and will the fm then have a reasonable (R/Q)?

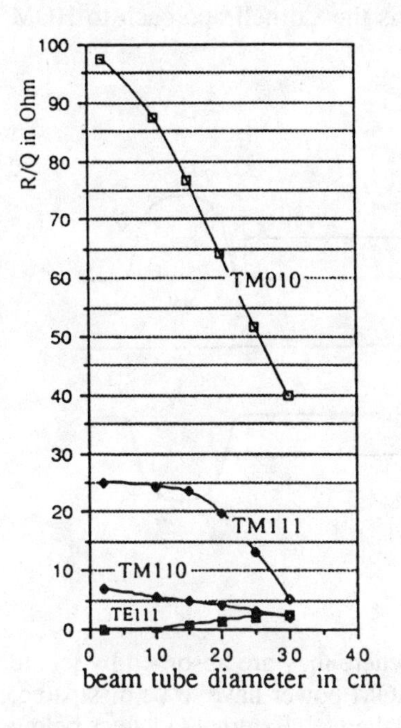

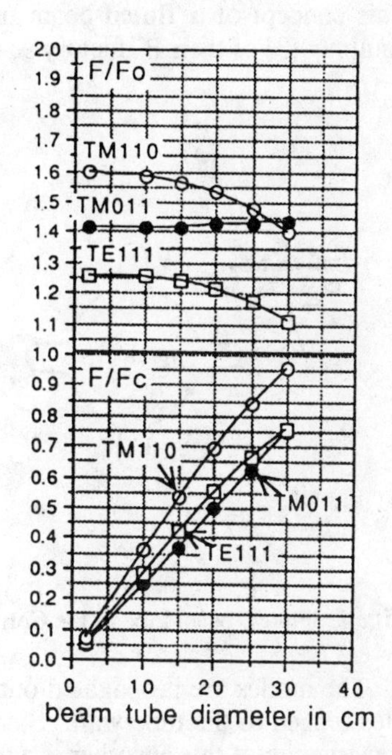

Fig. 2. (R/Q) as function of beam-tube diameter **Fig. 3.** Normalized mode frequencies

Using a cavity code like URMEL such questions can easily be studied, especially if the cavity form is that of a simple pill box of length $\lambda/2$ and with attached beam tubes. During such a study the beam-tube diameter has been varied but the fm frequency kept constant at 400 MHz in correcting the pill-box diameter. Results [2] are plotted in Figs. 2 and 3

In increasing the diameter of the beam tube the (R/Q)s of most modes decrease, that of the fundamental mode included. The main dipole mode TM_{111} looses R/Q two times faster than the fundamental one, and this is just the effect which we wish to observe. But there is a notable exception. The dipole mode TE_{111} which comes first above the fundamental one develops (R/Q) and at the same time (upper part of Fig. 3) its frequency falls and approaches that of the fm. The wider the beam tube the more the TE_{111} mode needs damping but the more difficult is its separation from the fm. Due to this lowering of their frequency both the TE_{111} and the TM_{110} mode also 'refuse' to propagate (lower part of Fig. 3).

The situation can be improved [3] if one decreases the length of the cell which for a one-cell cavity can be done ad libitum (but not for a multicell where the increasing iris thickness would impair cell-to-cell coupling).

For the pill-box example a length of $0.35\ \lambda$ would be most advantageous moving the (R/Q) of fm and TM_{011} mode up and down respectively and increasing the TE_{111}, TM_{011} and TM_{111} frequencies. But still the TE_{111} mode is not propagating.

Another idea was needed to achieve damping of all modes by the beam tubes themselves. In fact, giving the beam tube the more complex form of a waveguide with four ridges [4] one can lower the cut-off frequency of dipole modes selectively. This concept of a 'fluted beam tube' (see Fig. 4) is the Cornell approach to HOM damping [3] of their B–factory sc cavity.

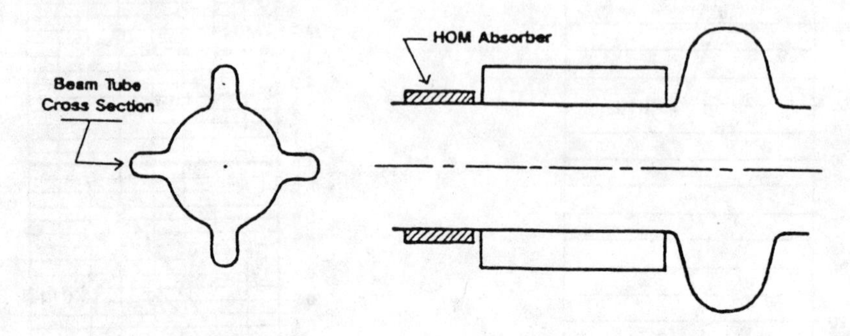

Fig. 4. 'Fluted' beam tube of the Cornell cavity

The modes are propagated out of the cryostat where they are absorbed by ferrite tiles brazed to the tube wall. Since kilowatts of HOM power have to be dissipated, construction of this absorber is a major technical challenge. External Q values below 100 are reached for all HOM as has been demonstrated by calculations [5] and model measurements.

Figure 5 compares the resulting cavity geometry with the typical shape of a copper cavity with 'nose cones' around the beam-tube opening to enhance the fm R/Q. The sc cavity has a three-fold advantage if loss factors are compared. Offering also three-times-higher accelerating gradients, a given voltage can be produced with about nine-times-less beam impedance.

The nose cones of a room temperature copper cavity prevent damping of lower-frequency HOM by the beam tubes but, without a cryostat, space is available to attach guides directly to the cavity body. A copper-cavity design [6] for the SLAC B–factory reaches loaded Q values of only 20 with three optimally positioned guides (see Fig. 6).

The question whether the HOM power within the guides should be absorbed by interior vacuum loads or transferred via broad-band transitions and vacuum windows to external coaxial loads is still the subject of R & D. At CEBAF, based on lossy ceramics, vacuum loads with temperature independent characteristics [7] have been developed and a workshop on this topic will be held there

2.2 Transmission Line Couplers

For a collider like LEP or TRISTAN the cross section of a guide would be far too generous for the HOM power to be transported. An appropriate size can be chosen if transmission lines are used but generally a filter has then to be foreseen which, exposed to high fm currents and voltages, is the most critical element of the design.

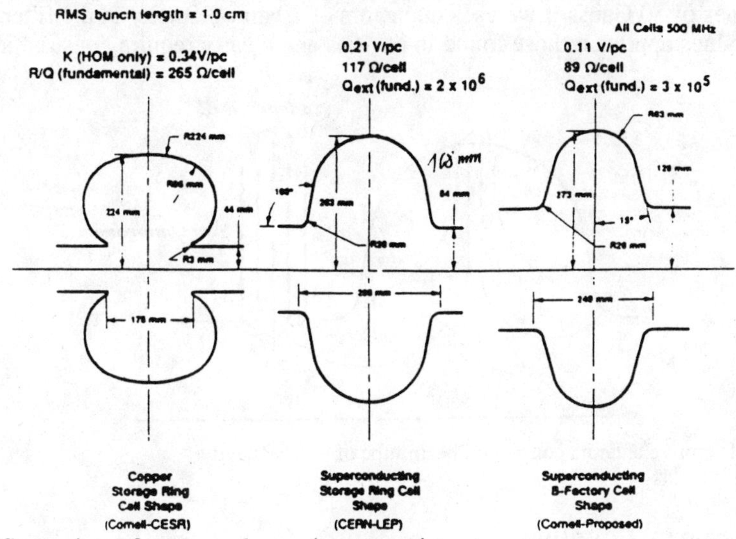

Fig. 5. Comparison of copper and sc cavity geometries

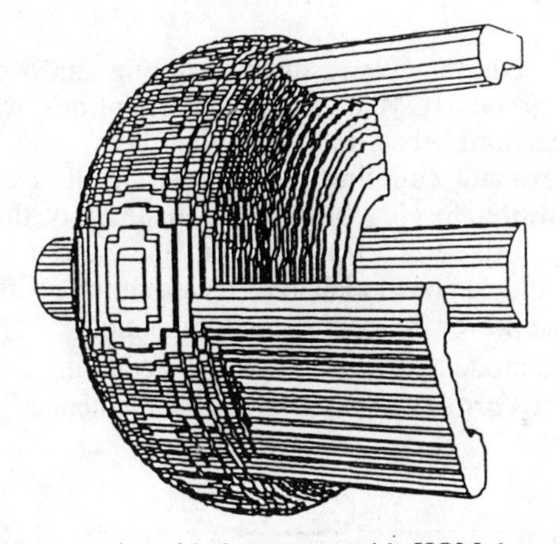

Fig. 6. The SLAC B-Factory cavity with three waveguide HOM dampers

2.2.1 Filter Requirements

This point is illustrated by taking a simple coaxial antenna coupler on the beam tube of the sc 4-cell LEP cavity (see Fig. 7) as an example. Using a 10-cm diameter 50-Ω line terminated without filter by a matched load, and an antenna penetration into the beam tube which produces the HOM damping required for LEP (Q_{ex}=15000), coupling to the fm corresponds to a Q_{ex} of 100000. With a fm (R/Q) of 230 Ω and assuming a fm accelerating gradient of 5 MV/m (V = 8.6 MV) this coupler would pick up a fm power of $P_{ex} = V^2/(2(R/Q)Q_{ex})$ = 1.6 MW! This is the incident power which the filter has to reject. Depending on its type, the current or voltage in its elements will be at least as big as those of the travelling wave transporting P_{ex}. These are calculated to be 12500 V and 250 A. Such a current will result in magneticfieldvalues of 50 Gauss if we use conductors of 20 mm-diameter in the filter circuit. These values approach those found in cavities and clearly require construction from sc materials.

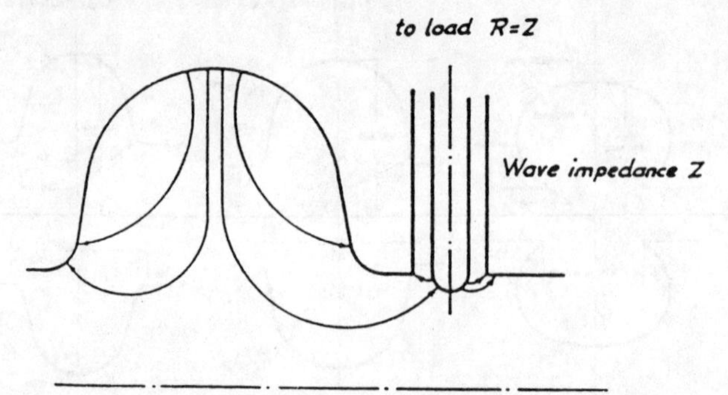

Fig. 7. Primitive antenna coupler on beam tube of sc LEP cavity

2.2.2 Resonant Coupling

To alleviate the filter requirements resonant coupling can be used. This makes it possible to achieve given HOM damping specifications with a reduced probe penetration and hence a diminished fm load on the filter.

The concept of resonant coupling [8] suggests itself if one inspects the field around the probe tip more closely. In part it is made up of the fringe or stray field which is unavoidably formed if a transmission line ends in an open circuit. The energy stored in this field can be modelled by a capacitor C_f parallelto the resistor $R = Z_0$ which represents the power flow out to the termination.

Modelling further a mode by C_m and L_m and the coupling by C (see Fig. 8) we see that part of the coupling current I is bypassed by C_f and hence P_{ex} is reduced.

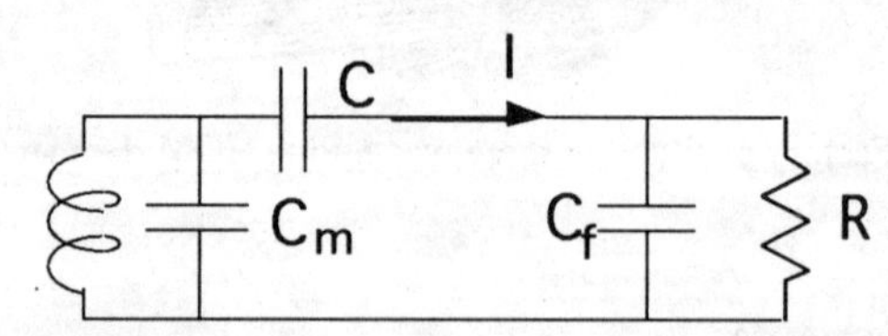

Fig. 8. Model of coupling to primittive antenna coupler of Fig. 7

But the full damping potential can be realized if C_f is compensated by a parallel inductance L (see Fig. 9). The coupler itself becomes a resonator although generally with low $Q = \omega C_f R$.

Within the loaded bandwidth of the mode we now can neglect the effect of C_f and L. Transforming the series connection of C and R into the equivalent parallel circuit of C and $R_p = 1/(\omega^2 C^2 R)$ and calculating the mode's external Q from

$$Q_{ex} = \omega(C_m + C)R_p \approx \omega C_m / (\omega^2 C^2 R) = \frac{C_m C_f}{C^2} \frac{1}{\omega C_f R} \tag{20}$$

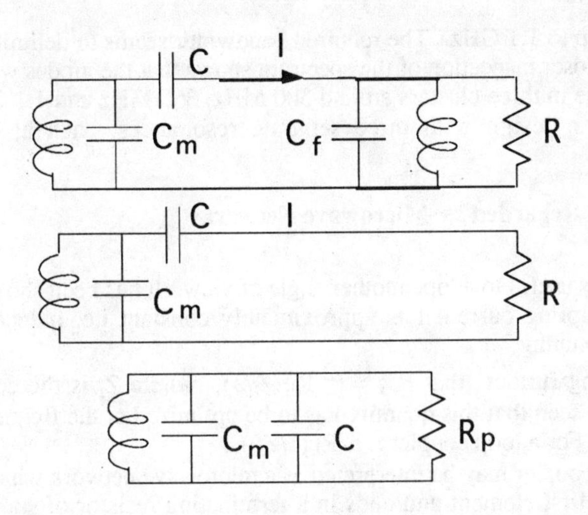

Fig. 9. Compensation of fringe field capacitance and two steps of simplification

we find that for a given coupling factor k = $C/\sqrt{C_m C_f}$, mode damping and coupler Q are proportional:

$$1/Q_{ex} = k^2 Q \tag{21}$$

The coupling factor has physical reality and for the coaxial-line coupler treated here, so does C_f. But C_m and C individually have no physical significance, only their ratio, and one is free to choose for instance $C_m = C_f$ and not necessarily $C_m = 1/(\omega_m(R/Q)_m)$)as we did to model beam loading.

Increasing R will, in proportion, augment damping of the mode to which the coupler is tuned, as long as the voltage across R remains small and the coupling current I does not change. This is no longer true if kQ approaches 1. As detailed analysis shows, at $kQ = 1$ the mode bandwidth reaches a maximum of kf. Higher coupler Q leads only to mode splitting, in the limit of infinite Q by a frequency difference δF = kF. This limit can be treated by a cavity code like MAFIA to determine k and hence Q_{ex} from Eq. (21).

If we now examine magnetic coupling by a loop we come to the same conclusion. The self inductance of the loop takes over the role of C_f but connected in series with the load resistor which therefore, due to voltage division, only sees a part of the induced voltage. Compensation requires a capacitance in series with the loop and the coupler's Q becomes $Q = \omega L_s/R$. Hence we *increase* damping by *decreasing* R.

This compensation technique always gives good results, in fact uncompensated coupler constructions have the compensation frequency zero and there a cavity has no HOM. Nevertheless one will rarely go to the limit $kQ \approx 1$ since one wants to dampen several modes with the same coupler.

For the 350 MHz LEP cavity, as a typical example of a sc cavity with rounded "spherical" form and wide beam tube, the HOM spectrum to be covered begins at 460

MHz to reach up to 1.1 GHz. The required bandwidth seems to delimit the coupler's Q to 1. But a closer inspection of the spectrum shows that the modes with significant R/Q values come in three clusters around 500 MHz, 660 MHz and 1.1 GHz requiring a HOM coupler design with three separate resonances, each at one of these frequencies.

2.2.3 Couplers Regarded as Microwave Networks

At this stage it is useful to adopt another angle of view. It has been shown that for $kQ < 1$ the coupling current I is approximately constant i.e. independent of the coupler's Q and tuning.

Remembering further that $P_{ex} = I^2\,\mathrm{Re}(Z_i/2)$, where Z_i is the coupler's input impedance, it is seen that this quantity has to be optimized at the frequencies of dangerous modes. (For a loop coupler it is $\mathrm{Re}(1/Z_i)$).

An antenna coupler may be interpreted as a microwave network which has a shunt capacitor C_f as first element and ends in a terminating resistor of some convenient standard value. (The network of a loop coupler starts with L_s as a series element.) Numerical methods using network analysis programs can help to solve the problem of $\mathrm{Re}Z_i$ optimisation.

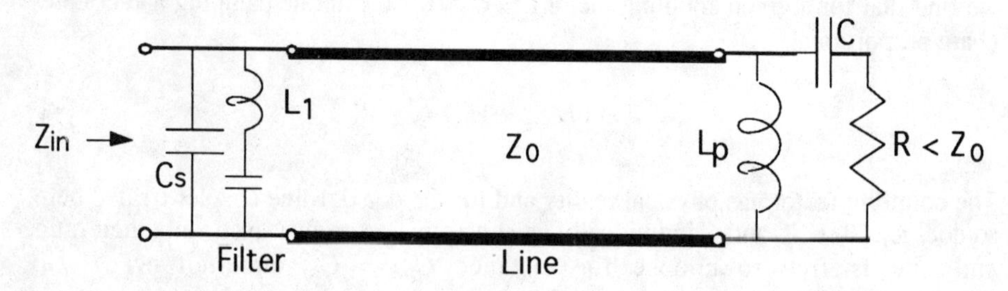

Fig. 10. A network model of the LEP HOM coupler (electric field coupling)

Figure 10 shows the network which models the electric-field coupling of the LEP cavity HOM coupler [8] A series LC-resonator parallel to C_f eliminates fm coupling. Its inductor forms a loop (see Fig. 11) arranged transverse to the cavity axis. Coupling to longitudinal modes is only via their E-field (as described by the network) but to dipole modes it is also magnetic (to their H_z component). This H_z coupling dominates for the TE_{111} modes. Due to their small frequency distance from the fm, they fall on the slope of the notch which suppresses E-field coupling at the fm frequency. An outline of the coupler's geometry is given in Fig. 11
The broadband sensitivity curve of Fig. 12 is obtained by measuring with a network analyzer the transfer σ_{12} between a small electric probe brought near to the coupler's front end and the load resistor. It shows the three coupler resonances and their alignment with the three groups of mode resonances. For comparison the insert gives the calculated $\mathrm{Re}Z_i$ of the network model.

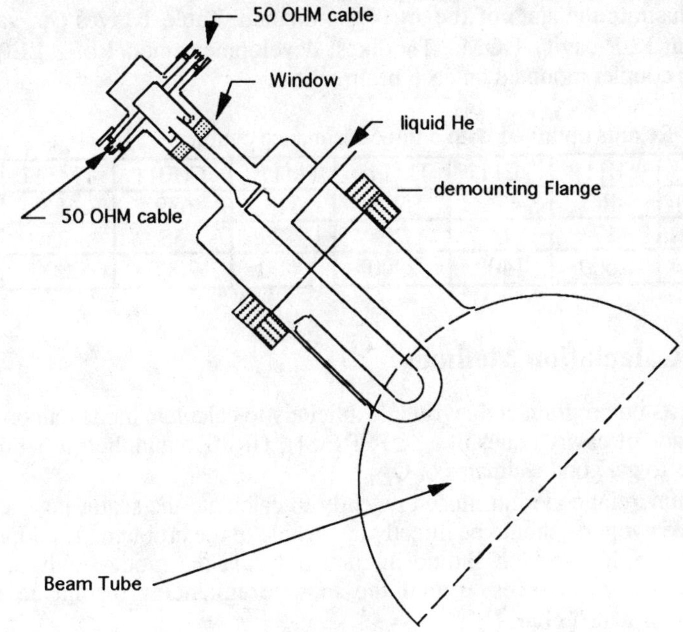

Fig. 11. Outline of the LEP HOM coupler's geometry

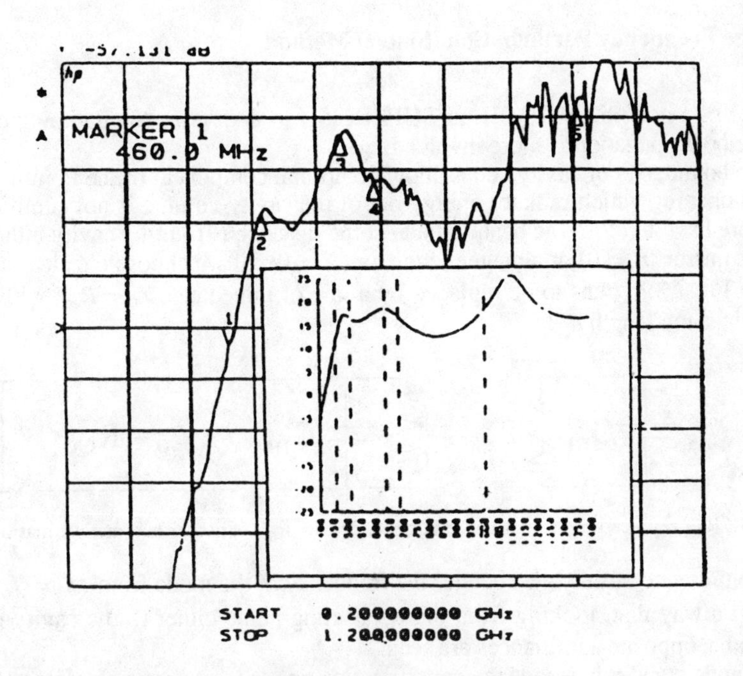

Fig. 12. Electric field sensitivity of the LEP HOM coupler

To illustrate the state of the art of the method, Table 1 gives Q_{ex} values of the significant LEP cavity HOM. The latest development model of a LEP coupler is used, one coupler mounted on each beam tube.

Table 1. Results obtained with a three resonance coupler

Mode	TE111	TE111	TM110	TM110	TM011	TM111	TM012
f (MHz)	462	476	506	513	639	688	1006
$(R/Q)(\Omega)$	17	14	20	12	55	25	22
Q_{ex}	2600	1400	2900	9000	2000	3000	5000

3 Q_{ex} Calculation Methods

There exists no program code written explicitely to calculate mode damping. But use can be made of cavity codes like SUPERFISH, URMEL and their three-dimensional offsprings to get good estimates of Q_{ex}.

A commercial code introduced recently to calculate the scattering parameters of microwave twoports should be directly applicable to the problem. Equipping a cavity with two couplers (which should interact only via the mode fields) a twoport is obtained. Its σ_{12} traces around the mode frequencies resonance curves the bandwidth of which gives Q_{ex}.

3.1 The Frequency Perturbation (Slater) Method

The use of a code like URMEL or SUPERFISH to determine the damping effect of beam tube propagation is more involved.

The boundaries of cavity codes allow no normal component of the Pointing vector i.e. an opening which radiates energy out of the cavity volume is not admitted. We therefore have to close the beam tube at some distance l from the cavity either by an electric mirror ($E_t = 0$) or a magnetic mirror ($H_t = 0$). If we choose the electric mirror then in Fig. 13 R_{ex} has to be replaced by a line of impedance $Z_0 = R_{ex}$ with a short circuit at some length d

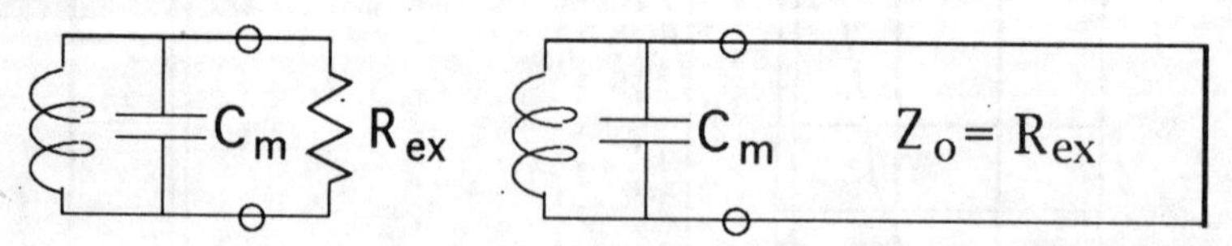

Fig. 13. R_{ex} replaced by line of wave impedance $Z_0 = R_{ex}$. Shortcircuit as termination.

The presence of this line changes the frequency of the mode from $\omega_m = (L_m C_m)^{-1/2}$ to ω_r in a way that, looking from the connecting plane either to the cavity or to the beam tube, opposite admittances are seen.

To understand this resonance condition imagine that, to create a voltage V, a current I is injected into the connecting terminals. At resonance, since the circuit has no losses, the current must become zero. With $\Omega_r = \omega_r/\omega_m - \omega_m/\omega_r$ and $\varphi_r = \omega_r d / v$

$$I = V\left[\omega_m C_m \Omega_r - 1/\left(Z_0 \tan\varphi_r\right)\right] = 0$$

But since
$$\omega_m C_m Z_0 = \omega_m C_m R_{ex} = Q_{ex}$$

it follows that
$$\Omega_r Q_{ex} = 1/\tan\varphi_r \tag{22}$$

To determine Q_{ex} one uses the code to calculate ω_r for a sequence of different beam tube lengths l and plots ω_r against l. But l and the parameter d may differ by a constant d_0. A three parameter fit varying Q_{ex}, ω_m and d_0. is needed to determine Q_{ex}.

3.2 The Open-circuit-voltage, Short-circuit-current Method

Imagine that, due to a special beam or auxiliary oscillator, a cavity mode is excited to some *constant* stored energy U and that, via a coupler and transmission line of impedance Z_0, power P_{ex} flows to a load $R = Z_0$. If U and P_{ex} can be determined then $Q_{ex} = \omega U/P_{ex}$

Remember now [9] that at constant stored energy we may interprete the coupler output by an active power source with the Thevenin equivalent of Fig. 14.

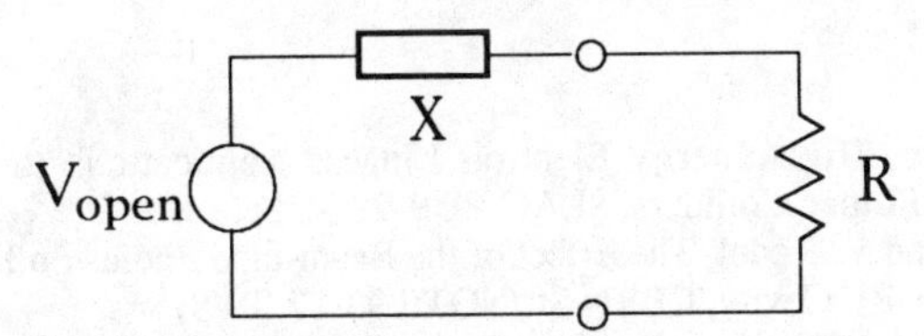

Fig. 14. Thevenin equivalent of the coupler output for constant stored cavity energy

Since the cavity and coupler circuit are without losses the internal impedance X must be a pure reactance with magnitude: $X = V_{open} / I_{short}$. Knowing V_{open} , I_{short} and U we can calculate Q_{ex}.

$$P_{ex} = \frac{1}{2} V_{open}^2 \frac{R}{R^2 + X^2} \tag{23}$$

$$Q_{ex} = \frac{\omega U}{P_{ex}} = 2\omega U \left(\frac{1}{R I_{short}^2} + \frac{R}{V_{open}^2} \right) \tag{24}$$

V_{open} and I_{short} can be 'measured' by a cavity code in terminating the line by a magnetic or electric mirror respectively and *normalizing to the same U*. In principle two program runs suffice but it is advisable to repeat the calculation for different line lengths. One will then find that V_{open} and I_{short} vary. But Q_{ex} according to Eq. (24) must remain invariant.

3.2.1 Application to a Waveguide

If the propagating power is not on a transmission line but on a guide, as in the beam tube example, one has to introduce generalized forms of voltage V_g, current I_g and wave impedance Z_g so that for a travelling wave $P = V_g I_g/2$ and $Z_g = V_g/I_g$. Longitudinal modes excite in the beam tube of radius a a TM_{01} wave for which

$$P = \frac{1}{2}\pi a^2 E_r(a) H_\varphi(a) \tag{25}$$

A convenient choice of V_g I_g and Z_g is then

$$V_g = \sqrt{\pi}\, a\, E_r(a) \qquad I_g = \sqrt{\pi}\, a\; H_\varphi(a) \tag{26}$$

$$Z_g = \frac{V_g}{I_g} = \frac{E_r}{H_\varphi} = \frac{\lambda}{\lambda_g} 120\pi \tag{27}$$

λ and λg are the free-field and guide wavelength respectively.

References

[1] P.B. Wilson, High Energy Electron Linacs: Applications to Storage Ring RF Systems and Linac Colliders, SLAC-PUB-2884, Feb. 82

[2] E. Haebel and V. Rodel, The Effect of the Beam-tube Radius on Higher-Order Modes in a Pill-Box RF Cavity, CERN SL/NOTE 93-17 (RFS)

[3] H. Padamsee et al., Accelerating Cavity Development for the Cornell B-Factory, Proc. of the IEEE Particle Accelerator Conference, San Francisco, 1991,, p. 786

[4] T. Kageyama, Grooved Beam Pipe for Damping of Dipol Modes in RF Cavities, KEK, pp. 91-133

[5] D. Moffat et al., Use of Ferrite-50 to Strongly Damp Higher-Order Modes, in Ref. [3]

[6] R. Rimmer et al. Higher-Order Mode Damping Studies on the PEP-II B-Factory RF Cavity, Proc. of the European Particle Accelerator Conference, Berlin, 1992, pp. 1289-1291

[7] I.E. Campisi et al., Higher-Order Mode Damping and Microwave Absorption at 2K, in Ref. [6], pp.1237-1239

[8] E. Haebel, Couplers, Tutorial and Update, Particle Accelerators, 1992, Vol. 40, pp. 141-159

[9] W. Hartung and E. Haebel, In Search of Trapped Modes in the Single-Cell Cavity Prototype for CESR-B, to be published in the Proc. of the Particle Accelerator Conference, Washington, 1993

The Beam-Beam Interaction in e^+e^- Storage Rings

Robert H. Siemann*
Stanford Linear Accelerator Center
Stanford, CA 94305, USA

1 Introduction

Colliders are designed for studying relatively rare, small impact parameter collisions that produce elementary particles. This is not the dominant interaction between the beams, however. That dominant interaction, the beam-beam interaction, is due to the electromagnetic fields of the beams.

The simplest and most pragmatic treatment of the beam-beam interaction is to parametrize the luminosity, L, in terms of the beam-beam tune shift, ξ. The tune shift is the shift in the vertical betatron tune of a small-amplitude particle due to the electromagnetic fields of the other beam. Expressed in terms of beam parameters

$$\xi = \frac{r_e}{2\pi} \frac{N\beta_y^*}{\gamma\sigma_y(\sigma_y+\sigma_x)} \tag{1.1}$$

where $r_e = 2.82\times10^{-15}$ m is the classical electron radius, N is the number of particles in a beam bunch, β_y^* is the vertical β-function at the interaction point, γ is the beam energy in units of rest energy, and σ_y and σ_x are the rms vertical and horizontal beam sizes at the collision point. The luminosity

$$L = \frac{1}{4\pi} \frac{N^2 f_c}{\sigma_y \sigma_x} \tag{1.2}$$

(f_c is the collision frequency) can be rewritten in a frequently used parametrization

$$L = \frac{N f_c \gamma \xi (1+\sigma_y/\sigma_x)}{2 r_e \beta_y^*} \quad . \tag{1.3}$$

---

* This work was supported by the Department of Energy contract DE-AC03-76SF00515.

It is known from experience that there is a soft limit on ξ, $\xi \lesssim 0.05$. The pragmatist chooses a value of ξ and then uses eq. (1.3) to trade-off luminosity, beam current (the total current is $I_T = eNf_c$), σ_y/σ_x, and β_y^*. A "conservative" design might be based on $\xi = 0.03$ while more aggressive ones might use $\xi = 0.05$. This approach has shortcomings primarily because it isn't based on an understanding of the underlying physics of the beam-beam interaction. Other beam parameters are assumed independent of and unaffected by ξ, and ξ is assumed independent of and unaffected by them. In fact, beam-beam performance depends on many other parameters including some of those in eq. (1.3), betatron and synchrotron tunes, bunch length, lattice errors, and radiation damping. Therefore, eq. (1.3) has limited applicability, and extrapolations into new regimes such as those being considered for heavy quark factories—small β_y^*, crossing angles, closely spaced bunches, unequal beam energies, etc—have uncertainties associated with them. Furthermore, a "conservative" choice of ξ may not lead to a conservative overall design. Other beam parameters or accelerator systems such as the RF and vacuum systems may be pushed unnecessarily.

This article is a personal perspective about the physics of the beam-beam interaction. This is an active area of research combining operational experience, experiments, computer models, and theory with the goal being to overcome the shortcomings above. This research hasn't progressed sufficiently to quantitatively explain beam-beam limits, but there are qualitative explanations of many of the features of the beam-beam interaction and clear directions for future developments.

2 Observations

Experimental aspects of the beam-beam interaction are the subject of several articles that give detailed observations.[1-6] Some of these papers synthesize data from several storage rings. This is a difficult task because there are numerous important parameters, and when colliders are compared many of these parameters are different. It is hard to know which of these differences are essential, which are secondary, and even if all the differences have been identified. Rather than repeating this type of quantitative analysis, this section stresses a common feature of the observations—there are two beam-beam limits.

Normally beams are injected into a collider with electrostatic separation at the interaction point. While the beams are still separated they are Gaussian with rms beam sizes at the interaction point of

$$\sigma_{x0} = \sqrt{\beta_x^* \varepsilon_x} \quad ; \quad \sigma_{y0} = \sqrt{\beta_y^* \varepsilon_y} \quad . \tag{2.1}$$

The emittances, ε_x and ε_y, are determined by the magnet lattice and the properties of synchrotron radiation.[7] Most electron storage rings have had $\sigma_{x0} >> \sigma_{y0}$. That is the natural relation between the sizes because i) the main dipoles bend in the horizontal plane leading to $\varepsilon_x >> \varepsilon_y$, and ii) a quadrupole doublet is the simplest interaction region configuration. If the quadrupole polarities are chosen to give

$\beta_x^* >> \beta_y^*$, the tune shifts in the two planes, ξ_x [given by eq. (1.1) with x and y interchanged] and ξ_y ($\equiv \xi$), can be made roughly equal :

$$\frac{\xi_x}{\xi_y} = \frac{\beta_x^*/\sigma_x}{\beta_y^*/\sigma_y} = \sqrt{\frac{\beta_x^* \varepsilon_y}{\beta_y^* \varepsilon_x}} . \tag{2.2}$$

Synchrotron light measurements show that the horizontal size doesn't change significantly when the beams are brought into collision, $\sigma_x = \sigma_{x0}$. However, above some current the vertical beam size becomes dependent on N rather than being determined by synchrotron radiation, $\sigma_y \neq \sigma_{y0}$. Figures 1 and 2 are a compilation of luminosity and tune-shift data for colliders when performance has been optimized. There are two distinct regimes. At low currents $L \propto N^2$ (I_T^2) indicating $\sigma_x\sigma_y$ is constant. The data do not show this regime for all colliders because for them it doesn't exist or is below the current range presented. At high currents $L \propto N$, and, since the horizontal size is unchanged, $\sigma_y \propto N$. TRISTAN has a single beam current limit and is the only collider not to reach this regime. The tune shift is derived from the luminosity measurements using eq. (1.3), so the tune shift plots are another way of presenting the same data. They show the tune shift reaching a limit. This is one of the beam-beam limits. It is associated with the beam core, and, as the figures show, the tune-shift limit varies from collider to collider.

The second beam-beam limit comes from changes of the beam distribution. The dominant effect is the appearance of "non-Gaussian" tails in the vertical—the number of particles with large vertical betatron amplitude is greater than that of a Gaussian distribution with the measured rms width of the core. The beam-beam interaction increases the population of the tail of the distribution even more than it increases the core size.

Figure 3 shows a beam distribution measurement. The non-Gaussian tails are clear. Systematic studies of the tails have never been made because of the difficulty of the measurements. The luminosity is insensitive to the tails, and synchrotron radiation monitors are plagued by unwanted reflections from the vacuum chamber that dominate the image beyond 2 to 3 σ_y. The best measurement technique is destructive—measuring the lifetime as a collimator is moved toward the center of the beam. Systematic studies are tedious because new beams must be injected after each measurement.

Particles with sufficiently large amplitudes hit the vacuum pipe causing experimental backgrounds and reducing the beam lifetime. The dynamic aperture due to magnet nonlinearities may play a role in that particles may fall outside the dynamic aperture before hitting the physical aperture. The second beam-beam limit has been reached when the lifetime or backgrounds become unacceptable. This beam-beam limit is associated with the beam tails. It isn't parametrized by a value of ξ; ξ is determined by the size of the core and has reached its limiting value below the second beam-beam limit.

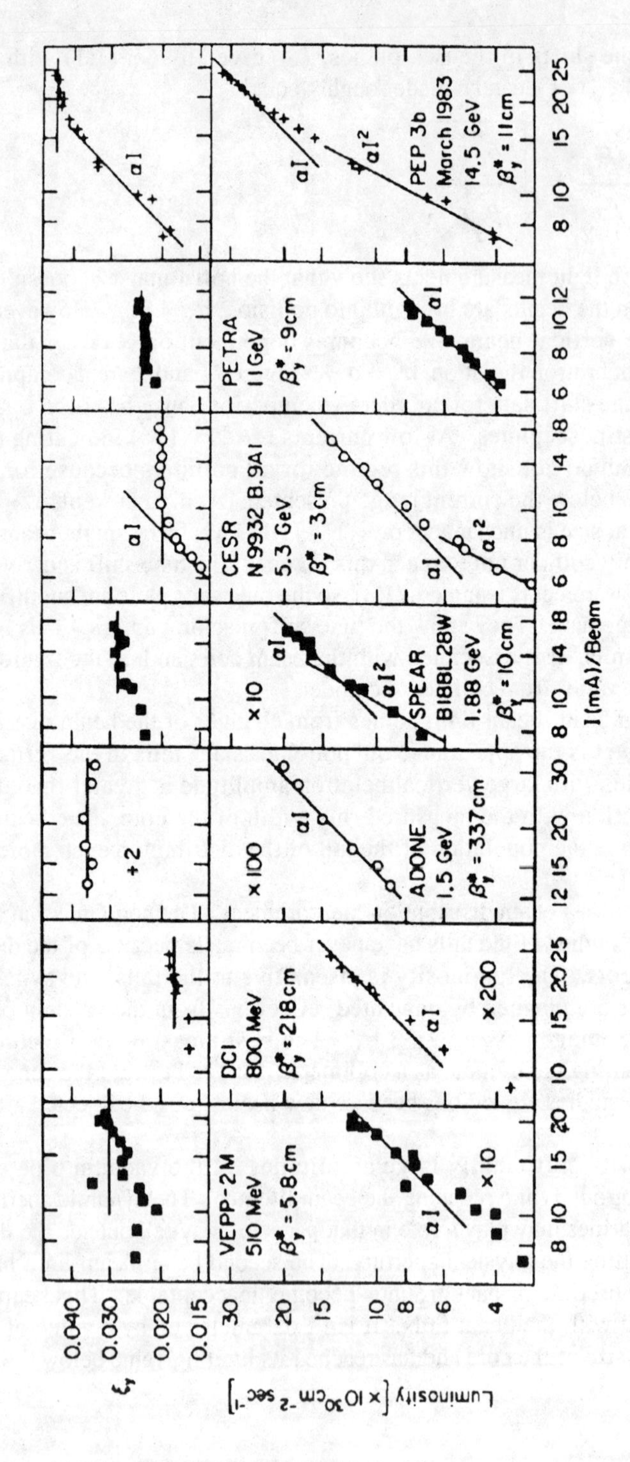

Figure 1: The beam-beam performance of e^+e^- colliders from an article by John Seeman (Ref. 1).

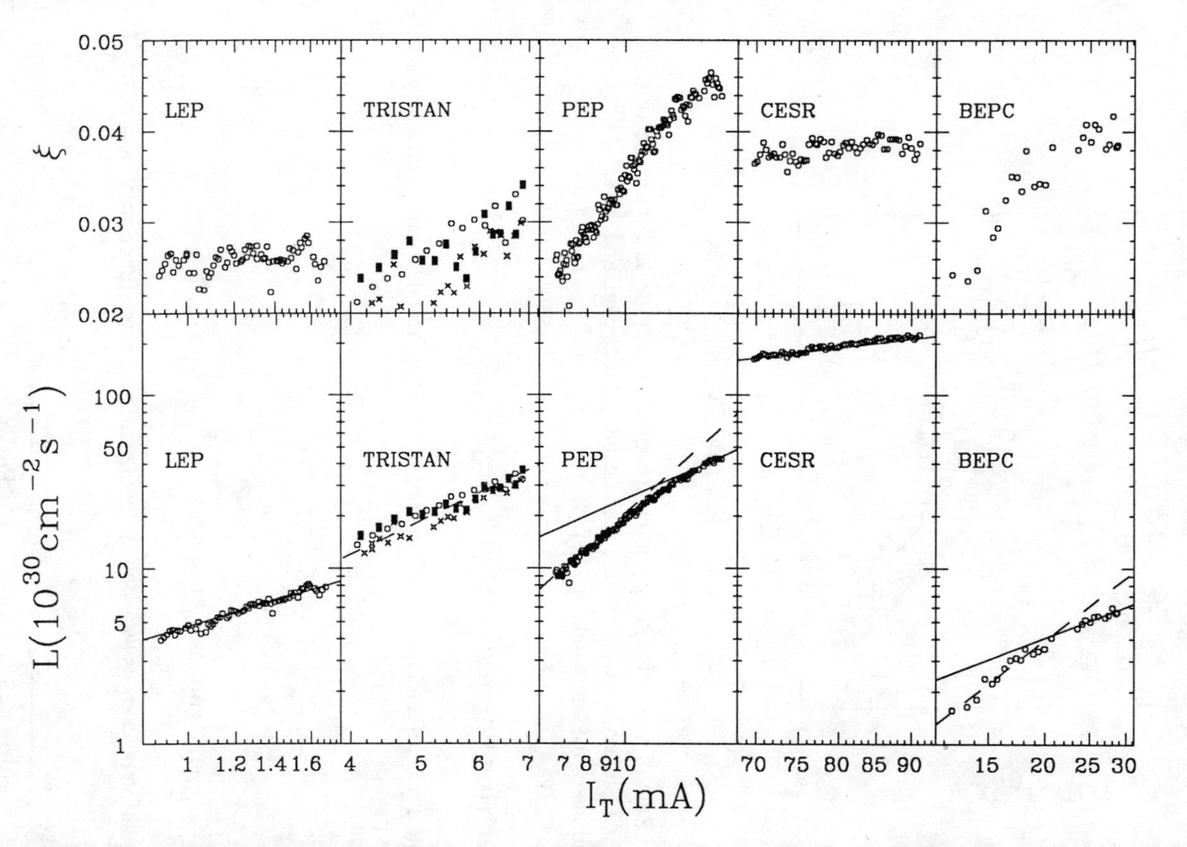

Figure 2: An update of beam-beam performance. The LEP data were provided by J. Jowett, the TRISTAN data by H. Fukuma and Y. Funakoshi, the PEP data by M. Donald, the CESR data by D. Rice, and the BEPC data by C. Zhang. The solid lines are eyeball fits to $L \propto I_T$, and the dashed lines to $L \propto I_T^2$. The different symbols for TRISTAN indicate measurements by different detectors.

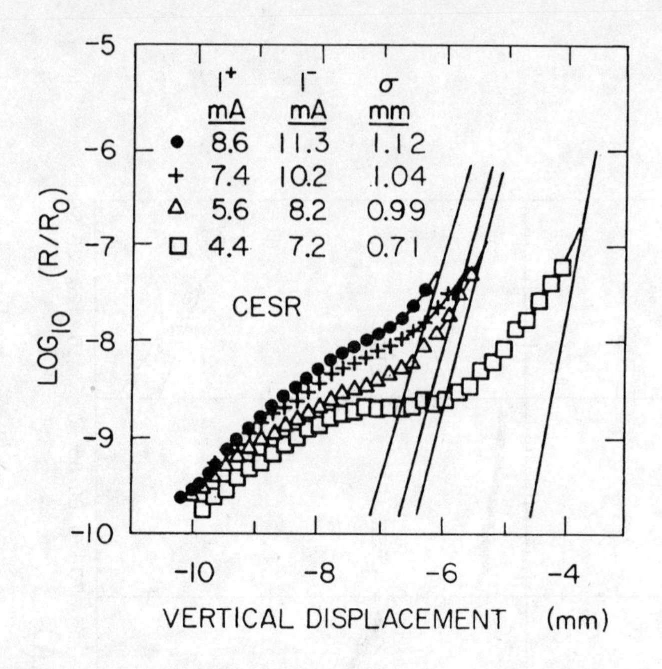

Figure 3: Rate of bremsstrahlung photons produced by a thin Be wire in the vertical tail of the CESR beam. R_0 is the rate at the center of the beam, and the solid lines are extrapolations of the Gaussian core (ref. 8).

3 Underlying Physics

The angles of a relativistic particle passing through an oncoming beam (Figure 4) change by[9]

$$\left.\begin{matrix}\Delta x' \\ \Delta y'\end{matrix}\right\} = \frac{-Nr_e}{\gamma\sigma_x}\sqrt{\frac{2\pi}{1-R^2}}\left.\begin{matrix}\mathrm{Im} \\ \mathrm{Re}\end{matrix}\right\} f_{BB} \; ;$$

$$f_{BB} = W(u+iRv) - \exp[-(1-R^2)(u^2+v^2)]\,W(Ru+iv) \; . \qquad (3.1)$$

In this equation $\sigma_x > \sigma_y$ is assumed, W is the complex error function,[10] $R = \sigma_y/\sigma_x$, and

$$u = \frac{x}{\sigma_x\sqrt{2(1-R^2)}} \; ; \; v = \frac{y}{\sigma_y\sqrt{2(1-R^2)}} \; . \qquad (3.2)$$

An example of $\Delta y'$ is plotted in Figure 5. At small y/σ_y, $\Delta y' \propto y$ while at large values it falls like $1/y$.

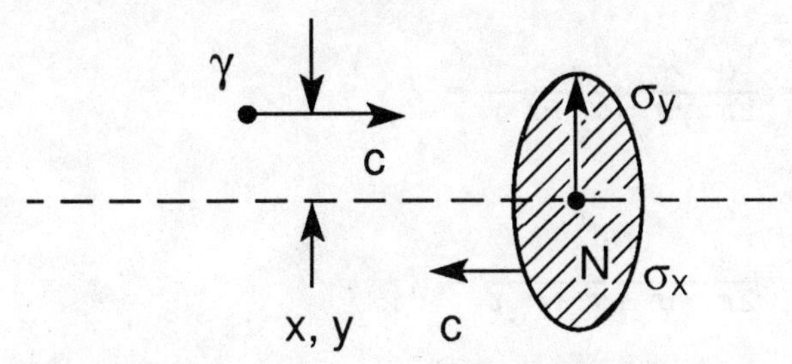

Figure 4: A relativistic particle of energy γ passing through an oncoming thin pancake of N oppositely charged particles. The transverse distributions of the pancake are Gaussian with rms widths σ_x and σ_y. The particle is displaced from the center of the oncoming beam by (x,y).

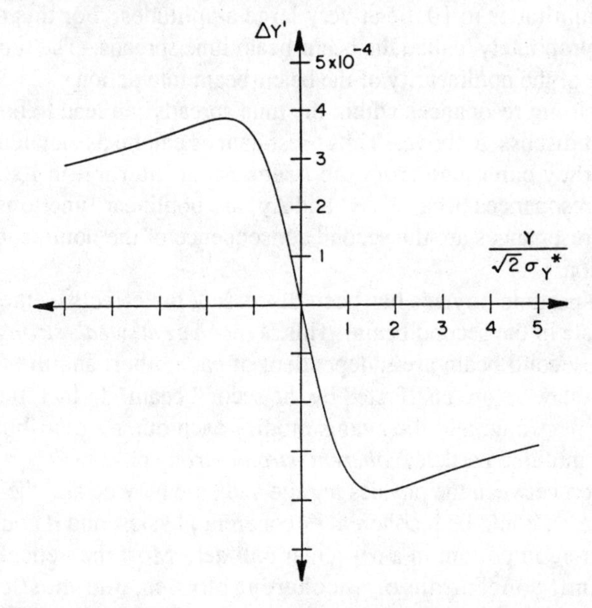

Figure 5: The vertical kick experienced by a typical electron in CESR (ref. 11).

The complex error function can be approximated for small arguments as $W(\zeta) \approx 1 + 2i\zeta/\pi^{1/2}$ giving

$$\left.\begin{matrix}\Delta x' \\ \Delta y'\end{matrix}\right\} = \frac{-2Nr_e}{\gamma(\sigma_x+\sigma_y)} \left.\begin{matrix}\mathrm{Im} \\ \mathrm{Re}\end{matrix}\right] \left[i\frac{x}{\sigma_x} + \frac{y}{\sigma_y} \right] . \tag{3.3}$$

This is a focusing quadrupole in both dimensions. Displacement, betatron amplitude and betatron phase are related by $z = A_z\cos(\psi_z)$ where z = x or y. Small-amplitude particles experience this linear focusing for all values of phase, and it changes their tunes by

$$\Delta Q_y = \frac{r_e}{2\pi} \frac{N\beta_y^*}{\gamma\sigma_y(\sigma_y+\sigma_x)} = \xi_y \equiv \xi \ , \tag{3.4}$$

and

$$\Delta Q_x = \frac{r_e}{2\pi} \frac{N\beta_x^*}{\gamma\sigma_x(\sigma_y+\sigma_x)} = \xi_x \ . \tag{3.5}$$

Particles with larger betatron amplitudes are outside the linear region of eq. (3.1) for a range of phases, and the focusing averaged over phase is weaker than for a small-amplitude particle. Very-large-amplitude particles have almost no change in tune due to the beam-beam interaction. The beam has a range of tunes from $\{\Delta Q_x, \Delta Q_y\} = \{\xi_x, \xi_y\}$ at small amplitudes to $\{0, 0\}$ at very large amplitudes. For this reason ξ_x and ξ_y are often, appropriately, called the beam-beam tune spreads. The tune spreads are one consequence of the nonlinearity of the beam-beam interaction.

Sufficiently strong resonances within the tune spreads can lead to beam-beam phenomena like that discussed above. Those resonances can be associated with the magnet lattice, or they can come from the beam-beam interaction itself which produces nonlinear resonances because $\Delta x'$ and $\Delta y'$ are nonlinear functions of x and y. These nonlinear resonances are the second consequence of the nonlinearity of the beam-beam interaction.

So far single-particle physics has been discussed, the effects of the fields of one beam on a particle in the second beam. This is *incoherent*, *weak-strong* physics. The particles in the second beam are independent of each other, and the fields and distribution of the first beam are unaffected by the second beam. In fact, the fields at the collision point are strong, and the beams modify each other's distributions and fields. This leads to multiple particle, *coherent*, *strong-strong* physics.

The connection between the physics regime and core blowup and the tune-shift limit is an open issue. It could be incoherent or coherent physics, and it could be that one or the other is more important in a particular collider. Most theoretical analyses of the tune-shift limit concentrate on incoherent physics, and most computer simulations make approximations that inhibit coherent physics. Experiments that would distinguish between them have not been performed, but there are some observations that can be explained only by coherent physics.

The large-amplitude particles leading to lifetime and background limitations are rare. These rare particles are independent of each other and cannot affect the distribution of the other beam. Clearly these particles are described by single-particle, incoherent physics.

4 The Incoherent Beam-Beam Interaction

4.1 Hamiltonian Formalism

Hamiltonian methods are extremely powerful. They can be used to calculate many aspects of the incoherent beam-beam interaction, and they provide a framework for addressing such topical questions as the effects of synchrotron motion, finite bunch length, unequal beam energies, optical errors, crossing angles, etc.

In the absence of the beam-beam interaction the Hamiltonian for the transverse motion of a single particle is[12]

$$H_0(x,p_x,y,p_y,s) = -eA_s(s) - \left[1+\frac{x}{\rho(s)}\right]\left[p^2-p_x^2-p_y^2\right]^{1/2} . \quad (4.1)$$

In this equation s is the usual coordinate along the reference orbit, ρ is the bending radius, A_s is the s-component of the vector potential describing the magnet lattice, $p = [E^2/c^2 - m^2c^2]^{1/2}$ is the total momentum, and p_x and p_y are the momenta conjugate to x and y. The ideal solution is to find constants of the motion. This is possible when the lattice is made up of dipoles and quadrupoles. In that case canonical transformations can be used to simplify H_0 to[12]

$$H_0 = \frac{2\pi Q_{x0}}{C} I_x + \frac{2\pi Q_{y0}}{C} I_y \quad (4.2)$$

where C is the accelerator circumference, Q_{x0} and Q_{y0} are the betatron tunes and it has been assumed for simplicity of illustration that there are no skew quadrupoles. The canonical transformation has been used to transform to "action-angle coordinates"; the actions, I_x and I_y, are constants of the motion because their conjugate coordinates, the angles ψ_x and ψ_y, do not appear in the Hamiltonian and $dI_z/ds = -\partial H_0/\partial\psi_z = 0$ (z = x or y).

The "smooth approximation" is a convenient description of transverse motion where the betatron phases advance at constant rates, $d\psi_z/ds = 2\pi Q_{z0}/C$, rather than at the actual instantaneous rates $d\psi_z/ds = 1/\beta_z(s)$. The transverse coordinates in terms of the action-angle coordinates of the Hamiltonian in eq. (4.2) are

$$z = \sqrt{2\beta_z(s) I_z}\cos[\psi_z+\chi_z(s)] \quad (4.3)$$

where

$$\chi_z(s) = \int_0^s \frac{d\zeta}{\beta_z(\zeta)} - Q_{z0}\frac{2\pi s}{C} . \quad (4.4)$$

The angle ψ_z is the betatron phase in the smooth approximation, $d\psi_z/ds = \partial H_0/\partial I_z = 2\pi Q_{z0}/C$, and $\chi_z(s)$ accounts for the difference between the approximate and actual phases. [There are alternate action-angle coordinates where the angle is the betatron phase rather than the phase in the smooth approximation, but using the choice in eqs. (4.2) to (4.4) allows one to account easily for the rapid phase advance near the interaction point where β is small.] Note that χ_z is periodic with period C:

$$\chi_z(s+nC) = \chi_z(s) \quad . \tag{4.5}$$

The Hamiltonian including the beam-beam interaction is

$$H(x,p_x,y,p_y,s) = H_0+V_{BB}(x,y,s) \tag{4.6}$$

where V_{BB} is the beam-beam potential. This potential is nonlinear, and it isn't possible to find constants of the motion. A perturbation method must be used. The steps in this method are:

1. Write the Hamiltonian, H, in terms of the action-angle coordinates of the unperturbed Hamiltonian, H_0.
2. Fourier analyze H with respect to s, ψ_x, and ψ_y.
3. Calculate the dependences of tunes on action from the average value of the perturbation.
4. Determine resonance conditions and resonance properties from the slowly varying terms of H.

A sample calculation is performed in the next section to illustrate the techniques and arrive at conclusions relevant to heavy quark factories. It is a generalization of previously published work[13,14] to finite-length, flat beams.

4.2 Bunch-Length Effects

4.2.1 Write the Hamiltonian, H, in Terms of the Action-Angle Coordinates of the Unperturbed Hamiltonian, H_0

The picture is that given by Figure 4 with the thin pancake replaced by an oncoming bunch with rms length σ_L. The beam-beam potential, assuming only one interaction point, is

$$V_{BB} = \frac{-Nr_e}{\gamma}\frac{2}{\sqrt{2\pi\sigma_L^2}}\sum_{n=-\infty}^{\infty} V_F(x,y,s)\exp\left[-2[s-(nC+c\tau)]^2/\sigma_L^2\right] . \tag{4.7}$$

The potential V_F comes from a solution of Poisson's equation[15]

$$V_F = \int_0^{\infty}\frac{dq}{\sqrt{(2\sigma_x^2+q)(2\sigma_y^2+q)}}\exp\left\{-\left[\frac{x^2}{2\sigma_x^2+q}+\frac{y^2}{2\sigma_y^2+q}\right]\right\} \quad ; \tag{4.8}$$

V_F has explicit s dependence because $\sigma_z^2 = \sigma_{z0}^2 + \varepsilon_z s^2/\beta_z^*$ near the collision point. This is important in the vertical dimension, and the calculation is valid for $\sigma_L/\beta_y^* < 1$. The modulation of the collision point due to synchrotron motion of tune Q_s and amplitude $\hat{\tau}$ is

$$\tau = \frac{\hat{\tau}}{2}\cos(2\pi nQ_s) \tag{4.9}$$

where eqs. (4.7) and (4.9) have a factor of one-half associated with them that arises from the relative motion of the particle and the opposing beam.

The transverse coordinates can be rewritten using eq. (4.3) to give

$$V_F(I_x,\psi_x,I_y,\psi_y,s) =$$

$$\int_0^{\infty} \frac{dq}{\sqrt{(2\sigma_x^2+q)(2\sigma_y^2+q)}} \exp\left\{-\left[\frac{2\beta_x I_x \cos^2(\psi_x+\chi_x)}{2\sigma_x^2+q} + \frac{2\beta_y I_y \cos^2(\psi_y+\chi_y)}{2\sigma_y^2+q}\right]\right\} \quad (4.10)$$

$$\cong \int_0^{\infty} \frac{dq}{\sqrt{(2\sigma_{x0}^2+q)(2\sigma_{y0}^2+q)}} \exp\left\{-\left[\frac{2\beta_x^* I_x \cos^2(\psi_x+\chi_x)}{2\sigma_{x0}^2+q} + \frac{2\beta_y^* I_y \cos^2(\psi_y+\chi_y)}{2\sigma_{y0}^2+q}\right]\right\} . \quad (4.11)$$

The approximation $\sigma_L/\beta_y^* < 1$ was used in going from eq. (4.10) to (4.11), and the only remaining s-dependence in eq. (4.11) is in χ_x and χ_y.

4.2.2 Fourier Analyze H with Respect to s, ψ_x, and ψ_y

This Fourier analysis will show the resonant structure of the beam-beam interaction. The Hamiltonian is periodic in ψ_x and ψ_y with period 2π, so

$$H = H_0(I_x,I_y) - \frac{Nr_e}{\gamma} \sum_{p,r=-\infty}^{\infty} \int_{-\infty}^{\infty} dk\, A_{pr}(I_x,I_y,k) \exp[i(p\psi_x+r\psi_y-ks)] . \quad (4.12)$$

As discussed below coefficient A_{pr} corresponds to resonances with horizontal order $|p|$ and vertical order $|r|$. It is

$$A_{pr} = \frac{1}{(2\pi)^3} \int_0^{2\pi} d\psi_x \int_0^{2\pi} d\psi_y \int_{-\infty}^{\infty} ds \exp[-i(p\psi_x+r\psi_y-ks)]$$

$$\times \frac{2}{\sqrt{2\pi\sigma_L^2}} \sum_{n=-\infty}^{\infty} V_F \exp\left[-2[s-(nC+c\tau)]^2/\sigma_L^2\right] . \quad (4.13)$$

A first result of making the approximation going from eq. (4.10) to eq. (4.11) is that the ψ integrals and the s integral can be factored by making a change of variables $\theta_z = \psi_z + \chi_z$

$$A_{pr}(I_x, I_y, k) = T_{pr}(I_x, I_y)\, \frac{1}{2\pi} \int_{-\infty}^{\infty} ds\, \exp[i(p\chi_x + r\chi_y + ks)]$$

$$\times \frac{2}{\sqrt{2\pi\sigma_L^2}} \sum_{n=-\infty}^{\infty} \exp\left[-2[s-(nC+c\tau)]^2/\sigma_L^2\right] ; \qquad (4.14)$$

$$T_{pr} = \frac{1}{(2\pi)^2} \int_0^{2\pi} d\theta_x\, e^{-ip\theta_x} \int_0^{2\pi} d\theta_y\, e^{-ir\theta_y} \int_0^{\infty} \frac{dq}{\sqrt{(2\sigma_{x0}^2+q)(2\sigma_{y0}^2+q)}}$$

$$\times \exp\left\{-\left[\frac{2\beta_x^* I_x \cos^2\theta_x}{2\sigma_{x0}^2+q} + \frac{2\beta_y^* I_y \cos^2\theta_y}{2\sigma_{y0}^2+q}\right]\right\} . \qquad (4.15)$$

The integral T_{pr} is zero when either p or r is odd.

The s integral can be performed by i) using the periodicity of χ_z [eq. (4.5)], ii) making a second use of the approximation $\sigma_L/\beta_y^* < 1$ to write $\chi_z(s) = s/\beta_z^* - 2\pi Q_z s/C$, and iii) using an integral from Gradshteyn and Ryzhik[16]

$$A_{pr} = \frac{1}{2\pi} T_{pr}(I_x, I_y) \exp\left[\frac{-k_{pr}^2\sigma_L^2}{8}\right]$$

$$\times \sum_{n=-\infty}^{\infty} \exp\left[\frac{ik_{pr}\hat{\tau}c}{2} \cos(2\pi n Q_s) + iknC\right] . \qquad (4.16)$$

The wavenumber k_{pr} is

$$k_{pr} = k + p(1/\beta_x^* - 2\pi Q_{x0}/C) + r(1/\beta_y^* - 2\pi Q_{y0}/C) . \qquad (4.17)$$

Finally, using a Bessel function sum[17] and the Poisson sum rule[18]

$$A_{pr} = \frac{1}{C} T_{pr}(I_x, I_y) \exp\left[\frac{-k_{pr}^2\sigma_L^2}{8}\right]$$

$$\times \sum_{m,n=-\infty}^{\infty} i^m J_m(k_{pr}\hat{\tau}c/2)\, \delta[kC - 2\pi(n - mQ_s)] \qquad (4.18)$$

where J_m is a Bessel function of order m. Sidebands spaced by Q_s have developed around the betatron resonances. The decomposition in eq. (4.18) makes sense only when a few of the sidebands are important.

Substituting back into eq. (4.12)

$$H = H_0(I_x, I_y) - \frac{Nr_e}{C\gamma} \sum_{\substack{m,n,p,\\ r=-\infty}}^{\infty} T_{pr}(I_x, I_y) \exp\left[\frac{-k_{pr}^2 \sigma_L^2}{8}\right]$$

$$\times\, i^m J_m(k_{pr}\hat{\tau}c/2) \exp\{i[p\psi_x + r\psi_y - 2\pi(n - mQ_s)s/C]\}. \quad (4.19)$$

The four-fold sum is the beam-beam perturbation. This all seems terribly messy, but the next steps show how this formality pays off.

4.2.3 Calculate the Dependences of the Tunes on Action from the Average Value of the Perturbation

The average value of the perturbation is given by the term with p = r = m = n = 0. All the other terms are oscillatory. When its phase varies rapidly, the effect of a term on the motion averages to zero quickly. A phase varies slowly if the tunes have special values leading to resonances, or if one or more of the tunes is low. Usually the fractional parts of the betatron tunes are not close to zero, but the synchrotron tune can be low. It is in hadron colliders where the effects of synchrotron motion are averaged over hundreds or thousands of turns. This leads to the possibility of adiabatic behavior where i) resonance conditions change slowly enough that particles trapped in a resonance stay trapped as resonance conditions change,* and ii) the decomposition in eq. (4.18) isn't the most illuminating approach.

The synchrotron tune in electron colliders is large enough that the synchrotron motion is averaged in tens of turns. The important terms in the perturbation are its average value and a few resonances. The phase advance is $d\psi_z/ds = \partial H/\partial I_z$. The derivative of H_0 gives $2\pi/C$ times the nominal tune, and the derivative of the average value of the perturbation gives the average phase advance from the perturbation. It is

$$\left\langle \frac{d\psi_z}{ds} \right\rangle \left[\equiv \frac{2\pi}{C} Q_z \right] = \left\langle \frac{\partial H}{\partial I_z} \right\rangle = \frac{2\pi}{C} Q_{z0} - \frac{Nr_e}{C\gamma} \frac{\partial T_{00}}{\partial I_z} \quad . \quad (4.20)$$

For example, in the vertical

---

* The synchrotron amplitude and tune enter the criteria for adiabatic motion, but the focus is on Q_s because $\hat{\tau}/\beta^* \sim 1$ in modern electron and hadron colliders. The differences between the "quasilinear" and "adiabatic" regimes are discussed in ref. 19.

$$Q_y = Q_{y0} + \frac{2Nr_e\beta_y^*}{\gamma(2\pi)^3}\int_0^{2\pi} d\theta_x \int_0^{2\pi} d\theta_y \cos^2\theta_y \int_0^{\infty} \frac{dq}{\sqrt{(2\sigma_{x0}^2+q)(2\sigma_{y0}^2+q)^3}}$$

$$\times \exp\left\{-\left[\frac{2\beta_x^* I_x \cos^2\theta_x}{2\sigma_{x0}^2+q} + \frac{2\beta_y^* I_y \cos^2\theta_y}{2\sigma_{y0}^2+q}\right]\right\} . \tag{4.21}$$

In the limit $I_x, I_y \to 0$,

$$Q_y = Q_{y0} + \frac{Nr_e\beta_y^*}{2\pi\gamma}\int_0^{\infty} \frac{dq}{\sqrt{(2\sigma_{x0}^2+q)(2\sigma_{y0}^2+q)^3}} . \tag{4.22}$$

Evaluating the integral

$$Q_y = Q_{y0} + \xi_y . \tag{4.23}$$

Equation (4.21) and the analogous one for Q_x [eq. (4.21) with x and y interchanged] give the dependences of tunes on actions. Usually the integrals have to be done numerically. Figure 6 shows the results of such a calculation.

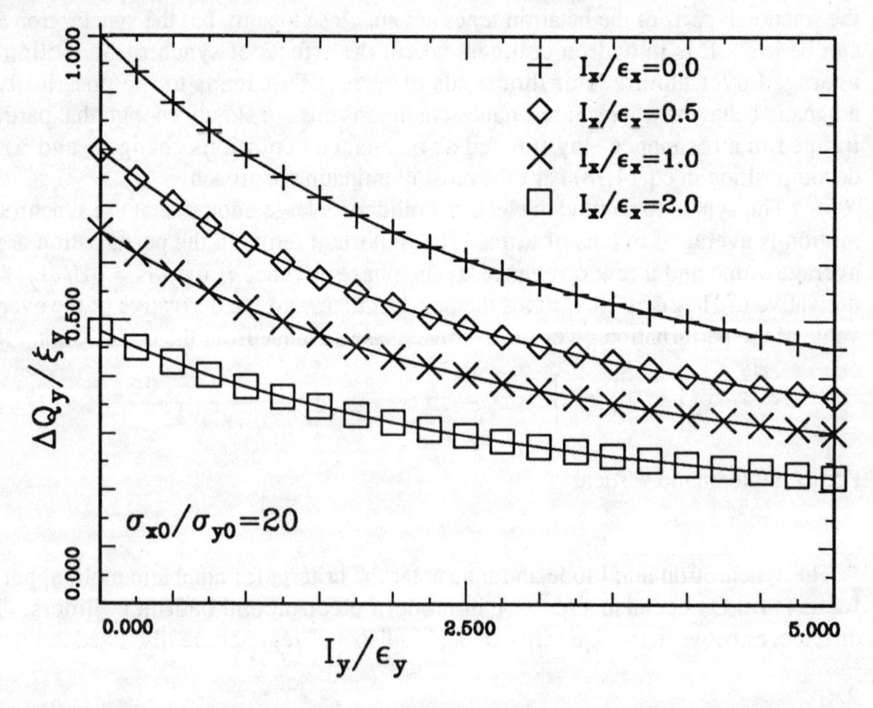

Figure 6: The vertical tune, $Q_y = Q_{y0} + \Delta Q_y$, for different values of the horizontal and vertical actions calculated from eq. (4.21).

4.2.4 Determine Resonance Conditions and Resonance Properties from the Slowly Varying Terms of H

A term in the Hamiltonian has a cumulative effect, resonant build-up, when its phase varies slowly. The resonance condition is

$$\frac{d}{ds}\left[p\psi_x + r\psi_y - 2\pi(n-mQ_s)\frac{s}{C}\right] \cong 0 , \qquad (4.24)$$

or

$$pQ_x + rQ_y + mQ_s = n . \qquad (4.25)$$

The tunes are the actual tunes, not the nominal tunes, and eqs. (4.21), the analogous one for Q_x, and (4.25) can be solved for the locus of resonant actions, $\{I_{xR}, I_{yR}\}$. This is illustrated in Figure 7. The resonance order equals $|p| + |r| + |m|$.

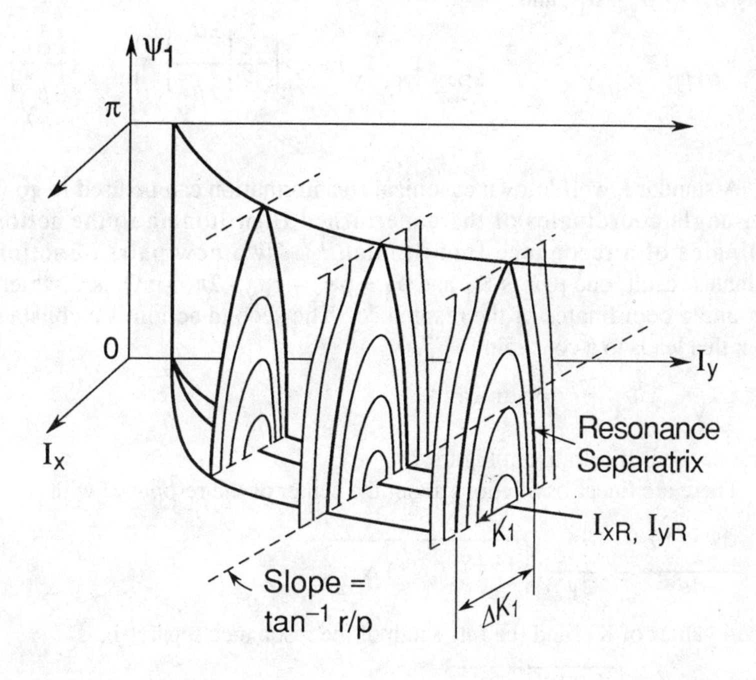

Figure 7: A nonlinear resonance illustrated in a three-dimensional phase space. The axes are the unperturbed actions I_x and I_y and the resonant angle $\psi_1 = p\psi_x + r\psi_y - 2\pi(n\text{-}mQ_s)s/C$. The resonance condition, eq. (4.24), is satisfied for $\{I_{xR}, I_{yR}\}$. The resonance action is K_1, and, from eq. (4.29), the resonance oscillation direction is $\tan^{-1}(r/p)$. This figure is adapted from ref. 20.

The Hamiltonian of a single resonance, assuming it is isolated from all the others, is

$$H_{prm} = H_0 - \frac{Nr_e}{C\gamma} T_{00} - \frac{2Nr_e}{C\gamma} T_{pr}(I_x, I_y) \exp\left\{-\frac{1}{2}\left[\frac{p\sigma_L}{2\beta_x^*} + \frac{r\sigma_L}{2\beta_y^*}\right]^2\right\}$$

$$\times J_m\left\{\left[\frac{p}{2\beta_x^*} + \frac{r}{2\beta_y^*}\right] c\hat{\tau}\right\} \cos[p\psi_x + r\psi_y - 2\pi(n - mQ_s)s/C] , \tag{4.26}$$

$$= H_0 - \frac{Nr_e}{C\gamma} T_{00} - \frac{2Nr_e}{C\gamma} F_{prm}(I_x, I_y, \hat{\tau}) \cos(p\psi_x + \ldots) . \tag{4.27}$$

Equation (4.25) has been used, the term with (-p, -r, -m, -n) has been combined with the term for (p, r, m, n), and the initial phase of the last term has been neglected. Usually $\beta_x^* >> \beta_y^* \sim \sigma_L$, and

$$F_{prm}(I_x, I_y, \hat{\tau}) \cong T_{pr}(I_x, I_y) \exp\left\{-\frac{1}{2}\left[\frac{r\sigma_L}{2\beta_y^*}\right]^2\right\} J_m\left\{\frac{rc\hat{\tau}}{2\beta_y^*}\right\} . \tag{4.28}$$

A standard, well-known canonical transformation can be used to go from the action-angle coordinates of the unperturbed Hamiltonian to the action-angle coordinates of a resonance Hamiltonian.[19] Two new pairs of action-angle coordinates result; one pair is K_1 and $\psi_1 = p\psi_x + r\psi_y - 2\pi(n\text{-}mQ_s)s/C$ which are the action-angle coordinates of the resonance. The second action is a constant of the motion; this leads to a constraint

$$pI_y - rI_x = \text{constant} . \tag{4.29}$$

The resonance is illustrated in Figure 7.

There are linear oscillations about the center of the resonance with

$$\frac{d\psi_1(\hat{\tau})}{ds} = \frac{Nr_e}{C\gamma} \sqrt{|2F^R_{prm}(\hat{\tau})\Lambda_{pr}|} \tag{4.30}$$

for small values of K_1, and the full width of the resonance separatrix is

$$\Delta K_1 = 4\sqrt{\left|\frac{2F^R_{prm}(\hat{\tau})}{\Lambda_{pr}}\right|} . \tag{4.31}$$

The quantity Λ_{pr} is proportional to the rate of change of tune with action

$$\frac{dQ_1}{dK_1} = \frac{C}{2\pi}\frac{d^2\psi_1}{dsdK_1} = \frac{Nr_e}{2\pi\gamma}\Lambda_{pr}$$

$$= \frac{Nr_e}{2\pi\gamma}\left[p^2\frac{\partial^2 T_{00}}{\partial I_x^2} + r^2\frac{\partial^2 T_{00}}{\partial I_y^2} + 2pr\frac{\partial^2 T_{00}}{\partial I_x \partial I_y}\right]\Bigg|_{I_{xR}, I_{yR}}, \qquad (4.32)$$

and

$$F^R_{prm}(\hat{\tau}) = F_{prm}(I_{xR}, I_{yR}, \hat{\tau}) \ . \qquad (4.33)$$

4.2.5 Discussion

The last section contains the results. The factors determining resonance properties are:

1. The strength of the perturbation, Nr_e/γ. The small-amplitude frequency depends on it, but the separatrix size is independent of it. Qualitatively, the resonance potential well stays the same size but gets deeper as the beam-beam strength increases.
2. The rate of change of the resonance tune, $pQ_x + rQ_y + mQ_s$, along the direction of oscillation in the $\{I_x, I_y\}$ plane is proportional to the "detuning", Λ_{pr}.
3. The remainder of the dependence on the resonant actions is given by $T_{pr}(I_{xR}, I_{yR})$. It must be calculated numerically; sample calculations are shown in Figure 8. When p or r is odd, $T_{pr} = 0$. Odd-order resonances can be introduced by an offset at the interaction point. When I_{xR}/ε_x and I_{yR}/ε_y are small, T_{pr} is small and the potential well is small and shallow. As p and r increase, T_{pr} decreases, reducing the importance of high-order resonances.
4. The form factor $\exp[-\frac{1}{2}(r\sigma_L/2\beta_y^*)^2]$ accounts for the nonlinear force acting over a range of vertical betatron phase. The resultant phase averaging increases with r, the vertical order of the resonance. The horizontal phase does not change over σ_L, and, therefore, p and β_x^* do not enter. It is likely that phase averaging is the mechanism contributing to good CESR performance with $\sigma_L/\beta_y^* \sim 1.1$.[21]
5. There are resonances with m = 0 involving betatron motion only and synchrobetatron resonances with $m \neq 0$ arising from the modulation of the collision point from synchrotron motion. $J_m(rc\hat{\tau}/2\beta_y^*)$ gives the dependence on synchrotron amplitude. The Bessel function $J_m(\zeta)$ has its first maximum at $\zeta \sim m$. The *mth* synchrobetatron resonance is important for $\hat{\tau} \geq 2\beta_y^* m/rc$, and particles with large synchrotron amplitudes have more synchrobetatron resonances.

"Difference" resonances have sign(p) = -sign(r), and, from eq. (4.29), $|p|I_y + |r|I_x$ is a constant. The energy associated with the transverse motion can be transferred between horizontal and vertical motions as long as the above sum remains constant. "Sum" resonances have sign(p) = sign(r) and the constant of motion is

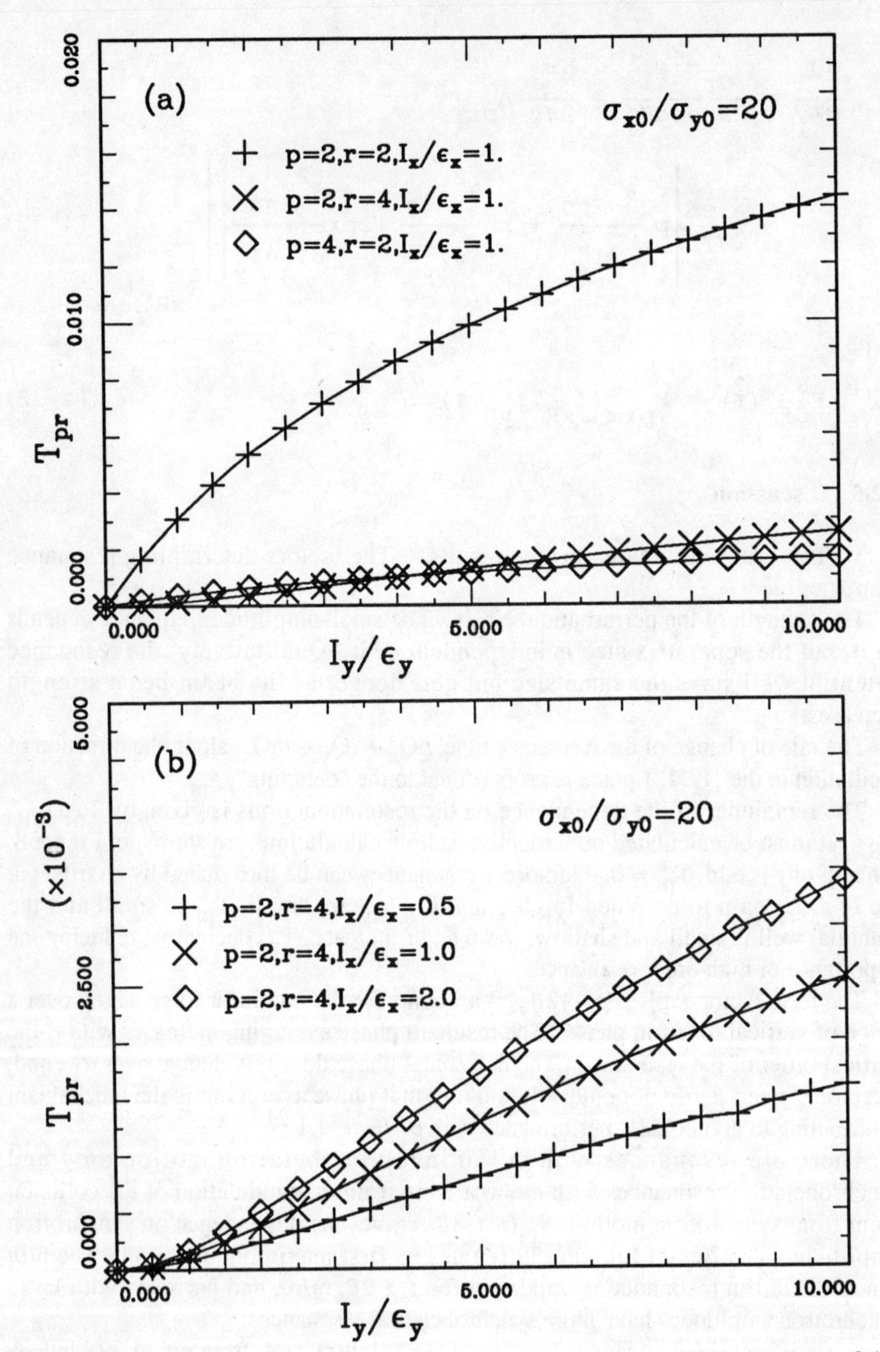

Figure 8: Plots of T_{pr} [eq. (4.15)] for different resonances and different values of the actions.

$|p|I_y - |r|I_x$. There is no restriction on the energy associated with transverse motion. The horizontal and vertical actions can grow without bound, and reduced lifetimes are probably associated with sum resonances. Radiation damping is outside the scope of this Hamiltonian analysis, but the naive expectation is that it limits the actions. In fact, it may enhance the ability of particles to reach large amplitudes through resonance streaming.[22]

The discussion so far has concentrated on single, isolated resonances. They explain the beam-beam performance in hadron colliders[23] and much of the beam core behavior in e^+e^- simulations,* but the interaction between resonances could be important, particularly for lifetime effects. Synchrobetatron sidebands are separated in tune by Q_s and in action by

$$\delta K_1 = \frac{2\pi\gamma Q_s}{Nr_e |\Lambda_{pr}|} . \tag{4.34}$$

Stochastic motion occurs when resonances overlap.[26] The Chirikov criterion is that there is resonance overlap and resulting chaotic motion when

$$\frac{\Delta K_1}{\delta K_1} > \frac{2}{\pi} \quad \text{or} \quad \frac{Nr_e}{\gamma} > \frac{Q_s}{\sqrt{2\left|F^R_{prm}(\hat{\tau})\Lambda_{pr}\right|}} . \tag{4.35}$$

The threshold Nr_e/γ for chaotic motion decreases with Q_s until the adiabatic regime with stable motion is entered. At this point the picture of separated synchrobetatron sidebands is not appropriate, and there is a transition from the quasilinear to the adiabatic regime.[19]

Particles with large synchrotron amplitudes have a number of sidebands, and the resonance overlap criterion suggests a connection between particles that are determining the lifetime and particles with large synchrotron amplitudes. Such connections have been seen in simulations,[27] but, as far as I know, there is no convincing connection between particles with large synchrotron amplitudes and particle losses. An alternative mechanism for reaching large amplitudes involves the interaction between nonlinear resonances and synchrotron radiation. It is the subject of the next section.

4.3 Resonance Streaming[22] and Phase Convection[20]

Nonlinear resonances can combine with the noise and damping from synchrotron radiation to produce non-Gaussian tails and, possibly, explain the reduced lifetime that is the second beam-beam limit. This possibility has motivated general studies of the interaction of nonlinear resonances, noise, and damping,[20,22]

---

* References 24 and 25 are reviews of beam-beam simulations with complete references.

and the results have been applied to collider lifetimes.[28] Only betatron motion has been considered, and this section has that restriction.

A particle subject to noise and deterministic forces from isolated nonlinear resonances and damping could reach large vertical amplitudes by a variety of routes. The "most probable" route is the one with the weakest net damping. One could imagine starting with a number of particles at the same value of I_y and different values of (I_x, ψ_x, ψ_y) and tracking them backwards in time in a system without noise. They all damp to the origin of phase space, but at different rates. The particle to damp slowest has backtracked along the most probable route to the starting value of I_y.[20] Resonances, through the mechanism of resonance streaming,[22] often determine the most probable route to large vertical amplitudes.

There are two extremes of the relative importance of a resonance versus synchrotron radiation. In one the time for damping and fluctuations to transport a particle across the resonance is short compared to the oscillation period. Resonant build-up cannot occur, and the resonance is not important. The dominant motion in the other extreme is oscillation about the resonance center $\{I_{xR}, I_{yR}\}$. Using eq. (4.25) the slope of the resonance center is

$$\frac{dI_{yR}}{dI_{xR}} = \frac{-\left[p\,\dfrac{\partial Q_x}{\partial I_x} + r\,\dfrac{\partial Q_y}{\partial I_x}\right]}{p\,\dfrac{\partial Q_x}{\partial I_y} + r\,\dfrac{\partial Q_y}{\partial I_y}} \,. \tag{4.36}$$

The derivatives are evaluated at $\{I_{xR}, I_{yR}\}$. Tunes decrease with increasing amplitude, so the partial derivatives are all negative. The slope is negative for sum resonances and can be positive or negative for difference resonances. The slopes can be large when the beams are flat and $I_y >> \sigma_{y0}^2$.

The different effects of damping from sum and difference resonances can be understood using Figure 9. Assume for the sake of illustration that the damping is only in the I_x direction. A decrease in I_x changes the centers of the resonance oscillations from A to B. In the case of the sum resonance,damping has shifted the oscillation center to a larger vertical action; the damping process has a component along $\{I_{xR}, I_{yR}\}$ that increases the vertical amplitude. This is resonance streaming. It occurs for sum resonances only, as contrasting the left- and right-hand sides of Figure 9 shows, and for flat beams, $\sigma_x/\sigma_y > 15$.[29]

The Fokker-Planck equation describes the evolution of the phase-space density in systems with damping and noise. When the potential is independent of time, there is a stationary solution similar to the Boltzmann distribution. When the potential is time (or s) dependent, as it is for the beam-beam interaction, there is no stationary solution. It is possible to calculate the distribution far from the core and independent of the initial distribution at large times when damping and noise are weaker than the resonances, however.[20] That density is affected strongly by resonances. It is enhanced by resonance streaming with resonances providing the most likely routes to large amplitude. Particles that fall into a sum resonance, stream to large amplitudes and then leave the resonance when noise and damping become dominant. Once they

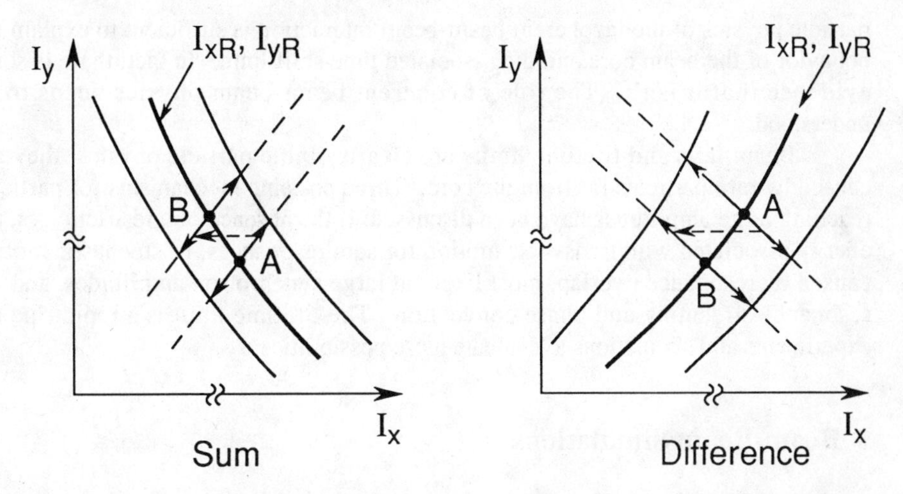

Figure 9: The intersections of resonance "tubes" such as shown in figure 7 with the I_x, I_y plane for sum and difference resonances. The resonance centers, the edges of the separatrices, and the directions of oscillation are shown.

leave the resonance they damp back to the origin of phase space. The result is a circulation of particles in phase space called "phase convection".[20]

The beam-beam interaction does not satisfy the approximations needed for calculating the density distribution far from the core (the beam tails), but a phenomenological model has been used to make estimates.[28] Phase convection and resonance streaming remain as central features of the results. In addition, these estimates show that nonlinearities of the magnet lattice are critical in determining the tails. Parametrizing that nonlinearity as if it were due to a single octupole, Gerasimov and Dikansky find that positive lattice nonlinearity, $dQ_{x0}/dI_x > 0$, enhances the tails strongly. This agrees with an experiment at VEPP-4.[3]

4.4 Concluding Remarks

This chapter started with a detailed calculation and ended with qualitative discussions about lifetime-limiting mechanisms. The calculation showed the relationship between β_y^* and two longitudinal parameters, σ_L and $\hat{\tau}$. Similar calculations following the same procedure can be used to show the effects of crossing angles, unequal beam energies, phase-advance errors between interaction points, etc.

It is difficult to go from the results of these calculations to quantitative statements about beam-beam limits. The value of the calculations is that they give insight into underlying physics, provide a framework for guiding and interpreting experiments and simulations, and identify possible methods for improving performance. Establishing direct, quantitative connections between calculations and performance is one of the themes of ongoing research into the beam-beam interaction. The situation is different for the two beam-beam limits. It isn't clear that the single-particle physics of the incoherent beam-beam interaction is sufficient to explain the behavior of the beam core and the associated tune-shift limit. In fact, there is some evidence that it isn't. The role of coherent beam-beam physics needs to be understood.

Beam tails and lifetime limits are clearly single-particle physics; they are caused by rare particles far from the core. Three possible mechanisms for particles reaching large amplitudes have been discussed: i) the absence of a restriction on the energy associated with transverse motion for sum resonances, ii) stochastic motion caused by resonance overlap, most likely at large synchrotron amplitudes, and iii) resonance streaming and phase convection. The lifetime limit is a topic ripe for experiments and simulations to evaluate these possibilities.

5 Beam-Beam Simulations

Simulations are an important part of beam-beam research. They are used for making performance estimates of existing or proposed colliders, and they are ideal for performing "experiments" that can be done with a degree of control and a variety of diagnostics that are impossible in real colliders.

Test particles representative of those in the beams are followed for a large number of turns with each turn consisting of transport between interaction points and collisions at the interaction points. Simulation of the arcs almost always includes betatron and synchrotron oscillations, radiation damping and quantum excitations. The techniques are standard.[30] Simulations have included chromaticity, lattice nonlinearities, lattice errors, collective effects, etc. depending on the physics under study. References 24 and 25 review much of this work.

The difference between simulations is the treatment of the collisions. This is also the area of recent progress. Simulations are weak-strong when test particles from only one beam, the weak beam, are tracked and the distribution of the opposing strong beam is unaffected by the test particles. This is single-particle, incoherent physics. The strong beam is usually Gaussian with the beam-beam impulse given by eq. (3.1). It is necessary to segment the strong beam longitudinally when $\sigma_L \sim \beta_y^*$.[14] This was discovered in a simulation experiment and later understood with a calculation similar to that in section 4.2.

Two beams are tracked in strong-strong simulations, and they modify each other's distributions. The test particles are representative of the beam, and their coordinates are used to determine the distribution and, from that, the electromagnetic fields at the interaction point and the beam-beam impulse. The most common and most straightforward procedure is to calculate the means and rms widths from test-particle coordinates, z_k (k = 1,..., K for each beam),

$$\bar{z} = \frac{1}{K} \sum_{k=1}^{K} z_k \, , \tag{5.1a}$$

$$\sigma_z^2 = \frac{1}{K} \sum_{k=1}^{K} (z_k - \bar{z})^2 , \tag{5.1b}$$

and use these in eq. (3.1) which gives the beam-beam impulse for a Gaussian distribution. This was considered a reasonable approximation because beam distributions remained roughly Gaussian in strong-strong simulations.

A number of effects that are important in operating colliders are seen in strong-strong simulations and are outside the scope of weak-strong simulations. First and foremost is blow-up of the vertical beam size leading to a tune-shift limit and luminosity proportional to the total beam current. Actual colliders have been modeled and the tune-shift limits from operation and simulation compared. The results are mixed. In general, agreement is found for well-established operating points, but the prediction of new, better operating points is poor. The well-established points have many hours of operator tuning invested in them. This tuning has gradually improved luminosity, presumably through the elimination of small errors that combine with the beam-beam interaction to determine performance. On the other hand, there is rarely enough accelerator studies time to tune extensively at exploratory operating points. It is impossible to know in any detail what errors are removed with tuning and include them in a model. At best one could select errors randomly and simulate an ensemble of colliders to determine the range of possible performance. My conclusion is that i) errors have been tuned out at well-established operating points, the tune-shift limit there is due to the beam-beam interaction alone, and, therefore, it can be explained by simulations; and ii) either there has been insufficient tuning at exploratory points or additional physics must be included in simulations. The assumption of the fields from a Gaussian distribution is one possibility that has been investigated recently.

A second effect seen in operation and in strong-strong simulations is the "flip-flop" effect where the two beams have substantially different vertical sizes. This is a hysteretic effect with small differences determining which beam is larger. It is difficult to reproduce actual performance in simulations because of the importance of small differences.

The third common effect is coherent centroid motion which is routinely observed in operations. There are two modes: the "0-mode" where the beam centroids are in phase at the collision point and the "π-mode" where they are 180° out of phase. These oscillations have limited amplitudes and are helpful as diagnostics for measuring ξ_x and ξ_y since the differences between the π- and 0-mode tunes are[31]

$$\frac{\Delta Q_x}{\xi_x} = \Lambda(r) = 1.330 - 0.370r + 0.279r^2$$

$$\frac{\Delta Q_y}{\xi_y} = (1 - r)\Lambda(r) \tag{5.2}$$

where $r = \sigma_y/(\sigma_x+\sigma_y)$. The coefficients result from coherent oscillations modifying the charge distribution, and they cannot be reproduced exactly in simulations that restrict the fields to be those of a Gaussian beam.

The increase in vertical beam size and the flip-flop are strong-strong, multiple-particle effects, but nonlinear motion of individual particles could account for changes in distributions without adding additional physics. However, coherent centroid oscillations cannot be explained within the framework of the incoherent beam-beam interaction.

The space-charge compensation experiments at DCI provide a second, strong piece of evidence that coherent beam-beam effects exist.[32] Those experiments indicate that coherent shape oscillations lead to a tune-shift limit. This is in sharp contrast to the harmless nature of coherent centroid oscillations. These experiments had four beams, an electron beam and a positron beam going in one direction colliding with an electron and a positron beam going in the opposite direction. The estimate was that the beam-beam potential was reduced by a factor of ten, and yet there was no striking improvement in performance. The tune-shift limit was set by ξ rather than its residual compensated value.

Is there any evidence of coherent shape oscillations in the more normal situation of two beams colliding? There is no experimental evidence. Seeing such oscillations requires imaging a beam on a single turn. Appropriate instruments have become available only recently,[33] and they have not been used in storage ring colliders. There is evidence for coherent oscillations in strong-strong simulations where the beam-beam impulse is calculated for a general distribution rather than using the expression for a Gaussian beam.[34]

Using the means and rms widths, eq. (5.1), together with eq. (3.1) for the impulse may not be a reasonable approximation. Strong-strong simulations are a relaxation calculation; the beam distributions and fields must be consistent with each other. Restricting the fields to those of a Gaussian beam indirectly restricts the beams to remain Gaussian. Relaxing that restriction has a good and a bad effect: it introduces new physics, but it makes the simulation sensitive to noise. Combining eqs. (5.1) and (3.1) is relatively noise free because only a few properties of the beam are extracted from the test-particle coordinates. Statistical techniques are needed to distinguish noise from real effects when a general expression for the beam-beam impulse is wanted. An adaptive, least-squares fitting procedure has been developed for nearly round beams, $\sigma_x = \sigma_y$.[35] Coherent shape oscillations are found using this procedure. Figure 10 is an example showing a coherent beam-beam resonance at $Q_{x0} = Q_{y0} = 5/6$. The beam shapes vary turn-by-turn with the extreme being one beam with a dense core and the other with a hollow core (Figure 11). It is impossible to represent such beam shapes with a Gaussian, and it isn't surprising that such turn-by-turn variations in sizes and shapes are not seen when fields from a Gaussian beam are used.

Whether coherent effects are important for flat beams awaits experiments and development of a beam-beam algorithm for flat beams.

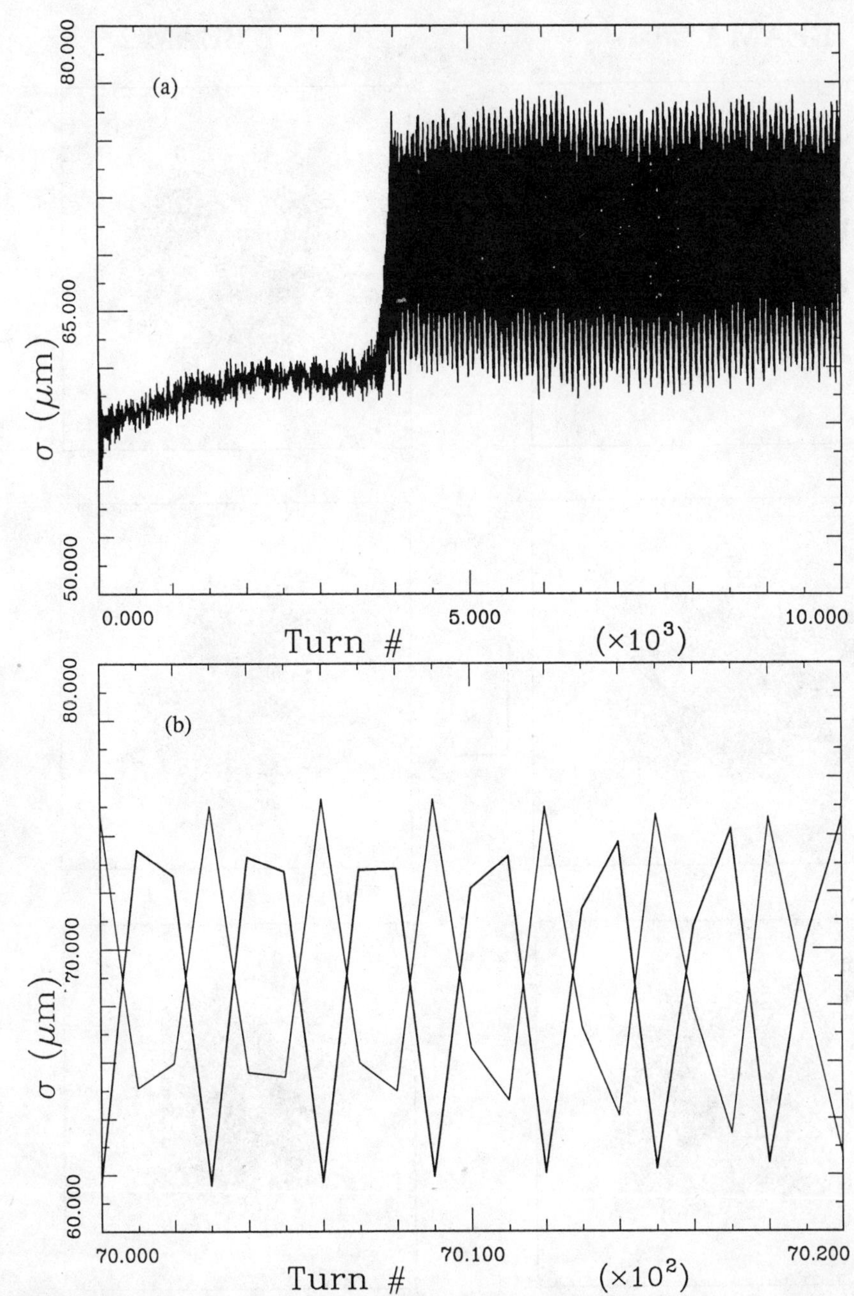

Figure 10: The rms beam sizes from a strong-strong simulation using a general expression for the beam-beam impulse. The parameters of the simulation are $Q_{x0} = Q_{y0} = 0.79$, $\sigma_{x0} = \sigma_{y0} = 55\ \mu m$, $\xi = 0.10$, and fractional energy loss per turn = 1×10^{-3}. a) Shows the onset of the instability for one of the beams. b) Shows the size variations of the two beams. They are anticorrelated and repeat every three turns. (From ref. 34).

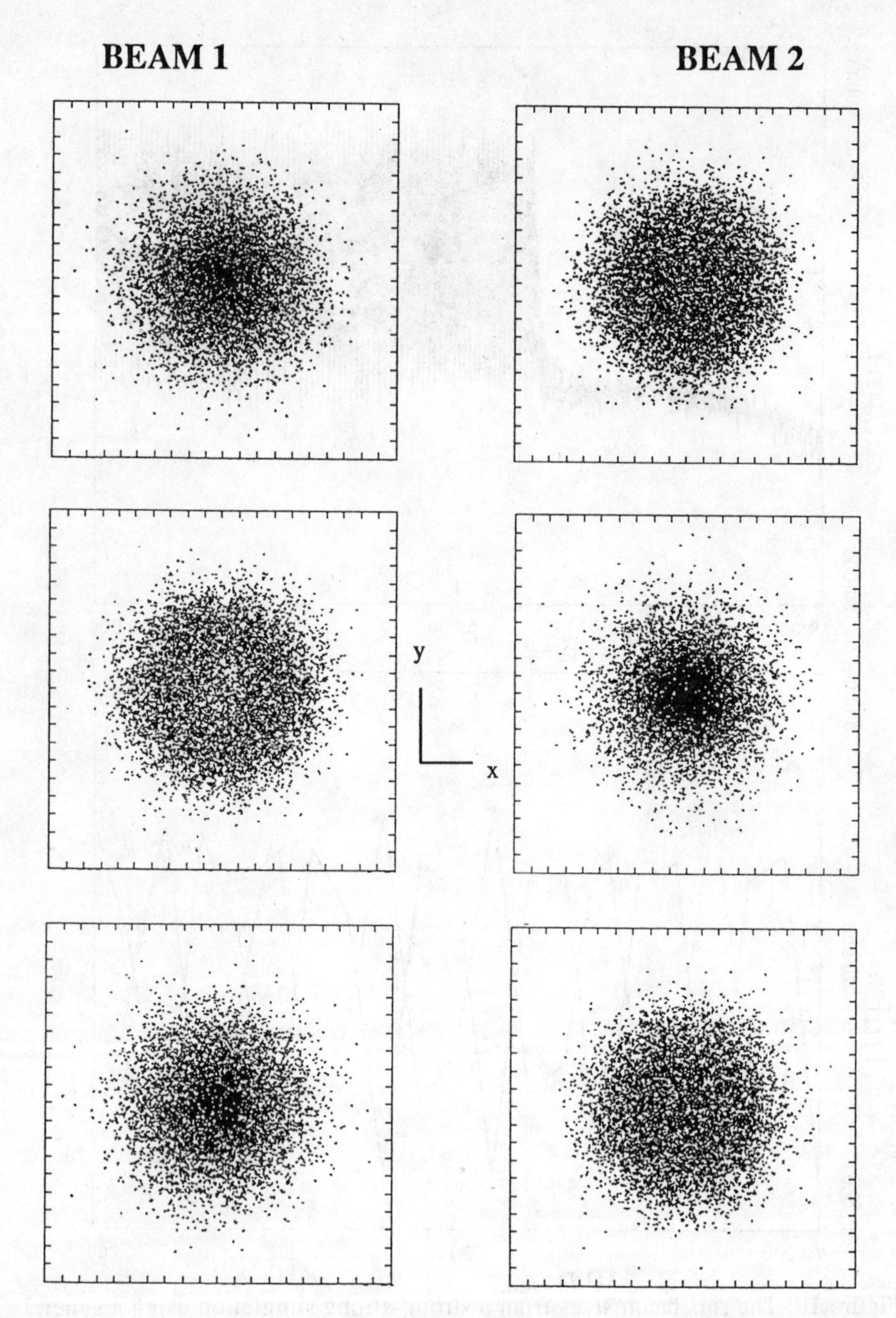

Figure 11: Scatter plots showing the beams on three successive turns for $Q_{x0} = Q_{y0} = 0.80$.

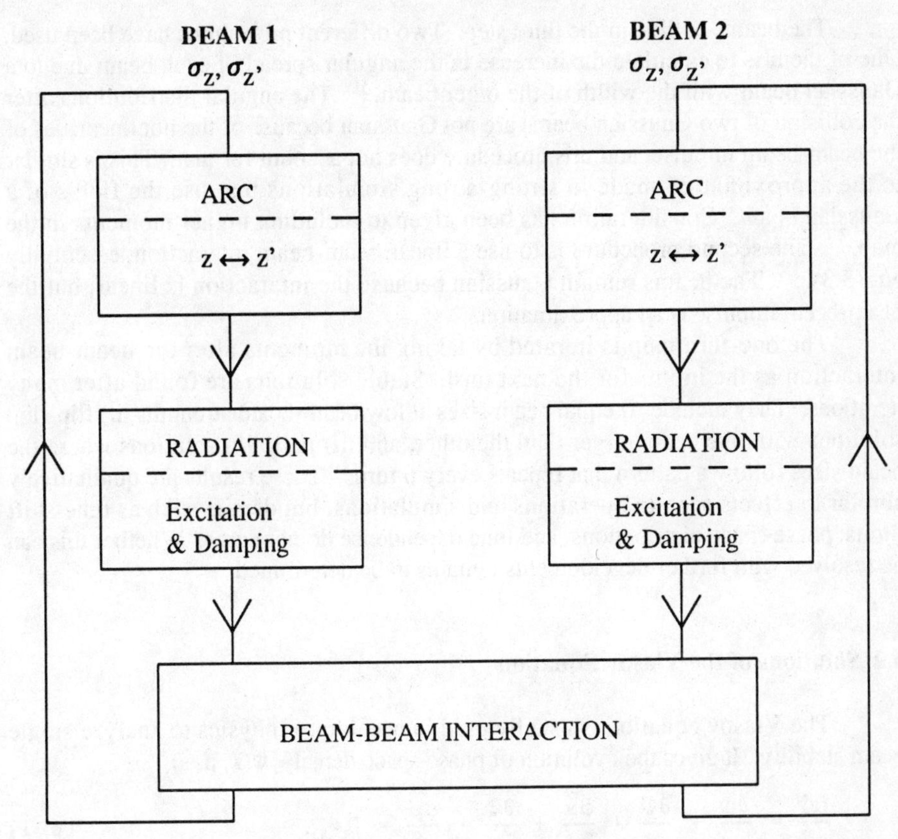

Figure 12: Schematic of an iterated map calculation of coherent beam-beam interactions.

6 The Coherent Beam-Beam Interaction

6.1 Iterated Maps of Moments

There are two different classes of coherent beam-beam interaction theories. One is based on iteration of a one-turn map for the beam moments and the other on solutions of the Vlasov equation.

The iterated map is shown schematically in Figure 12. The rms widths and angular spreads of the beams, the second moments of the distributions, are mapped for a single turn. The first elements of the map are the arcs transporting the beams between collision points. Tune-dependent linear combinations of the moments at the beginning of the arc give values at the end. The second step accounts for radiation. The moments are reduced by a fractional amount for damping, and random terms are added to model quantum excitations.

The beams collide in the third step. Two different procedures have been used. One of them is to calculate the increase in the angular spread of each beam due to a Gaussian beam with the width of the other beam.[36] The angular distributions after the collision of two Gaussian beams are not Gaussian because of the nonlinearities of the beam-beam impulse, and this procedure does not account for that. This is similar to the approximation made in strong-strong simulations that use the fields of a Gaussian beam. Consideration has been given to including higher moments in the map.[37] The second procedure is to use a linear beam-beam interaction, essentially eq. (3.3).[38] The beams remain Gaussian because the interaction is linear, but the beam-beam impulse is an approximation.

The one-turn map is iterated by taking the moments after the beam-beam interaction as the inputs for the next turn. Stable solutions are found after many iterations. They include i) equal beam sizes at low beam-beam strengths, ii) flip-flop solutions with one beam larger than the other, and iii) period-n solutions where the beam sizes follow a pattern that repeats every n turns. These results are qualitatively similar to effects seen in operations and simulations, but details such as tune-shift limits, phase-space distributions, and tune dependence do not agree. Whether this can be resolved with further developments remains to be determined.

6.2 Solutions of the Vlasov Equation

The Vlasov equation is usually used in accelerator physics to analyze single-beam stability. It gives the evolution of phase-space density, $\Phi(\vec{x}, \vec{p}, s)$,

$$\frac{d\Phi}{ds} = \frac{\partial\Phi}{\partial s} + \frac{\partial\Phi}{\partial\vec{x}} \cdot \frac{d\vec{x}}{ds} + \frac{\partial\Phi}{\partial\vec{p}} \cdot \frac{d\vec{p}}{ds} = 0 \quad . \tag{6.1}$$

The products are vector dot products. The forces acting on a particle come from external sources and from other particles. As a consequence the derivatives $d\vec{x}/ds$ and $d\vec{p}/ds$ can depend on Φ and the Vlasov equation is quadratic in Φ. It can be linearized by adding a small perturbation ϕ to the equilibrium distribution, Φ_0,

$$\Phi(\vec{x},\vec{p},s) = \Phi_0(\vec{x},\vec{p},s) + \phi(\vec{x},\vec{p},s) \quad . \tag{6.2}$$

When $\phi << \Phi_0$, terms of order ϕ^2 can be ignored leaving an equation linear in ϕ. Instabilities are unstable perturbations that are sought by analyzing the resultant equation for growing solutions. The characteristics of unstable perturbations are determined, but it isn't possible to tell whether these instabilities grow forever or are limited by some nonlinearity. That is outside the approximation used to linearize the Vlasov equation.

The Vlasov equation has been linearized and solved by two pairs of authors, Dikansky and Pestrikov[39] and Chao and Ruth,[40] for the one-dimensional beam-beam interaction. There are two examples of the one-dimensional beam-beam interaction: i) very flat beams, $\sigma_x >> \sigma_y$, where the vertical beam-beam force depends on y only, and ii) round beams, $\sigma_x = \sigma_y$, where only the radial coordinate matters. Dikansky and Pestrikov have analyzed both cases and found similar results. Chao and Ruth did their calculation for flat beams. I am more familiar with their work, and, for that reason only, the next section follows it.

6.3 Coherent Instabilities in the One-Dimensional Beam-Beam Interaction[40]

The Vlasov equation when only the vertical coordinate is considered is

$$\frac{d\Phi_k}{ds} = \frac{\partial\Phi_k}{\partial s} + y'\frac{\partial\Phi_k}{\partial y} + y''\frac{\partial\Phi_k}{\partial y'} = 0 \; . \tag{6.3}$$

The index k = 1,2 indicates that there is an equation for each beam. The distributions Φ_k are normalized to unity

$$\int_{-\infty}^{\infty}\int_{-\infty}^{\infty} \Phi_k(y,y')\,dy\,dy' = 1 \; , \tag{6.4}$$

and their projections onto the y-axis are

$$\rho_k(y) = \int_{-\infty}^{\infty} \Phi_k(y,y')\,dy' \; . \tag{6.5}$$

The second derivative, y'', depends on the storage ring lattice and the beam-beam interaction which, in turn, depends on the distribution of the other beam. The beam-beam impulse for a particle in beam 1 comes from Gauss' law. It is

$$\begin{aligned}\Delta y' &= \frac{-4\pi N r_e}{\gamma L_x}\left[\int_{-\infty}^{y} \rho_2(\zeta)\,d\zeta - \int_{y}^{\infty} \rho_2(\zeta)\,d\zeta\right] \\ &= \frac{-4\pi N r_e}{\gamma L_x}\int_{-\infty}^{\infty} \rho_2(\zeta)\,\Theta(y-\zeta)\,d\zeta\end{aligned} \tag{6.6}$$

where

$$\Theta(\zeta) = \begin{cases} 1 & \zeta > 0 \\ -1 & \zeta < 0 \end{cases} \; , \tag{6.7}$$

and L_x is the horizontal width of the beam. A Gaussian beam has $L_x = (8\pi)^{1/2}\sigma_x$ near x = 0.

The equilibrium distribution is the same for the two beams, $\Phi_1 = \Phi_2 = \Phi_0$, and it satisfies

$$\frac{\partial\Phi_0}{\partial s} + y'\frac{\partial\Phi_0}{\partial y} - F(s,y)\frac{\partial\Phi_0}{\partial y'} = 0 \; . \tag{6.8}$$

with

$$F(s,y) = K(s)y$$

$$+ \frac{4\pi N r_e}{\gamma L_x} \sum_{n=-\infty}^{\infty} \delta(s-nC) \int_{-\infty}^{\infty}\int_{-\infty}^{\infty} \Phi_0(\zeta,y')\Theta(y-\zeta)\,dy'd\zeta \quad . \tag{6.9}$$

The first term in F gives the focusing of the magnet lattice, and the second term comes from the beam-beam interaction. This equation is quadratic in Φ_0, and solving it is difficult. However, a solution is not required, and that simplifies the situation. There is more on this below.

Following eq. (6.2), eq. (6.3) is linearized by substituting $\Phi_k = \Phi_0 + \phi_k$. The approximate equation for ϕ_1 is

$$\frac{\partial \phi_1}{\partial s} + y'\frac{\partial \phi_1}{\partial y} - F(s,y)\frac{\partial \phi_1}{\partial y'}$$

$$- \frac{\partial \Phi_0}{\partial y'} \frac{4\pi N r_e}{\gamma L_x} \sum_{n=-\infty}^{\infty} \delta(s-nC) \int_{-\infty}^{\infty}\int_{-\infty}^{\infty} \phi_2(\zeta,y')\Theta(y-\zeta)\,dy'd\zeta \;\tilde{=}\; 0 . \tag{6.10}$$

Terms involving only the equilibrium distribution make no contribution because of eq. (6.8), and one term proportional to

$$\frac{\partial \phi_1}{\partial y'} \int_{-\infty}^{\infty}\int_{-\infty}^{\infty} \phi_2(\zeta,y')\Theta(y-\zeta)\,dy'd\zeta \tag{6.11}$$

has been neglected. There is a similar equation for ϕ_2 that is coupled to eq. (6.10). They can be uncoupled by introducing

$$\phi_{\pm} = \phi_1 \pm \phi_2 \tag{6.12}$$

which satisfy two independent equations

$$\frac{\partial \phi_{\pm}}{\partial s} + y'\frac{\partial \phi_{\pm}}{\partial y} - F(s,y)\frac{\partial \phi_{\pm}}{\partial y'}$$

$$\mp \frac{\partial \Phi_0}{\partial y'} \frac{4\pi N r_e}{\gamma L_x} \sum_{n=-\infty}^{\infty} \delta(s-nC) \int_{-\infty}^{\infty}\int_{-\infty}^{\infty} \phi_{\pm}(\zeta,y')\Theta(y-\zeta)\,dy'd\zeta \;\tilde{=}\; 0 . \tag{6.13}$$

The equilibrium distribution enters in three ways. An approximation is used to simplify the calculation making it possible to characterize instabilities but making it inconsistent and of limited quantitative value—three different Φ_0's are used, one for each way it enters. First, Φ_0 must satisfy eq. (6.8) to eliminate the terms involving only Φ_0 from the linearized Vlasov equation. Finding that Φ_0 is analogous to determining the longitudinal distribution of a single beam taking account of potential well distortion. It isn't necessary to know this Φ_0 for characterizing instabilities, but it does affect thresholds.[41]

Second, Φ_0 gives the contribution of the beam-beam interaction to the focusing through F(s,y), eq. (6.9). Assuming a stable equilibrium distribution exists and is such that it produces a linear focusing force, F(s,y) becomes

$$F(s,y) = F(s)y \ . \tag{6.14}$$

Since F is linear in the displacement just like a quadrupole lattice, a β-function can be found and action-angle coordinates, I and ψ, exist. The β-function accounts for the focusing of Φ_0 as well as the magnet lattice. The action is a constant of the motion, dI/ds = 0, and Φ_0 is a function of I only, $\Phi_0(I,\psi) = \Phi_0(I)$. Equation (6.13) can be rewritten in terms of the action-angle coordinates using these facts plus i) β is a minimum at the interaction point so that $\beta' = d\beta/ds = 0$ there, ii) $d\psi/ds = 1/\beta$, iii) $y'' = - F(s)y$, and iv) the chain rule. It becomes

$$\frac{\partial \phi_\pm}{\partial s} + \frac{1}{\beta}\frac{\partial \phi_\pm}{\partial \psi} \pm \frac{d\Phi_0}{dI}\sqrt{2I\beta}\,\sin\psi$$

$$\times \frac{4\pi N r_e}{\gamma L_x}\sum_{n=-\infty}^{\infty}\delta(s-nC)\int_{-\infty}^{\infty}\int_{-\infty}^{\infty}\phi_\mp(\zeta,y')\Theta(y-\zeta)\,dy'\,d\zeta = 0 . \tag{6.15}$$

There is a rough analogy in transverse single-beam stability calculations to the simplification of eq. (6.14); it is the assumption that the deflecting fields leading to the transverse impedance are linear in displacement.

The third appearance of Φ_0 is as a weighting factor $d\Phi_0/dI$ for the beam-beam contributions of $\phi_\pm$. The "water-bag" model with constant phase-space density out to a boundary is used:

$$\Phi_0(I) = \frac{1}{\pi\varepsilon}H(\varepsilon/2 - I) \tag{6.16}$$

where ε is the vertical emittance,

$$H(\zeta) = \begin{cases} 1 & \zeta < 1 \\ 0 & \zeta > 1 \end{cases} , \tag{6.17}$$

and the normalization is

$$2\pi\int_0^{\infty}\Phi_0(I)\,dI = 1 \ . \tag{6.18}$$

With this distribution the last term in eq. (6.15) is a delta-function, and the perturbation is localized to $I = \varepsilon/2$.

Fourier analyzing $\phi_\pm$

$$\phi_\pm = \delta(I-\varepsilon/2)\sum_{m=-\infty}^{\infty} g_m^{\pm}(s)\,e^{im\psi} \ . \tag{6.19}$$

Substituting into eq. (6.15)

$$\frac{dg_m^{\pm}}{ds} + \frac{im}{\beta} g_m^{\pm} \pm \frac{2Nr_e}{\pi \gamma L_x} \sqrt{\frac{\beta}{\varepsilon}} \sum_{n=-\infty}^{\infty} \delta(s-nC) \sum_{k=-\infty}^{\infty} M_{mk} g_k^{\pm} = 0 . \tag{6.20}$$

The matrix element M_{mk} of matrix **M** is

$$M_{mk} = \int_0^{2\pi} d\psi \sin\psi e^{-im\psi} \int_0^{2\pi} d\psi' e^{ik\psi'} \Theta(\cos\psi - \cos\psi')$$

$$= \begin{cases} \dfrac{-32im}{\left[(m+k)^2-1\right]\left[(m-k)^2-1\right]} & m+k \text{ even} \\ 0 & m+k \text{ odd} \end{cases} . \tag{6.21}$$

Equation (6.20) gives the evolution of the perturbation. Follow it for one turn assuming only one interaction point. The Fourier components are independent between collisions, and g_m, the perturbation with phase-space periodicity m, advances at -m times the betatron phase advance:

$$g_m(s) = g_m(0) \exp\left[-im \int_0^s \frac{ds}{\beta}\right] . \tag{6.22}$$

The total betatron phase advance for a complete turn is $2\pi Q$. The betatron tune, Q, includes the focusing from Φ_0. Let $g_m(0)$ denote the value just after the interaction point and $g_m(C)$ the value one turn later taking into account all but the last term in eq. (6.20):

$$g_m(C) = g_m(0) \exp(-i2\pi mQ) = R_{mm} g_m(0) . \tag{6.23}$$

This defines the elements of a diagonal matrix **R**. The perturbations are coupled to each other at the interaction point. The change during a collision is

$$\Delta g_m^{\pm} = \pm \frac{2Nr_e}{\pi \gamma L_x} \sqrt{\frac{\beta}{\varepsilon}} \sum_{k=-\infty}^{\infty} M_{mk} g_k^{\pm}(C) . \tag{6.24}$$

The one-turn transformation of the vector $\vec{g} = (\ldots g_2, g_1, g_0, g_{-1}, g_{-2}, \ldots)^T$ is

$$\vec{g}^{\pm} = \left[I \pm \frac{2Nr_e}{\pi \gamma L_x} \sqrt{\frac{\beta}{\varepsilon}} \, \mathbf{M} \right] \mathbf{R} \vec{g}^{\pm} = \mathbf{T} \vec{g}^{\pm} . \tag{6.25}$$

There is a coherent beam-beam instability when one of the eigenvalues of **T** has an absolute value greater than unity; this occurs when $|Tr(\mathbf{T})| > 2$. Different Fourier components are unstable at different tunes. Consider only g_m and g_{-m} to learn the instability condition. For these two components

$$\mathbf{T} = \begin{bmatrix} 1 \pm i\alpha_m & i\alpha_m \\ -i\alpha_m & 1 \mp i\alpha_m \end{bmatrix} \begin{bmatrix} \exp(-im2\pi Q) & 0 \\ 0 & \exp(im2\pi Q) \end{bmatrix}$$

$$= \begin{bmatrix} (1 \pm i\alpha_m)\exp(-im2\pi Q) & i\alpha_m \exp(im2\pi Q) \\ -i\alpha_m \exp(-im2\pi Q) & (1 \mp i\alpha_m)\exp(im2\pi Q) \end{bmatrix} \tag{6.26}$$

where

$$\alpha_m = \frac{32m}{4m^2-1}\left[\frac{2Nr_e}{\pi\gamma L_x}\sqrt{\frac{\beta}{\varepsilon}}\right] . \tag{6.27}$$

The motion is stable if

$$\left|\frac{1}{2}\mathrm{Tr}(\mathbf{T})\right| = \left|\cos(2\pi Qm) \pm \alpha_m \sin(2\pi Qm)\right| < 1 . \tag{6.28}$$

Instabilities occur near $Q \simeq n/2m$. In terms of $\delta Q = Q - n/2m$ eq. (6.28) is

$$\left|\cos(2\pi m\delta Q) \pm \alpha_m \sin(2\pi m\delta Q)\right| < 1 . \tag{6.29}$$

Chao and Ruth perform more detailed calculations including several values of $|m|$, multiple interaction regions, and multiple bunches, but the important results have been obtained above. These are:

1. There are coherent beam-beam resonances for $Q = n/2m$ corresponding to perturbations with phase-space periodicity m. The resonances are even order only.
2. At resonance the betatron tune of the lattice is less than n/2m because of the focusing from the equilibrium distribution, Φ_0. Treating the beam-beam interaction as a thin quadrupole producing a tune shift ξ, the lattice and betatron tunes are related by

$$\cos(2\pi Q) = \cos(2\pi Q_0) - 2\pi\xi\sin(2\pi Q_0) \tag{6.30}$$

which for the water-bag model implies

$$\xi = \frac{4Nr_e}{\pi L_x \gamma\sqrt{\varepsilon/\beta}} . \tag{6.31}$$

3. From eq. (6.29) ϕ_- is unstable for $\delta Q < 0$. Since $\phi_- = \phi_1 - \phi_2$, the beams are anticorrelated when ϕ_- is unstable. The beams are correlated when ϕ_+ is unstable which occurs when $\delta Q > 0$.
4. The full width of the resonance is

$$\Delta Q = \frac{32\xi}{\pi(4m^2-1)} . \tag{6.32}$$

The resonances become narrower as m increases (Figure 13).

A number of approximations have been made, and Landau and radiation damping have not been included. They should determine the important resonances.

Simulations are the appropriate way to judge the validity of this calculation and to make quantitative predictions.

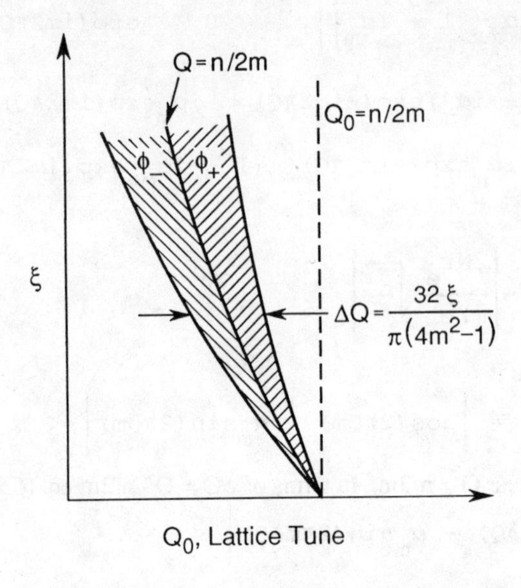

Figure 13: The results of the coherent beam-beam calculation of Chao and Ruth.[40]

6.4 Coherent Beam-Beam Simulations, Revisited

The coherent resonance shown in Figures 10 and 11 agrees with all but one of the results of the Vlasov equation solution. This particular resonance is Q = 5/6, and the beams are anticorrelated as expected if ϕ_- was unstable. The resonance has a stopband with a width about a factor of two smaller than eq. (6.32) (see Figure 14). The only substantive disagreement is that an instability with the beams correlated has never been seen. It could be that ϕ_+ is unstable initially, but the limiting behavior has different characteristics.

Other resonances have been searched for. Resonances with Q = n/2 and Q = n/4 are seen even when the fields of a Gaussian beam are used, and there is a coherent resonance at Q = 7/8. There are no odd-order resonances. The behavior at Q = 4/6 is close to that at Q = 5/6 suggesting that both are sixth-order resonances.

The simulation shows several features that are outside the Vlasov equation calculation. Nonlinearities limit the instability, and there are damping effects. The stopband does not extend to $\xi = 0$, but the width shrinks to zero at $\xi \sim 0.06$, presumably as a result of Landau damping. The instabilities are sensitive to radiation damping. The sixth-order resonance is present when the fractional energy loss per turn is as large as 10^{-3}, but it must be reduced to 10^{-5} to see the eighth-order resonance. The former is large compared to the energy loss in heavy quark factories while the latter is comparable to the energy loss in DCI where coherent phenomena are thought to have limited the improvement from space-charge compensation.

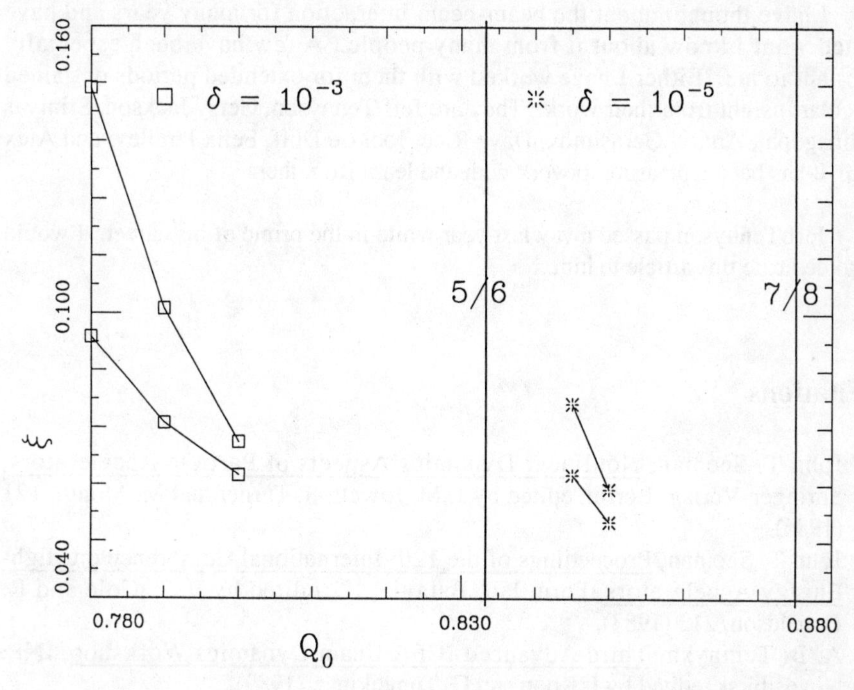

Figure 14: Onset and offset values of ξ as a function of Q_0 for sixth-order (squares) and eighth-order (asterisks) coherent resonances. In each case the region of coherent motion is between the lines. δ is the fractional energy loss per turn. (From ref. 34).

6.5 Concluding Remarks

The coherent beam-beam interaction has many of the qualitative features of the tune-shift limit. Although the simulation suggests that the tune-shift limit associated with it is higher than that achieved in operating colliders, the simulation was done for the special case of a nearly round beam with essentially a one-dimensional beam-beam interaction. A field calculating algorithm for flat beams has proven more difficult to develop, but work is continuing there. Perhaps the tune-shift limit will be lower for flat beams and a two-dimensional beam-beam interaction.

The most direct evidence about the role of the coherent beam-beam interaction should come from colliders operating at the beam-beam limit, however. The signature is clear: beam distributions changing turn-by-turn with the changes of the two beams correlated. Such measurements are possible and should be performed.

7 Acknowledgements and Dedication

I have thought about the beam-beam interaction for many years and have learned what I know about it from many people. A few have been especially important to me. Either I have worked with them for extended periods or gained particular insight from their work. They are Jeff Tennyson, Gerry Jackson, Srinivas Krishnagopal, Andrei Gerasimov, Dave Rice, Joel Le Duff, Felix Izrailev, and Alex Chao. It has been a pleasure to work with and learn from them.

Jeff Tennyson passed away last year while in the prime of his career. I would like to dedicate this article to him.

8 Citations

1. John T. Seeman, Nonlinear Dynamics Aspects of Particle Accelerators, Springer-Verlag, Berlin, edited by J. M. Jowett, S. Turner and M. Month, 121 (1986).
2. John T. Seeman, Proceedings of the 12th International Conference on High-Energy Accelerators, Fermilab, Batavia, IL, edited by F. T. Cole and R. Donaldson, 212 (1983).
3. A. B. Temnykh, Third Advanced ICFA Beam Dynamics Workshop, INP, Novosibirsk, edited by I. Koop and G. Tumaikin, 5 (1989).
4. A. Piwinski, Third Advanced ICFA Beam Dynamics Workshop, INP, Novosibirsk, edited by I. Koop and G. Tumaikin, 12 (1989).
5. D. H. Rice, Third Advanced ICFA Beam Dynamics Workshop, INP, Novosibirsk, edited by I. Koop and G. Tumaikin, 17 (1989).
6. P. M. Ivanov *et al*, Third Advanced ICFA Beam Dynamics Workshop, INP, Novosibirsk, edited by I. Koop and G. Tumaikin, 26 (1989).
7. Matthew Sands, The Physics of Electron Storage Rings—An Introduction, SLAC-121 (Nov, 1970).
8. G. Decker and R. Talman, IEEE Trans Nucl Sci **NS-30**, 2188 (1983).
9. M. Bassetti and G. Erskine, CERN-ISR-TH/80-06 (1980).
10. W. Gautschi, Handbook of Mathematical Functions, Nat. Bureau of Standards, Washington, Ninth printing, edited by M. Abramowitz and I. A. Stegun, 295(1970).
11. G. P. Jackson, PhD Thesis, Cornell Univ (1988).
12. Ronald D. Ruth, AIP Conf Proc **153**, 150 (1987).
13. F. M. Izrailev and I. B. Vasserman, Proc of the 7th All Union Conference on Charged Particle Accelerators, Dubna, USSR, 288 (1980).
14. S. Krishnagopal and R. Siemann, Phys Rev **D41**, 2312 (1990).
15. S. Kheifets, PETRA Note 119, DESY (Jan. 1976).
16. I. S. Gradshteyn and I. M. Ryzhik, Table of Integrals, Series and Products (Academic Press, New York, Fourth Edition, 1965) eq. 3.323.2.

17. I. S. Gradshteyn and I. M. Ryzhik, Table of Integrals, Series and Products (Academic Press, New York, Fourth Edition, 1965) eq. 8.511.4.
18. P. M. Morse and H. Feshbach, Methods of Theoretical Physics (McGraw-Hill Book Co, New York, 1953) p. 483.
19. A. L. Gerasimov, F. M. Izrailev and J. L. Tennyson, AIP Conf Proc **153**, 474 (1987).
20. A. L. Gerasimov, Physics Letters A **135**, 92 (1989)
A. L. Gerasimov, Physica **41D**, 89 (1990).
21. David H. Rice, Part Accel **31**, 107 (1990).
22. Jeffrey Tennyson, Physica **5D**, 123 (1982).
23. L. R. Evans, AIP Conf Proc **127**, 243 (1985).
24. S. Myers, Nonlinear Dynamics Aspects of Particle Accelerators (Berlin: Springer-Verlag, 1986, edited by J. M. Jowett, S. Turner & M. Month), p. 176.
25. R. H. Siemann, Third Advanced ICFA Beam Dynamics Workshop, INP, Novosibirsk, edited by I. Koop and G. Tumaikin, 110 (1989).
26. B. V. Chirikov, Physics Reports **52**, 263 (1979).
27. A. B. Temnykh, private communication.
28. A. L. Gerasimov and N. S. Dikansky, Nucl. Instr. Meth. **A292**, 209 (1990).
A. L. Gerasimov and N. S. Dikansky, Nucl. Instr. Meth. **A292**, 221 (1990).
A. L. Gerasimov and N. S. Dikansky, Nucl. Instr. Meth. **A292**, 233 (1990).
29. J. L. Tennyson, private communication.
30. R. H. Siemann, AIP Conf Proc **127**, 368 (1985).
31. K. Yokoya *et al*, KEK Preprint 89-14 (1989).
32. J. Le Duff *et al*, Proc of 11th Inter Conf on High-Energy Accel, 707 (1980).
J. Le Duff and M. P. Level, "Experiences Faisceau-Faisceau sur DCI", LAL/RT/80-03 (Orsay, 1980).
33. M. Minty *et al*, SLAC-PUB-5993 (1992).
34. S. Krishnagopal and R. Siemann, Phys Rev Lett **67**, 2461 (1991).
35. S. Krishnagopal and R. Siemann, LBL-31094, SLAC/AP-90 (1991).
36. Kohji Hirata, Phys Rev Lett **58** 25; **58**, 1798 (E) (1987).
Kohji Hirata, Phys Rev **D37**, 1307 (1988).
37. Kohji Hirata, Third Advanced ICFA Beam Dynamics Workshop, INP, Novosibirsk, edited by I. Koop and G. Tumaikin, 46 (1989).
38. M. A. Furman, K. Y. Ng, A. W. Chao, SSC-174 (1988).
M. Furman, Third Advanced ICFA Beam Dynamics Workshop, INP, Novosibirsk, edited by I. Koop and G. Tumaikin, 52 (1989).
39. N. S. Dikansky and D. V. Pestrikov, Part Accel **12**, 27 (1982).
40. A. Chao and R. Ruth, Part Accel **16**, 201 (1985).
41. Katsunobu Oide and Kaoru Yokoya, KEK Preprint 90-10 (1990).

The B-Factory Project at KEK

Shin-Ichi Kurokawa

National Laboratory for High-energy Physics, KEK
1-1 Oho, Tsukuba, Ibaraki, 305, Japan

Abstract

The B-Factory project at KEK aims to construct an accelerator complex to detect the CP-violation effect of B-mesons. It is a two-ring electron-positron collider of 3.5 x 8 GeV in the existing TRISTAN tunnel. The design peak luminosity is 10^{34} $cm^{-2}s^{-1}$, which will be realized in two steps: from a small-angle collision with a luminosity of 2×10^{33} $cm^{-2}s^{-1}$ to a large-angle crab-crossing scheme with the final luminosity of 10^{34} $cm^{-2}s^{-1}$.

1 Introduction

A high-luminosity, asymmetric, two-ring electron-positron collider for B-physics (B-Factory) is now being planned as the extension of TRISTAN at KEK [1]. The machine is operated at the ϒ(4S) resonance for detecting the CP-violation effect at bottom quarks. CP-violation studies benefit considerably from having a moving center of mass for the BB system: B mesons, produced in an asymmetric collider, are boosted along the beam direction and travel a few hundred μm before decaying. By the use of a precision vertex detector, we can identify the B and $\overline{B}$ mesons and their decay vertices and observe their time evolution. The energy asymmetry of 3.5 × 8 GeV was chosen for the KEK B-Factory in order to minimize the necessary integrated luminosity for detecting the CP-violation effect. The required luminosity is at least 2×10^{33} $cm^{-2}s^{-1}$, an order of magnitude beyond the present highest luminosity collider, CESR [2].

In designing the accelerators for the B-Factory, we follow the guidelines shown below:

(1) The infrastructure of TRISTAN [3] should be used to a maximum.

(2) We envision only one interaction region in order to concentrate our effort on achieving the maximum luminosity for the detector.

(3) The 2.5 GeV electron linac will be upgraded to 8 GeV in order to inject 3.5 GeV positrons and 8 GeV electrons directly into the B-Factory rings.

We plan to increase the luminosity of the B-Factory in two steps. We first build a 3.5 GeV x 8 GeV two-ring collider with a small-angle (± 2.8 mrad) crossing scheme with the luminosity of 2×10^{33} $cm^{-2}s^{-1}$ (step 1). In this step we cannot fill the whole bucket with beam, since we need a length for separation of electrons and positrons to avoid spurious collisions; therefore, every fifth bucket is filled with beam. Three-meter bunch spacing in this case is long enough to install beam separation equipment, such as separation dipole magnets. In the second step, we fill every bucket with beam by introducing a large-angle crossing (± 10–20 mrad). The luminosity will be increased by factor 5 i.e. to 10^{34} cm^{-2}s^{-1}. This two-step realization of the final luminosity is justified by the following fact: large-angle crossing requires a new scheme called crab-crossing [4] to avoid a geometrical loss of the luminosity and to prevent the beam-beam tune shift limit from decreasing due to synchrobetatron resonances. The crab crossing scheme needs special cavities (crab cavities) [5], which will require some 3–5 years of R & D work.

The machine parameters for both steps are essentially unchanged except for the bunch spacing and the total current. The same lattice is used for both steps with minor changes of the interaction region.

As shown in Fig. 1 the detector will be installed in the Fuji Experimental Hall of TRISTAN, which is presently occupied by the VENUS detector, one of the three detectors now working at TRISTAN. The superconducting solenoid magnet and outer-layer ion structure of VENUS will be used for the B-Factory detector with some slight modifications; the inner part of the detector will be completely renewed. Electrons and positrons are injected from the upgraded linac to the B-Factory rings in straight sections on both sides of the collision point. Figure 2 illustrates the cross sections of the tunnel for the B-Factory and TRISTAN.

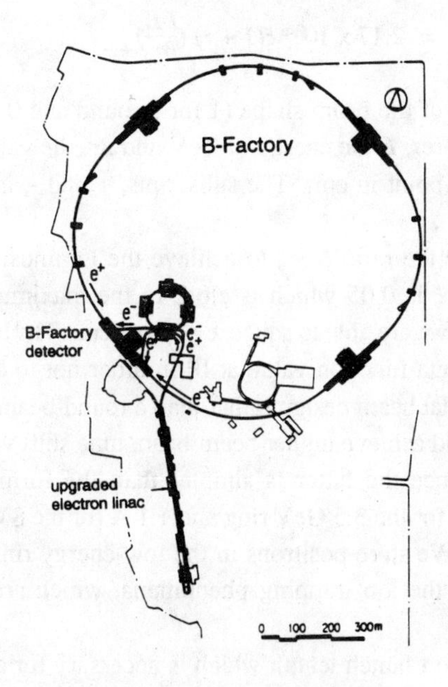

Fig. 1. Layout of the B-factory within the KEK site

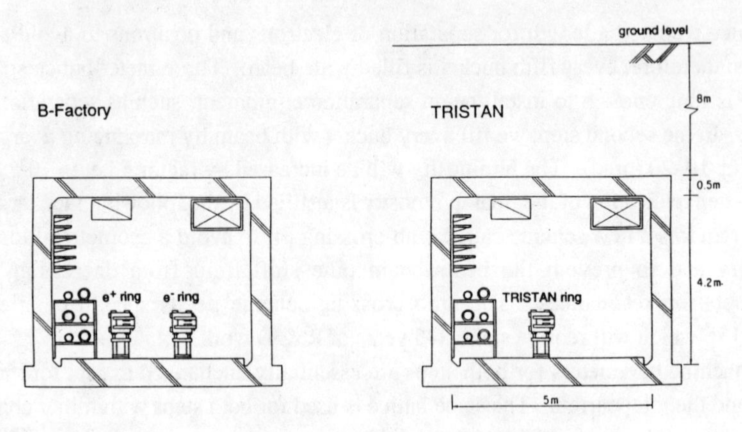

Fig. 2. Cross sections of the tunnel for the B-Factory and TRISTAN

2 Design Principle

On the basis of the assumption that horizontal and vertical beam-beam tune shifts of both beams are equal to a single specified value, ξ, and that both beams have the same cross sections at the interaction point, IP, the luminosity is given by the following expression in a unit of $cm^{-2}s^{-1}$:

$$L = 2.17 \times 10^{34} \xi(1 + r) \left(\frac{I \cdot E}{\beta^*_y}\right)_{+,-} \quad , \qquad (1)$$

where r is the aspect ratio of the beam shape (1 for a round and 0 for a flat beam), I the circulating current in amperes, E the energy in GeV and β^*_y the value of the vertical beta function at the interaction point in cm. The subscripts, + and -, mean that this formula applies to either ring.

We first try to increase the ratio ξ/β^*_y to achieve the luminosity goal with the least current. We assume that ξ is 0.05 which is close to the maximum value achieved by existing machines and that we are able to squeeze β^*_y to 1 cm. The bunch length should be less than one half of the beta function value at IP in order not to cause reduction of the luminosity. We choose a flat beam design rather than a round-beam one, since there is no clear evidence that we could achieve higher beam-beam tune shift values for round beams than for flat ones, and since the latter is simpler than the former. Even with these parameters, we need 2.6 A for the 3.5 GeV ring and 1.1 A for the 8 GeV ring to obtain the 10^{34} $cm^{-2}s^{-1}$ luminosity. We store positrons in the low-energy ring and electrons in the high-energy one to avoid the ion trapping phenomena, which are more severe at low energies.

In order to achieve a short bunch length which is necessary for a small β^*_y, we need a large RF voltage V_c, hence a large number of RF cavities. At the same time the number of the cavities should be kept to a minimum to avoid severe beam instabilities. We solve this by increasing the number of bunches in the beams [6]. If the total current I is given, the number of particles per bunch N_b is expressed as:

$$N_b = \frac{IS_b}{ec} \quad , \tag{2}$$

where S_b is the bunch spacing, e the charge of an electron and c the speed of light. The horizontal emittance ε_x is given by:

$$\varepsilon_x = \frac{r_e N_b}{2\pi\gamma\xi(1+r)} = \frac{r_e I S_b}{2\pi\gamma e c \xi(1+r)} \quad , \tag{3}$$

where r_e is the electron classical radius and γ the beam energy in a unit of the electron mass. The necessary RF voltage V_c is given by the formula,

$$V_c = \frac{E T_0 c^2 \sigma_\varepsilon^{2/3} \varepsilon_x^{2/3}}{e h \omega_0 (2R)^{2/3} \sigma_z^2} \quad , \tag{4}$$

where T_0 is the revolution period, h the harmonic number, ω_0 the revolution frequency, σ_ε the beam energy spread, R the average ring radius, and σ_z the bunch length. By combining this formula with Eq. (3), we get

$$V_c \propto \frac{S_b^{2/3}}{\sigma_z^2} \quad . \tag{5}$$

The high V_c necessary for the short bunch length can be compensated by decreasing the bunch spacing.

3 Lattice Design

3.1 Beam Parameters

The main parameters of the B-Factory accelerators are given in Table 1. The values in parentheses correspond to those for the first step. The high-energy ring, HER, and the low-energy ring, LER, have the same circumferences, emittances, and the beta functions at IP. This gives similar sizes of the normal cells, the betatron tunes, and the energy spreads to both rings. The radiation damping time of the low-energy ring becomes twice as long as that of the high-energy ring if only the radiation from the arcs is taken into account.

3.2 Chromaticity Correction

It is desirable to inject beams into the B-Factory without changing the optics at injection from that of collision. To this end, we are studying a non-interleaved sextupole chromaticity correction scheme expecting that this provides sufficient dynamic apertures at injection [7]. Between a pair of sextupoles no other sextupoles exist and the betatron phase advance is π in both the horizontal and vertical planes. The merit of this scheme is based on the cancellation of the geometric aberrations of the sextupole by a -I transformation in a pair. Figure 3 shows an example of the dynamic aperture. The dynamic aperture is larger than that required at injection shown with × signs in the figure. The aperture differs a little with different synchrotron tunes.

Table 1. Main parameters of the KEK B-Factory

Energy	E	3.5		8.0	GeV
Circumference	C		3018		m
Luminosity	L		1×10^{34} (2×10^{33})		$cm^{-2}s^{-1}$
Tune shifts	ξ_x/ξ_y		0.05/0.05		
Beta function at IP	β^*_x/β^*_y		1.0/0.01		m
Beam current	I	2.6 (0.52)		1.1 (0.22)	A
Natural bunch length	σ_z		0.5		cm
Energy spread	σ_ε	7.8×10^{-4}		7.3×10^{-4}	
Bunch spacing	S_B		0.6 (3.0)		m
Particles/bunch	N	3.3×10^{10}		1.4×10^{10}	
Emittance	$\varepsilon_x/\varepsilon_y$		$1.9 \times 10^{-8}/1.9 \times 10^{-10}$		m
Synchrotron tune	ν_s	0.064		0.070	
Betatron tune	ν_x, ν_y	~39		~39	
Momentum compaction	α	8.8×10^{-4}		1.0×10^{-3}	
Energy loss/turn	U_0	0.91		4.1	MeV
RF voltage	V_c	20		47	MV
RF frequency	f_{RF}		508		MHz
Harmonic number	h		5120		
Energy damping decrement	T_0/τ_E	2.6×10^{-4}		5.1×10^{-4}	
Bending radius	ρ	15.0		91.3	m
Length of bending magnet	l_B	0.42		2.56	m

Values in parentheses are for the first step.

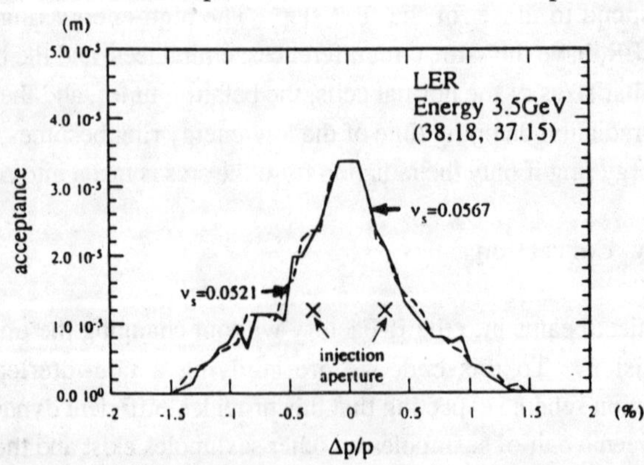

Fig. 3. Dynamic aperture for LER by the non-interleaved sextupole correction scheme. The signs × show the required apertures at injection.

4 Insertion Design

4.1 Separation Scheme

The insertion layout around IP is shown in Fig. 4 and Fig. 5 for the first step (small-angle crossing scheme) [8]. Electrons and positrons cross at an angle of ± 2.8 mrad. The experiment at CESR shows that with this small crossing angle, reduction of the beam-beam tune shift is negligible [9]. In order to decrease the synchrotron radiation from the incoming beam and to make enough separation quickly between the two orbits, the optics is no longer symmetric with respect to IP.

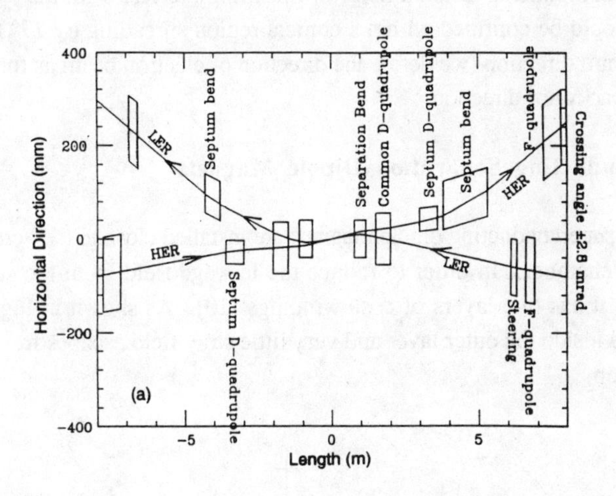

Fig. 4. Layout of the interaction region for a small-angle crossing

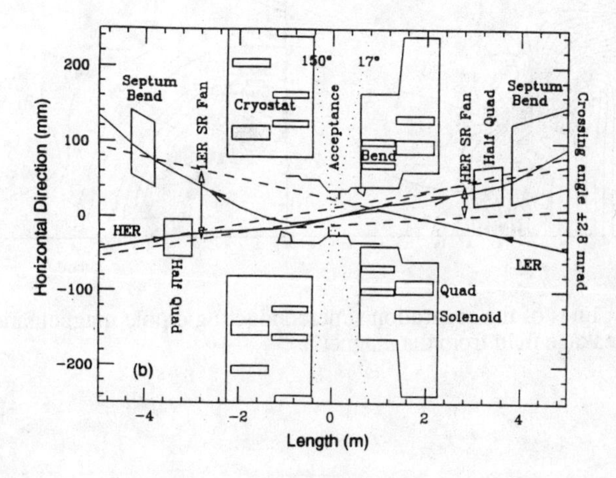

Fig. 5. Detailed layout of the interaction region

Beam separation is achieved by means of superconducting dipole magnets (separation bend). On both sides of IP the orbit of the incoming beam goes through the center of a defocusing superconducting quadrupole. Septum defocusing quadrupoles are half quadrupole magnets and hence can be inserted close enough to IP. These magnets focus vertically and deflect the HER beam outwards helping the orbit separation. The outgoing beam orbit is deflected away from the other by septum magnets.

The deflection angle is determined by the condition that synchrotron radiation from incoming particles on the LER center orbit has to be confined within 25 mm from IP in the transverse direction. The size of the IP duct is 60 mm, 35 mm outside and 25 mm inside, and 30 mm in full height.

Accelerator components inside a detector must not interfere with the experiment. The components should be confined within a conical region spreading by 17° from the center axis in the forward direction (we define the direction of electron beam as forward direction) and 30° in the backward direction.

4.2 Superconducting Separation Dipole Magnets

The separation superconducting dipole magnets are installed close to the vertex detector and precision drift chamber. In order to reduce the leakage field from the superconducting dipole magnet, it has two layers of cosθ windings [10]. As shown in Fig. 6a, almost all flux is confined inside the outer layer and very little stray field escapes from the magnet as shown in Fig. 6b.

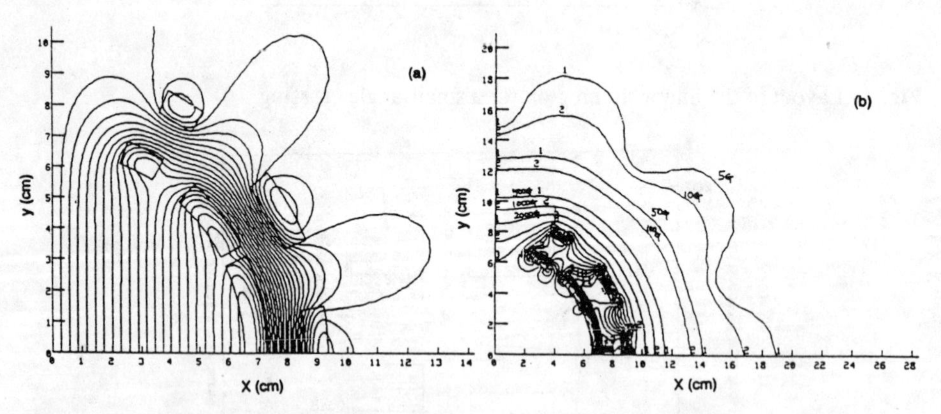

Fig. 6. (a) Flux lines of the separation superconducting dipole magnet, and (b) distribution of leakage field from the magnet

4.3 Background Consideration

Two sources of background to the detector are identified: one is synchrotron light radiated in both dipole and quadrupole magnets near the interaction region, IR; the other involves electrons and positrons which have lost some part of their energy due to the interaction with residual gas (these electrons and positrons are called "spent electrons"). The IP beam pipe and the masks have been designed according to the following guidelines:

(1) Photons radiated from the beam within the dipole and quadrupole magnets should not directly hit the IP beam pipe.
(2) Masks should intercept photons scattered by vacuum pipes to prevent them from hitting the IP beam pipe.
(3) Masks for spent electrons should be located at some distance from the IP beam pipe to prevent any shower leakage from entering the detector.

In the present design a simulation shows that the number of photons which hit the IP beam pipe is 10^{-3} per beam crossing. This number is sufficiently small for the detector. The rates of spent electrons which enter the detector through the IP beam pipe were estimated on the assumption of a 10^{-7} Pa vacuum around IR to be 0.8 kHz for 8 GeV at 0.26 A and 8 kHz for 3.5 GeV at 0.55 A; these rates are within a manageable range for the detector [11].

5 Coupled-Bunch Instabilities and RF System

Large currents, many bunches and short distance between bunches cause strong coupled-bunch instabilities both in the transverse and longitudinal directions. Three sources of coupled-bunch instability are identified. These are (1) higher-order modes (HOM) of RF cavity (transverse and longitudinal), (2) accelerating mode of RF cavity (longitudinal) and (3) resistive wall of vacuum chambers (transverse).

5.1 Normal Conducting RF Cavity

To prevent the coupled-bunch instabilities due to HOM,s we are studying a damped cavity [12], which was first proposed by R.B. Palmer [13] for linear colliders to reduce the Q values of HOM's. The basic idea of the damped cavity is that the HOM field is guided to waveguides through slots cut on disks of a disk-loaded type structure; the cutoff frequency of the waveguide is set higher than the fundamental accelerating mode frequency.

Figure 7 shows a schematic drawing of the two-cell damped cavity of the B-Factory. Table 2 lists the parameters of the damped cavity calculated by the code MAFIA [14]. The first prototype damped cavity has been completed and a low-power test is now under way. Table 5.1 also shows the measured values of some of HOMs.

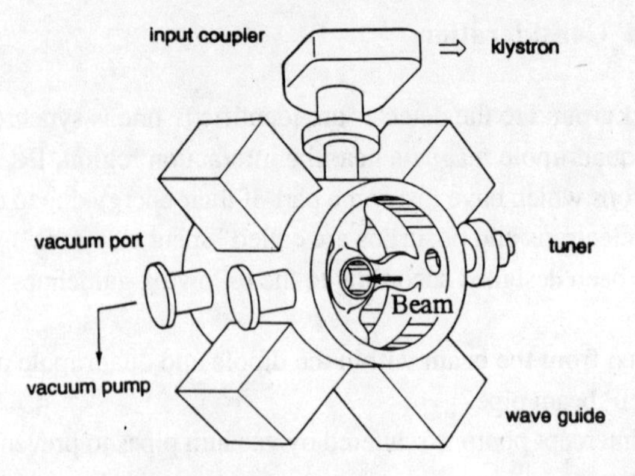

Fig. 7. Schematic drawing of the two-cell damped cavity of the B-factory

Table 2. Parameters of damped cavity

	Q_0	R	
TM010 (fundamental)	27,800	18.7	MΩ/m
	Qext	R/Q	
TM110			
0 mode	13.4	73.5	Ω/m
π mode	29.9[a] (41.1)[b]	462	Ω/m
TM011			
0 mode	uncoupled	1.29	Ω
π mode	10.6[a] (13.9)[b]	115	Ω

[a]Calculated values; [b]Measured values.

5.2 Superconducting RF Cavity

Owing to the high field gradient and the high loaded-Q value of a superconducting cavity, coupled-bunch instabilities excited by the fundamental accelerating mode of cavities are not severe (see section 5.3); therefore, it is worthwhile doing extensive R&D work on superconducting cavities in the hope that they can be used for the B-Factory. We also need to prove that superconducting cavities can work under high currents and strong synchrotron radiation fields, that higher modes can be successfully absorbed, and that the necessary power can be fed to cavities through input couplers and windows.

On the basis of the Cornell design [15], we designed the B-Factory superconducting cavity shown in Fig. 8. The main difference between our design and that of Cornell is that we use a simple larger-diameter beam pipe instead of a complicated fluted beam pipe and that we try to use an antenna-type input coupler which we are using for TRISTAN superconducting cavities instead of the wave-guide type coupler for the Cornell cavity.

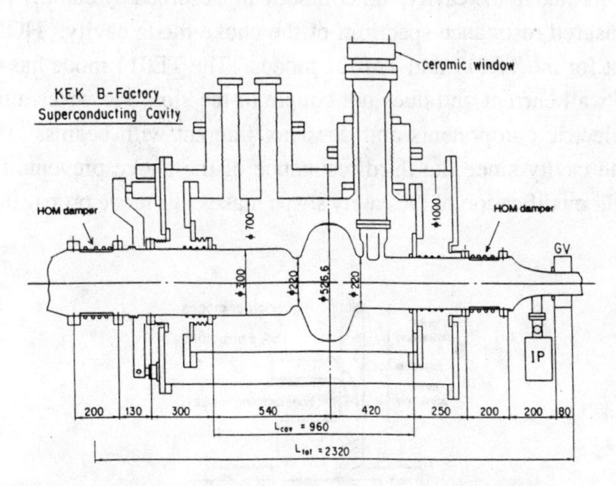

Fig. 8. Schematic drawing of the B-factory superconducting cavity

As the first step of R&D, we made a prototype Nb cavity in order to test whether we can achieve a sufficiently high field gradient with this type of cavity. The shape of this prototype cavity was not optimized, since we utilized one cell of TRISTAN cavity and attached a large-diameter beam pipe to it in order to speed up the production. Figure 9 shows the result of a vertical test; the maximum accelerating field obtained was 9.7 MeV/m with a Q value larger than 10^9. These values are good enough for the B-Factory. We have already optimized the shape of the B-Factory cavity; full-size aluminium and niobium model cavities are being produced. We plan to have a beam test at the TRISTAN accumulation ring in late 1994.

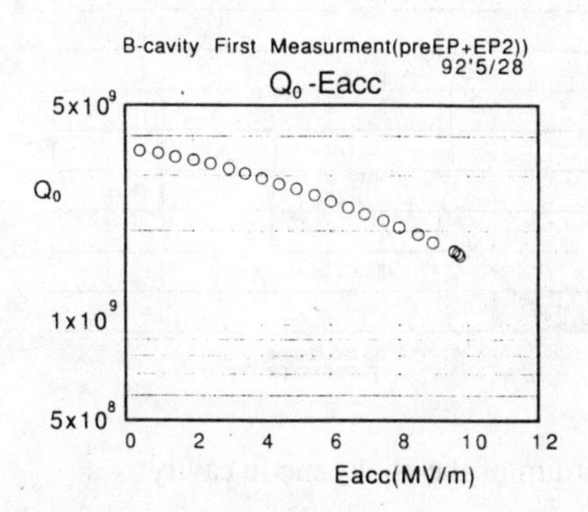

Fig. 9. The result of a vertical test of the prototype superconducting cavity

5.3 Choke-Mode Cavity

The choke-mode cavity proposed by T. Shintake of KEK [16] might be applicable to the B-Factory. As shown in Fig. 10, all modes that affect beams escape from a gap attached to the side of a pill-box cavity. Due to the choke structure only the accelerating mode is reflected and confined in the cavity; other modes are absorbed by dummy loads. Figure 11 shows the measured resonance spectrum of the choke-mode cavity. HOM's are heavily damped except for the TE011 and TM021 modes. The TE011 mode has no longitudinal component of wall current and does not couple to the slot. However, this mode has no longitudinal electric components and does not interact with beams. TM021 mode is confined in the cavity since the third resonance of the choke prevents this mode from escaping. Slight modification of the cavity shape makes this mode propagate to the slot.

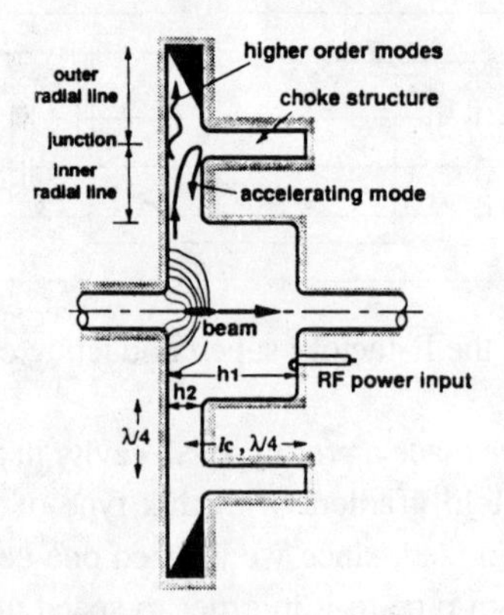

Fig. 10. Conceptual illustration of the choke-mode cavity

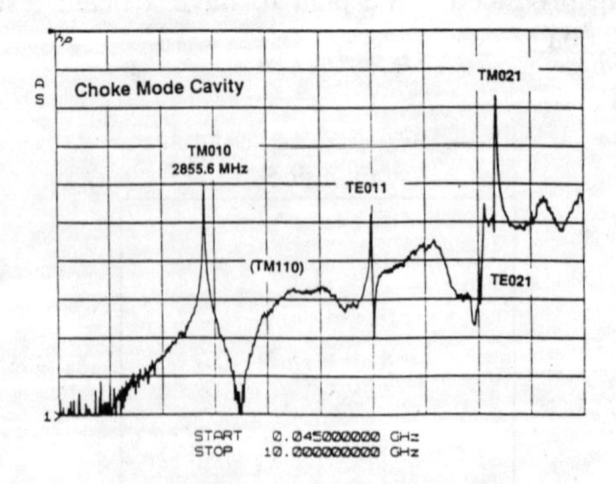

Fig. 11. Resonance spectrum of the choke-mode cavity

5.4 Coupled-Bunch Instability due to Fundamental Mode

Extremely heavy beam loading to the cavity, together with the small revolution frequency, leads to a quite violent longitudinal coupled-bunch instability [17]. In Fig. 12 the real part of the fundamental mode impedance of the KEK B-Factory cavity is depicted as a function of frequency. In the figure f_{RF} (508 MHz) designates the RF frequency, f_0 (101 kHz) the revolution frequency and f_s (7 kHz) the synchrotron frequency, respectively. Resonance frequencies for the instability exist around every higher revolution harmonic. The impedance at the right (left) side bands of the harmonics contributes to excitation (damping) of coupled-bunch modes. As the beam current increases, the resonant frequency of accelerating mode should be moved down to minimize the reflection power from the cavities. The amount of frequency detuning Δf is given by

$$\frac{\Delta f}{f_{RF}} = \frac{I}{2\,\dfrac{V_c}{N_{\text{cell}}}}\,\frac{R_s}{Q_0}\,\sin\phi\ ,$$

where I denotes the total beam current, V_c the total RF voltage, N_{cell} the number of cells, R_s the shunt impedance per cell, Q_0 the unloaded Q value and ϕ the synchronous phase. At LER of the KEK B-Factory the detuning frequency is more than three times as large as the revolution frequency; therefore, the resonant frequency of the cavity passes the resonant frequency of the instability three times as the beam current increases. The growth time becomes as small as 4.8 μs in the worst case.

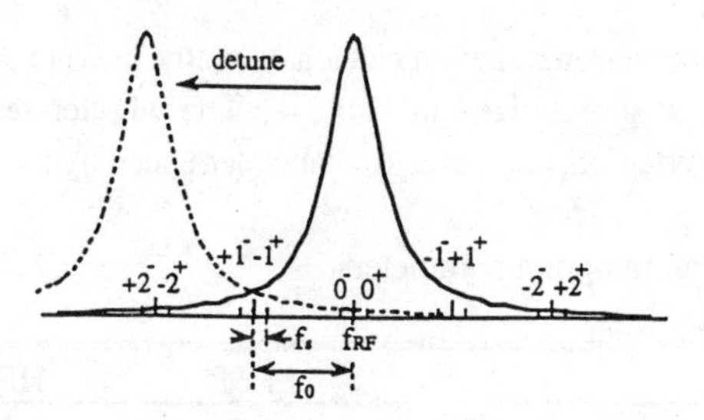

Fig. 12. Impedance of the fundamental mode of the cavity and those of coupled-bunch instabilities. f_{RF}, f_o and f_s are the RF, the revolution and the synchrotron oscillation frequency. $0^-, 0^+, 1^-, 1^+$, ... denote the coupled-bunch instability modes: suffix + means excitation and - damping mode.

The most straightforward way to avoid this instability is to employ superconducting cavities. The detuning frequency for superconducting cavities is much smaller than that for normal conducting cavities because of a high field gradient and a low R_s/Q_0 value. Due to this small detuning frequency and the large loaded-Q value of superconducting cavities, the growth time of the coupled-bunch instability becomes as long as 23 ms (loaded Q of 2.5×10^5 is assumed). Table 3 summarizes the comparison between the normal conducting and superconducting cases. Another possible cure for this instability is the so-called RF feedback system [18].

Table 3. Comparison of normal conducting (NC) and superconducting (SC) cavities

	NC	SC	
Rs/Q_0	197	85	Ω/cell
V_C	22	22	MV
N_{cell}	56	8	
f_{RF}	508	508	MHz
I (A)	2.6	2.6	A
Δf	331	20	kHz
f_{rev}	101	101	kHz
P_{window}	148	450	KW
t_{gr}	4.8×10^{-6}	2.3×10^{-2}	s

6 Vacuum

6.1 Design Guidelines

As a design goal of the vacuum system we set a pressure of 10^{-7} Pa at full current of the second step. This pressure guarantees a long beam life and a low rate of spent electrons to the detector. Additionally, this low pressure is desirable to prevent any ion-trapping phenomena in HER.

The amount of synchrotron radiation calculated on the basis of the machine parameters of the KEK B-Factory is summarized in Table 4. The wiggler section is exposed to very intense synchrotron radiation, compared to that experienced by a normal cell.

Table 4. Synchrotron radiation parameters

		LER	HER	
Energy	E	3.5	8.0	GeV
Beam currents	I	2.6	1.1	A
[Arc section]				
Bending radius	ρ	15	91	m
Length of arc section	l_a	2130	2130	
Critical energy	ε_c	6.3	12.5	keV
Max. power density	P_{max}	4.6	6.3	kW/m
Av. photon density	N_{av}	3.3×10^{18}	3.3×10^{18}	phot./s/m
[Wiggler section]				
Bending radius	ρ	8.3	44.4	m
Length of wig. section	l_w	140	140	m
Critical energy	ε_c	11.5	25.6	keV
Max. power density	P_{max}	30	33	kW/m
Av. photon density	N_{av}	5×10^{19}	1×10^{19}	phot./s/m

6.2 Pumping System and Pressure Distribution

Distributed type pumping is better than lumped-type pumping to achieve a low average vacuum pressure. Since even in HER less than 40% of the vacuum chamber is within magnetic field, the use of distributed ion pumps is precluded. We choose NEGs as main pumps for the B-Factory. Figure 13 shows the layout of the pumping system of one cell of HER. If we assume a photo-desorption coefficient η of 10^{-6}, and the pumping speed of NEGs is 100 l/s, we get the distribution of effective pumping speed and the pressure distribution shown in Fig. 14.

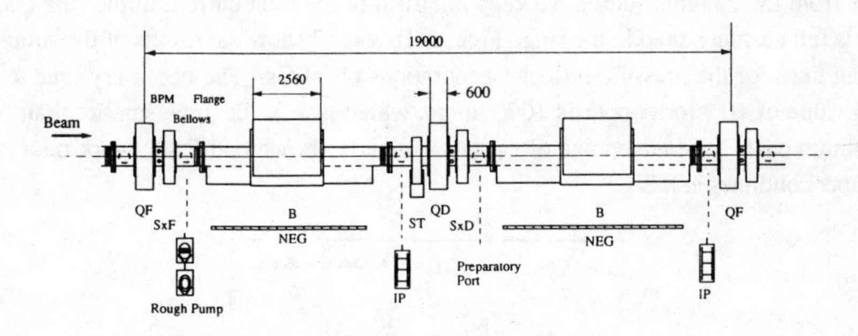

Fig. 13. Layout of the pumping system of one cell of HER

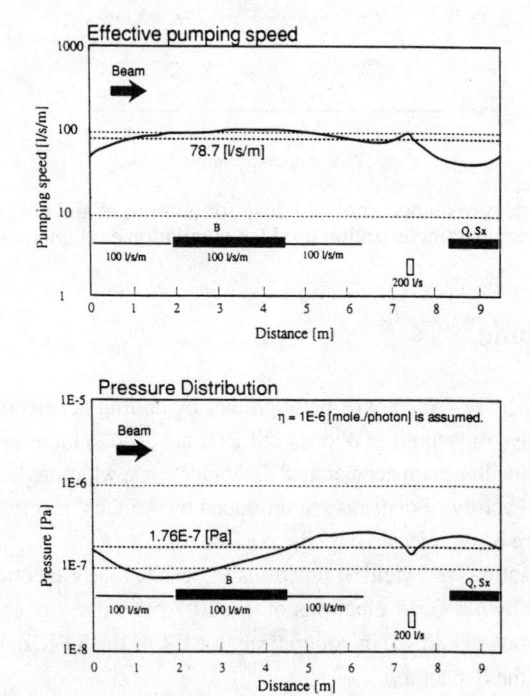

Fig. 14. Distribution of effective pumping speed and pressure within one cell of HER

6.3 Chamber Material

Since the maximum power density for the arc section is less than 10 kW/m (see Table 4), copper and aluminum can be used as the chamber material for the B-Factory; however, copper has various advantages over aluminum. First, its self-shielding property makes lead shielding unnecessary. Second, copper has a smaller η value than aluminum as shown in Fig. 15, which depicts the result obtained by A.G. Mathewson et al. [19]. We parameterized the dependence of η on photon dose according to these data and made a simulation by assuming that: (1) by the accumulation of dose the pumping speed of NEG decreases from 400 l/s/m to 70 l/s/m, when we regenerate NEGs; (2) In the first 1000 hours from the commissioning we keep one fifth of the total current in the ring and only after is full current stored in the ring. Figures 16 and 17 show the results of the simulation. Zigzag lines for the pressure reflect regenerations of NEGs. The necessary time to reach the η value of 10^{-6} for copper is 1000 hours, which is a factor three smaller than for the aluminum case. The advantage of copper is clearly shown and R&D work on a copper chamber continues at KEK.

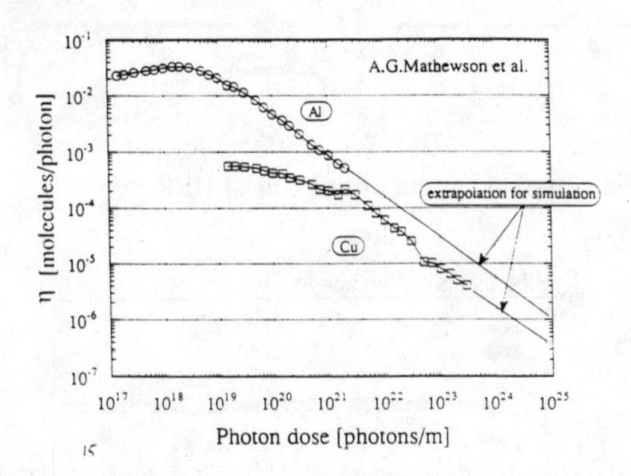

Fig. 15. Dependence of η on accumulated dose for aluminum and copper [19]. Lines are drawn according to the parameterization used for simulation explained in Sec. 6.3.

7 Linac Upgrade

The present KEK 2.5 GeV linac will be upgraded by adding accelerating structures and changing 20 MW klystrons to 60 MW ones. SLEDs are used to increase the field gradient. After this upgrade, the linac can accelerate 8 GeV electrons, which will be injected directly into HER of the B-Factory. Positrons are produced by 4.5 GeV electrons and accelerated up to 3.5 GeV before being injected directly to LER.

If we assume a normalized yield of positrons to be 2%/1 GeV electron, the intensity of positrons produced by 4.5 GeV electrons of 4×10^{10} per pulse amounts to 3.6×10^{9} per pulse; this corresponds to 880 sec injection time to LER of the KEK B-Factory. This time is short enough for the B-Factory.

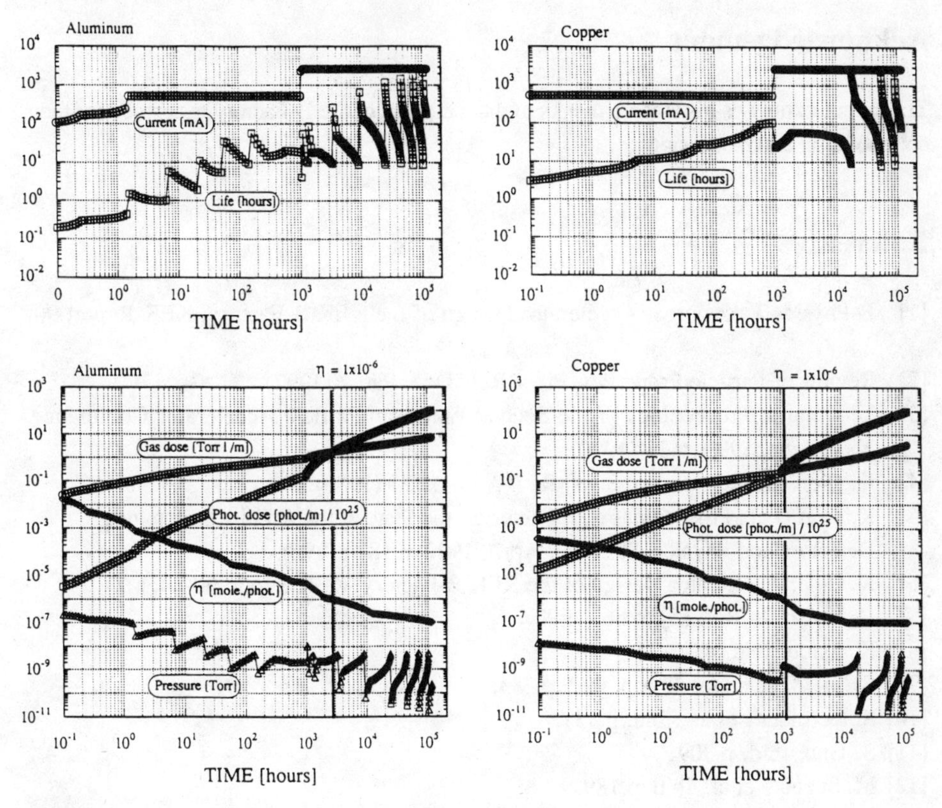

Fig. 16. Result of the simulation for aluminium

Fig. 17. Result of the simulation for copper

8 Summary

The B-Factory project at KEK is regarded as the third phase of TRISTAN. After having pursued the energy frontier by increasing the beam energy from 25 GeV to 32 GeV, the TRISTAN has stepped into its second phase from February 1990. In the second phase we put the stress on accumulating as large an integrated luminosity as possible with a modest energy (at present TRISTAN is operated at 29 GeV). The goal is to accumulate the 300 pb^{-1} integrated luminosity. By the summer of 1994 this goal will be reached. We envision that construction of the B-Factory will start from April 1994 and by the end of 1998 the commissioning of the B-Factory will take place. Various R&D works for the B-Factory are now extensively performed.

Acknowledgement

The author expresses his sincere thanks to the members of the B-Factory task force for their collaboration and discussions.

References

[1] B-Physics Task Force, Accelerator Design of the KEK B-Factory, KEK Report 90-24.
[2] Review of Particle Properties, Phys. Rev. D45, Part 2 (1992).
[3] Y. Kimura, Proceedings of the Second European Particle Accelerator Conference, Nice, June, 1990, p.23 (1990).
[4] K. Oide and K. Yokoya, Phys. Rev. A40, p.315 (1989).
[5] K. Akai et al., Proceedings of B Factories, the State of the Art in Accelerators, Detectors and Physics, Stanford, April, 1992, p.181 (1992).
[6] K. Oide, Accelerator Design of the KEK B-Factory, KEK Report 90-24, p.27.
[7] H. Koiso, In Ref. [5], p.86 (1992).
[8] K. Satoh, ibid. p.261.
[9] CESR Accelerator Group, ibid. p.283.
[10] S. Kurokawa et al., ibid. p.331.
[11] S. Uno, ibid. p.309.
[12] M. Suetake et al., ibid. p.189.
[13] R. B. Palmer, SLAC-PUB-4542 (1989).
[14] K. Klatt et al., Proceedings of the 1986 Linear Accelerator Conference, SLAC, 1986, Report 303.
[15] CESR Accelerator Group, In Ref. [5], p.172 (1992).
[16] T. Shintake, Jpn. J. Appl. Phys. 31 p.1567 (1992).
[17] P.W. Wilson, in these Proceedings.
[18] F. Pedersen, in these Proceedings.
[19] A.W. Mathewson et al., Proceedings of 2nd Topical Conference on Vacuum Design of Synchrotron Light Sources, Argonne, Nov. 1990.

Detectors for ϕ, τ-charm and B Factories

Jasper Kirkby

CERN, Geneva, Switzerland

Abstract. We provide an overview of the physics goals and detector designs for ϕ, τ-charm and B Factories.

1 Introduction

1.1 The Particle Factory

In order to make progress in particle physics, it is now widely recognized that the next generation of particle accelerators must pursue two complementary lines: the "high-energy frontier" and the "high-precision frontier". Their common goals are to find answers to the open questions facing the Standard Model and to search for experimental clues towards a more fundamental theory. The basic approach of experiments at the high-energy frontier is to search for new particles and new interactions, whereas experiments at the high-precision frontier search for subtle discrepancies with the Standard Model, by examining previously-discovered particles with increased precision and sensitivity.

The "high-energy" path has been followed with great success by accelerators since their inception. The next-generation machines—the LHC and SSC hadron colliders with their high rates and energies of several 10's of TeV—pose great challenges for the detectors. The primary detector requirement is high selectivity; signals as small as $\mathcal{O}(100)$ events per year must be extracted from $\mathcal{O}(10^{15}-10^{16})$ interactions.

The "high-precision" approach has emerged more recently, because of technical and scientific reasons. The former are a consequence of continual improvements in the technology and performance of both accelerators and detectors since the original experiments took place at (what are now) low energies. This has opened up the possibility of exploring old experimental territory with greatly-improved sensitivity and precision, in a search for surprises. The scientific interest in this approach has also grown in recent years due both to the discovery that there appear to be only a limited number of fundamental fermions in Nature (the present three families), and also to the on-going improvements in the precision of the Standard Model predictions.

The high-luminosity machines that are required to explore the high-precision frontier are known as particle "factories" (a word coined from industry to describe the mass-production of particles). At such machines the primary detector requirement is high precision, implying careful optimization of both the detector and the machine for the particle(s) under study and, in most cases, high-resolution general-purpose detectors.

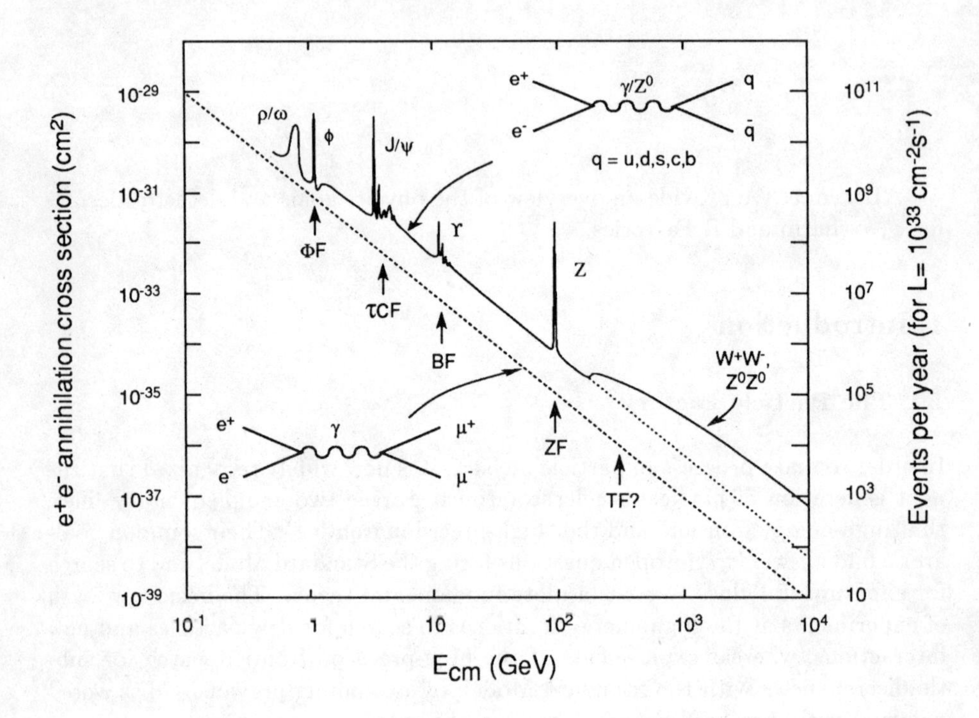

Fig. 1. The e^+e^- annihilation cross-section in the centre-of-mass energy range 1 GeV $< E_{\rm cm} <$ 10 TeV.

1.2 Choice of Accelerator and Energy

The primary requirements of a particle factory are to produce specific particles in copious quantities and with low backgrounds. High-luminosity e^+e^- colliders are well-suited to these requirements, and their design is now a subject of considerable world-wide interest, as this meeting[1] confirms.

The optimum energy to study a particular particle at an e^+e^- collider generally corresponds to the region near its pair-production threshold. In such regions are generally found the highest cross-sections, lowest backgrounds and other favourable experimental conditions. Other energies of interest correspond to production of the narrow vector resonances ($\phi, J/\psi, \Upsilon$, Z^0, ...) which, in addition to being interesting particles in their own right, constitute high-rate secondary sources of lighter particles. The energies of interest for e^+e^- collider particle factories are summarized in Table 1. The production cross sections and typical event rates for these machines are indicated in Fig. 1. As shown in Fig. 2, particle factories are expected to have 10 to 100 times the luminosities of present machines.

Table 1. Summary of the energies of interest for e^+e^- collider particle factories.

E_{cm} (GeV)	Particle resonance	Particle threshold	Accelerator
1	ϕ	$s\bar{s}$	ϕ Factory
3-5	J/ψ, ψ'...	$\tau\bar{\tau}$, $c\bar{c}$	τc Factory
9-11	$\Upsilon(1S)$, $\Upsilon(2S)$...	$b\bar{b}$	B Factory
91	Z^0	-	Z Factory
??	θ	$t\bar{t}$	T Factory

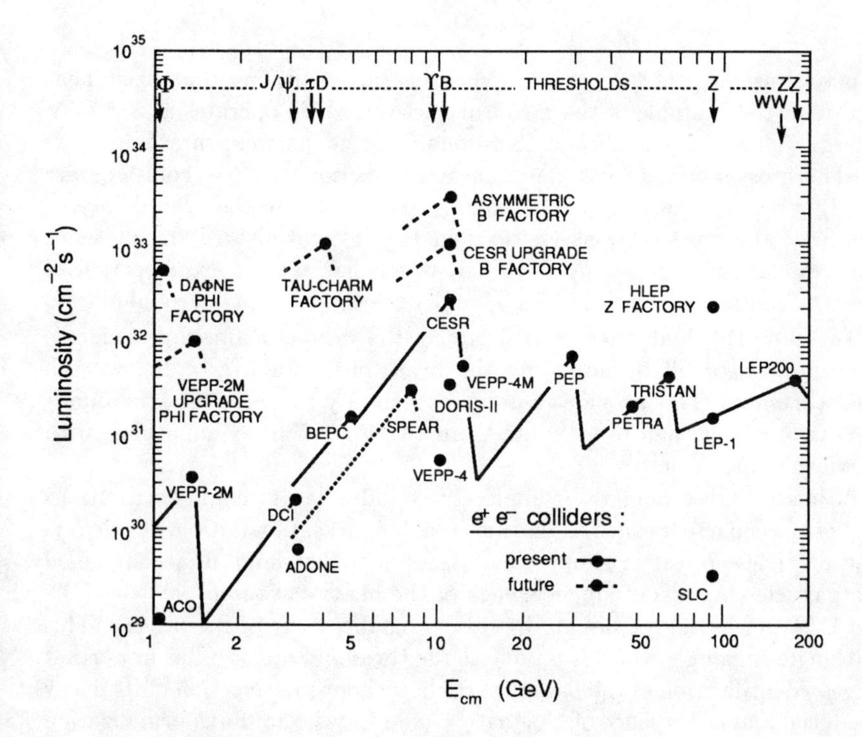

Fig. 2. The luminosity of present and future e^+e^- colliders in the centre-of-mass energy range $1 \leq E_{cm} \leq 200$ GeV. The solid line represents the present envelope of the maximum luminosity. The dashed lines indicate the design luminosities of future factories.

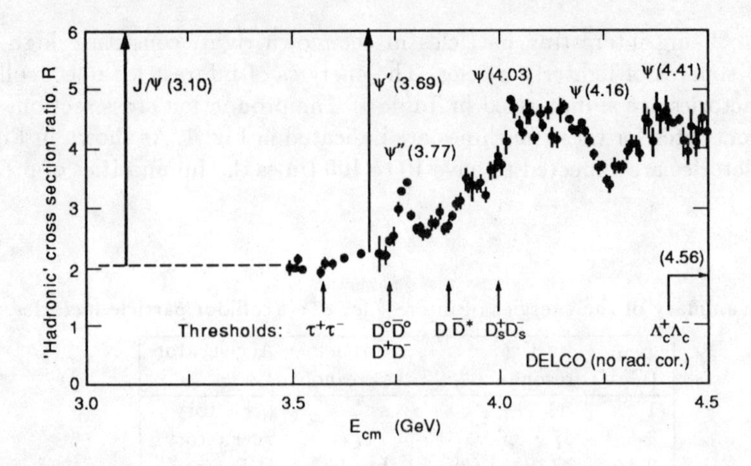

Fig. 3. An example of a heavy-flavour threshold region: the energy range of the τ-charm Factory. The ratio $R = \sigma(e^+e^- \to$ 'hadrons') / $\sigma(e^+e^- \to \mu^+\mu^-)$, where 'hadrons' include both $q\bar{q}$ and $\tau^+\tau^-$ events. The data are from DELCO at SPEAR.

We may illustrate the favourable experimental conditions that exist near threshold with the example of the τ-charm Factory, which operates at 3–5 GeV total energy. This region is rich with resonances and particle thresholds (see Fig. 3). The cross-sections for τ and charm production in e^+e^- collisions are higher in this region than at any other energy. This is due to the overall $E_{\rm cm}{}^{-2}$ dependence of the cross-sections (as seen in Fig. 1) and also to the presence of charm resonances such as the $\psi''(3.77)$, which has a peak cross section of 5 nb (over a continuum level of 13 nb) and decays with equal probability to pure $D^0\bar{D}^0$ or D^+D^- final states. Furthermore, the various τ and charm signals can be turned on or off by adjusting the beam energy above or below each particular threshold. This provides a unique possibility to measure backgrounds *experimentally* rather than to estimate them by Monte Carlo simulations, with their inevitable uncertainties.

Operating near threshold also completely excludes background contributions from higher-mass particles such as, in this case, b quarks. In addition, the heavy-flavoured particles appear in simple final states near threshold, unaccompanied by extra particles. As a consequence, each of the heavy-flavoured particles (D^0, $D^\pm$, and $D_s^\pm$) can be tagged simply by observing the decay of its partner. This so-called single-tagging—which is unique to the threshold region—has important advantages: identification of the parent particle without pre-selection of its decay mode, reduced biases, presence of kinematic constraints, exact flux normalization and reduction of backgrounds. Single-tagging exists also for τ leptons since they can be produced below $c\bar{c}$ and $b\bar{b}$ thresholds, and so the $\tau^+\tau^-$ events can be cleanly isolated with simple selection criteria on one of the two τ's. Finally, there are also several kinematic advantages which result from the low particle velocities near threshold (such as monochromatic spectra for two-body decays).

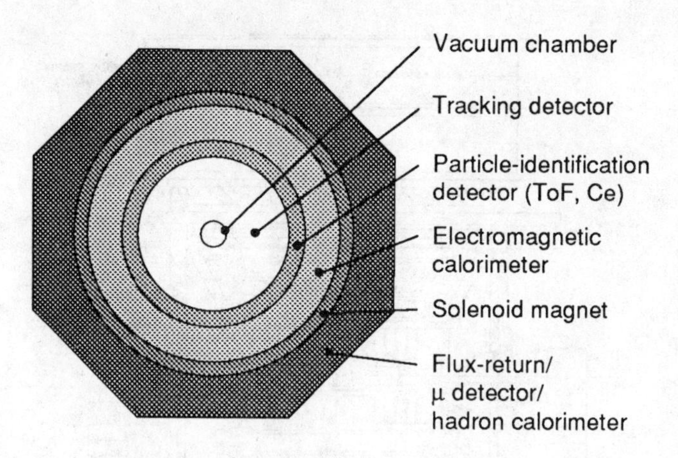

Fig. 4. The transverse layout of the universal e^+e^- collider detector.

2 Detector Overview

2.1 Evolution of the Universal e^+e^- Detector

The universal e^+e^- detector, as illustrated in Fig. 4, has a familiar layout: cylindrical "onion" layers of sub-detectors and a solenoid magnet, arranged in order of increasing density with increasing radius. Almost all present and future collider detectors follow this basic layout, possibly with minor adjustments such as the location of the solenoid coil.

It is interesting to recall the origins of this style of detector. The most successful early pioneer of the "4π general-purpose detector" was the Mark I detector (Fig. 5)[2], which operated at SPEAR between 1973 and 1978. Mark I had all the by-now-standard features: a large-acceptance cylindrical tracking chamber in a solenoid magnet, particle identification (time-of-flight), electromagnetic calorimetry and μ identification. The discoveries of this detector are the stuff of text books on the history of science: the "ψ" part of the J/ψ, the charmonium system, the τ lepton, the charmed mesons, and the first evidence for hadronic jets.

Mark I dramatically demonstrated the importance of large acceptance for e^+e^- collider detectors, both to maximize the rate and also to permit a complete analysis to be made of all the particles in each event. However, by today's standard, the performance of Mark I was relatively crude with, for example, only 65% solid angle coverage, 30% energy resolution for γ's, and 20% probability for a hadron to be mis-identified as an e or μ. This allowed room for important contributions from special-purpose detectors such as the Crystal Ball (emphasizing γ's), DASP (particle identification and momentum resolution), DELCO (e identification), PLUTO (μ identification) and others.

Today we have 15-20 year's improvement in detector technologies since the original experiments, resulting in general-purpose detectors that can essentially

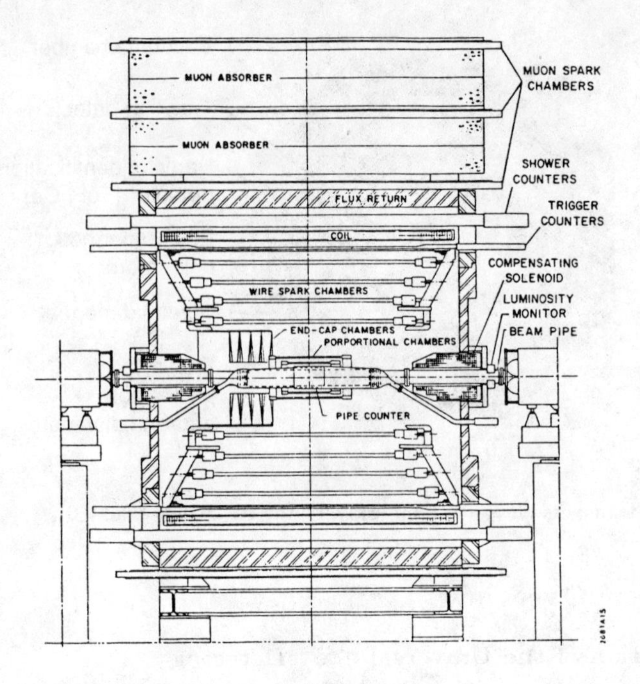

Fig. 5. The Mark I detector (1973-1978) at SPEAR.

"do everything" with little compromise in performance. An impressive example of such a detector is the CLEO II detector (Fig. 6)[3] operating at CESR, which represents one of the first detectors (along with L3 at LEP and the Crystal Barrel at LEAR) to combine high-resolution charged *and* neutral measurements. The next-generation detectors for particle factories will benefit from further technological advances and have even better performance. Faced with such high overall performance in general-purpose detectors, it has become difficult to identify a role for special-purpose detectors—although this may easily change in future as the physics develops and so too, perhaps, does the need for dedicated experiments.

2.2 General Design Considerations

Physics at particle factories is characterized by high precision—small systematic errors and sensitivities to rare decays—which demands critical attention to all aspects of the performance, reliability and monitoring of the detector, trigger, on-line and off-line computing. The detectors themselves are designed with large, uniform sub-systems to minimize the uncertainties in acceptance and performance at boundaries. The minimum number of sub-systems are used that can do the job. Due to the complex nature and poor access of the detectors, a high reliability and, in some cases, redundancy is required for the detector sub-systems. The need for internal redundancy applies also to the detection capabilities. This

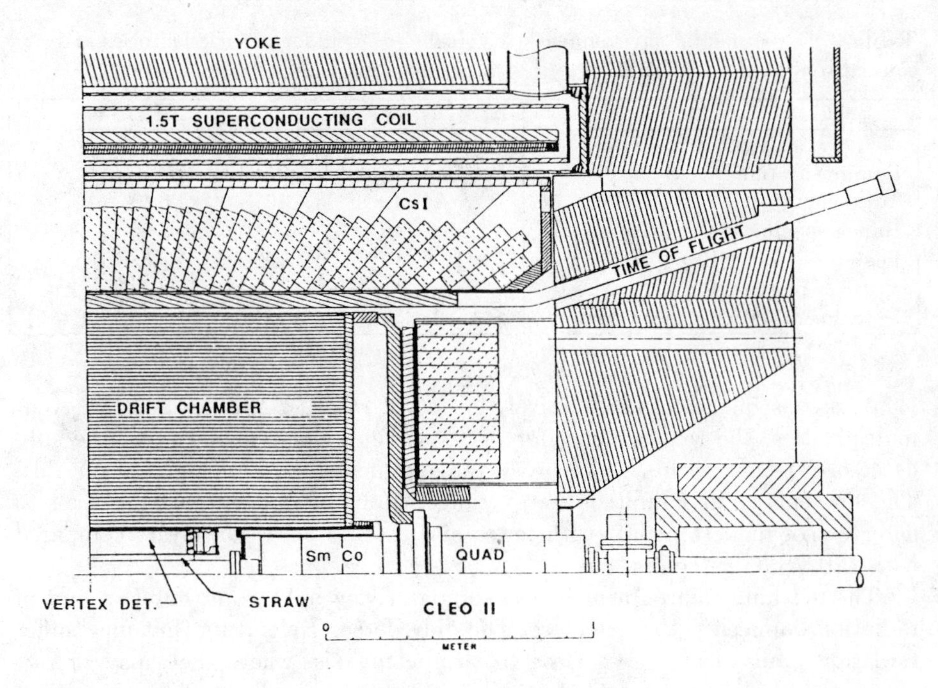

Fig. 6. The CLEO II detector at CESR.

improves the overall performance and, most importantly, provides cross-checks of the performance of the individual detector sub-systems; in many experiments, a thorough understanding of the detector performance is vital in achieving the desired level of systematic errors.

At particle factories, particularly close co-ordination is required between the designs of the machine and detector. The physics requirements shape not only the design of the detector but also the design of the machine in aspects such as luminosity, choice of beam energies, continuous injection, monochromator optics, polarized beams, stable beam-energy and its precise calibration, etc. The machine and detector are intimately connected and each places physical constraints on the other. The most obvious example is the presence of the final-focus elements—the $\mu\beta$ quadrupole magnets—which project deep inside the detectors. For example, at the τ-charm Factory, the radius of the superconducting $\mu\beta$ quadrupole cryostats is 25 cm and their inner faces are only 80 cm from the interaction point. The $\mu\beta$ quadrupole magnets are necessary to achieve tight focusing of the beams at the collision point, and are present in all factory designs.

2.3 Experimental Environment

The experimental environment at e^+e^- collider particle factories is rather mild compared with other future machines (Table 2), and can be accommodated with modest extrapolations of current experimental techniques.

Table 2. Experimental environment at typical e^+e^- collider particle factories and, for comparison, at the LHC/SSC.

	ϕF	τcF	BF	LHC/SSC
Luminosity ($cm^{-2}s^{-1}$)	5×10^{32}	10^{33}	3×10^{33}	$10^{33}(10^{34})$
Total energy (GeV)	1	4	10	14-40 TeV
Bunch spacing (ns)	3	40	10	15
Peak event rate (Hz)	2×10^3	$10^3 - 10^4$	10^2	$10^8(\times 10)$
$< n_{ch} >$ per event	3	4	8	100
Detector radiation dose (Gy/year)	$\lesssim 1$	$\lesssim 1$	$\lesssim 1$	$10^2 - 10^4(\times 10)$

As a consequence of the relatively low event rates (< 10 kHz) and low event multiplicities, the requirements for response time and granularity of the sub-detectors are not severe, and a broad range of detector technologies is suitable. The low energy and limited energy range of the secondary particles lead to detectors of modest depth, with fewer and thinner detection layers compared with LHC/SSC detectors.

The machine environment is also comparatively mild from the viewpoint of radiation damage to the detectors. The only detector elements that may suffer radiation damage are those close to the beam axis, where they may receive abnormal doses from unstable beam conditions such as during beam injection. The low event rates, together with the presence of circulating electrons rather than protons, mean that the detector can constitute the local shielding around the interaction point, resulting in good access to the local electronics of the detector.

Regarding machine-detector backgrounds, there are two basic types: synchrotron photons and beam-gas/wall interactions. The former contribute to the occupancy of the detector elements and to the aging of drift-chamber wires, especially in the region close to the vacuum chamber. Synchrotron radiation background is generally easier to control and of less concern than beam-gas/wall interactions since the latter lead to tracks in the detector and hence to events that are not easily rejected by simple triggers. However, experience has shown that the machine backgrounds at e^+e^- colliders can be sufficiently attenuated by suitable masks in the vacuum chamber near the interaction region and by adjustable aperture limits elsewhere in the ring. Of course, as yet there is no experience of the backgrounds from an *asymmetric* collider with transverse magnetic fields close to the interaction point, as will be the case for B factories. Here the main concern will be off-energy beam particles that are swept into the detector, but calculations[4] indicate that the backgrounds can be kept to acceptable levels by careful design of the masking.

2.4 Data Acquisition

The low event rates and multiplicities at e^+e^- particle factories (Table 2) result in data-flow rates that can be handled with present techniques: a 1 kHz event rate generates a data rate of about 20 Mbyte s^{-1}. However, in view of the short bunch-spacing, deadtimeless level 1/2 triggers are required, i.e. event data must be stored in pipelines while trigger decisions are being made. Furthermore, in order to avoid saturating the storage media and the off-line analysis, powerful on-line processors are required to analyze, select and compress the events of interest before writing them to tape or disk (an event rate of 20 Mbyte s^{-1} would fill an 8 mm cassette in only 100 s).

The trigger and data acquisition systems for factory detectors follow the same approach as those for experiments at future hadron colliders: pipelined front-end electronics, multi-level triggers and on-line processing of the events. Although the requirements are more modest for factory detectors than for hadron collider detectors, they are nevertheless sizeable by today's standards: on-line processing powers of $\mathcal{O}(10^4)$ MIPS (provided by a large "farm" of parallel processors) and data storage requirements of $\mathcal{O}(10^{14})$ byte per year. The processor farm and data storage requirements are doubled after including the off-line needs, which include re-analysis of the recorded data and the generation and analysis of Monte Carlo event samples comparable in size to those of the real data.

In view of the huge data sets that factories will generate each year, a challenging aspect of the experimental programme will be data storage, access and analysis. Although the bulk of the data will probably be stored on robot-controlled cartridges, a large amount of data [$\mathcal{O}(10)$ Tbyte per year] must be available on disk for rapid access. High-speed links are necessary to allow access to the central data library from computers at the collaborating institutes. A 2 Mbyte s^{-1} link is sufficient for most purposes except bulk data transfer, which is most efficiently executed by physically transporting cassettes containing selected data from the central library. High-speed network connections between the detector and the collaborating institutes are also important for remote trouble-shooting of the detector by experts.

3 ϕ Factory Detectors

3.1 Physics Goals

The main interest in the ϕ is as a pure source of $K_L^0 K_S^0$ events for investigating CP violation. The branching ratio $\phi \rightarrow K_L^0 K_S^0$ is large (34%) and the neutral kaons always appear as a $K_L^0 K_S^0$ pair since they are produced in a pure quantum state ($J^{PC} = 1^{--}$). The peak ϕ cross section is 4.4 μb, which corresponds to an event rate of 2.2 kHz at the DAΦNE reference luminosity of 5×10^{32} $cm^{-2}s^{-1}$[5, 6]. A ϕ Factory at this luminosity produces 7.6×10^9 $K_L^0 K_S^0$ pairs per year (10^7 s $\times L_{peak}$). In addition, large samples of K^+K^- (1.1×10^{10} events per year) and $\gamma\eta$ (2.8×10^8 events per year) are generated. These have additional physics interest and also provide important calibration signals for the detector.

The primary physics motivation[7] for a ϕ Factory is to search for **direct CP violation**, i.e. evidence that CP violation occurs in direct $\Delta S = 1$ decays and not only in $\Delta S = 2$ transitions (involving $K^0 \leftrightarrow \bar{K}^0$ mixing). Indirect CP violation (due to mixing) is characterized by the parameter ϵ which describes the mass eigenstates K_L^0 and K_S^0 in terms of the CP eigenstates K_1 and K_2 ($K_S^0 = K_1 + \epsilon K_2$ and $K_L^0 = K_2 + \epsilon K_1$). The parameter ϵ' characterizes direct CP violation, and a non-zero value would lead to a difference in the amplitude ratios $|\eta_{+-}|$ and $|\eta_{00}|$:

$$|\eta_{+-}| = \epsilon + \epsilon' = \frac{\mathrm{Ampl}(K_L^0 \to \pi^+\pi^-)}{\mathrm{Ampl}(K_S^0 \to \pi^+\pi^-)}$$

$$|\eta_{00}| = \epsilon - 2\epsilon' = \frac{\mathrm{Ampl}(K_L^0 \to \pi^0\pi^0)}{\mathrm{Ampl}(K_S^0 \to \pi^0\pi^0)}$$

Unfortunately, the Standard Model predicts extremely small values (few $\times 10^{-4}$) for the ratio ϵ'/ϵ. The present experimental error on ϵ'/ϵ from fixed-target experiments is 7×10^{-4} and the goal of the ϕ Factory is to measure this ratio to a precision of 10^{-4}. This will require a measurement of each of the branching ratios $\mathrm{Br}(K_L^0/K_S^0 \to \pi^+\pi^-/\pi^0\pi^0)$ to 5×10^{-4} accuracy, which could be statistically achieved with a sample of 5×10^{10} ϕ decays (requiring 10 fb^{-1}, or two years at a luminosity of 5×10^{32} $\mathrm{cm}^{-2}\mathrm{s}^{-1}$).

Further physics goals[7] of a ϕ Factory include:

- **Other searches for CP violation,** e.g. $K_S^0 \to \pi^0\pi^0\pi^0$ ($\mathrm{Br} \simeq 2 \times 10^{-9}$), which is just observable, corresponding to ≈ 30 events in one year.
- **Rare K_S^0 decays.** By tagging a K_L^0, a ϕ Factory effectively provides a pure K_S^0 beam of up to $\approx 10^{10}$ K_S^0 per year. This unique capability will lead to a substantial improvement in the measurements of many rare K_S^0 decays, such as $\pi^0\gamma\gamma$, $\pi^0\nu\bar{\nu}$, $l^+l^-\gamma$, $\pi^0 l^+l^-$, etc.
- **Rare $K^\pm$ decays,** e.g. $\pi^\pm\gamma\gamma$ (present Br limit $< 10^{-6}$) and $\pi^\pm\mu^+\mu^-$(flavour-changing neutral current decay; present Br limit $< 2 \times 10^{-7}$), etc.
- **Radiative ϕ decays,** e.g. $\phi \to \eta'\gamma$ ($\mathrm{Br} \simeq 1.2 \times 10^{-4}$) and $\phi \to f_0(975)\gamma$. The experimental sensitivity is $10^{-6} - 10^{-7}$.
- **Rare η decays,** e.g. searches for C violation in the electromagnetic interaction, e.g. $\eta \to l^+l^+\pi^0$and $\eta \to 3\gamma$.

3.2 Detector Requirements

The design of a ϕ Factory detector is driven by the requirements of the CP violation experiment to measure precisely the branching ratios for $K_S^0 \to \pi^+\pi^-$, $K_L^0 \to \pi^+\pi^-$, $K_S^0 \to \pi^0\pi^0$ and $K_L^0 \to \pi^0\pi^0$. The parent K_L^0/K_S^0 is distinguished by its observed decay length; the mean decay length ($\beta\gamma c\tau$) for a K_L^0 produced in ϕ decay is 343 cm, and for a K_S^0 it is 5.9 mm. A reconstructed decay having a vertex < 2 cm from the interaction point indicates a K_S^0 and tags a K_L^0 on the opposite side; a decay length > 20 cm indicates a K_L^0 and tags a K_S^0.

A large-radius tracking detector is therefore required in order to accept a sufficiently high fraction of the K_L^0 decays. The $\pi^\pm$ are quite soft (50-250 MeV/c) and so a relatively-weak magnetic field is optimal. The electromagnetic calorimeter is the most demanding component of the detector. The vertex position of each $\pi^0\pi^0$ decay must be measured to a precision $\sigma \leq 1$ cm. This is necessary to maintain acceptable event losses for tagged K_S^0, which are required to have an observed decay vertex inside a fiducial volume of 6 cm radius. (A major concern of the CP violation experiment is to achieve a precise understanding of the detection efficiencies for the four decays $K_L^0/K_S^0 \rightarrow \pi^+\pi^-/\pi^0\pi^0$, which are different and do not cancel.) In order to efficiently reject backgrounds from $3\pi^0$ decays, the calorimeter must be hermetic and fully efficient down to γ energies of 20 MeV. Finally, the energy resolution must be fairly good: equivalent to a Pb-glass calorimeter or better. The main features of a ϕ Factory detector are summarized in Fig. 7.

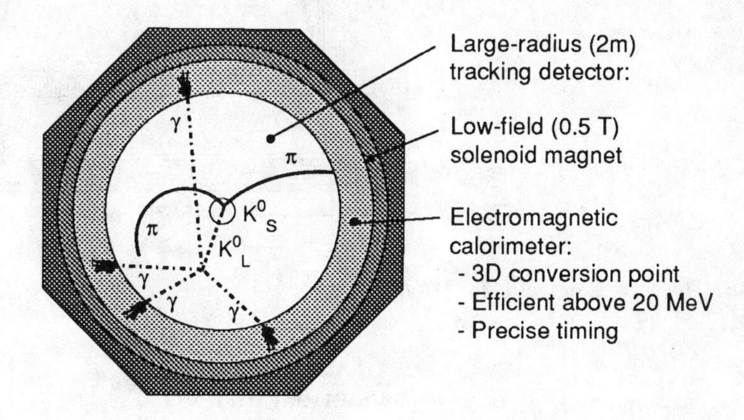

Fig. 7. Main features of a ϕ Factory detector.

3.3 Detector Example

The solutions to these requirements can be illustrated by the choices of the KLOE detector (Fig. 8)[8] at DAΦNE:

- **Tracking detector:** This is a 2 m-radius drift chamber filled with a He-based gas. A warm solenoid generates an axial magnetic field of 0.67 T. The drift chamber is filled with a He-based gas in order to reduce the multiple Coulomb scattering of the soft particles and also to minimize the regeneration of K_S^0 inside the tracking volume. Similarly, a large-radius (8 cm) vacuum chamber is used in order to ensure that all accepted K_S^0 decays are free of regeneration backgrounds. The expected momentum resolution is $\sigma_p/p = 0.3\%$ and the $K^0 \rightarrow \pi^+\pi^-$ mass resolution is $\sigma_M = 0.7$ MeV/c^2.
- **Electromagnetic calorimeter:** This is a Pb/SCIFI sampling calorimeter (Fig. 9) using very thin (0.5 mm) Pb layers and 1 mm diameter plastic

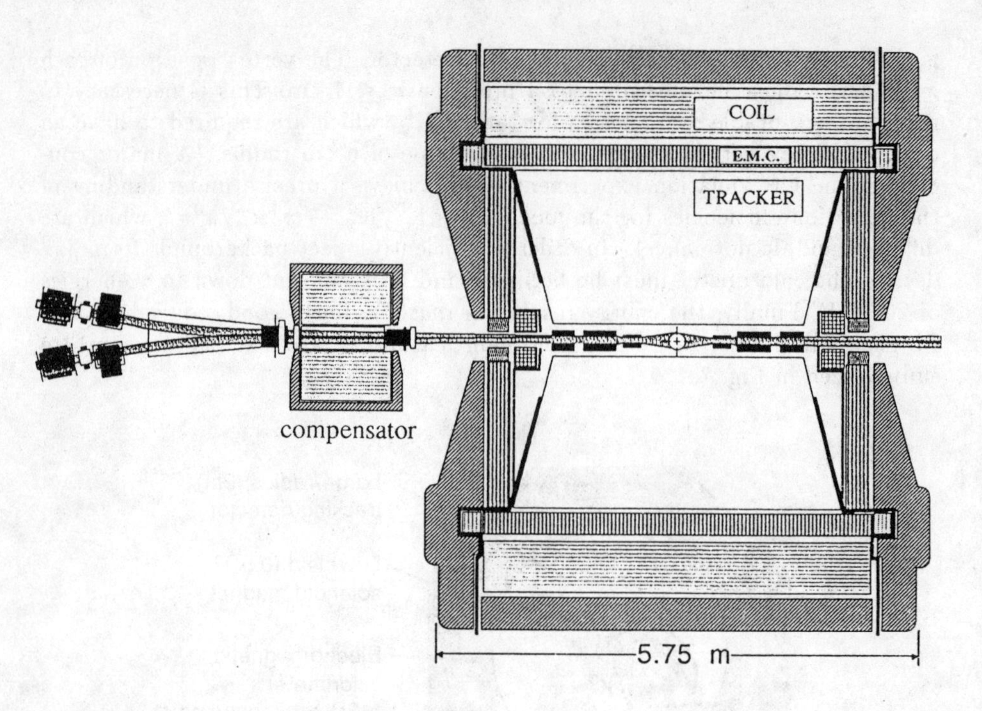

Fig. 8. The KLOE detector for DAΦNE.

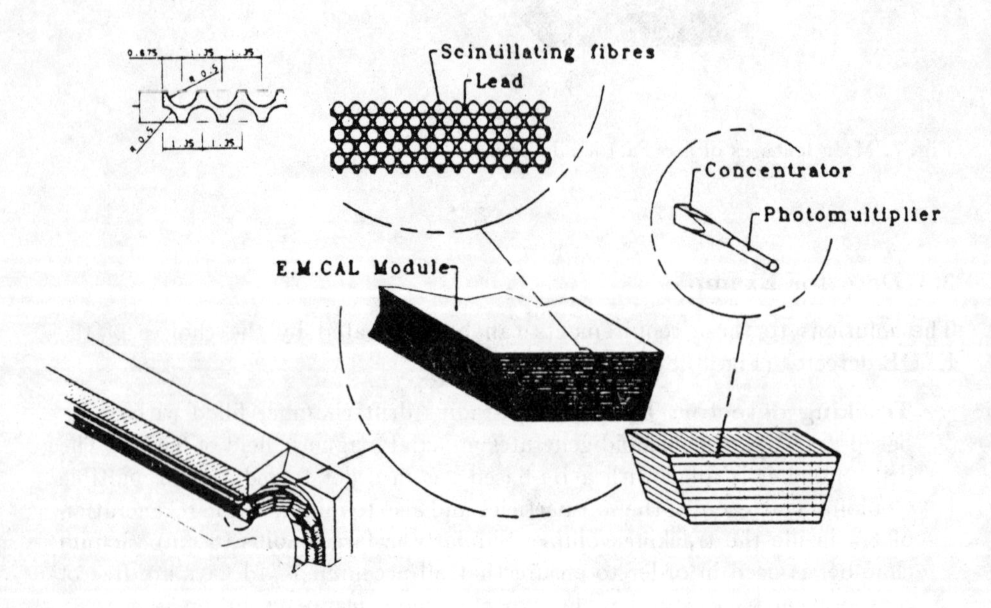

Fig. 9. The Pb/SCIFI electromagnetic calorimeter for KLOE.

scintillating fibres, which are read out with mesh phototubes located inside the magnetic field. The resulting sampling fraction is $\simeq 12\%$ and $X_0 = 1.6$ cm. This device measures the $\pi^0\pi^0$ vertex position by precise γ timing: a timing resolution $\sigma = 300$ ps is expected for the worst case, $E_\gamma \simeq 20$ MeV. By combining the 4 γ's in a $\pi^0\pi^0$ decay, an overall decay vertex precision $\sigma \leq 1$ cm can be achieved. The device is efficient for γ's down to 20 MeV and has an acceptable energy resolution ($\sigma_E/E \leq 5\%/\sqrt{E(\mathrm{GeV})}$). The resulting $\mathrm{K}^0 \to \pi^0\pi^0$ mass resolution is $\sigma_M = 3.4$ MeV/c^2.

4 τ-charm Factory Detectors

4.1 Physics Goals

The physics interest in a τ-charm Factory encompasses the τ and ν_τ leptons, charm (D^0, $\mathrm{D}^\pm$, $\mathrm{D_S^\pm}$, $\Lambda_c^\pm$, etc.) and charmonium (J/ψ, ψ', χ_c, η_c, etc.). The rates for $\tau^+\tau^-$, $\mathrm{D^0\bar{D}^0}$, $\mathrm{D^+D^-}$ or $\mathrm{D_S^+D_S^-}$ events are $1 - 3 \times 10^7$ per year ($L_{peak} = 10^{33}$ cm^{-2}s^{-1}; 10 fb^{-1} year^{-1}), depending on the beam energy (refer to Fig. 3). With standard optics, $\mathcal{O}(10^{10})$ J/ψ or ψ' events are generated per year (and up to 10^{11} J/ψ events per year may be possible with monochromator optics[9, 10]).

The primary physics goals[11, 12, 13] of a τ-charm Factory are probably the following:

1. **A search for finite ν_τ mass.** Finite mass neutrinos—especially the ν_τ—are a favourite explanation for the dark matter of the Universe and for the lack of solar neutrinos (since ν oscillations would then be possible). Two types of measurement will be made at a τ-charm Factory:
   - **Direct ν_τ mass measurement [$m > \mathcal{O}$(1 MeV/c²)].** This involves measurements of the end-point of the $5\pi^\pm$ mass spectrum in $\tau^\pm \to 5\pi^\pm\,\nu_\tau$, and of the end-point of the $K^-K^+\pi^\pm$ mass spectrum in $\tau^\pm \to K^-K^+\pi^\pm\nu_\tau$. With a two-year (20 fb^{-1}) sample at 3.67 GeV, the expected upper limit sensitivity (95% CL) is about 3 MeV/c^2, to be compared with the present limit of 31 MeV/c^2.
   - **Indirect ν_τ mass measurement [$\mathcal{O}$ (100 eV/c²)$< m < \mathcal{O}$ (1 MeV/c²)].** Any ν_τ heavier than $\mathcal{O}(100$ eV/c$^2)$ would have to decay in order to agree with the observed density of the universe. The decays $\nu_\tau \to \nu_l$ X ($\mathrm{l} = e, \mu$) may then occur, where X is a massless, weakly interacting, spin zero Goldstone boson (Majoron, familon, flavon, etc.). The τ lepton would then also be able to decay through same process $\tau^- \to \mathrm{l^-X}$, and calculated branching ratios are in the range $10^{-6} - 10^{-2}$. At a τ-charm Factory these decays have the distinctive signature of monochromatic leptons. The sensitivity limit is Br $\simeq 10^{-5}$.
2. **Precise measurements of the pure leptonic D decays.** Pure leptonic decays of $\mathrm{D}^\pm_{(s)}$ mesons can be rigorously calculated in the Standard Model and are important in determining the weak decay constants f_D and f_{D_s}

which occur in calculations of second-order-weak processes (mixing, CP violation, etc.). In addition, the ratio of the $D^{\pm}$ and $D_s^{\pm}$ pure leptonic decay rates will provide a precise determination of the ratio of the CKM matrix elements V_{cd} and V_{cs}. The largest pure leptonic branching ratios are $Br(D_s^+ \to \tau^+\nu_\tau) \simeq 3.3\%$, $Br(D_s^+ \to \mu^+\nu_\mu) \simeq 3.6 \times 10^{-3}$, and $Br(D^+ \to \mu^+\nu_\mu) \simeq 3.5 \times 10^{-4}$. Whereas none of these decays has been cleanly isolated so far, they can each be measured to a few per cent precision at a τ-charm Factory.

3. **A high-statistics study of gluonic matter and its spectrum.** This involves a systematic search for gluonia and hybrids in the gluon-rich environment of charmonium decays: $J/\psi \to \gamma gg$, ggg; $\eta_c \to gg$, and $\chi_c \to gg$.

Further physics goals[11, 12, 13] of a τ-charm Factory include:

- **A search for insights into the generation puzzle**—the reason why there appear to be three families of elementary particles in Nature. This programme involves a number of precise measurements of fundamental processes for which the Standard Model makes rigorous predictions. They include the following:
  - **$D^0\bar{D}^0$ mixing.** This process is highly suppressed in the Standard Model and so many of its extensions lead to enhanced rates which may be "background-free". Signatures of mixing are like-sign dileptons from dual semileptonic decays ($l^{\pm}l^{\pm}X$) or dual identical hadronic decays, such as $(K^+\pi^-)(K^+\pi^-)$ or $(K^-\pi^+)(K^-\pi^+)$. The expected rate for $D^0\bar{D}^0$ oscillations is $\simeq 10^{-5}$, which can be measured with one year's data at a τ-charm Factory.
  - **Precision τ branching ratios.** The goal is to measure all of the τ branching ratios with high precision—reducing the present errors by a factor of 10 or more. In particular, the one-prong decays—which are accurately known in the Standard Model—will be measured to $\mathcal{O}$ (0.1%) precision. Precise measurements of τ branching ratios may also provide the most accurate measurement of the strong coupling constant α_S.
  - **Current structure of the τ-ν_τ-W vertex.** Here the aim is to measure the nature of the weak current in τ decays to a precision comparable to present μ decay experiments. (The latter have provided the most precise measurements so far of the pure V-A nature of the weak decay current). The experiment involves a precise determination of the decay parameters—ρ_l, η_l, ξ_l, δ_l ($l = e, \mu$), and ξ'_μ—in the leptonic channels $\tau^- \to e^-\bar{\nu}_e\nu_\tau$ and $\tau^- \to \mu^-\bar{\nu}_\mu\nu_\tau$.
  - **A study of the D semileptonic branching ratios,** and the determination of the CKM matrix elements V_{cd} and V_{cs}. Many of the semileptonic branching ratios are measurable to an accuracy better than 1%, to be compared with the present errors of 12% for $D^0 \to K^-e^+\nu_e$ and 50% for $D^0 \to \pi^-e^+\nu_e$. With precise data on a broad range of semileptonic decays, the theoretical (hadronic form factor) uncertainties may be largely removed and precise values obtained for V_{cd} and V_{cs}.

– **Searches for CP violation.** It may be possible to observe direct CP violation by measuring differences in the decay parameters of hyperons and anti-hyperons produced in the processes

$$J/\psi \to \Lambda\,\bar{\Lambda} \to (p\pi^-)\,(\bar{p}\pi^+), \quad \text{and}$$

$$J/\psi \to \Xi^-\;\Xi^+ \to (\Lambda\pi^-)\,(\bar{\Lambda}\pi^+) \to (p\pi^-\pi^-)\,(\bar{p}\pi^+\pi^+)$$

Since the expected differences are small [e.g. $\mathcal{O}(10^{-4})$ in the decay asymmetry parameter α], very large statistics ($> 10^{11}$ J/ψ decays) are required. These may be reached if a monochromator optics can be successfully implemented.

– **Searches for physics beyond the Standard Model.** New physics may be found in almost any of the previously-described experiments. An additional approach to this question is to search for rare τ, D and J/ψ decays. The Br sensitivity is $\simeq 10^{-7} - 10^{-8}$.

4.2 Detector Requirements

The physics programme of a τ-charm Factory imposes broad requirements on the performance of the detector. There are experiments for which a primary concern is the quality of the momentum resolution (e.g. direct ν_τ mass measurement), of the lepton identification (e.g. indirect ν_τ mass measurement), of the hermeticity (e.g. pure-leptonic $D^\pm$ and $D_s^\pm$ decays), of the photon measurements (e.g. gluonium spectroscopy), and of the π/K separation (e.g. $D^0\bar{D}^0$ mixing). Simply stated, a high-performance *general-purpose* detector is required for a τ-charm Factory. The main features of a τ-charm Factory detector are summarized in Fig. 10.

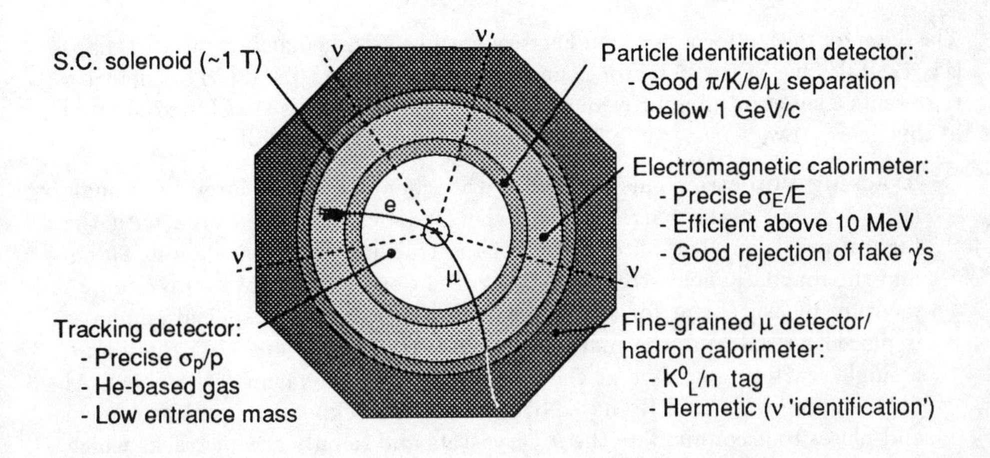

Fig. 10. Main features of a τ-charm Factory detector.

Table 3. Comparison of the performances of current and future detectors at τcF energies. A BF detector will have similar overall performance as a τcF detector. The symbol '$\oplus$' denotes addition in quadrature.

	Mark III (SPEAR) / BES (BEPC)	τcF (& BF)
Charged particles:		
Momentum res: σ_p/p(GeV/c)	$1.5\%p \oplus 1.5\%/\beta$ [MkIII] $0.7\%p \oplus 1.3\%/\beta$ [BES]	$0.3\%p \oplus 0.3\%/\beta$
p^{π}_{min}(MeV/c) for efficient tracking	80	50
Ω(barrel) ($\times 4\pi$ sr)	70%	90%
Photons:		
Energy resolution: σ_E/E(GeV)	$17\%/\sqrt{E}$	$2\%/E^{\frac{1}{4}} \oplus 1\%$
Angular resolution: $\sigma_{\theta,\phi}$ (mr)	10 [MkIII] 5 [BES]	$4/\sqrt{E}$
2γ angular separation: $\Delta\theta_{2\gamma}$ (mr)	20	50
E^{γ}_{min}(MeV) for efficient detection	100	10
Particle identification:		
h $\rightarrow$ e rejection	4% at 0.5 GeV/c	0.1% (10^{-5} inc. RICH)
h $\rightarrow$ μ rejection	5% at 1.0 GeV/c	1%/p(GeV/c) + 1%
$\pi \rightarrow$ K rejection	3σ at 0.7 GeV/c	3σ at 1.0 GeV/c (10^{-4} inc. RICH)
K^0_L/n detection efficiency	60%	95%
E^{ν}_{min}(MeV) for efficient ν tagging	-	$\simeq 100$

4.3 Detector Example

The detector design for a τ-charm Factory has been studied at several workshops [11, 12, 13]. One of the detector concepts is shown in Fig. 11[13]. This detector represents a substantial improvement over previous detectors that have operated in this energy range (see Table 3). The main features are as follows:

- **Tracking detector:** This involves a drift chamber with a large solid angle (90%$\times 4\pi$ sr) in the barrel region, where particles are measured with the highest precision. The outer radius of the tracking detector is about 85 cm and the magnetic field strength is 1-1.2 T. In order to achieve a precise momentum measurement for the (typical) low-energy tracks, special emphasis is placed on reducing the material in the tracking chamber to a minimum: a single scattering surface at the entrance (1 mm Be vacuum chamber), Al field-shaping wires and He-based drift gas. The tracking detector has conical end-plates to accommodate the $\mu\beta$ cryostats and to only accept tracks which come from the interaction point. This design minimizes the occupancy from background particles and improves the selectivity of the tracking trigger.
- **Electromagnetic calorimeter:** A CsI(Tl) [or CsI(Na)] crystal calorimeter is chosen in order to achieve high resolution measurements of γ's, and

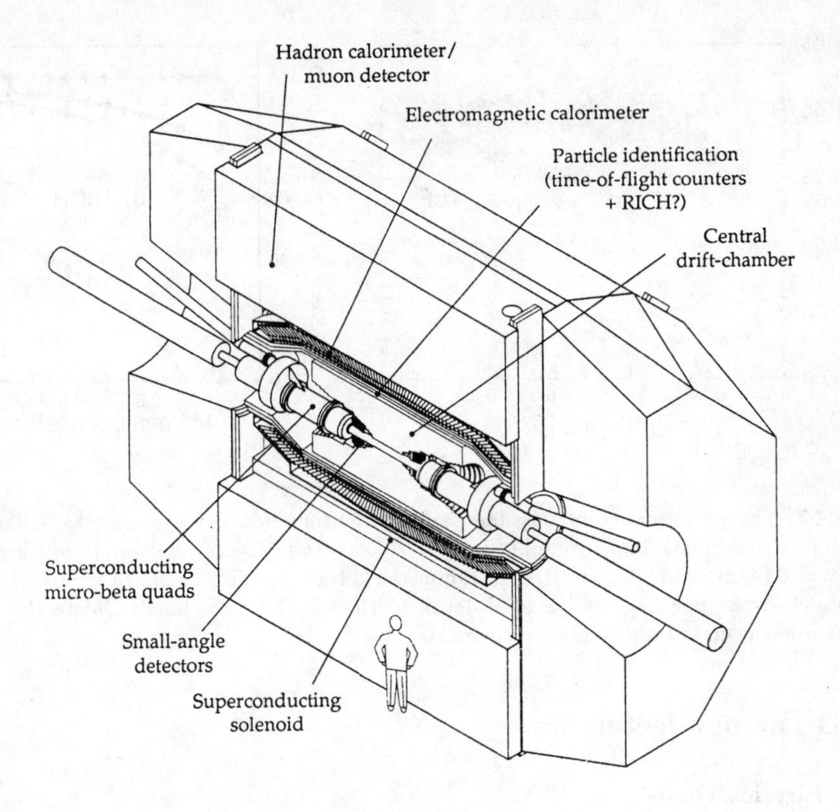

Fig. 11. A detector concept for a τ-charm Factory.

high efficiency down to low energies of $\mathcal{O}(10)$ MeV. A novel feature of this calorimeter is that each crystal is split into two longitudinal sections with separate photodiode readouts.

- **Particle identification:** High-quality particle identification in the region below 1 GeV/c is achieved by a combination of dE/dx measurements in the central drift chamber, time-of-flight measurements based on scintillating-fibre counters and, possibly, a fast RICH using either a solid NaF or liquid Freon (C_6F_{14}) radiator (see Fig. 12).
- **Neutrino "identification":** The presence of ν's is inferred on an event-by-event basis by ensuring that the detector is hermetic to all other particles. Charged particles and photons are measured with high efficiency and precision in the tracking detector and electromagnetic calorimeter (barrel and small-angle sections). The hermeticity is completed by means of a fine-grained outer hadron calorimeter/μ detector to tag the presence of neutral hadrons. Demanding upper limits are placed on the inefficiency for γ's in the electromagnetic calorimeter (<1%), and for K^0_L/n's in the hadron calorimeter (<5%).

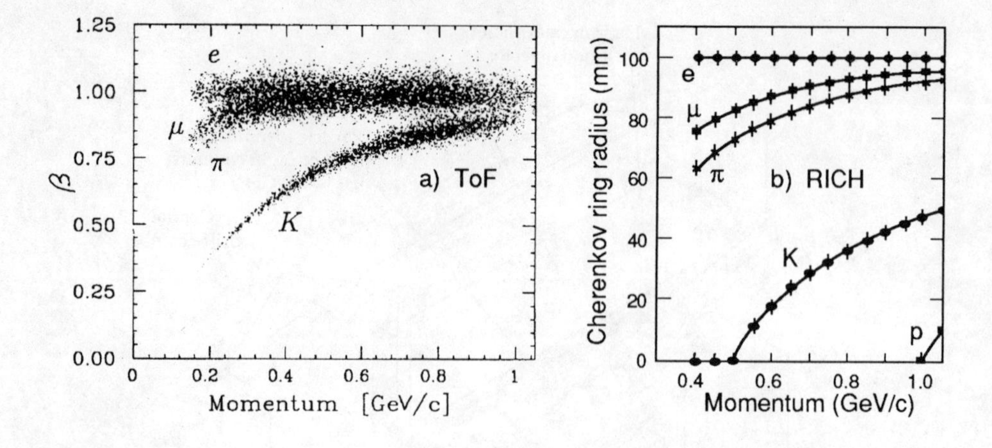

Fig. 12. The performance of possible particle identification detectors for a τ-charm Factory detector: a) Time-of-flight; $\sigma_{ToF} = 120$ ps (two layers) and minimum flight path = 0.95 m, and b) Fast RICH; 10mm-thick NaF radiator, a mean of about 20 photo-electrons per ring, and a total depth of 10 cm. The error bars indicate the 1σ measurement error on a single Cerenkov ring.

5 B Factory Detectors

5.1 Physics Goals

The physics interest of a B Factory[3],[14]–[21] is centered on bottom-flavoured particles (B_d^0, B_s^0, $B^\pm$, Λ_b^0, etc.), but there also exists an extensive programme of two-photon physics and of heavy-flavour physics which includes bottomonium (Υ, χ_b, η_b, etc.), the τ lepton and charm particles.

The primary physics goal is to investigate CP violation in the $B^0\bar{B}^0$ system. Near $b\bar{b}$ threshold there are several resonances where the B mesons are produced with enhanced cross sections and in simple final states. The most important is the $\Upsilon(4S)$ resonance at 10.6 GeV, which has a peak cross section of 1.1 nb (over a continuum level of 2.6 nb) and decays with equal probability to pure $B_d^0\bar{B}_d^0$ or B^+B^- final states.

The CP-violation experiments involve observing a first decay at time t_1 that identifies whether the particle is a B^0 or $\bar{B}^0$, e.g. a B^0 semi-leptonic decay to μ^+X or e^+X. This tags the presence of an accompanying pure $\bar{B}^0$ at time t_1, which subsequently propagates and may decay at time t_2 to a pure CP eigenstate, e.g. $J/\psi\ K_S^0$ ($\rightarrow l^+l^-K_S^0$). There are four possible different sequences for these decays depending on whether the tag or the CP eigenstate appears first, and whether it is a B^0 or $\bar{B}^0$. CP violation leads to differences between the rates of these time-ordered sequences. The experimental goal is to reach sufficient sensitivity to check whether the Standard Model description of CP violation—as a phase incorporated into the CKM mixing matrix—is correct. As well as the decay $B_d^0 \rightarrow$

$J/\psi\ K_S^0$, this will require measurements of other decays to CP eigenstates such as $B_d^0 \rightarrow \pi^+\ \pi^-$ and $B_s^0 \rightarrow \rho\ K_S^0$. Unfortunately, all of these branching ratios are very small [e.g. Br ($B_d^0 \rightarrow J/\psi\ K_S^0$) = 3×10^{-4}] and consequently, large luminosities are required. Calculations indicate that an integrated luminosity $L \simeq 50\ \mathrm{fb}^{-1}$ is required to cover most of the range of CP-violating effects allowed by the Standard Model. With this luminosity, the final number of reconstructed events is only $\approx 300 - 600$ from a parent sample of $3 \times 10^7\ B_d^0\bar{B}_d^0$ events. This leads to a minimum required luminosity for an asymmetric B Factory of $3 \times 10^{33}\ \mathrm{cm}^{-2}\mathrm{s}^{-1}$ (30 fb^{-1} per year).

Further physics goals[14]–[19] of a B Factory include:

- **$B_s^0\bar{B}_s^0$ mixing.** This measurement also needs an asymmetric machine and a precision vertex detector. The technique is to operate an asymmetric collider at the Υ(5S), which has a mass of 10.9 GeV/c^2 and lies above $B_s^0\bar{B}_s^0$ threshold. The signature of mixing is like-sign dilepton events which, because of the expected rapid mixing, will display an oscillating intensity vs. decay time difference $\Delta(t) = t_2 - t_1$.
- **Rare B decays.** Decays such as $B \rightarrow K^*\gamma$ and $B \rightarrow K^*l^+l^-$ involve so-called "loop" or "penguin" higher-order diagrams and are an interesting area in which to search for signs of new physics. None has been observed so far, and expected branching ratios are in the region $\approx 10^{-6}$, which is well within the experimental sensitivity of a B Factory. Other interesting classes of rare B decays are flavour-changing neutral currents (e.g. $B^0 \rightarrow \mu^+\mu^-$ and $B^0 \rightarrow e^+e^-$) and lepton-number-violating decays (e.g. $B \rightarrow \mu^+e^-$). The experimental sensitivity to these decays is Br $\simeq 10^{-8}$.
- **Hadronic and semi-leptonic B decays.** Measurements of semi-leptonic decays such as $B \rightarrow D^*l\nu$ and $B \rightarrow \rho l\nu$ will provide a precise determination of the CKM matrix elements V_{cb} and V_{ub}.
- **τ and charm physics.** A B Factory with $L = 3 \times 10^{33}\ \mathrm{cm}^{-2}\mathrm{s}^{-1}$ generates statistically comparable τ and charm data samples as a τ-charm Factory with $L = 10^{33}\ \mathrm{cm}^{-2}\mathrm{s}^{-1}$. Consequently, a B Factory will perform, with similar statistical precision, many of the experiments described previously for the τ-charm Factory. However, the backgrounds will generally be larger (e.g. b backgrounds may contribute) and systematic rather than statistical errors may limit the final precisions of certain experiments. Nevertheless, in general there will exist the important possibility of verifying precision τ-charm results on two different machines with independent systematic errors.
- **Υ physics.** Bound states of heavy quark-antiquark pairs provide an interesting testing ground for many aspects of QCD.
- **Two-photon physics.** Two-photon physics at a B Factory covers the mass region up to ≈ 3 GeV/c^2 and will therefore provide valuable information on aspects of light quark spectroscopy. In particular, since they contain no electric charge, glueballs should be strongly suppressed. The two-photon results should therefore complement the study of radiative J/ψ decays at a τ-charm Factory, where glueballs should be produced abundantly.

5.2 Detector Requirements

Measurements of the CP asymmetries in $B^0\bar{B}^0$ events require knowledge of the decay time of the B^0 and $\bar{B}^0$. Although the decay *time* is very small ($\simeq 1$ ps) and not directly measurable, it can be translated into a measurable decay *distance* by boosting the $\Upsilon(4S) \rightarrow B_d^0\bar{B}_d^0$ events [the mean B flight path is only 30 μm from an $\Upsilon(4S)$ at rest]. This has led to the design requirement of an *asymmetric* collider, i.e. having unequal beam energies, operating in the 10 GeV centre-of-mass energy region. Beam collision energies of about 9.33 GeV on 3 GeV represent the optimum compromise between sufficient boost and loss of detection efficiency due to increased collimation along the beam axis. At these beam energies, the average separation of the B_d^0 and $\bar{B}_d^0$ decay vertices is $\Delta z =$ 210 μm[19]. This separation can be measured by precision tracking techniques such as silicon strip detectors, which can provide a vertex separation resolution $\sigma(\Delta z) = 50$ μm.

Both the CP violation experiments and the rest of the physics programme at a B Factory impose broad requirements on the performance the detector. The overall requirements are similar to those of a τcF detector—i.e. a high-resolution general-purpose detector—but with the important additions of a precision vertex measurement and particle identification over an extended energy range (up to $\simeq 5$ GeV). Furthermore, in order to preserve a uniform acceptance in the centre-of-mass, the detector is generally extended along the direction of the high-energy (e^-) beam. The main features of a B Factory detector are summarized in Fig. 13.

5.3 Detector Example

There have been several conceptual designs prepared for B Factory detectors[3], [14]–[19]. Their overall performance is similar to a τcF detector (Table 3). As an example, the HELENA detector[19] for the proposed B Factory at DESY is shown in Fig. 14. The main features of this detector are as follows:

- **Vertex detector:** This comprises three layers of double-sided silicon strip detectors arranged in a forward and barrel part which cover $\theta = 12^o - 150^o$ (Fig. 15). The innermost layer is located as close as possible to the beam axis in order to achieve the best vertex precision; the vacuum chamber has a radius of only 25 mm. A vertex precision $\sigma_z = 30$ μm is expected for a strip pitch of 25 μm. A drawback of silicon strip detectors is their mass, which adversely affects the measurements of low-momentum charged particles; the barrel vertex detector represents $0.85\%/\sin\theta$ radiation lengths (which is comparable to the total material in the large central tracking chamber).
- **Tracking detector:** This is a cylindrical drift chamber of outer radius $\approx$1 m. As with a τcF detector, the main emphasis of the tracking detector is to reconstruct low momentum tracks with high precision. Also similar is the use of conical end-plates but, in this case, with an asymmetric geometry. However, rather than the traditional solenoid, the analyzing magnet involves a pair of Helmholtz coils which provide a magnetic field $B = 1$ T. The

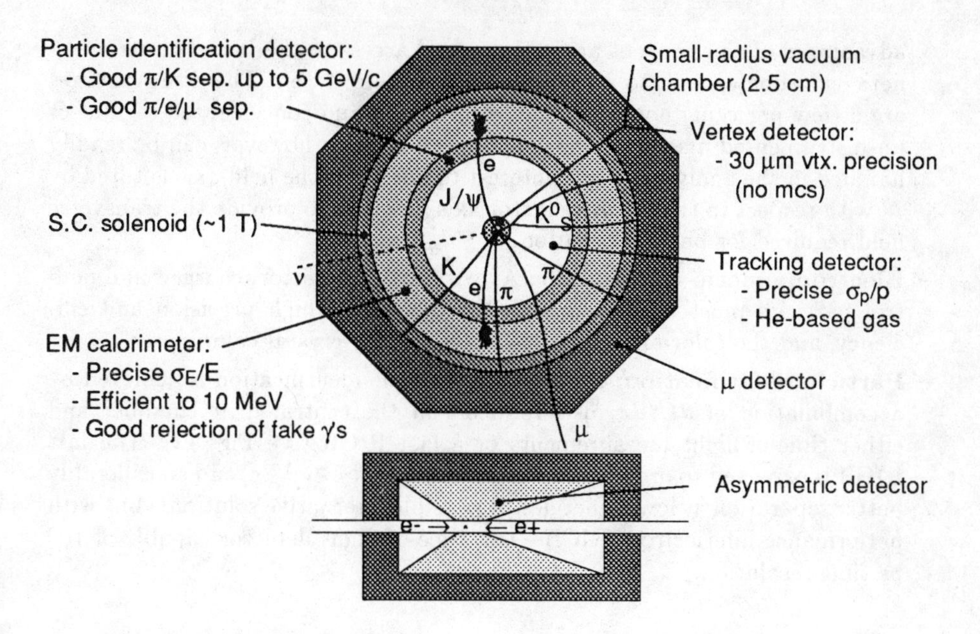

Fig. 13. Main features of a B Factory detector.

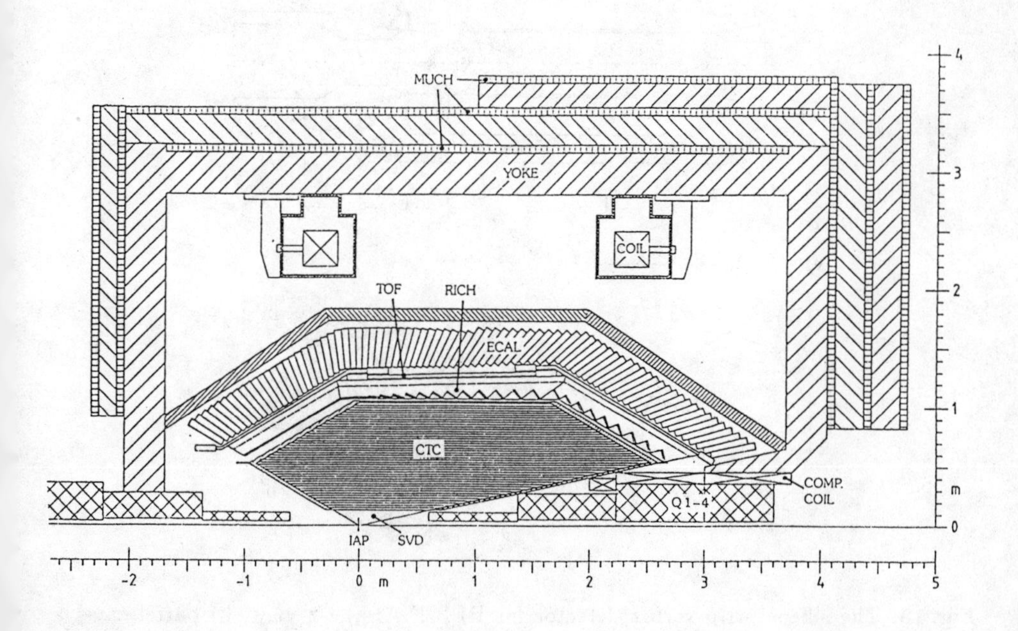

Fig. 14. The DESY B Factory detector, HELENA.

advantages of this magnet are an improved access to the barrel electromagnetic calorimeter and, to a lesser extent, a reduced cost. The disadvantages are a (few per cent) non-uniformity of the field and concentrated regions of un-instrumented material. The field non-uniformity, however, can be readily handled in the analysis. It is of interest to note that the field axis is tilted by 7° with respect to the beam/detector axis, in order to provide the transverse field required for beam separation.

- **Electromagnetic calorimeter:** As with a τcF detector, charged and neutral particles must be measured with comparable high precision and efficiency, and the calorimeter of choice is a CsI(Tl) crystal calorimeter.
- **Particle identification:** High-quality particle identification is achieved by a combination of dE/dx measurements in the central drift chamber and either time-of-flight measurements or a fast RICH (see Fig. 12). The fast RICH is expected to provide 3σ π-K separation at 5 GeV/c, and considerably better separation at lower energies. A possible alternative solution—but with performance inferior to a RICH—is a time-of-flight detector capable of 100 ps time resolution.

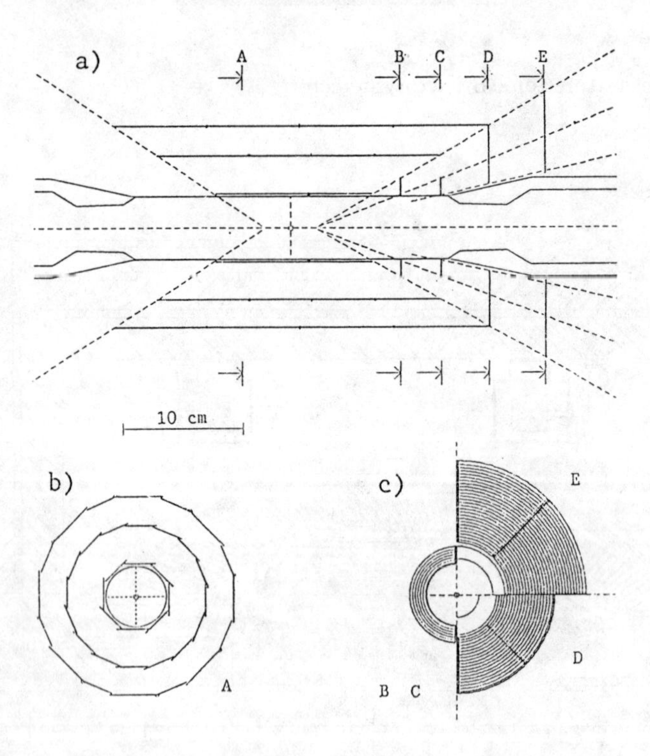

Fig. 15. The silicon strip vertex detector for HELENA: a) r-z view, b) barrel cross section at A, and c) forward modules B-E.

6 Conclusions

Progress in particle physics will require particle factories to complement the high-energy colliders. All factories have the same basic goal: to investigate and extend the Standard Model by precision experimental studies of the known quarks and leptons. High statistics and low backgrounds are the primary requirements for these experiments and so the luminosities required for particle factories are high: 10 to 100 times the presently-achieved values. However, to take full advantage of these high statistics will require very careful control of the experimental systematic errors. This, in turn, means that both the machines and the detectors must be carefully optimized for specific physics.

The required performances for the detectors, triggers and data acquisition systems can be met with present technologies or their reasonable extrapolations. The main challenges will be calibrations and the control of detector systematics. Detectors at e^+e^- collider particle factories will attempt to carry out experiments with characteristic precisions that have previously been almost exclusively the realm of fixed-target experiments (with several generations of experience). Few, if any, of the experiments will be easy.

The experiments at ϕ, τ-charm and B factories both complement and overlap each other, providing vital independent cross-checks of the results. In this way the particle factories should together be able to provide a sharp and reliable picture of the precise nature of the present fundamental particles and their interactions.

7 Acknowledgements

I would like to thank Jonathan Dorfan, Paolo Franzini, David Hitlin, Luciano Maiani, Fumihiko Takasaki and Maurice Tigner for providing me with material for this paper. I am indebted for many discussions with my colleagues in Europe, Russia, the United States and elsewhere who are developing the τ-charm Factory. Finally, I would like to thank Mel Month, Stuart Turner, Ted Wilson and all the staff of the Joint US-CERN Accelerator School for a most informative and enjoyable meeting.

References

1. *Frontiers of Particle Beams; Factories with e^+e^- Rings*, Benalmadena, Spain, 29 October - 4 November 1992, these proceedings.
2. A.M. Boyarski et al., *An Experimental Survey of Positron-Electron Annihilation into Multiparticle Final States in the Center of Mass Energy Range 2 GeV to 5 GeV*, SLAC Proposal SP-2, (1971).
3. M. Ogg et al., *Detector for a B-Factory*, CLNS 91-1047 (1991).
4. R.D. Ehrlich, *Backgrounds at e^+e^- B Factories*, in [1].
5. C. Biscari, *ϕ Factory Design*, in [1].
6. Proc. *Workshop on Physics and Detectors for DAΦNE*, INFN - Laboratori Nazionali di Frascati, Rome, Italy, 9-12 April 1991, ed. G. Pancheri (1991).

7. *The DAΦNE Physics Handbook*, INFN - Laboratori Nationali di Frascati, Rome, Italy, eds. L. Maiani, G. Pancheri and N. Paver (1992).
8. A. Aloisio et al., *KLOE; a General Purpose Detector for DAΦNE*, KLOE Note No. 1 (1992).
9. A. Zholents, *Polarized J/ψ Mesons at a Tau-Charm Factory with a Monochromator Scheme*, CERN SL/92-27 (1992); A. Faus-Golfe and J. Le Duff, *A Versatile Lattice for a Tau-Charm Factory that Includes a Monochromatization Scheme*, LAL-Orsay preprint LAL/RT 92-01 (1992).
10. J.M. Jowett, *Lattice and Interaction Region Design for τ-charm Factories*, in [1].
11. Proc. *Tau-Charm Factory Workshop*, SLAC, California, USA, 23-27 May 1989, ed. L.V. Beers, SLAC-Report-343 (1989).
12. Proc. *Workshop on JINR c-tau Factory*, JINR, Dubna, Russia, 29-31 May 1991, eds. V.A. Bednyakov and G.A. Chelkov (Dubna, 1992).
13. Proc. *Meeting on the Tau-Charm Factory Detector and Machine*, Sevilla, Spain, 29 April-2 May 1991, eds. J. Kirkby and J.M. Quesada (Univ. Sevilla, 1992).
14. *Proposal for an Electron Positron Collider for Heavy Flavour Particle Physics and Synchrotron Radiation*, Paul Scherrer Institute, Villigen, Switzerland, PSI-PR-88-09 (1988).
15. *Feasibility Study for a B-Meson Factory in the CERN ISR Tunnel*, CERN, Geneva, Switzerland, ed. T. Nakada, CERN 90-02 (1990).
16. Proc. *Workshop on Physics and Detector Issues for a High-Luminosity Asymmetric B Factory at SLAC*, January-June 1990, ed. D. Hitlin, SLAC-373 (1991); *An Asymmetric B Factory Based on PEP; Conceptual Design Report*, SLAC-372 (1991).
17. S. Stone et al. (CESR-B), *Physics rationale for a B Factory*, CLNS 91-1043 (1991); *CESR-B; Conceptual Design for a B Factory Based on CESR*, CLNS 91-1050 (1991).
18. Proc. *Workshop on Physics and Detectors for KEK Asymmetric B-Factory*, KEK, Tsukuba, Japan, 15-18 April 1991, eds. H. Ozaki and N. Sato, KEK Proceedings 91-7 (1991); F. Takasaki et al., *Progress Report on Physics and Detector at KEK Asymmetric B Factory*, KEK Report 92-3 (1992).
19. H. Albrecht et al., *HELENA, a Beauty Factory in Hamburg*, DESY 92-041 (1992).
20. M.S. Zisman, *B Factory Design*, in [1].
21. S. Kurokawa, *KEK B-Factory Project*, in [1].

The Role of a B Factory in the U.S. Program

M.S. Witherell

Physics Department, University of California, Santa Barbara, CA 93106-9530, USA

One of the most important physics issues in particle physics today is understanding the source of CP violation. The experimental signature of the Standard Model explanation of CP violation is the existence of large, predictable asymmetries in the decay of the $B^0(\overline{B}^0)$ into CP eigenstates. The asymmetric $e^+ e^-$ B factory was conceived to provide a thorough test of these predictions. In this talk I discuss the particle physics and the detector issues connected with such a machine.

1 Introduction

At the present stage of particle physics, we are in the remarkable and unprecedented situation of having a mathematically consistent theory, the Standard Model of Particle Physics, which describes all of the experimental data available. In the Standard Model, there are three generations of quarks, each containing two quarks, and three generations of leptons. We have a gauge theory of the electroweak interactions, $SU(2) \times U(1)$, and a gauge theory of the strong interactions, QCD, all with associated gauge bosons. Finally, there is the Higgs scalar boson, whose interactions are responsible for the masses of the other particles.

The aims of particle physics are to complete the experimental verification of the Standard Model, to search for physics outside it, and, if there is new physics, to learn how the theory must be extended to accommodate it. In the next decade, three experimental goals stand out as providing particularly decisive tests of the Standard Model. One is the search for the Higgs particle, or if there is none, the alternative cause of spontaneous symmetry breaking. The Higgs search will be carried out at the highest energy hadron colliders, the SSC and LHC. Another goal is the search for the top quark, and the measurement of its mass, which is now underway at the Tevatron collider. The third experimental goal, and the subject of this talk, is the complete study of CP violation phenomena in the B meson system.

There is only one manifestation of CP violation which we can directly measure, that in the K^0 system, which was first seen 28 years ago. We believe that this CP violation arises in the cross coupling between quark generations. Of the 18 observable fundamental parameters in the Standard Model, 10 are directly related to quark properties: six quark masses and 4 parameters in the quark mixing matrix, or Cabibbo-Kobayashi-Maskawa (CKM) matrix. The only way to account for CP violation within the Standard Model is as a consequence of the properties of the CKM matrix.

2 The CKM Matrix

The W^+ boson decays into the three lepton generations, $e^+\nu_e$, $\mu^+\nu_\mu$, and $\tau^+\nu_\tau$, with equal rates, described with the universal weak coupling constant g. On the other hand, quark flavor is not conserved, and the W can decay into any of the nine possible $Q\overline{q}$ combinations. (Here Q is a u, c, or t quark and q is a d, s, or b quark. Of course the W boson cannot decay to the heavier top quark because of energy conservation, but the same coupling applies for the top quark decaying into a W.) For the quark decays, the amplitude is given by $g\ V_{Qq}$, where the nine parameters V_{Qq} are the elements of the CKM matrix.

This extra factor in the weak coupling of quarks comes about because the quark mass eigenstates are not the weak eigenstates. We can define the weak eigenstates d', s', and b' such that the W couples only to $u\overline{d'}$, $c\overline{s'}$, and $t\overline{b'}$. Then the CKM matrix represents a rotation between the d, s, b basis and the d', s', b' basis:

$$V_{Qq} = \begin{pmatrix} V_{ud} & V_{us} & V_{ub} \\ V_{cd} & V_{cs} & V_{cb} \\ V_{td} & V_{ts} & V_{tb} \end{pmatrix}. \tag{1}$$

Because V represents a rotation, it is a unitary matrix. It is almost diagonal, but the off-diagonal elements represent very important physics. It takes four separate parameters to describe the matrix, and of these one is a CP-violating phase. Thus with three quark generations there is just enough freedom to allow for CP violation.

With the single phase in the CKM matrix, we have a natural explanation for CP violation within the Standard Model. We have no convincing evidence to confirm the explanation however. Either CP violation does come from this phase, and is an intrinsic part of the Standard Model, or it is the first sign of physics outside the Standard Model. This is the question that must be resolved by studying B decays. Eventually, precise measurements of the CKM matrix should lead us to relationships among the quark masses and mixings, and from that a deeper understanding of the CP phase.

If we apply the unitarity condition to the first and third columns in the CKM matrix, we get the expression

$$V_{ud}V_{ub}^* + V_{cd}V_{cb}^* + V_{td}V_{tb}^* = 0. \tag{2}$$

Using the fact that the diagonal elements are approximately unity, we get a triangle relation between the off-diagonal elements:

$$V_{ub}^* + V_{td} = -V_{cd}V_{cb}^*. \tag{3}$$

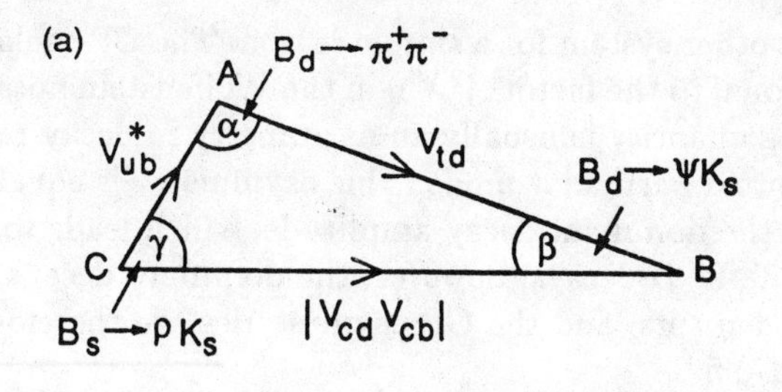

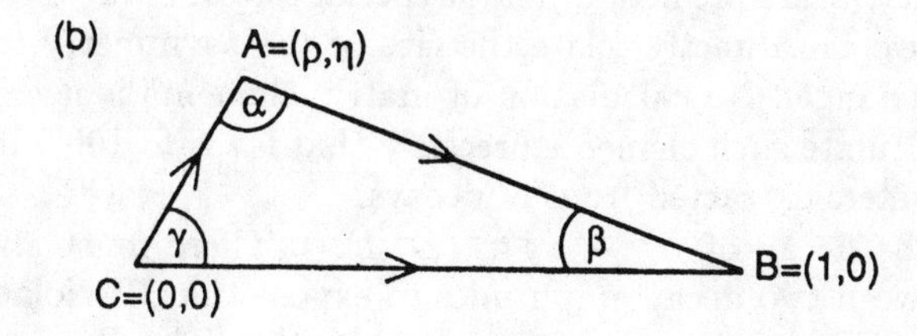

Fig. 1. (a) Representation in the complex plane of the triangle formed by the CKM matrix elements using the relation $V_{ub}^* + V_{td} = |V_{cd}V_{cb}|$. (b) The rescaled triangle, with vertices at A(ρ, η), B(1,0), and C(0,0). Each angle is related to the CP-violation parameter for one class of B decay.

This triangle in the complex plane is shown in fig. 1(a). The amplitude for all CP-violating processes is proportional to the area of this triangle.

The Wolfenstein parameterization for the CKM matrix gives a useful way of displaying the four free parameters:

$$\begin{pmatrix} 1-\lambda^2/2 & \lambda & A\lambda^3(\rho - i\eta) \\ -\lambda & 1-\lambda^2/2 & A\lambda^2 \\ A\lambda^3(1-\rho-i\eta) & -A\lambda^2 & 1 \end{pmatrix}. \tag{4}$$

It uses an expansion in powers of $\lambda = V_{us} = 0.22$. The smallest matrix elements are those representing the transitions from the first to third generations, V_{ub} and V_{td}. The imaginary phase which causes CP violation is included in these elements. By dividing the lengths of the triangle in fig. 1(a) by the base, one gets the triangle in fig. 1(b). This triangle has unit base, and it is completely described by the two parameters ρ and η.

3 CP Violation in *B* Decays

A striking experimental signature of the Standard Model explanation for CP violation is the existence of large, predictable asymmetries in B decays. They are larger than in any other system for a simple reason. The CP-violating amplitude is always proportional to the factor $A^2\lambda^6\eta$ in the Wolfenstein notation. The measured CP-violating quantity is usually an asymmetry in decay rates for particle and antiparticle into a particular mode. This asymmetry is equal to the ratio of this amplitude to the dominant decay amplitude, which leads to an asymmetry of about 10^{-3} in K. In B decays, however, the dominant decays are suppressed by smaller CKM elements, and the CP asymmetries are therefore much larger, in the range of 0.1-0.5.

The asymmetries are predictable for the special case of $B^0(\overline{B}^0) \rightarrow$ CP eigenstate. For this case we can directly relate the size of the asymmetry to the angles of the unitarity triangle. No calculation of matrix elements is necessary. It is our inability to calculate such elements precisely that leads to 100% uncertainties in the CP parameters extracted from K decays.

Consider the decay of $B^0 \rightarrow \pi^+\pi^-(D^+D^-)$. There must always be an interference between two decay amplitudes to expose the CP-violating phase. For these decays to CP eigenstates, mixing provides this. The B^0 can decay directly to the final state, or it can mix into a $\overline{B}^0$ and then decay. Particle-antiparticle mixing is a second-order weak interaction proportional to V_{td}^2 which is proportional to $(1-\rho-i\eta)^2$.

Calculating the decay rate for this mode from the sum of the mixed and unmixed amplitudes, one gets

$$\frac{d\Gamma}{dt} \propto e^{-\Gamma t}[1 \pm \sin(2\beta) sin(x\Gamma t)]. \tag{5}$$

The sign of the interference depends on whether the initial state is B^0 or $\overline{B}^0$. The amplitude of the oscillation is $\sin(2\beta)$ for decays to final states such as D^+D^-, $\sin(2\alpha)$ for $\pi^+\pi^-$ and related decays.

For e^+e^- colliders running at the $\Upsilon(4s)$ the situation is further complicated by Bose symmetry. The two B mesons produced stay in a particle-antiparticle state until the time of the first decay. For example, at time t, the $\overline{B}^0$ decays into a tagging mode which can measure that it is not a B^0. At that instant, the other particle must be a B^0. Some time later, it decays into a CP mode. Then the time dependence is

$$\frac{d\Gamma}{dt} \propto \exp(-[t_1+t_2]/\tau) \times [1 \pm \sin(2\beta)\sin(x[t_2-t_1]/\tau)]. \tag{6}$$

The time evolution of the mixing is governed by the difference in the two decay times. Integrating over all times, the asymmetry integrates to zero. It is therefore necessary to measure the difference in decay times of the two B mesons.

This is the physical phenomenon that motivates the design of the B factory. In order to measure this $\Delta t = t_2 - t_1$ in a reaction with no accompanying particles, we need to have the B mesons moving in the same direction at relativistic velocities. Then the distance between the two decay vertices is proportional to the decay time difference. For a machine of approximately 3 GeV on 9 GeV, this distance is about $180\mu m$.

4 The Experimental Status of the CKM Matrix

Our knowledge of the CKM matrix can be summarized by giving the allowed range for the two parameters ρ and η. It is determined by five measurements: 1) the top quark mass, which can only be inferred indirectly from electroweak measurements, especially at LEP; 2) V_{cb} from $b \to c$ semileptonic decays; 3) CP violation parameters from $K_l \to 2\pi$; 4) the rate for B^0–$\overline{B}^0$ mixing; and 5) V_{ub} from $b \to u$ semileptonic decays. The effect of all of these measurements is shown in figure 2. The allowed region is shaded.

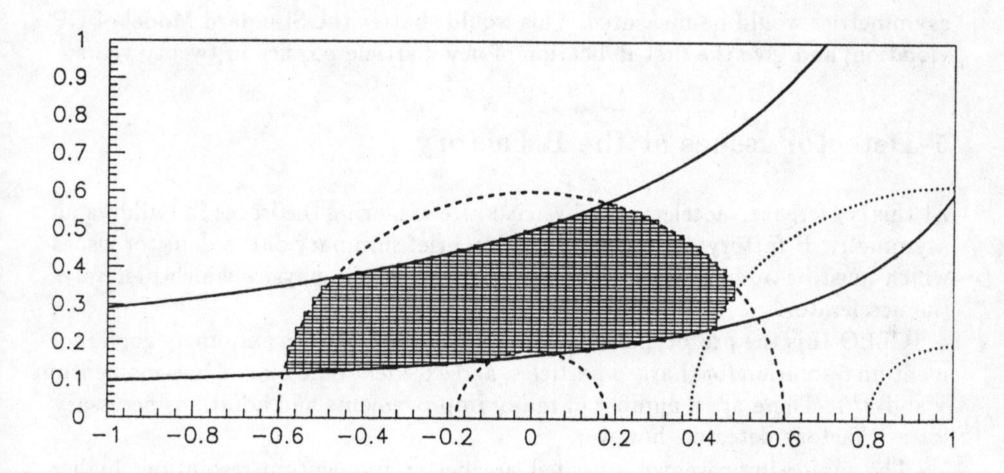

Fig. 2. Constraints on the coordinates of point A of the rescaled unitarity triangle, for $m_t = 150$ GeV and $V_{cb} = 0.044$. The coordinates are ρ (horizontal) and η (vertical). The constraints come from measurements of $|V_{ub}/V_{cb}|$ (dashed circles), x_d (dotted circles), and ϵ (solid hyperbolas). The crosshatched region is allowed.

As one can see from this plot, the allowed region is very large on the relevant scale. Remember that the angles α and β are just the angles of the triangle formed with the points (0,0), (1,0), and (ρ, η). The amplitudes for the CP asymmetries are therefore $\sin(2\beta) = 0.15 - 1.00$ and $\sin(2\alpha) = 0.0 - 1.0$. The expected asymmetries can be very large, but they are not well predicted until the non-CP quantities are better known.

There will be great progress in narrowing down the allowed region over the next eight years or so. The top quark mass will be well measured by CDF and D0. The B decays which determine V_{cb} and V_{ub} will be measured better by CLEO, and even more precisely at a B factory. Mixing of B_s mesons might be measured at the Tevatron or at LEP. The theoretical interpretation of B^0 mixing should become more precise as lattice gauge calculations are improved. Precise measurement of direct CP violation in K decays at Fermilab and CERN should add an additional constraint. Thus new experimental information from every laboratory will improve our knowledge of the CKM matrix.

In a plausible scenario of the year 2001, the allowed area in (ρ, η) space from these measurements would be narrow indeed, an order of magnitude smaller than it is today. We could make sharp predictions for the CP asymmetries, approximately ± 0.2 on $\sin(2\beta)$ and ± 0.1 on $\sin(2\alpha)$. With a run at an asymmetric B factory of $100 fb^{-1}$, one could measure these asymmetries with values like $\sin(2\beta) = 0.37 \pm 0.06$ and $\sin(2\alpha) = 0.93 \pm 0.11$. These measurements would provide a stunning experimental verification of the source of CP violation, and end a 35-year search. The other possibility is that some completely different asymmetries would be measured. This would shatter the Standard Model of CP violation, and give the first indication of new particle physics in twenty years.

5 Detector Issues at the B Factory

At this conference, accelerator physicists are exploring the issues in building an asymmetric B factory. I am going to give a brief summary of the detector issues which must be addressed for the experiment to do the physics which motivates the accelerator.

CLEO-II is the prototype for a B factory detector. It has extremely good momentum resolution for charged particles, and excellent detection of low-momentum γ and π^0's. There are a number of major improvements and additions necessary for a B factory detector, however.

The major improvements needed are better momentum resolution, higher rate capability, and larger acceptance at forward angles. The last item is required because of the asymmetric energy configuration. the additions are a silicon vertex detector, to measure the time evolution of the CP asymmetries, and a detector which identifies hadrons at high energy, probably some form of Cerenkov counter. There is an intense R & D program under way at many laboratories throughout the high energy physics community to develop these detectors needed to do B factory physics.

The single most important feature of the B factory detector is probably the vertex detector, for it is the time evolution of CP asymmetries which motivates the asymmetric design of the B factory. If one cannot measure the vertices well enough, there is no point building the machine. There are silicon detectors operating at many experiments around the world. The B factory detector must have good resolution and segmentation in both dimensions, and thus use either crossed-strip or pixel technology. The new features which are being developed for B factory work are a low-mass mounting scheme, faster readout electronics, and good performance at forward angles.

Figure 3 shows the vertex detector now being constructed for installation in CLEO in 1994. The low-mass mounting and crossed-strip silicon detectors needed for the B factory are implemented in that detector. For the first time, this detector will face many of the practical design problems common with the B factory. In the design of the detector for the luminosity upgrade of CESR, CLEO proposes to extend this with silicon detectors out to a radius of 12 cm. This will provide excellent measurement of angles, as well as precise measurement of impact parameters and vertices. The silicon then operates not only as a vertex detector, but as a full tracking detector, with full responsibility for low-momentum tracks.

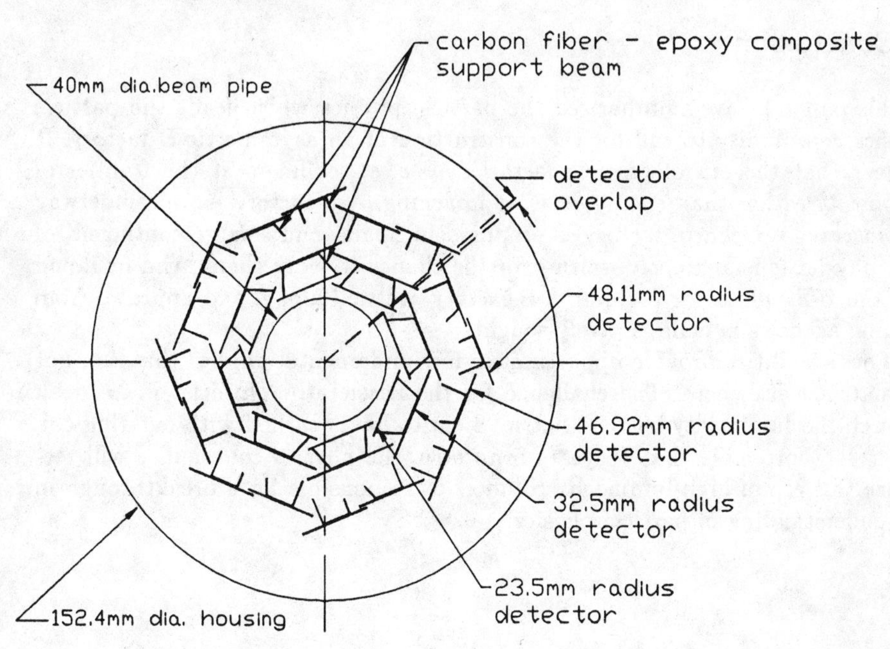

Fig. 3. Cross section of CLEO silicon vertex detector, taken in the r–ϕ plane, near $z = 0$. The structure is supported at the ends.

There are a number of important machine issues at the interface of the detector and the accelerator. For stability of machine components near the interaction region, it is desirable to have a support pipe, of a low-mass composite, which passes through the detector. On the other hand, it compromises the tracking resolution of the detector. One must carefully compare the costs and benefits of such a pipe before deciding whether it is justified.

The assembly of the inner parts of the detector and the machine components in the region of the experiment are inextricably intertwined. In addition, the background limits for the silicon and drift chamber are critical constraints. Probably more than at any previous machine, there has been constant interaction between the accelerator physicists responsible for delivering a high-luminosity machine and high energy physicists who are building the experiment.

There are some conclusions one can draw about the status of the detector for the B factory. Most of the requirements for a B factory detector can be met with components that exist today, or with reasonable extrapolation. There are a few well-defined topics for detector R & D: silicon vertex detector, low-mass drift chamber, particle identification, electronics and data acquisition, and background limits are probably the most important. In general, however, the machine is a greater challenge than the detector.

6 Conclusion

In this paper I have summarized the particle physics which leads the particle physics community to call for the construction of an asymmetric B factory. It is clear that the era of particle factories — e^+e^- colliders at the frontier of luminosity rather than energy — is approaching. A ϕ factory is well underway at Frascati. A τ-charm factory is proposed in Spain, and a large contingent of U.S. physicists is anxiously waiting for the chance to work there. And in Japan and the U.S., plans for building a B factory are well along, and approval from funding agencies is being actively sought.

These facilities provide opportunities for fundamental physics that can not be matched elsewhere. The challenge for the accelerator physicist is to break through the luminosity barrier that we seem to have reached with one-ring colliders. It is probable that 10 years from now the physics community will recognize this era of high-luminosity colliders as responsible for a breakthrough in our understanding of particle physics.

List of Participants

BACONNIER, Y.	CERN, Geneva, Switzerland
BAGLIN, C.	CERN, Geneva, Switzerland
BARBADILLO-RANK, M.	University of Seville, Spain
BISCARI, C.	INFN, Frascati, Italy
CALVINO, F.	ETSEIB, Barcelona, Spain
CASTRO, P.	CERN, Geneva, Switzerland
CHAUTARD, F.	SSCL, Dallas, TX, USA
DE RUJULA, A.	CERN, Geneva, Switzerland
DELAHAYE, J.-P.	CERN, Geneva, Switzerland
DONALD, M.H.R.	SLAC, Stanford, CA, USA
EHRLICH, R.	Cornell University, Ithaca, NY, USA
ERICKSON, R.	SLAC, Stanford, CA, USA
FAUGIER, A.	CERN, Geneva, Switzerland
FAUS-GOLFE, A.	LAL, Orsay, France
FEIKES, J.	DESY, Hamburg, Germany
FERNANDEZ-FIGUEROA, C.	CERN, Geneva, Switzerland
FILIPPOV, A.N.	INP, Novosibirsk, USSR
FURMAN, M.	LBL, Berkeley, CA, USA
GALLUCCIO, F.	INFN, Naples, Italy
GARREN, A.	LBL, Berkeley, CA, USA
GROSSE WIESMANN, P.	CERN, Geneva, Switzerland
GUIGNARD, G.	CERN, Geneva, Switzerland
HAEBEL, E.	CERN, Geneva, Switzerland
JIMINEZ-ARTACHO, E.	Universidad Politecnica de Madrid, Spain
JOHNSON, K.	LANL, Los Alamos, NM, USA
JOWETT, J.	CERN, Geneva, Switzerland
JUDKINS, J.	SLAC, Stanford, CA, USA
KEIL, E.	CERN, Geneva, Switzerland
KIRKBY, J.	CERN, Geneva, Switzerland
KUROKAWA, S.	KEK, Ibaraki-Ken, Japan
KREJCIK, P.	SLAC, Stanford, CA. USA
KRUSCHE, A.	CERN, Geneva, Switzerland
LEDUFF, J.	LAL, Orsay, France
LUIJCKX, G.	NIKHEF, Amsterdam, Netherlands
MAAS, R.	NIKHEF, Amerstam, Netherlands
MANGLUNKI, D.	CERN, Geneva, Switzerland
MASULLO, M.	INFN, Naples, Italy
MATSUMOTO, S.	KEK, Ibaraki, Japan
MEOT, F.	LNS, Gif-sur-Yvette, France
MISTRY, N.	Cornell University, Ithaca, NY, USA
MONTH, M.	Fermilab, Batavia, IL, USA
MORTON, P.	SLAC, Stanford, CA, USA
MUNOZ , M.	ETSEIB, Barcelona, Spain
PALKOVIC, J.	SSCL, Dallas, TX, USA
PEDERSEN, F.	CERN, Geneva, Switzerland
PHINNEY, N.	SLAC, Stanford, CA, USA
PLASS, G.	CERN, Geneva, Switzerland
PONCET, A.	CERN, Geneva, Switzerland
QUESADA, J.-M.	University of Seville, Spain

RIUNAUD, J-P.	CERN, Geneva, Switzerland
RIVKIN, L.	PSI, Villigen, Switzerland
ROY, G.	CERN, Geneva, Switzerland
RUBIN, D.L.	LNS, Ithaca, NY, USA
RUJULA, DE A.	CERN, Geneva, Switzerland
SAGAN, D.	Cornell University, Ithaca, NY, USA
SANCHIS LOZANO, M.	IFIC, Burjasot, Spain
SHAPOSHNIKOVA, E.	CERN, Geneva, Switzerland
SHEPPARD, J.C.	SLAC, Stanford, CA. USA
SIEMANN, R.	SLAC, Stanford, CA, USA
SMITH, S.	Daresbury Lab., Warrington, UK
STINSON, G.	University of Alberta, Edmonton, Canada
STRUBIN, P.	CERN, Geneva, Switzerland
TAVARES, P.F.	CERN, Geneva, Switzerland
TRANQUILLE, G.	CERN, Geneva, Switzerland
TURNER, S.	CERN, Geneva, Switzerland
WANG, D.	CERN, Geneva, Switzerland
WANG, J.-M.	BNL, Upton, NY, USA
WANG, X.	SSCL, Dallas, TX, USA
WILLMOTT, C.	CERN, Geneva, Switzerland
WILSON, E.J.N.	CERN, Geneva, Switzerland
WILSON, P.	SLAC, Stanford, CA, USA
WITHERELL, M.	Univ. of California, Santa Barbara, CA, USA
WU, Y.	NIKHEF, Amsterdam, Netherlands
ZISMAN, M.S.	LBL, Berkeley, CA, USA
ZOBOV, M.	INFN, Frascati, Italy
ZOLFAGHARI, A.	MIT, Middleton, MA, USA

Printing: Druckhaus Beltz, Hemsbach
Binding: Buchbinderei Schäffer, Grünstadt